Climate Change Biology

Second Edition

Climate Change Biology

Second Edition

Lee Hannah

The Betty and Gordon Moore Center for Science and Oceans,
Conservation International;
The Bren School of Environmental Science and Management
University of California, Santa Barbara

AMSTERDAM • BOSTON • HEIDELBERG • LONDON
NEW YORK • OXFORD • PARIS • SAN DIEGO
SAN FRANCISCO • SINGAPORE • SYDNEY • TOKYO
Academic Press is an imprint of Elsevier

Academic Press is an imprint of Elsevier
32 Jamestown Road, London NW1 7BY, UK
525 B Street, Suite 1800, San Diego, CA 92101-4495, USA
225 Wyman Street, Waltham, MA 02451, USA
The Boulevard, Langford Lane, Kidlington, Oxford OX5 1GB, UK

First edition 2011
Second edition 2015

Notices

Knowledge and best practice in this field are constantly changing. As new research and experience broaden our understanding, changes in research methods, professional practices, or medical treatment may become necessary.

Practitioners and researchers must always rely on their own experience and knowledge in evaluating and using any information, methods, compounds, or experiments described herein. In using such information or methods they should be mindful of their own safety and the safety of others, including parties for whom they have a professional responsibility.

To the fullest extent of the law, neither the Publisher nor the authors, contributors, or editors, assume any liability for any injury and/or damage to persons or property as a matter of products liability, negligence or otherwise, or from any use or operation of any methods, products, instructions, or ideas contained in the material herein.

British Library Cataloguing in Publication Data
A catalogue record for this book is available from the British Library

Library of Congress Cataloging-in-Publication Data
A catalog record for this book is available from the Library of Congress

ISBN: 978-0-12-420218-4

For information on all Academic Press publications visit our website at http://store.elsevier.com/

Printed in the United States of America

Contents

Acknowledgments

Monica Pessino, Molly Thompson, and Karoleen DeCastro at Ocean o' Graphics UCSB made the original figures for this book to come to life and helped with countless revisions. Thanks to Patrick Roehrdanz, Jacob Skaggs, and Trinidad Pizano for their long hours tracking down the reproduced figures. Lynn Scarlett graciously contributed insights on her experiences during the endangered species listing for polar bears. Thanks to Andy Richford for seeing the importance of the subject and believing in the project from the beginning. Candice Janco and Sean Combs provided great editorial support. Finally, my sincere appreciation to my colleagues at Conservation International for their advice, encouragement, and belief in the importance of training a new generation of conservation professionals in an era of change.

USING THE FIGURES

The figures and boxes in this book are designed to provide a window into the primary literature. Many of the figures and all of the 'spotlight' boxes have been selected to represent classic climate change biology references. The sources for those figures and boxes are given in *author, date* format in the caption or at the end of the box, with the citations given in the reference list at the end of the book. More detailed explanations of many of the figures are available at http://booksite.elsevier.com/9780124202184. Instructors can assign readings from the primary literature to help build understanding of how a field evolves from primary research. Interested students can use the figure references and online explanations as access points to the peer review literature for research assignments or deeper understanding of concepts discussed in the text.

SECTION 1

Introduction

CHAPTER 1

A New Discipline: Climate Change Biology

The sun warms the Earth. Gases in the atmosphere capture heat and reradiate it back to the surface. This "greenhouse effect" transforms the Earth from a cold, rocky ball into a living planet. But how does this system operate, and how are human actions affecting this natural process?

These questions have been largely ignored in biology and conservation in the past. The recognition that human change is occurring in the climate—and that natural change is inevitable—is leading to a revolution in biology. A new discipline is emerging, melding well-established fields of inquiry such as paleoecology with new insights from observations of unfolding upheaval in species and ecosystems. The scope of the discipline encompasses all of the effects of human greenhouse gas pollution on the natural world. This is climate change biology.

The changes are too big to ignore. Extinctions have begun, and many more are projected. Species are moving to track their preferred climates, the timing of biological events cued to climate is shifting, and new plant and animal associations are emerging, whereas well-established ones are disappearing. Biologists are seeing change everywhere, and nowhere is change more important than in dealing with climate.

Climate change biology is the study of the impact of climate change on natural systems, with emphasis on understanding the future impacts of human-induced climate change. To understand future change, the discipline draws on lessons from the past, currently observed changes, biological theory, and modeling. It encompasses many existing disciplines, including paleoecology, global change biology, biogeography, and climatology. Climate change biology uses insights from all of these disciplines but not all of the results of these disciplines. For instance, paleoecological data that help us understand how biological systems will respond to anthropogenic climate change are a major part of climate change biology, but many aspects of paleoecology may remain outside the realm of the new discipline. Climatology is relevant to climate change biology but most climate studies fall outside the discipline. However, when climatologists conduct studies specifically to unlock biological mysteries, climatology is part of climate change biology. The practitioners of climate change biology therefore

Climate Change Biology. http://dx.doi.org/10.1016/B978-0-12-420218-4.00001-9

come from a broad range of biological and physical sciences, and their inclusion within the discipline is defined by their interest in understanding biological responses to climate change, particularly future changes due to human influences on the Earth's atmosphere (Figure 1.1).

SPOTLIGHT: BIRTH OF A DISCIPLINE

Rob Peters and Thomas E. Lovejoy founded climate change biology when both were with the World Wildlife Fund in the late 1980s. Lovejoy famously met with Steve Schneider (then director of the National Center for Atmospheric Research) and said, "I want to talk about how what you do affects what I do." Lovejoy describes the ensuing discussion as an "aha" moment for both scientists. Peters took the "aha" idea and turned it into the discipline of conservation in the face of climate change. A great poker and street hockey player, Peters was no stranger to getting to the spot before others. His papers with various coauthors in the late 1980s and early 1990s outlined much of early thinking on the subject. They were the first in their field. The classic 1985 article, "The Greenhouse Effect and Nature Reserves," framed the issues to be confronted succinctly (Peters and Darling, 1985). It even opened with a passage from Shakespeare (Macbeth): "I look'd toward Birnam, and anon, methought, the wood began to move."

Peters, R.L., Darling, J.D.S., 1985. The greenhouse effect and nature reserves. BioScience 35, 707–717.

FIGURE 1.1 Earth's atmosphere.
The atmosphere of the Earth is an amazingly thin layer of gases. At its thickest, the atmosphere is approximately 100 km deep, which is less than 1/100 of the Earth's diameter (12,700 km). Viewed from this perspective, the atmosphere appears as a thin, vulnerable shroud around the Earth. Alterations to this gossamer protective layer may have major consequences for life. *Source: Reproduced with permission from NASA.*

Climate change biology explores the interactions of biological systems with the climate system, as well as the biological dynamics driven by climate change. The interactions are not small. The climate system is in many respects driven by biology. Atmosphere and climate are themselves the products of eons of biological processes. Biological by-products are the very gases that capture the warmth of the sun and transform the planet. Everything from the color of plants across vast areas to the cycling of moisture between plants and the atmosphere help determine climate. The cycle is completed as the interactions of climate with biology determine where plants and animals can live, in turn influencing where, how far, and how fast they will move.

A GREENHOUSE PLANET

Water vapor and carbon dioxide (CO_2) are the two most abundant greenhouse gases in the atmosphere. They are both transparent to visible light arriving from the sun, but each traps heat coming from the Earth's surface (Figure 1.2). Both occur naturally, but CO_2 is also released by human burning of fossil fuels. The increase in atmospheric CO_2 concentrations resulting from human pollution is projected to cause major alterations to the Earth's climate system and global mean temperature in the twenty-first century.

Evidence spanning millions of years, and particularly from the past million years, suggests that greenhouse gases are a critical component of the Earth's climate system. Warm periods have been repeatedly associated with high levels of atmospheric CO_2 during the ice ages of the past 2 million years. Deeper in time, periods of high CO_2 concentrations or methane release have been associated with global warm periods.

GREENHOUSE EFFECT

Some gases in the Earth's atmosphere "trap" heat. Sunlight warms the Earth's surface, which then radiates long-wave radiation. Some of this radiation is absorbed and reemitted by gases such as CO_2 and water vapor. Part of the reemitted radiation is directed back at the Earth, resulting in a net redirection of long-wave radiation from space and back to Earth. This warms the lower reaches of the atmosphere, much as glass in a greenhouse traps heat from the sun, and so is known as the greenhouse effect.

The concentration of CO_2 in the atmosphere increased more than 30% in the twentieth century. This increase is due primarily to the burning of fossil fuels. Beginning with coal at the outset of the industrial revolution, and transitioning to oil and natural gas as economies advanced, the power for our electricity, industry, and transport has been drawn heavily from fossil fuels. Fossil fuels

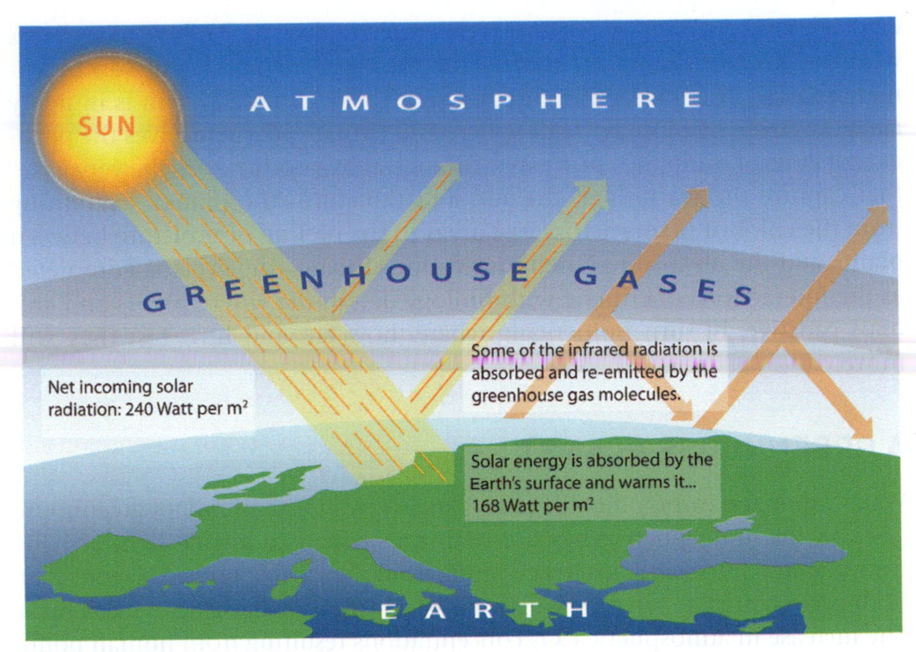
SUN
ATMOSPHERE
GREENHOUSE GASES
Some of the infrared radiation is absorbed and re-emitted by the greenhouse gas molecules.
Net incoming solar radiation: 240 Watt per m²
Solar energy is absorbed by the Earth's surface and warms it... 168 Watt per m²
EARTH

FIGURE 1.2 **The greenhouse effect.**
Solar radiation reaches the Earth, warming the surface. The surface then radiates long-wave radiation back toward space. Greenhouse gases absorb and reemit some of this long-wave radiation. The net effect is that some radiation that would have escaped to space is reradiated within the atmosphere, causing warming. *Source: From Climate Change 2001: The Scientific Basis. Intergovernmental Panel on Climate Change, 2001.*

are rich in carbon, and burning them both releases their stored energy and combines their carbon with oxygen to produce CO_2.

Rising CO_2 levels have direct effects on the growth of plants and on seawater chemistry while indirectly leading to global warming. These direct and indirect effects have profound implications for biological processes and the survival of species.

BOUNDARIES OF LIFE

Every species has climatic and physical tolerances that determine where it can live. Most species also initiate internal processes based on climatic cues. These two factors combine to determine much of the biology of how species interact, including how individual pairs of species share space and react to one another and how multiple-species assemblages come to exist together.

For example, coral reefs grow where the combination of water temperature and seawater chemistry falls within a relatively narrow range of suitable conditions. Water temperature must be above approximately 10 °C for reef-building, shallow-water corals to survive. At between 28 and 31 °C, depending on region and species, corals suffer high mortality. These same corals require dissolved calcium carbonate levels of 0.3Ω(Ω measures the degree of saturation of seawater with aragonite, a form of calcium carbonate) in order to produce their calcium carbonate skeletons and build reefs. Coral reefs are therefore found only in warm waters, primarily in the tropics, where seawater saturation with calcium carbonate is sufficient.

Species interactions with one another are often critical in mediating the effects of climate change. For example, the exact mechanism of coral mortality at high temperatures, is driven by a species interaction. Zooxanthellae are algae symbionts of corals that live within the coral polyp. These symbionts photosynthesize, nourishing the coral, while the coral provides a skeletal structure that keeps the zooxanthellae near the surface, where light for photosynthesis is abundant. At very high temperatures, the symbiosis breaks down and the coral expels the zooxanthellae. Without the photosynthetic pigments of their algal symbionts, the coral turns white or "bleaches." Some bleached corals regain their zooxanthellae and recover, but many die. Reef-building corals therefore have both thermal and ocean chemistry limits to their distribution, with species interactions, in this case in the form of symbiosis, determining the exact upper thermal limit of survival.

The combination of factors that determines where a species can survive is familiar to ecologists as the concept of niche. As with corals, most species respond directly to temperature or other climatic variables in both direct and indirect ways. Earlier definitions of niche, including those of Joseph Grinnell, who created the term in 1917, placed emphasis on species interactions as determinants of survival. Although some species interactions are nonclimatic, many others, such as the coral's interplay with zooxanthellae, are inextricably linked to climate. Later ecologists refined the concept, giving even greater emphasis to environmental variables. G. Evelyn Hutchinson defined the concept of niche as a composite (multidimensional hypervolume) of the environmental gradients across which a species could live. Many of the environmental variables relevant to this definition of niche are climatic, including temperature, precipitation, and rainfall seasonality. A polymath, Hutchinson also once said, "I sincerely hope that all of the things we are doing to the Earth's atmosphere cancel each other out."

Rapid, human-induced climate change is driving major movements in niche space. Thousands of range shifts have been recorded in plants, birds, mammals, amphibians, and insects. These range shifts result when climatic gradients shift as a consequence of global warming. Species' climatic tolerances do not change (or do not change as rapidly as climate is changing), so they must track suitable climate to survive. In today's landscapes heavily dominated by human uses such as agriculture and cities, tracking suitable climate can be a major problem for species.

In one of the earliest documented cases of shifting ranges, Edith's checkerspot butterfly was found to be shifting northwards and upslope. Similar shifts have now been found in hundreds of butterflies as well as many other invertebrates and many vertebrates such as birds and fish. Range shifts in plants are being recorded from the Cape of Good Hope in Africa to the Alps in Europe.

SHIFTING INTERACTIONS

As ranges shift, ecology is reinvented. The concept of community seems outmoded because species move in response to their own unique climatic tolerances, not as groups of organisms. Species that have coexisted throughout human memory turn out to be members of relatively ephemeral assemblages when viewed on geologic timescales.

For example, drought is driving the dieback of pinyon pines (*Pinus edulis*) across huge portions of the southwest United States. Pine dieback has affected more than 1,200,000 ha of pinyon–juniper woodland, making the pinyon–juniper association look temporary for much of the area it used to characterize. Referring to a pinyon–juniper "community" still has descriptive value because where it still exists, the association is home to many species in common. But for ecological purposes it has become clear that junipers and pinyons do not exist as an interdependent unit: They have simply shared similar climatic conditions within their respective tolerances, but those conditions are now diverging.

The same is true for species and their food. Switching of long-standing prey preferences or primary patterns of herbivory have now been seen owing to climate change in many areas. Edith's checkerspot butterfly populations in California's Sierra Nevada mountains have switched from feeding predominantly on blue-eyed Mary (*Collinsia parviflora*) to English plantain (*Plantago lanceolota*) owing to mismatching of caterpillar emergence and nectar availability due to climate change.

CHEMISTRY OF CHANGE

In the oceans and on land, greenhouse gas pollution also has direct effects on biological systems. CO_2 dissolves in seawater to produce acid and reduce the amount of calcium carbonate held in the water (saturation state). Reduced saturation state makes it more difficult or impossible for creatures to secrete calcium carbonate shells or skeletons. Consequences may include extinction, reduced abundance, or range shifts for species as diverse as squid, shelled sea creatures, and corals. Acidification can have direct effects by altering the pH of seawater. Ocean surface waters already have about 30% more H^1 ions (less basic; pH change from 8.1 to 8.0) due to dissolving of CO_2 pollution during the past two centuries (Figure 1.3).

FIGURE 1.3 Ocean chemistry and marine life.
Marine organisms such as these Pacific white-sided dolphins (*Lagenorhynchus obliquidens*) are already experiencing ocean acidification. The pH of seawater varies significantly by region and by depth, and it is increasing owing to human CO_2 emissions. CO_2 from human fossil fuel combustion enters the atmosphere and then dissolves in seawater, making it more acidic. Surface waters already contain about 30% more hydrogen ions than they did in preindustrial times. *Source: Courtesy of NOAA.*

On land, CO_2 stimulates plant growth because it is one of the principle inputs to photosynthetic pathways. This effect is not uniform for all species, and it may favor plants using the C_3 photosynthetic pathway. Global vegetation patterns may therefore be influenced by direct CO_2 effects as well as by warming. The complex, long-term effects of CO_2, either in the oceans or on land, are yet to be fully understood.

LINKAGES BACK TO CLIMATE

Biological systems have thermal properties and emit gases that in turn change climate. The amount of the sun's energy reflected (albedo) or absorbed changes greatly when vegetation changes. The replacement of tundra with coniferous forest owing to climate warming is darkening boreal latitudes, increasing heat absorption and causing further warming. The moisture transpired by trees in one area of the Amazon condenses in the atmosphere and falls as rain in other areas of the Amazon. Conversion of rain forest to savanna breaks this cycle and can lead to descent into mega-drought. The climate system is therefore influenced by what happens to biological systems, completing the chain of causation.

Natural CO_2 fluxes are large relative to emissions from fossil fuel burning, but the human emissions are enough to disturb the natural balance of the carbon cycle and increase atmospheric concentrations. How much and how fast they increase depends in large part on what is happening in other parts of the (natural) carbon cycle. Understanding the sinks, sources, and fluxes of the carbon cycle is a priority for understanding the full linkages between climate and biology.

CLIMATE CHANGE BIOLOGY

Climate change biology is a field of growing interest to a new generation of biologists, conservationists, students, and researchers. Understanding the impacts of greenhouse gases on biology and understanding the influence of biology on climate are in their infancy. The body of knowledge of past change is large, however, and the chronicle of changes currently under way is substantial and rapidly growing.

Progress in the field will help us understand how organisms respond to climate change, what conservation measures can be designed to lessen the damage, and how the interplay between the biosphere and the climate will determine the health of human and natural systems for centuries to come. The outcomes for nature are not divorced and separate from human health and happiness but, rather, are integral to achieving long-term human development in the face of climate change. Human development depends on healthy natural systems. People increasingly turn to nature for inspiration and to the outdoors for recreation, as well as relying on myriad natural systems for provision of food and materials. Maintaining healthy natural systems is an immense challenge when those systems change rapidly, as they are today.

Exploring what we have learned so far sets the stage for fuller understanding. Principles are emerging, such as the ephemeral nature of communities, that will provide a solid foundation for learning in the future. There will certainly be surprises—we are putting the planet through the largest, fastest climatic change since the rise of human civilization—but the early identification of, and learning from, those surprises is part of the excitement of a new field of inquiry. Management responses will at first be based on current fragmentary understanding, but with time, management lessons will emerge and help refine the science underpinning our early assumptions.

The structure of this book follows these principles. An overview of the climate system concludes this introductory section. The second section explores the impacts of human-induced climate change on nature that are currently being observed. The third section turns to the past for lessons about climate change in terrestrial, marine, and freshwater biological systems. Based on these insights, theory and modeling of future potential change are explored in Section 4. The last two sections of the book explore how the insights of climate change biology can be applied to the design of more dynamic conservation systems and how international policy and greenhouse gas reduction efforts influence biology and conservation.

Training the next generation of climate change biologists begins now. The first generation of climate change biologists were generally researchers specialized in other areas of ecology or biology, either finding new relevance for their observations or discovering the importance of climate change to their field of inquiry. The next generation will often be interested first in climate change and then in the tools of other subdisciplines as a means of exploring climate change questions. As it takes its place in more established circles, climate change biology is ready to provide answers of immense importance to people and nature, for generations to come.

CHAPTER 2

The Climate System and Climate Change

This chapter introduces the basics of the climate system and climate change. How do we know climate is changing? How are future changes simulated? What causes natural and anthropogenic climate change? These questions are answered here, forming a foundation for climate change understanding that is needed to explore biological responses.

THE CLIMATE SYSTEM

The Earth's climate system is composed of the atmosphere, the oceans, and the Earth's land surface (Figure 2.1). The dynamic elements of the system are hydrology and the movement of gases, including water vapor. Elements external to the climate system but very important in determining its behavior include the sun, variations in the Earth's orbit in relation to the sun, and the shape and position of continents and oceans.

The atmosphere traps energy by capturing and reradiating radiation that would otherwise escape into space. Long-wave radiation (heat) given off by the land surface and oceans is absorbed by greenhouse gases in the atmosphere. This energy is then reradiated in all directions, the net effect being a trapping of a portion of the energy in the Earth's atmosphere near the surface. Clouds in the atmosphere can reflect incoming solar energy, cooling the surface. During the day, this effect can outstrip the warming effect of the water vapor in the clouds, whereas at night the warming effect of clouds dominates. The main constituents of the atmosphere are nitrogen (78%) and oxygen (21%). Water vapor and CO_2 are minor constituents of the atmosphere but potent greenhouse gases.

The oceans are the second major component of the climate system. From a climatic standpoint, the greatest importance of the oceans is as vast reservoirs of water and dissolved gas. The oceans contribute most of the water vapor found in the atmosphere. Warmer oceans give off more water vapor. They also produce larger and more severe storms such as hurricanes. The oceans absorb CO_2, reducing its concentration in the atmosphere.

Climate Change Biology. http://dx.doi.org/10.1016/B978-0-12-420218-4.00002-0

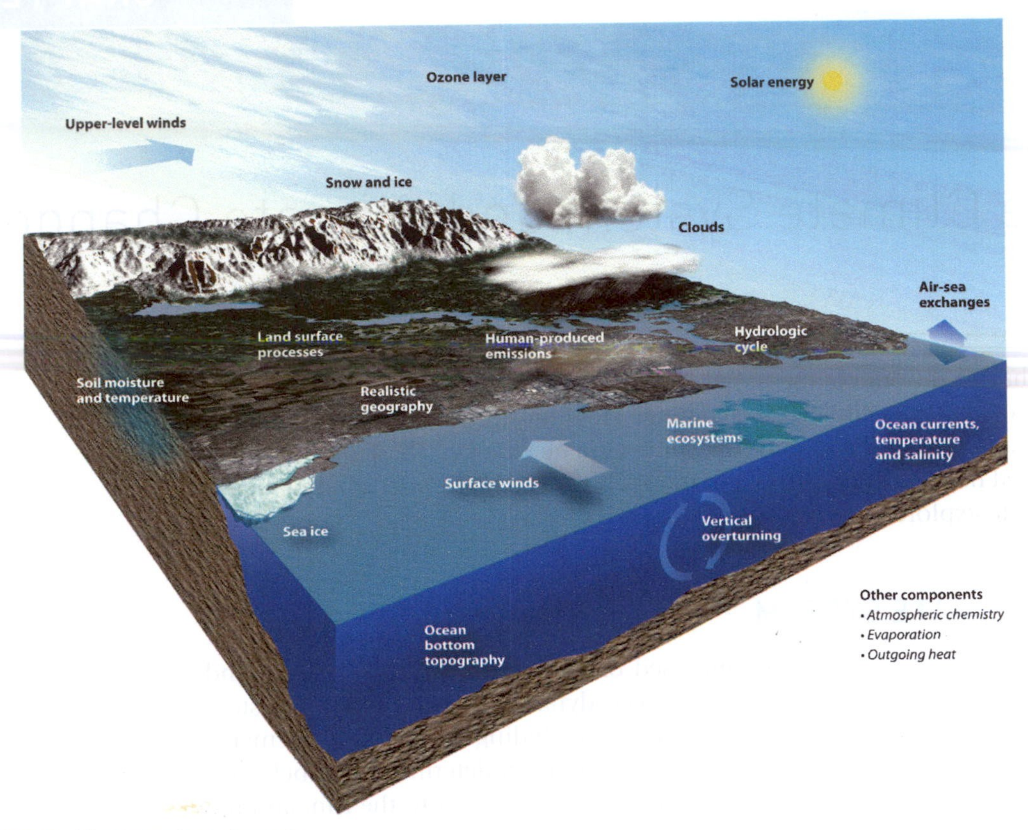

FIGURE 2.1 Climate system elements.
The land surface, oceans, and atmosphere are the major elements of the climate system. Human-driven change in the climate system acts largely through additions of greenhouse gases to the atmosphere. *Copyright UCAR, reproduced with permission of UCAR/NCAR.*

The land surface consists of vegetation, exposed soil and rock, human structures, and snow and ice. The reflective properties of these surfaces make a large difference in how the planet warms. Dark surfaces absorb solar energy and reradiate it as heat that may be trapped by greenhouse gases in the atmosphere. Light surfaces reflect sunlight back into space in wavelengths not trapped by greenhouse gases, so they have a cooling effect.

Snow and ice are particularly important parts of the Earth's surface in the climate system because they reflect the sun well. White surfaces reflect solar energy, cooling the Earth's surface. Glaciers, snowpack, and sea ice all measurably cool the Earth by reflecting sunlight. Increases in average global temperature reduce the area of ice and snow by melting, thus reducing the resultant reflectivity of the planet and producing a positive feedback loop in the climate system as the Earth warms still further (Figures 2.2–2.4).

POSITIVE FEEDBACKS IN CLIMATE

There is a positive feedback on warming when snow and ice melt. Snow and ice reflect light. Their reflectance is measured as the albedo. Light-colored materials have a high albedo, and dark materials have a low albedo. The near white of snow and ice gives them a high albedo, whereas the dark waters revealed when ice melts or the dark needles of forest conifers revealed when snow melts have a low albedo. This means that water or forest absorbs more energy from sunlight than does snow or ice, accelerating warming.

FIGURE 2.2 **Upsala glacier, patagonia, 1928 (top) and 2004 (bottom).**
Source: Top: © Archivo Museo Salesiano/De Agostini. Bottom: © Greenpeace/Daniel Beltrán.

Hydrology is the movement of water within and between elements of the climate system. Because water vapor has powerful heating (greenhouse gas) and cooling (daytime clouds) effects, the movement of water is of unparalleled importance in the climate system. Water moves through the hydrologic cycle, evaporating from the oceans, condensing as clouds, and then raining out over land to form fresh water that flows to the sea. Increases in global temperature can accelerate this hydrologic cycle by speeding up evaporation from the ocean surface.

EVOLUTION OF THE EARTH'S CLIMATE

The atmosphere as we know it was made possible by life. The atmosphere, in turn, made higher life-forms possible. The Earth was formed 4.5 billion years ago, and within approximately 1 billion years single-celled life appeared. Microbial photosynthesis over hundreds of millions of years produced enough

FIGURE 2.3 Boulder glacier, glacier national park, 1932.
Source: Reproduced with permission from Archives and Special Collections, Mansfield Library, The University of Montana.

FIGURE 2.4 Boulder glacier, 1988.
Source: Reproduced with permission from Archives and Special Collections, Mansfield Library, The University of Montana.

oxygen to make it a major component of the atmosphere. Much of this photosynthesis occurred in microbial mats, some of which formed structures known as stromatolites, which are stony accretions that are dominant in the fossil record for billions of years. By approximately 600 million years ago, oxygen buildup was sufficient to support the formation of an ozone layer in the upper atmosphere. Sunlight bombarding the upper atmosphere split oxygen atoms to create free oxygen radicals, some of which recombined with oxygen to form ozone. At this point, even though atmospheric oxygen levels were still only a fraction of modern levels, the major characteristics of modern atmosphere were in existence—oxygen, nitrogen, water vapor, and an ozone layer.

The ozone layer allowed terrestrial life to emerge. Previously, life had been possible only in the oceans, where the water column shielded organisms from damaging UV radiation. With the emergence of the ozone layer, UV radiation was screened out in the upper atmosphere, allowing life-forms to emerge onto land. Photosynthetic organisms were still dominant, allowing the continuing buildup of oxygen in the atmosphere.

The interaction of the atmosphere, water, and continental configurations began to govern climate. Major changes in climate were associated with the periodic formation of supercontinents, glacial episodes, and volcanism. At least three supercontinents have existed in the past billion years of Earth history. Rodinia existed from approximately 1 billion years ago to 750 million years ago. Pannotia was formed approximately 600 million years ago and lasted for 50–60 million years. The most recent supercontinent, Pangaea, was formed approximately 250 million years ago and later broke into its constituent components of Gondwanaland and Laurasia. Among several episodes of volcanism, the greatest was the massive outpouring that formed the Siberian Traps 250 million years ago.

The Earth's climate alternated between "icehouse" and "greenhouse" conditions once the modern atmosphere had evolved. Major icehouse episodes in deep time occurred between 800 and 600 million years ago and again at about 300 million years ago. The Earth has generally been warmer than in the present since emerging from icehouse conditions approximately 280 million years ago, but there have been remarkable increases and decreases in temperature within that time span as well.

There have been four major warm periods and four major cool or cold periods during the past 500 million years (Figure 2.5). During cool or cold phases, there is polar ice and substantial ice on land, and the global mean temperature is low. In the warm periods, there is little or no polar ice or ice on land. The warm periods generally are associated with high atmospheric CO_2 levels, whereas the icehouse periods are associated with low CO_2. Warm greenhouse conditions dominated for most of deep time (100 million to 1 billion years ago) but were punctuated by several icehouse episodes. More recently,

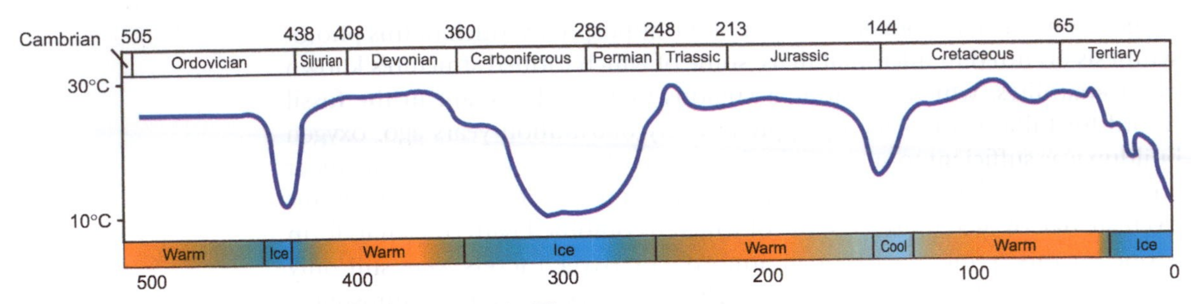

FIGURE 2.5 Global temperature during the past 500 million years.
Global mean temperature has fluctuated between icehouse and hothouse conditions during the past billion years. Four major hothouse periods have seen a largely ice-free planet, whereas four major icehouse periods have had major polar or continental ice sheets. The current climate is in a warm phase within an icehouse period. *Source: Reproduced and redrawn with permission from Christopher R. Scotese.*

a gradual cooling has dominated, leading to the ice ages of the past 2 million years. The current 10,000-year interglacial period is one of several brief warm blips in the predominantly icehouse conditions of the past 2 million years.

During the past 100 million years, a slight cooling trend gradually reversed approximately 80 million years ago and then was interrupted by a dramatic, brief warm period approximately 55 million years ago (Figure 2.6). During this warm spike, global mean temperature rose several degrees very rapidly and then dropped again only a few million years later. This spike, known as the Paleocene-Eocene Thermal Maximum (PETM), was followed by gradual warming that led to a longer warm period known as the Early Eocene Climatic Optimum.

Cooling dominated from 50 to 30 million years ago, leading to ice formation in both the northern and the southern polar regions approximately 40 million years ago. This ice cover was sporadic at first and then became continuous in Antarctica in a rapid cooling event approximately 34 million years ago. Slight warming kept the ice cover in the Northern Hemisphere sporadic until approximately 2 million years ago, when the Pleistocene ice ages began.

Climate dynamics have been particularly pronounced during the past 2 million years as the Earth has plunged into, and more briefly back out of, glacial periods (see Fig. 2.6). Glacial conditions have dominated this period, with warm greenhouse intervals coming at roughly 100,000-year intervals and lasting only a few thousand years each. This period has been characterized by much climatic variability, including very rapid climate "flickers"—sudden shifts to warmer or colder conditions that occurred in less than 1000 years.

Glacial/interglacial transitions are driven by orbital forcing of climate. When conditions are right for land ice to last through many summers in the large landmasses of the Northern Hemisphere, an ice age is initiated. As solar input

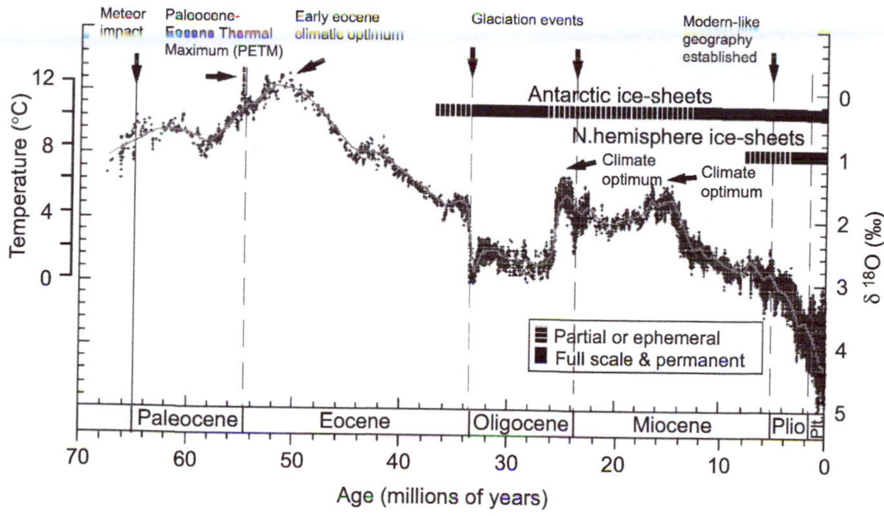

FIGURE 2.6 Earth temperature change during the past 70 million years.
The past 70 million years have seen the planet cool from the thermal maximum at the Paleocene–Eocene boundary 54 million years ago to current ice age conditions. Polar ice formed in the Southern Hemisphere beginning approximately 35 million years ago and in the Northern Hemisphere approximately 8 million years ago. Northern Hemisphere polar ice became permanent approximately 2 million years ago, initiating the ice ages. *Source: Zachos et al. (2001) Reprinted with permission from AAAS.*

to northern landmasses changes with variations in the Earth's orbit, the northern land ice melts, initiating a warm, greenhouse interval.

The last glacial period gave way to the current warm period starting approximately 10,000 years ago (the Holocene). After several late-glacial climate flickers, notably the Younger Dryas, the climate became more stable as recorded in Greenland and Antarctic ice cores, though pollen records suggest substantial climate variation in both temperate and tropical areas. Orbital forcings are unusual in this period and may result in an interglacial period considerably longer than those typical of the past 500,000 years. It is onto this unusually warm, stable climate that human greenhouse gas emissions are pushing additional climatic warming.

NATURAL DRIVERS OF CHANGE

Energy from the sun drives the climate system. The sun's warmth is unevenly distributed across the planet, which sets winds and ocean currents in motion, transporting heat from the equator to the relatively cooler poles. Energy from the sun drives the hydrologic cycle as well, evaporating water from the oceans and freshwater bodies. Natural forcings such as volcanic eruptions modify summer climate on shorter timescales (Figures 2.7 and 2.8).

SPOTLIGHT: FORCING THE SYSTEM

The climate system is forced by both natural and human-driven processes. Orbital forcing is particularly important in driving natural change. It includes variations in the Earth's orbit that result in relatively more or less solar radiation reaching the Earth. The Earth's orbit is not perfectly round; the tilt of the Earth on its axis varies in its orientation to the sun, and the tilt itself wobbles—it changes with time. All of these factors result in changes in incoming solar energy and drive changes in the climate system. Volcanic activity ejects large amounts of particulates into the atmosphere, causing cooling, and is another forcing external to the climate system (Figures 2.7 and 2.8). Finally, most recently and most dramatically, human pollution of the atmosphere with greenhouse gases has resulted in radiative forcing of the climate system—changes that affect the reradiation of the energy of the sun, warms the atmosphere and results in climate surprises.

Source: Schneider, S. H., 2004. Abrupt non-linear climate change, irreversibility and surprise. Global Environmental Change, 14 (3), 245–258.

FIGURE 2.7 Pinatubo eruption.
The summit caldera on August 1, 1991, a month and a half after the June 15 explosive eruption that ejected so much ash into the atmosphere that it altered global climate. *Photo by T. J. Casadevall, U.S. Geological Survey.* From http://www.beringia.com/climate/content/volcanoes.shtml.

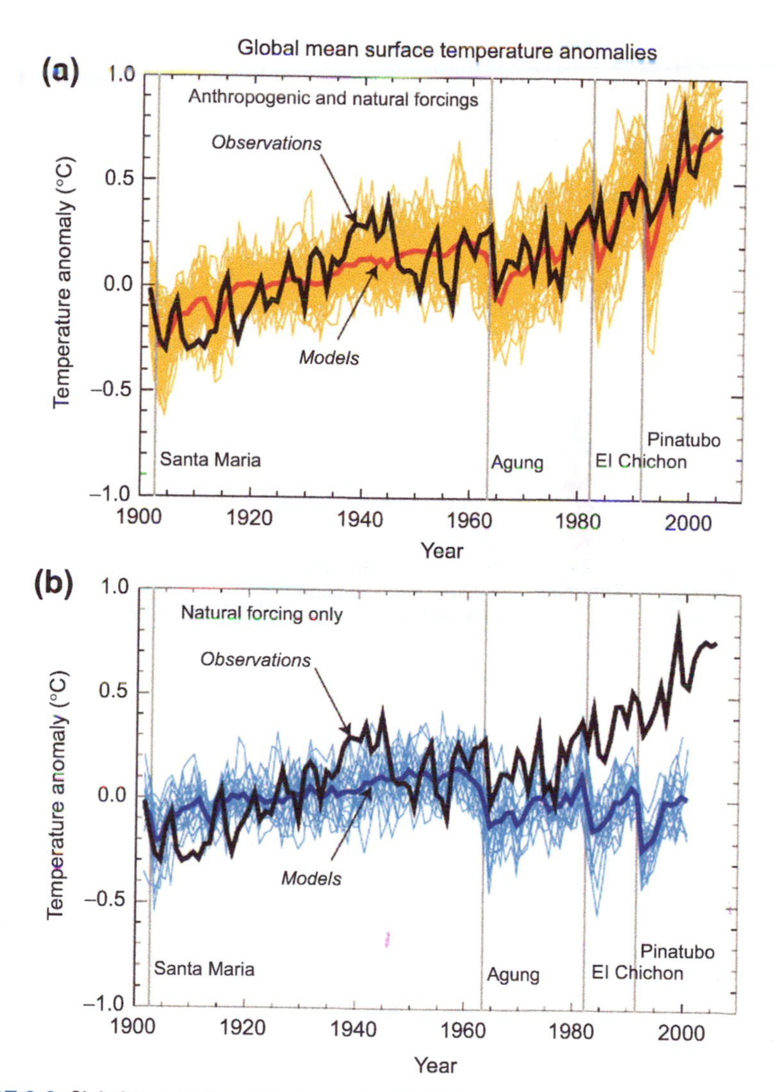

FIGURE 2.8 Global temperature change due to natural and human forcing. Global temperature cooled measurably in the years immediately after the Mount Pinatubo eruption (bold line). This global temperature trace indicates major volcanic events that drove decreases in global temperature. It is coupled with mean temperature projections from global climate models (general circulation model—GCM), computer simulations (colored lines) showing that the actual temperature record can be fully reproduced only when human forcings, primarily burning of fossil fuels and deforestation, are included in the GCM simulations. *Source: From Climate Change (2007): The Physical Science Basis. Working Group I Contribution to the Fourth Assessment Report of the Intergovernmental Panel on Climate Change. Figure TS.23. Cambridge University Press.*

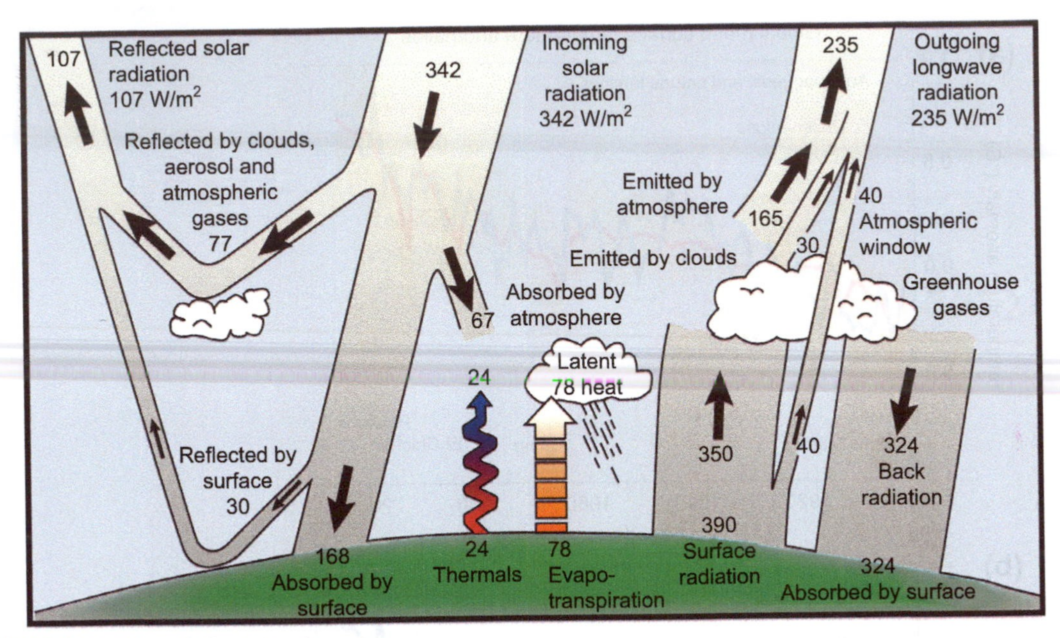

FIGURE 2.9 Earth's radiation balance.
Approximately 342 W/m² of solar energy reaches the Earth's surface. 107 W/m² is reflected into space, whereas 235 W/m² is emitted from the Earth as long-wave radiation. *Source: From Climate Change (2007): The Physical Science Basis. Working Group I Contribution to the Fourth Assessment Report of the Intergovernmental Panel on Climate Change. Cambridge University Press.*

The energy reaching the top of the Earth's atmosphere is estimated to average 342 W per square meter (W/m²). Some of that energy is reflected back into space by fine particles in the atmosphere, clouds, or the Earth's surface, leaving approximately 235 W/m² to warm the atmosphere and the surface of the Earth (Figure 2.9). Over the whole Earth, this is an immense amount of energy—approximately 150 million times more energy than is produced by the world's largest power station.

The exact amount of energy reaching the Earth varies, however, as does the distribution of that energy to various parts of the world. Changes in the orbit of the Earth bring it closer to the sun or farther away or tilt one part of the planet closer to the sun. The energy output of the sun may vary as well, up to several tenths of 1%. These variations in orbit affect the amount of energy reaching the Earth, changing the sun's warming effect and hence changing climate.

There are three main types of orbital variation affecting the Earth's climate (Figure 2.10). The first, called eccentricity, relates to the shape of the Earth's orbit around the sun. The path that the Earth carves in space varies from nearly circular to strongly egg shaped (elliptical). When the orbit is elliptical,

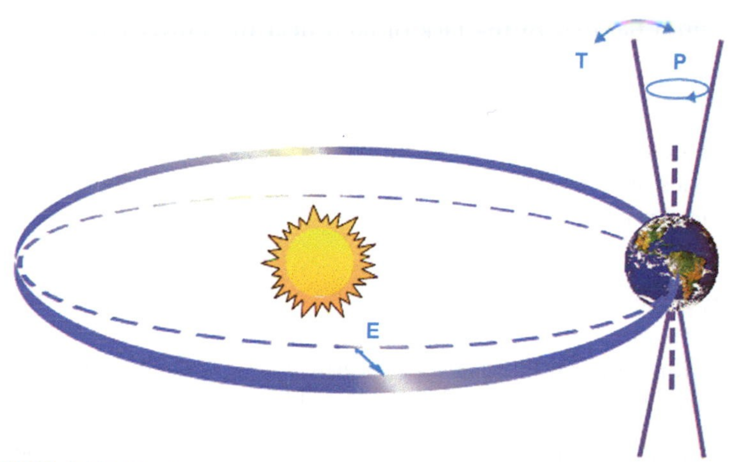

FIGURE 2.10 Orbital forcings.

Three major orbital forcings affect the amount of solar radiation incident on Earth. The Earth's orbit is oval rather than round, resulting in more radiation reaching the Earth when it is closer to the sun. This effect is eccentricity (E). The tilt of the Earth (T; also referred to as obliquity) varies, which affects the amount of radiation reaching the Northern Hemisphere. Finally, the time of year during which the Northern Hemisphere is tilted toward the sun varies, which is called precession of the equinoxes (P). These forcings are often referred to as Milankovitch forcings, for the Serbian physicist who recognized that the amount of solar radiation received in summer in the Northern Hemisphere determined the timing of ice ages. *Source: From Climate Change (2007): The Physical Science Basis. Working Group I Contribution to the Fourth Assessment Report of the Intergovernmental Panel on Climate Change. Cambridge University Press.*

the Earth will be much closer to the sun in some parts of its orbit, changing seasonal heating. The Earth wobbles on its axis as it circles the sun, giving rise to the second and third types of orbital forcing. The amount of tilt varies and is referred to as obliquity. The direction of tilt slowly rotates and is called precession.

All three of these orbital forcings affect the distribution of heating between seasons or between hemispheres more strongly than they affect the overall amount of solar energy reaching the Earth. Their effect on climate is therefore due to amplifications and dynamic effects rather than to changes in raw energy input.

The most pronounced of these amplifications are the ice ages, which are driven by the unequal amounts of land in the Northern and Southern Hemispheres. North America and Eurasia have huge landmasses near the poles. When the Northern Hemisphere receives less heat, particularly in summer, ice may form on this land. The ice reflects sunlight and cools the entire planet. When the Northern Hemisphere receives more heat, the ice melts and the planet warms. Warming and cooling of the Southern Hemisphere has no such effect because of the lack of land near the poles. There is very little land near the poles in South America and Africa because both taper as they approach the South Pole.

Precession determines which hemisphere tilts toward the sun in summer. Precession varies on a 23,000-year cycle. When the Northern Hemisphere is tilted toward the sun in summer, summers are very hot and ice cannot build up on the large northern landmasses.

Obliquity is the amount of tilt in the Earth's axis. The Earth wobbles on its axis like a spinning top. When the tilt is strong toward the Northern Hemisphere, it is difficult for continental ice sheets to form. There is a 41,000-year periodicity to obliquity. Obliquity is sometimes referred to as tilt.

Eccentricity is the shape of the Earth's orbit around the sun. This shape varies from more circular to more oval with two cyclic periods—100,000 and 400,000 years. The more circular orbit results in more even distribution of solar energy. The more oval orbit can result in less solar energy reaching the Northern Hemisphere's large, ice-prone landmasses and can help trigger a glacial period.

Glacial periods start with cool summers. Combinations of orbital forcings that lead to cool summers allow ice to be retained through the warm season and continental ice sheets to form in North America and Europe. The Northern Hemisphere landmasses are particularly important because they offer enough high-latitude landmass for the formation of continental ice sheets. A similar dynamic for the Southern Hemisphere does not exist because there is little landmass to hold ice in South America or Africa at high latitudes. In the late 1800s, scientists believed that cold winters led to ice ages. Milutin Milankovitch, a Serbian geophysicist and engineer, recognized that cool summers were the key to ice buildup. Cycles in orbital forcing—Milankovitch cycles—bear his name in recognition of his contribution to understanding their role in the ice ages.

Recent research points to a role for the Southern Hemisphere in the formation and termination of glacial periods as well, also driven by Milankovitch forcings. Low obliquity (tilt) brings cool summers to both hemispheres, which favors ice buildup in the north and intensified circumpolar current in the south. The intensification of the circumpolar current reduces upwelling of CO_2-rich water. The reduction in atmospheric CO_2 cools the planet, facilitating continental ice sheet buildup in the north. The Southern Hemisphere may also push the Northern Hemisphere along as glacial periods end—high obliquity results in warmer summers in both hemispheres. This begins to melt the continental ice sheets in the north, whereas in the south it intensifies circumpolar currents and winds, pumping CO_2-rich water to the surface and warming the planet.

MAJOR FEATURES OF PRESENT CLIMATE

Energy from the sun drives circulation patterns in both the oceans and the atmosphere. Atmospheric circulation is driven by the principle that warm air is less dense than cool air and therefore rises. Ocean circulation is driven by both temperature and salinity. Warm water rises, cool water sinks, and salty water is more dense than fresh water, leading salty water to sink and less salty water to rise.

The Earth receives more heat from the sun at the equator than it does at the poles. A pot that is off center on the stove also receives heat unequally, resulting in water roiling to the top where the heat is received and moving out to the cooler edges of the pot. The sun's heat received at the Earth's equator acts in the same way, causing the Earth's atmosphere to roil—warm air rises and builds up in the tropics, pushing toward the cooler poles. As air masses move from the tropics toward the poles, they cool, descend, and eventually return to the tropics in a giant loop. This movement of heat, known as heat transport, creates large, systematic patterns of circulation in the atmosphere.

This heat imbalance sets up gradients that drive heat transfer from the equator toward the poles. Warm air and water rise, pooling at the equator, setting up circulation patterns typified by rising warm air or water near the equator and sinking cold air or water near the poles, with movement in between.

In the atmosphere, these circulation patterns are known as Hadley cells (Figure 2.11). There are two Hadley cells between the equator and each pole. Hadley cells have both vertical and horizontal structures. Viewed in cross-section, air masses in a Hadley cell rise at the equator, move toward the pole, and then descend. From above, the circulation is clockwise, as moving air is deflected by the Coriolis effect imparted by the Earth's rotation.

In counterpoint to the Hadley cells, in the tropics there are East–West-oriented circulation cells. These circulation patterns arise when pressure differences across ocean basins drive surface winds in one direction, balanced by transfers aloft in the opposite direction. Over the Pacific Ocean, the circulation is known as Walker cell circulation or the "Southern Oscillation." It drives easterly surface winds across the Pacific. Breakdown in Walker cell circulation in the tropical Pacific results in an El Niño event.

Trade winds are surface winds caused by air movement and Hadley cells being deflected by the Coriolis effect. The trade winds are easterly, meaning that they blow from the east. They move westward along the equator in both the Northern and the Southern Hemisphere. Where the trade winds converge along the equator, a zone of uplift and cloud formation results, which is known as the Intertropical Convergence Zone (ITCZ). The trade winds are balanced by return flows in the midlatitudes by west-to-east blowing winds known as the westerlies.

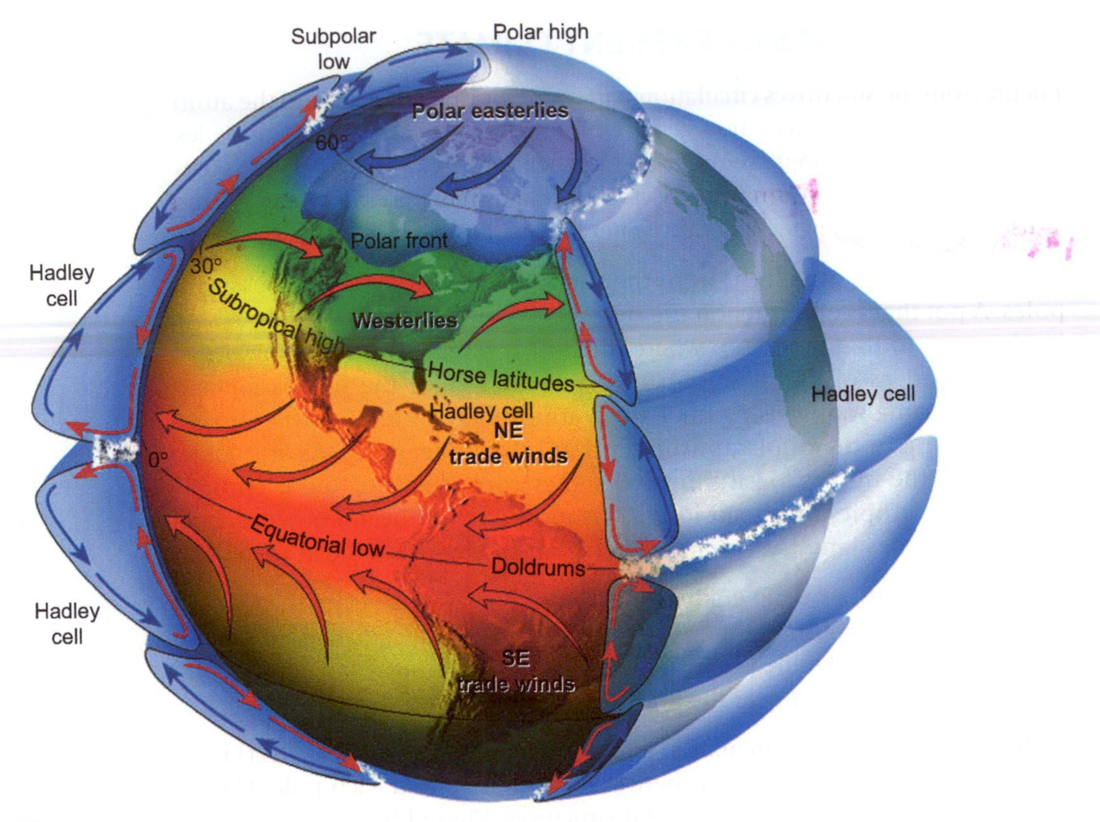

FIGURE 2.11 Hadley cells.
Warm air rises in the atmosphere, cools, and descends. This phenomenon results in the formation of major vertical circulation features in the atmosphere known as Hadley cells. Clouds form in the tropics where Hadley cells meet near the equatorial low - the Intertropical Convergence Zone (ITCZ). The ITCZ migrates north and south, resulting in two rainfall peaks in most parts of the tropics. *Source: Lutgens et al. (2001). Reproduced with permission from Pearson Publishers.*

Major ocean circulation patterns follow the wind patterns, forming large gyres with east-to-west flow along the equator and west-to-east flow at the midlatitudes. However, ocean current direction varies from wind direction by 15–45° progressively with depth, an effect known as the Ekman spiral. When surface ocean currents strike continents, they deflect and follow the shoreline, forming boundary currents.

Upwelling results when along-shore winds move ocean water. The wind-driven surface movement is deflected by the Ekman spiral, resulting in transport of water away from the coast. This moving water has to be replaced, so water from depth is drawn to the surface. The movement of this cold, nutrient-rich water from depth to the surface is referred to as upwelling (Figure 2.12).

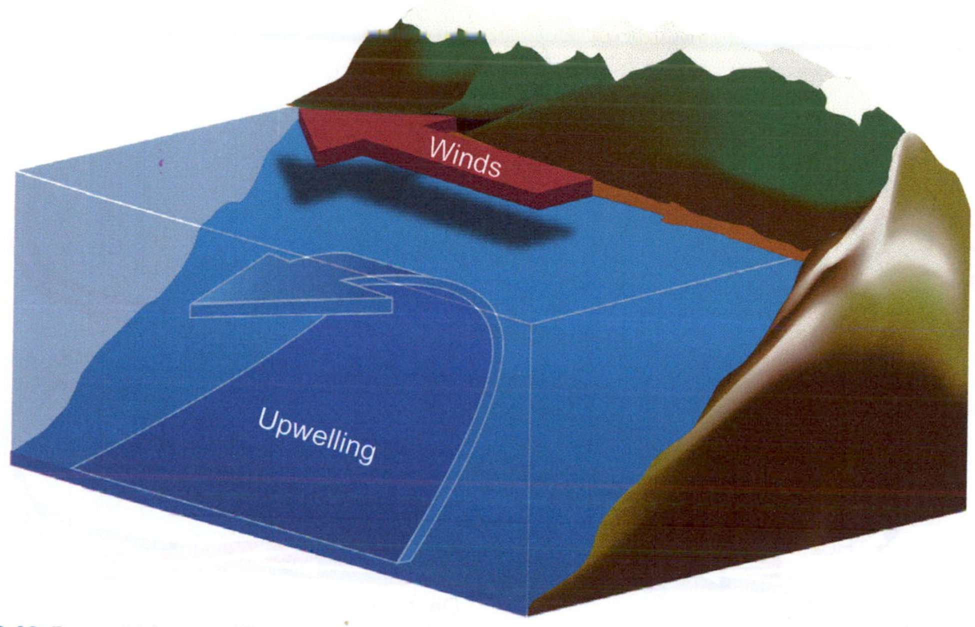

FIGURE 2.12 **Forces driving upwelling.**
Alongshore winds create water movement that is deflected by Eckman forces. Replacement water rises from the depths, creating upwelling. *Source: From Wikimedia Commons.*

In the oceans, the equator-to-pole circulation is the thermohaline circulation (Figure 2.13). It is more complex because it must work its way around landmasses and because it involves salinity as well as warmth. Warm water at the equator evaporates, leaving behind water that is both warmer and more salty, and hence more dense. This salty warm water moves toward the poles, where it cools and sinks, renewing the circulation.

The influence of the thermohaline circulation is especially strong in the North Atlantic, bringing in massive quantities of heat from the equator. This portion of the thermohaline circulation is known as the Gulf Stream. When the Gulf Stream shuts off, it robs heat from two major landmasses near the poles, greatly accelerating ice buildup. Glacial periods seem to end when the Gulf Stream strengthens, pumping energy northward to melt the ice sheets. Whereas the onset of glacial periods seems to be more gradual, the end of a glacial period can be dramatically rapid. Climate flickers such as the Younger Dryas can be initiated when the thermohaline circulation shuts down during the glacial/interglacial transition. Changes in the thermohaline circulation are therefore an important trigger for climate change.

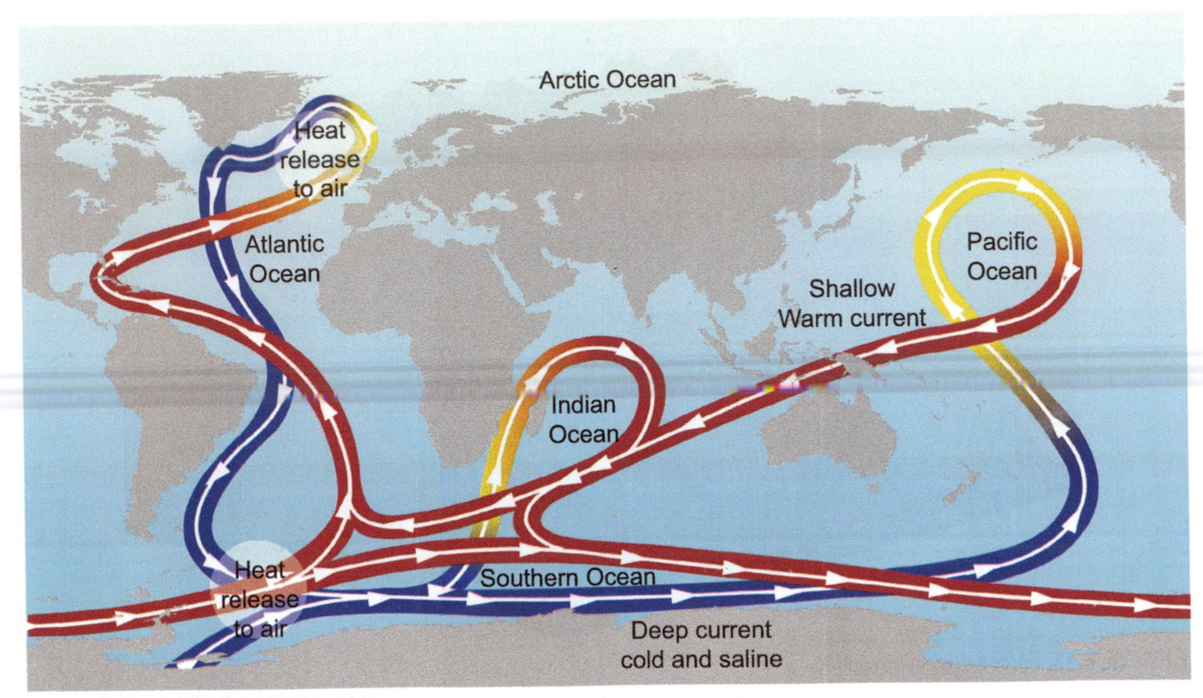

FIGURE 2.13 **Thermohaline circulation.**
Major circulation features in the oceans are established when seawater warms at the equator, evaporating and becoming more saline, and then moves near the surface (red) toward the poles, where it cools and sinks. It then moves near the bottom (blue) back to the equator, to rise and begin the process anew on timescales of hundreds of years. Because it involves both temperature and salinity, this feature is termed thermohaline circulation. *Source: Lovejoy and Hannah (2005). Reproduced with permission from Yale University Press.*

STABLE STATES OF THE SYSTEM

The circulation patterns of the Earth's climate system change over time. Like freeway traffic that either moves freely or backs up clear across town, atmospheric circulation may exhibit dramatically different patterns at different times, frequently switching back and forth among two or more relatively stable states. As the Earth spins, its rotation sets up waves in atmospheric circulations, much as water in a river rapid sets up standing waves. In such systems, it is natural that a wave crest or "high" in one region will be connected to wave troughs or "lows" in neighboring regions.

El Niño events are among the best known of these multiple-state patterns (Figure 2.14). During El Niño events, ocean circulation patterns change across the Pacific Ocean. Rain patterns shift and atmospheric circulation changes in response to alterations in ocean water temperatures. These effects are felt in the Pacific but are also reflected in other, far distant parts of the globe.

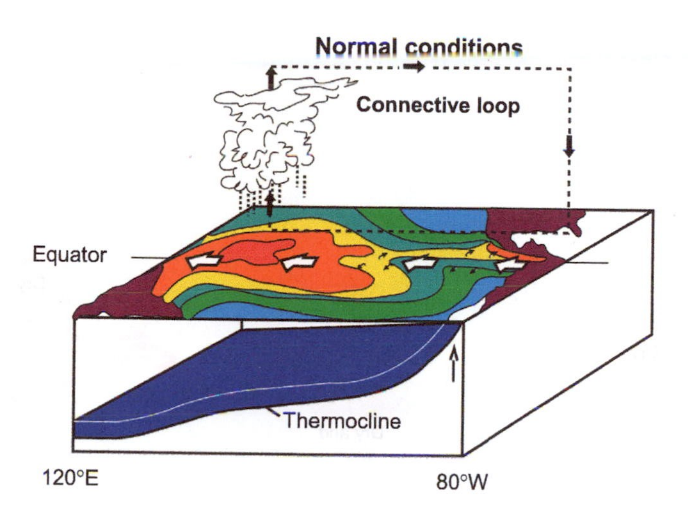

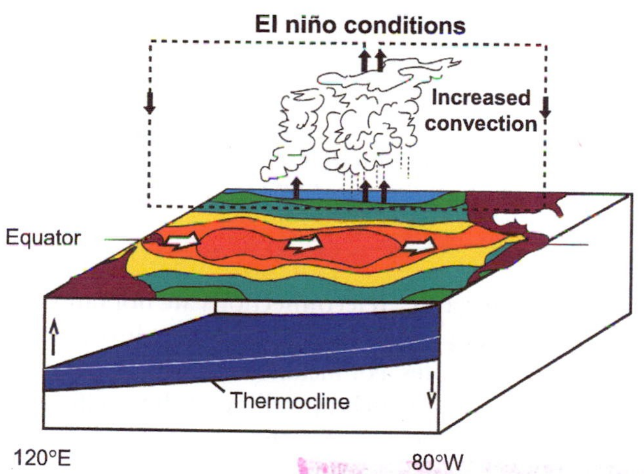

FIGURE 2.14 **El Niño.**
Periodically, the gross circulation of the southern Pacific Ocean changes, in a phenomenon known as El Niño. Under El Niño conditions, the thermocline becomes more shallow and upwelling is reduced along western South America. This results in pooling of warm water in the central Pacific and changes in precipitation and convection patterns. *Source: Lovejoy and Hannah (2005). Reproduced with permission from Yale University Press.*

Thus, El Niño years are associated with less upwelling of deep ocean water and enhanced rainfall in the Pacific but also with decreases in rainfall and drought in Africa. These long-distance effects are the result of global circulation patterns sitting next to, and driving, one another, almost like gears. What happens in one circulation cell is passed on to the next and may result in consequences in faraway places. Such long-distance, linked impacts are called

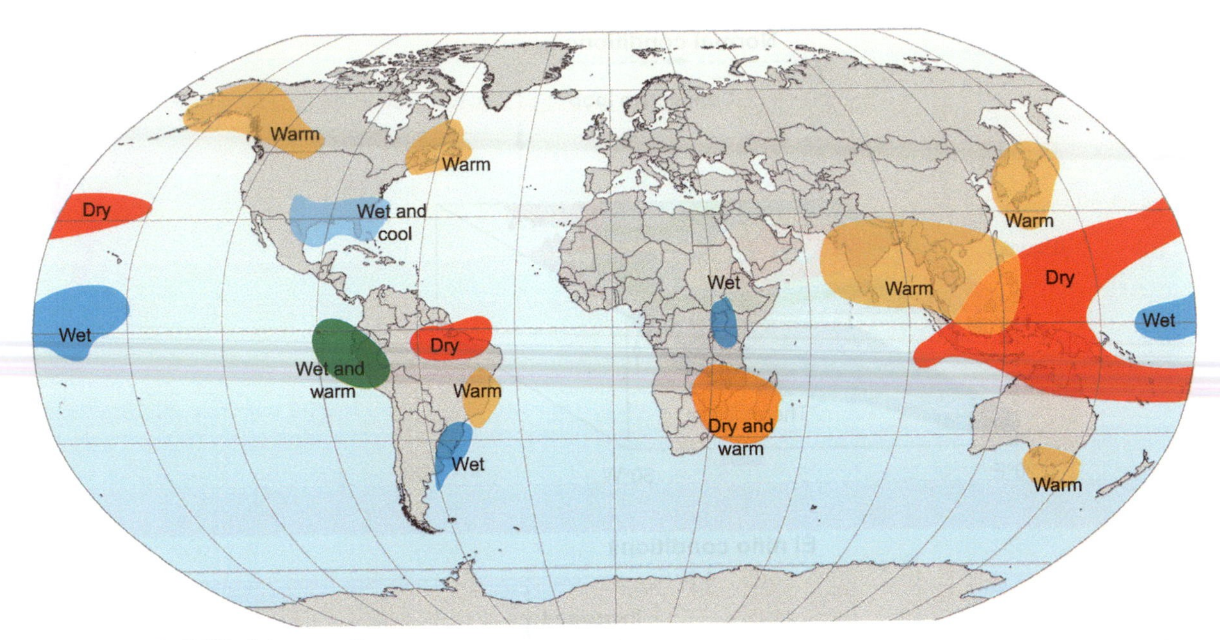

FIGURE 2.15 El Niño teleconnections.

Changes associated with an El Niño event include drying, warming, changes in precipitation, and cooling in different combinations in regions as widely separated as South Africa and Eastern Asia. *Source: Reproduced with permission from Mark Bush. Ecology of a Changing Planet, 3rd edition: Benjamin Cummings.*

"teleconnections (Figure 2.15)." Teleconnections are not random; they tend to be linked to complementary "sister" states. They often involve coupled changes in ocean and atmospheric states. For instance, the complement to El Niño conditions are La Niña events, in which upwelling in the Pacific is enhanced and rainfall reduced. The oscillation between these two conditions is known as the El Niño/Southern Oscillation or ENSO.

Other large-scale modes of atmospheric variability include the North Atlantic Oscillation and the Pacific Decadal Oscillation. The Pacific Decadal Oscillation affects the North Pacific Ocean and switches states approximately every 10 years, as its name implies. The North Atlantic Oscillation is dominated by two modes, one in which arctic air pounds Europe and another in which European weather is considerably more pleasant. The thermohaline circulation is an excellent example of a teleconnection because what happens in the North Atlantic may affect climate across the entire planet.

HUMAN-DRIVEN CHANGE: RISING CO_2

The rise of CO_2 due to fossil fuel burning and deforestation has been traced in a simple study on the Mauna Loa volcano in Hawaii. Air intakes atop the

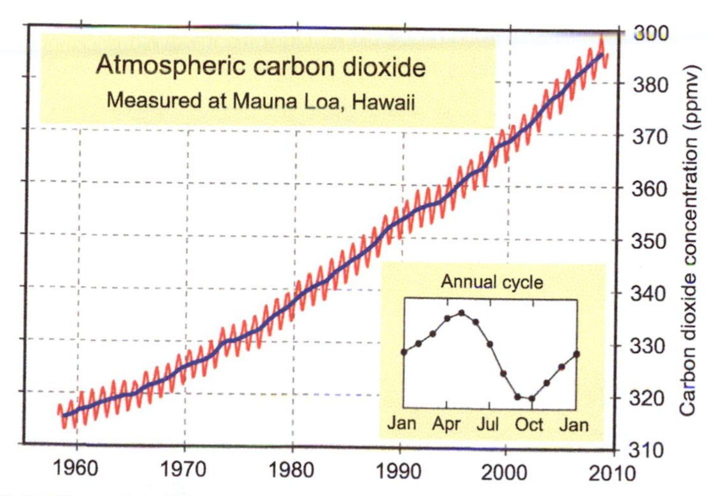

FIGURE 2.16 Mauna Loa CO_2 record.
The CO_2 record from Mauna Loa clearly shows strongly rising atmospheric CO_2 concentrations since approximately 1960. Superimposed on a multiyear increase is a much smaller "sawtooth" annual cycle, which results from the release and uptake of CO_2 from vegetation. *Source: From Climate Change (2007): The Physical Science Basis. Working Group I Contribution to the Fourth Assessment Report of the Intergovernmental Panel on Climate Change.*

mountain capture samples that are then analyzed for CO_2 content. Mauna Loa was chosen because its island location and high elevation place it far away from short-term contamination from any city air pollution. The record of CO_2 at Mauna Loa is therefore pure: it shows what is happening in the atmosphere very plainly—and plainly CO_2 is rising dramatically (Figure 2.16).

CHARLES DAVID KEELING

Rising atmospheric CO_2 was first measured by Charles David Keeling at the Mauna Loa observatory on the island of Hawaii. Keeling worked at the Scripps Institute of Oceanography in San Diego and, along with Roger Revelle, the director of the institute, concluded that direct measurement of changing CO_2 was needed. In the late 1950s, Keeling settled on the remote slopes of Mauna Loa to escape local variation in CO_2 caused by urban emissions or vegetation. The program begun by Keeling continues today and has provided incontrovertible evidence of the effect of human pollution on the atmosphere.

The Mauna Loa record is so sensitive that it clearly shows the pulse of the seasons. Each spring, plants come to life, sucking CO_2 from the atmosphere. Then each fall, leaves fall and decompose, releasing CO_2 to the atmosphere as they decay. This cycle is balanced across the equator. As Northern Hemisphere plants die and release CO_2 in the fall, plants in the Southern Hemisphere are taking up CO_2 with the flush of new spring growth. However, landmasses in the north

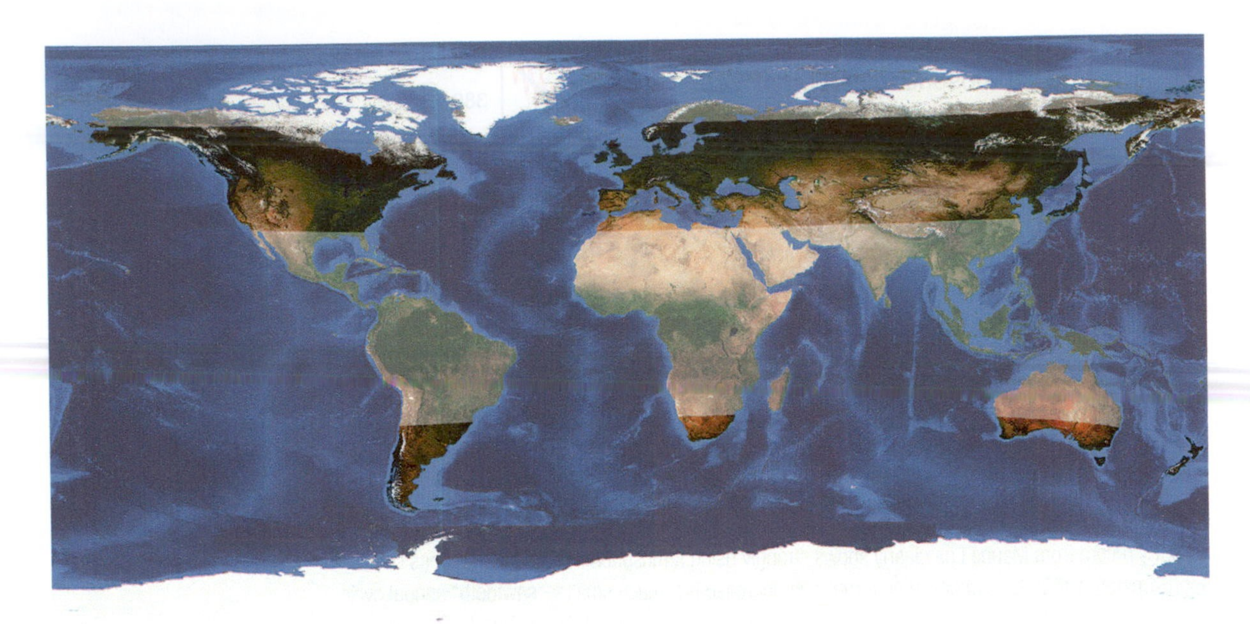

FIGURE 2.17 **Northern and Southern Hemisphere landmasses.**
Large amounts of land surface in the high latitudes of the Northern Hemisphere result in a large fall flush of CO_2 into the atmosphere and a large, measurable uptake of CO_2 from the atmosphere in the spring. The Northern Hemisphere effect dominates the same effect in the Southern Hemisphere because the Southern Hemisphere has little land at comparable latitudes. The dominance of landmasses in the Northern Hemisphere is also important in the formation of glacial periods. *Source: Courtesy of NASA.*

are far greater than those in the south (Figure 2.17), so Southern Hemisphere processes balance only a small part of the seasonal cycle in the north. A net global uptake of CO_2 occurs in the Northern Hemisphere in spring, with net CO_2 release in the Northern Hemisphere fall. This seasonal seesaw is reflected in the Mauna Loa record. The short-term drop in annual CO_2 during the northern spring is reflected in an annual blip of dropping CO_2 in the Mauna Loa record. A steep blip of increased CO_2 accompanies the northern fall each year in the Mauna Loa measurements.

CO_2 concentrations in the Earth's atmosphere are clearly and steadily rising. Each year, observers at Mauna Loa note slightly higher levels in both the spring highs and the fall lows—the annual seesaw in atmospheric CO_2 is slowly ratcheting up. This rise has been noted each year since about 1940, exactly as expected given the large amounts of oil, gas, and coal that are burned for fuel each year.

Fossil fuel use worldwide more than quadrupled in the period of the Mauna Loa record—from just over 2 Pg (as carbon) in 1960 to more than 8 Pg annually (a petagram (Pg) is 10^{15} g, or 1000 million metric tons) today. Clearing of forests and other land use changes contribute approximately one-fourth of total emissions of CO_2, for a global total of more than 8000 million metric tons a year.

ISN'T WATER A GREENHOUSE GAS?

Water is a greenhouse gas, one that absorbs in the same part of the spectrum as CO2 (see figure). Since water is much more abundant in the atmosphere than CO2, doesn't this mean that adding CO2 would have little effect on warming? Early scientists understood these gas absorption properties and as a result underestimated the greenhouse effect for nearly half a century.

What the scientists of the early 1900s didn't realize was that the Earth loses heat not at the surface where water vapor is abundant, but at the top of the atmosphere, where water vapor is essentially absent. At the top of the atmosphere, the absorbtion of CO_2 is dominant, and it doesn't saturate much at levels (4 times to 10 times preindustrial concentration) likely to be produced by human pollution.

So water is a greenhouse gas, but not where it matters – in the upper atmosphere. In the upper atmosphere where the Earth loses heat to space, the effect of CO_2 dominates. This is why the Earth is warming in response to human pollution.

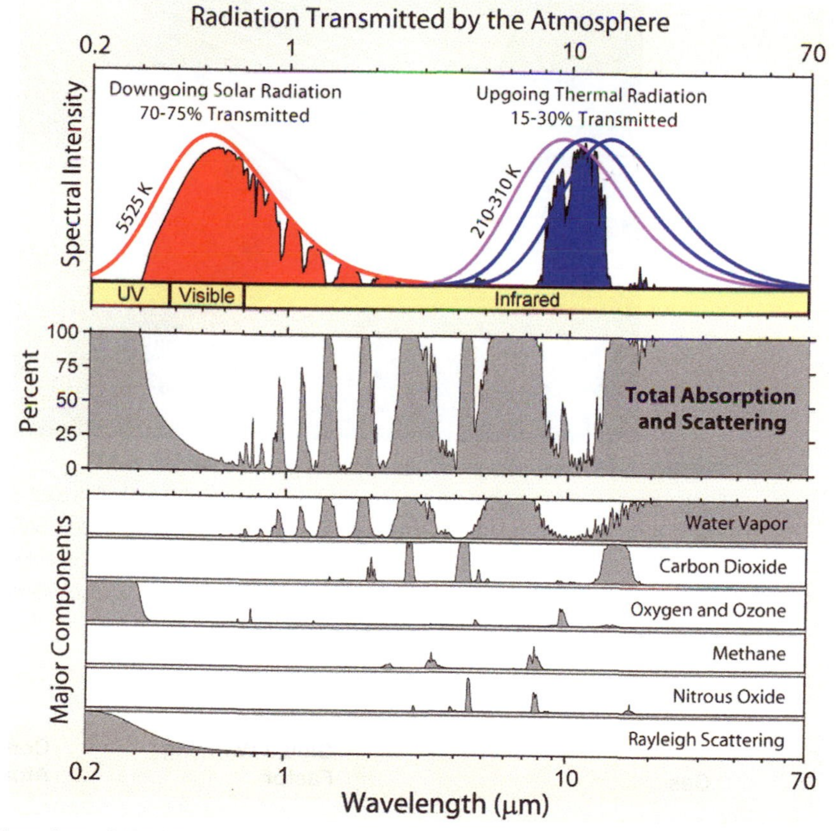

Incoming radiation from the sun (red curve) and outgoing radiation from the surface of the Earth (blue curve) are both attenuated by absorption of photons by gases in the Earth's atmosphere. The resulting actual incoming (solid red) and outgoing (solid blue) radiation are the result of subtraction of the effects of various gases (labeled) and particles (Rayleigh scattering) in the atmosphere (gas absorption properties are shown in gray) from the theoretical total. *Figure courtesy of Global Warming Art, reproduced with permission.*

However, CO_2 is not the only greenhouse gas. Methane, water vapor, and other gases have warming effects (Figure 2.18), some of them much more pronounced than that of CO_2 (Table 2.1). Methane is particularly important because although it is a minor constituent of the atmosphere (its concentrations are measured in parts per billion), it has a strong warming effect—it is a potent greenhouse gas. Human activities produce methane, although in much

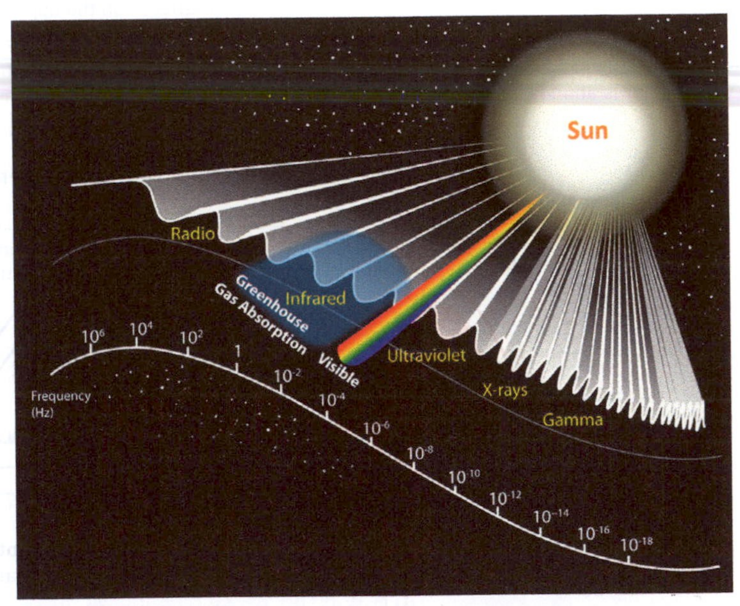

FIGURE 2.18 Electromagnetic absorption of greenhouse gases.
Electromagnetic radiation from the sun reaches the Earth as visible light, ultraviolet, or infrared radiation. This radiation strikes the Earth and is reradiated as longer wavelength radiation. Greenhouse gases such as CO_2 and methane absorb radiation in the portion of the spectrum that is just longer in wavelength than visible light. They then reradiate this energy, warming the atmosphere. *Source: University of California.*

Table 2.1 Greenhouse Gases, Potency, and Concentration

Gas	Global Warming Factor	Concentration in Atmosphere (ppb)
Carbon dioxide (CO_2)	1	379,000
Methane (CH_4)	21	1760
Nitrous oxide (N_2O)	310	320
Chlorofluorocarbons (CFCs)	5000–14,000	<1

ppb, parts per billion.
Source: IPCC.

smaller quantities than CO_2. Many types of farming result in methane emissions, with releases from decaying vegetation in flooded rice fields being the greatest source. Methane concentration in the atmosphere has increased from 700 ppb in preindustrial times to more than 1700 ppb today. Human activities do not strongly affect atmospheric water vapor concentrations directly, but water vapor concentrations are affected indirectly by temperature. Finally, some gases that are found in small amounts in the atmosphere and in human emissions are very strong greenhouse gases and may play a significant role in affecting global climate.

Because CO_2 in the atmosphere is currently increasing, we expect the climate to warm. This effect has been measured—global mean temperature is rising (Figure 2.19). Global mean temperature increased nearly 1 °C in the 100 years ending in 2005 (0.74 ± 0.18 °C (1.33 ± 0.32 °F)). The oceans generally warmed less than land so that most terrestrial regions, particularly continental interiors, have warmed in excess of the global mean (Figure 2.20). Some regions have cooled, whereas most have warmed. Cooling and warming trends are sometimes found in close proximity. Areas of Antarctica, for instance, have warmed as much as 2.5 °C, whereas other areas of the continent have cooled slightly.

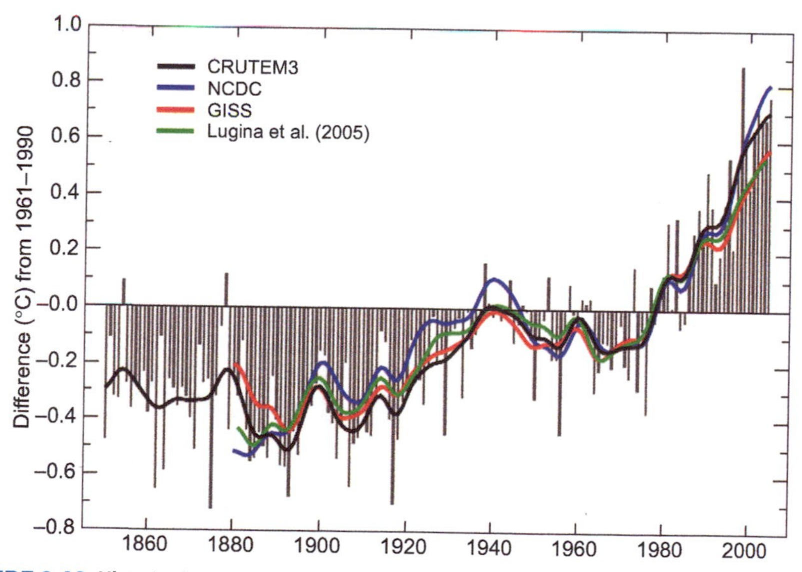

FIGURE 2.19 Historic rise in global mean temperature.
Units are deviation in degrees Celsius from the reference period 1961–1990. Colored lines represent temperature reconstructions using different methods. Bars indicate values from the instrumental record. *Source: From Climate Change (2007): The Physical Science Basis. Working Group I Contribution to the Fourth Assessment Report of the Intergovernmental Panel on Climate Change.*

CLIMATE CHANGE OR GLOBAL WARMING?

Early descriptions of climate change often used the term "global warming." This phrase is less used today because scientists have documented so many manifestations of the effects of greenhouse gases in the atmosphere—including increases and decreases in precipitation, warming, and even short-term cooling—that the broader term "climate change" is preferred.

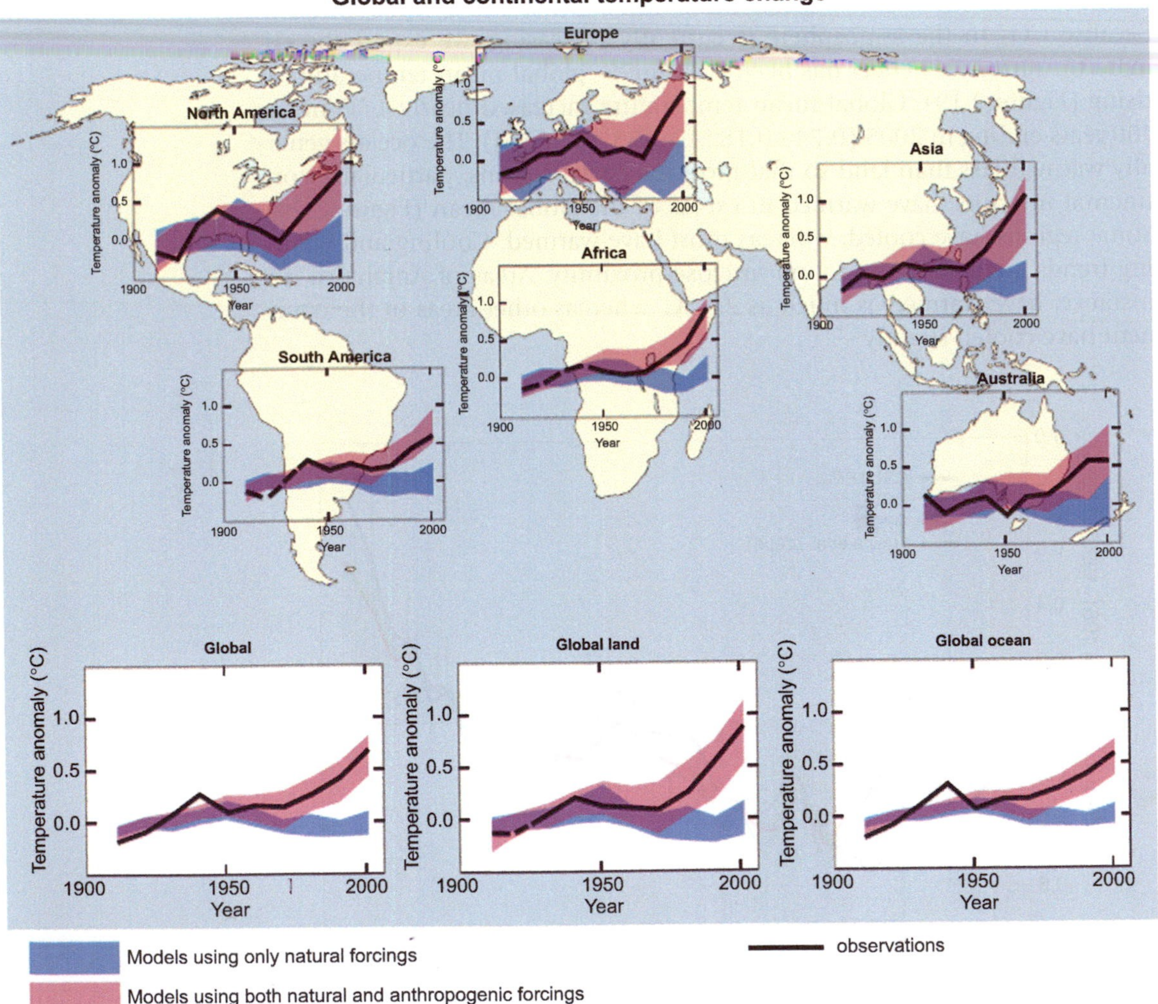

FIGURE 2.20 Rise in mean temperature by continent matches human-driven climate change.
Bold lines indicate historical record. Pink shading indicates the range of values from GCM simulations using both human and natural forcings on climate. Blue shading indicates range of values from GCMs forced with only natural (no human greenhouse gases) forcings. *Source: From Climate Change (2007): The Physical Science Basis. Working Group I Contribution to the Fourth Assessment Report of the Intergovernmental Panel on Climate Change.*

RAPID CLIMATE CHANGE

Rapid climate changes have been very common in the transition to current warm conditions and for the past 2 million years. In fact, rapid climate shifts appear as far back as we have good methods for detecting them. There may be multiple causes for rapid change, but several mechanisms are emerging as especially important.

Shutdown of the thermohaline circulation (Figure 2.21) is one factor that clearly drives rapid climate change. Meltwater from land ice in Greenland and North America enters the North Atlantic during warming (Figure 2.22), causing the waters of the Gulf Stream to become less salty. This less saline water is less dense and thus cannot sink and complete the return trip to the equator. The thermohaline circulation shuts down, stopping transport of heat from the equator. The net result of the shutdown is colder conditions throughout the North Atlantic, especially in Europe.

THE POWER OF THE GULF STREAM

The Gulf Stream is a mass of warm water transported from the tropical Atlantic northward by the thermohaline circulation. The arrival of this warm water affects the climate of the North Atlantic, making northern Europe significantly warmer than it would be without the Gulf Stream influence. Without the thermohaline circulation, northern Europe would be plunged into a cold spell—an example of how climate change might even result in local cooling.

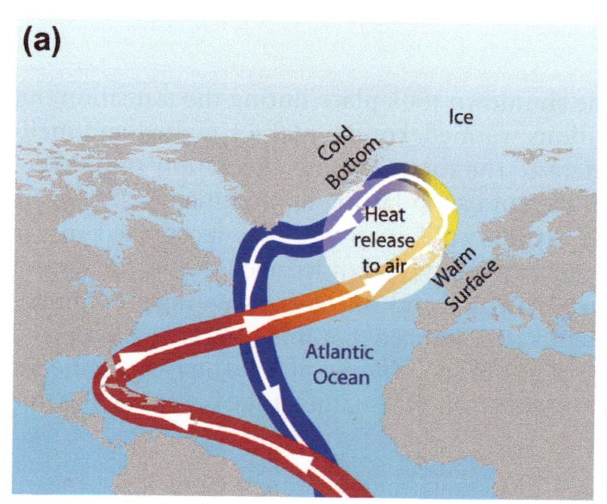

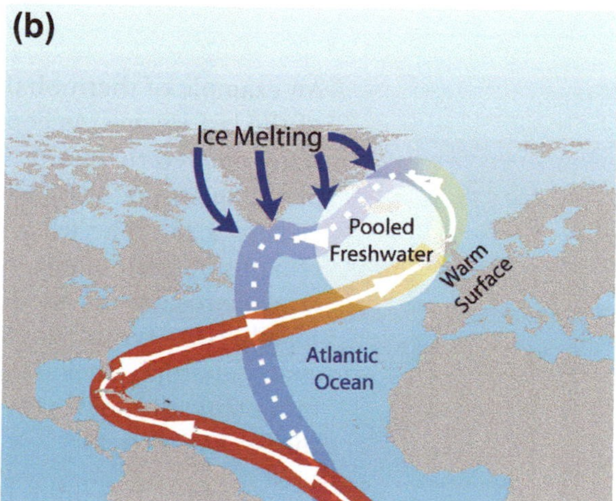

FIGURE 2.21 Shutdown of thermohaline circulation.
Thermohaline circulation is driven by dense water cooling and sinking (a). When polar ice melts (b), freshwater pulses in the North Atlantic can reduce contact of the Gulf Stream with ice and reduce its salinity. This leads to warmer, less saline water that is less likely to sink. If the freshwater pulse is strong enough, it can shut down thermohaline circulation.

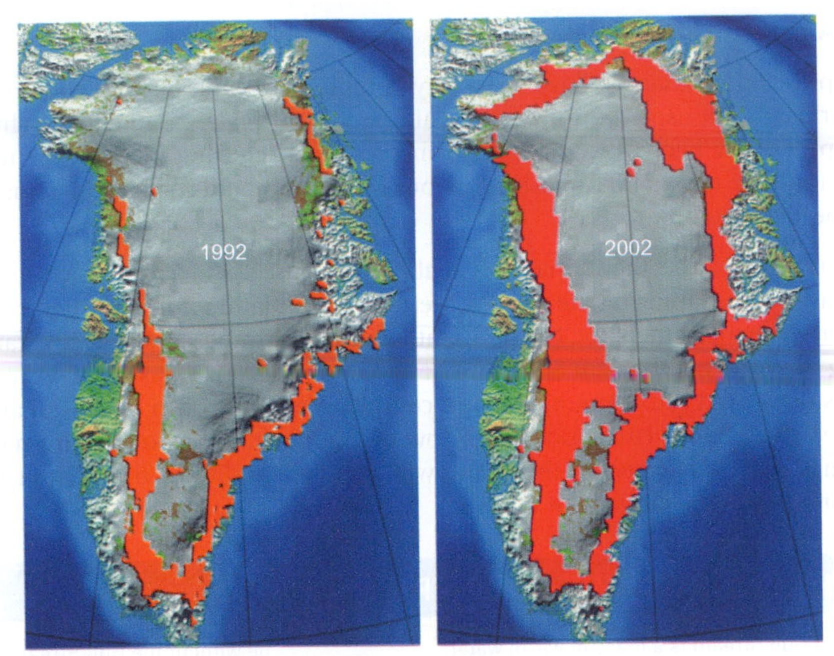

FIGURE 2.22 Recent greenland ice melting.
Red indicates areas of ice melt. Melt zones increased with warming in the latter half of the twentieth century. Greenland melt increases sea level rise, in contrast to the melting of sea ice (e.g., in Antarctica), which does not increase sea level because the ice is already displacing seawater. Continued acceleration of melting could result in shutdown of thermohaline circulation. *Source: Arctic Climate Impact Assessment.*

An example of thermohaline shutdown took place during the transition out of the last ice age. As conditions warmed, continental ice melted and meltwater entered the North Atlantic. The thermohaline circulation shut down, plunging Europe into a sudden cold snap lasting approximately 1000 years (Figure 2.23). The existence of this cold snap was first recognized in the remains of fossil plants. An arctic plant typical of ice age Europe, the mountain reinrose, was found in a narrow band of deposits dating to approximately 11,000 years ago. Scientists recognized that this indicated a brief cold snap in Europe. They named the cold snap for the plant. The Latin name for mountain reinrose is *Dryas octopetala*, and the name of the cold snap became the "Younger Dryas."

The Younger Dryas holds important lessons for the future because it was caused by warming that led to ice melt and thermohaline shutdown. An important question about future climate change due to greenhouse gas emissions is whether warming could again shut down thermohaline circulation.

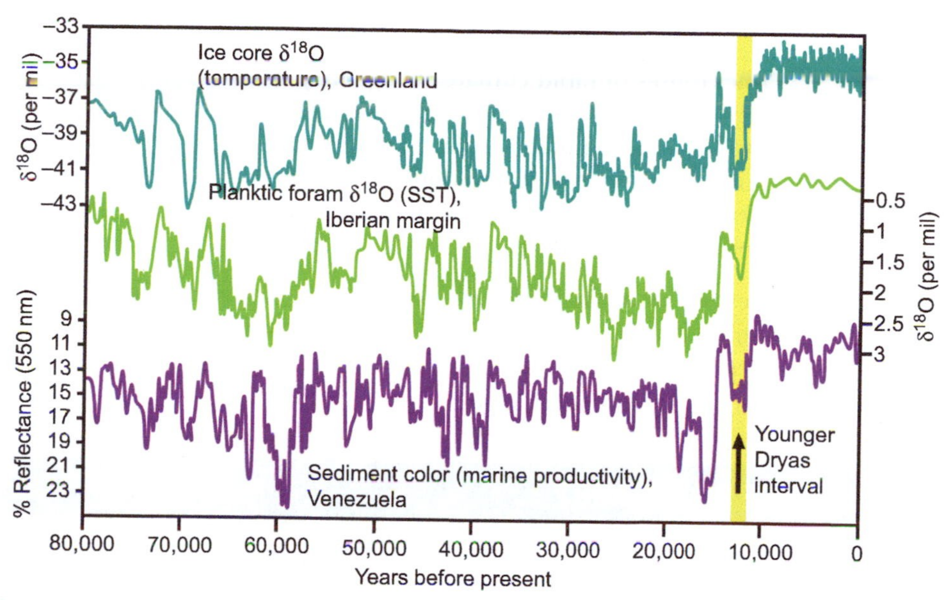

FIGURE 2.23 The Younger Dryas in context.

Temperature variation from three proxies shows very frequent, short climate flickers. All three proxies are in agreement on the frequency and relative timing of these flickers, indicating that they are global phenomena. The Younger Dryas event was a long-lasting climate flicker spanning about 1000 years that was particularly pronounced in Europe because it resulted from a shutdown of the Gulf Stream portion of thermohaline circulation. Note the relative stability of the warm climate of the past 10,000 years after the Younger Dryas. Human civilization has evolved in a period of atypical climatic stability. *Source: Reproduced with permission from Yale University Press.*

SPOTLIGHT: THE SEESAW EFFECT

Climate change does not unfold evenly on the two sides of the equator. Often, effects in one hemisphere are accompanied by changes opposite in sign in the other hemisphere or show up later in the other hemisphere. Alterations in northern sea ice extent are often an initiating event in these interhemispheric teleconnections, so that effects may be seen first in the Northern Hemisphere and then seen later or reversed in the Southern Hemisphere. This has been termed the "seesaw" effect.

Researchers see the seesaw effect in ice core proxies for temperature change. The "wiggles" of temperature change are indicated by oxygen isotope proxies in these cores. The wiggles in Antarctic ice cores do not match those in Greenland ice cores because of the seesaw effect.

Large, rapid changes in temperature are seen in the Greenland ice core record. Are these changes global? The answer seems to be no. Shackleton (2001) reviews research in this field and sees a seesaw of delayed change in the Southern Hemisphere rather than a global synchronous response. Rapid change in the Antarctic ice core seems to be offset from rapid changes in the Greenland ice core by thousands of years. Deep water temperatures in the Atlantic seem to follow the Antarctic pattern, whereas surface water temperatures follow the Greenland pattern.

The cause of the seesaw is most likely decoupling of climate connectivity across the equator. Circulation features such as Hadley cells originate at the equator, so there may be a delay in transmitting large changes across this boundary. Alternatively, the circumpolar current in the Antarctic could be a barrier to change, and the Antarctic climate may be out of synch with the rest of the planet. It has even been suggested that large abrupt change may originate in the tropics. Whatever the cause, Greenland and Antarctic ice core records clearly seesaw.

Source: Shackleton, N., 2001. Paleoclimate. Climate change across the hemispheres. Science 291, 58–59.

Other causes of rapid climate change may be associated with sudden releases of greenhouse gases such as CO_2 or methane. In the past, such releases may have occurred naturally from massive seabed deposits of methane hydrates, emissions from volcanic eruptions, or decay of vegetation associated with asteroid impacts. In the future, massive human emissions of greenhouse gases may have similar effects.

These emissions have massive effects on the global carbon cycle and are driving major changes in climate. Carbon cycle changes are important because they affect the balance that determines concentrations of CO_2 in the atmosphere and, hence, climate change.

SPOTLIGHT: HOTHOUSE OR ICE AGE?

Previous interglacial periods have typically lasted approximately 10,000 years. The current warm climate has lasted more than 10,000 years: Are we headed for another ice age?

The answer seems to be "no." The orbital forcings that create interglacial periods are in an unusual configuration that has not been seen for approximately 400,000 years. The last time the Earth's orbit was in a similar configuration, there was an unusually long interglacial period. That interglacial period is known as Marine Isotopic Stage 11 (MIS 11).

One clue about MIS 11 comes from Antarctic ice cores. Raynaud et al. (2005) unraveled folds in the Vostoc Antarctic ice core to examine MIS 11 more closely. Once they had corrected for a fold in the ice, it became clear that MIS 11 was unusually long and that greenhouse gas concentrations were high. Another piece of evidence came from a simple three-state model of ice ages. This modeling showed that the current combination of orbital tilt and eccentricity is likely to lead to an interglacial period of 20,000–30,000 years.

Humans are now pumping greenhouse gases into the atmosphere that will accentuate an already warm climate. If orbital conditions were typical, we might expect that warming to delay the advent of the next ice age. However, with the next ice age tens of thousands of years away, human warming is likely to take climate to temperatures not seen in millions of years.

Source: Raynaud, D., Barnola, J.M., Souchez, R., Lorrain, R., Petit, J.R., Duval, P., et al., 2005. Palaeoclimatology: the record for marine isotopic stage 11. Nature 436, 39–40.

THE VELOCITY OF CLIMATE CHANGE

Another way to look at the rapidity of climate change is the velocity of temperature change. The velocity of climate change is proportional to the distance that has to be traveled over the surface of the Earth to maintain a certain temperature as climate changes (Figure 2.24). This speed varies depending on terrain. On a perfectly flat smooth Earth, maintaining a constant temperature requires shifting north or south with latitude, toward the cooler poles to offset warming. This latitudinal shift sets the velocity of climate change in areas with little topography. In mountainous areas, a similar temperature change can be achieved by shifting upslope, and these distances are much shorter. In the real world, with topography, flat areas have a high velocity of climate change and mountains have a low velocity of climate change.

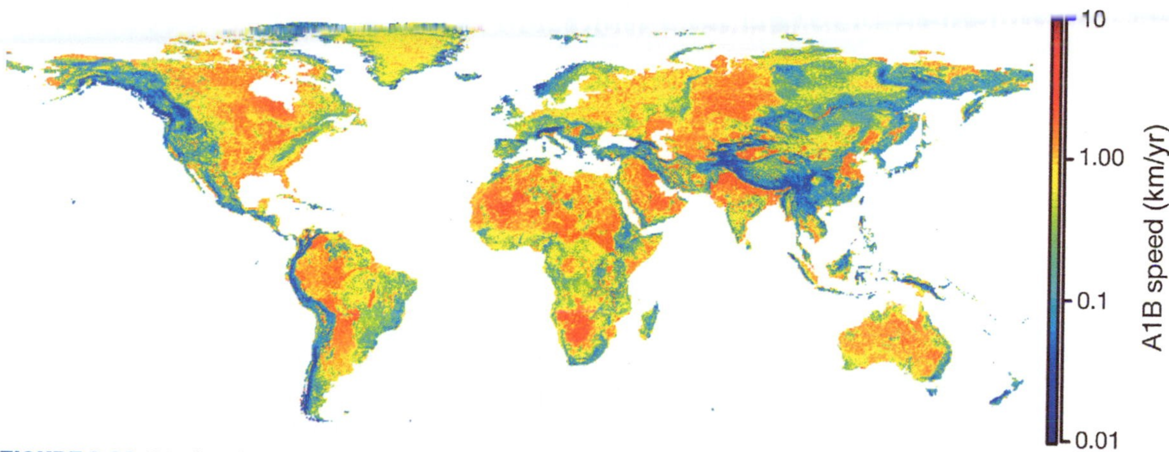

FIGURE 2.24 Velocity of climate change.
A global map of the velocity of temperature change shows the variation in speed of temperature change across the surface of the Earth. Velocity of climate change is the product of atmospheric temperature change and topography. Flat areas require latitudinal movement to track atmospheric temperature change and have a high velocity of climate change. Mountainous areas can track temperature change upslope across much shorter distances and so have a low velocity of climate change. *Source: Loarie et al (2009). Reprinted with permission from Nature.*

Velocity of climate change matters to living things, especially plants. Plants are mobile only when seed disperses, which makes keeping up with the velocity of climate change a problem. Plant dispersal may have trouble keeping pace with temperature shifts where the velocity of climate change is high. In the mountains, the velocity of climate change reduces the distances that plants have to cover to keep pace with climate change, but gravity may work against some dispersal mechanisms. Most obviously, gravity-dispersed seeds will move downhill, in the direction opposite of temperature shifts. Understanding the velocity of climate change and its interplay with species' dispersal is therefore important to understanding both past and future range shifts.

MODELING THE CLIMATE SYSTEM

Climate change models allow the simulation of the effects of the buildup of greenhouse gases centuries into the future, based on current understanding of atmospheric physics and chemistry. The typical horizontal resolution of a global climate model is 100–200 km. Combining global and regional models allows finer-scale examination of regional details of change to horizontal resolutions of 10–50 km. Most global models are run on supercomputers, whereas some regional models may be run on desktop computers (often taking six to eight months for a single realization).

STRUCTURE OF GENERAL CIRCULATION MODELS

General circulation models (GCMs) use a system of mathematical equations to simulate the movement of mass and energy from one part of the atmosphere to another. They divide the atmosphere and ocean into a series of three-dimensional cells, each of which transfers mass and energy to its neighbors based on the outcome of the equations within the cell. These are in principle the same type of model used to predict weather, but they are run on a broader (global) scale and for centuries rather than days.

Global climate models simulate climate changes across the entire planet. These models are often referred to as GCMs because they simulate general atmospheric circulation patterns. GCMs represent atmospheric and ocean circulation in a series of equations describing physical properties of gases and fluids. Each set of equations is solved for a volume of air or water, typically with dimensions of hundreds of kilometers. The atmosphere and oceans are represented by thousands of these cubes, distributed 10–20 layers thick across the face of the planet and down into the oceans. Energy and water vapor (or liquid) are passed between the cubes, allowing simulation of ocean currents and circulation in the atmosphere. This process is similar to that for models used to forecast weather, except it is applied over broader spatial scales to capture global effects and on longer timescales to capture climate instead of weather. Because of these broader spatial and temporal scales, the resolution of GCMs must be much coarser than that of weather models to stay within the computational limits of modern computers.

EVOLUTION OF GCMS

Models of global climate began as mathematical descriptions of atmospheric circulation. They were known as general circulation models or GCMs. As the models became more complex, layers of ocean were added. These models were known as coupled atmosphere–ocean GCMs (AOGCMs). Today, most GCMs are AOGCMs. More advanced models incorporate the effects of vegetation change on climate, effectively joining models of the biosphere to the ocean–atmosphere models. These most advanced models are known as earth system models.

Regional climate models (RCMs) are very similar in structure to GCMs, but they capture finer-scale resolution of change in a particular region (Figure 2.25). The equation-based processing, cubes, and layers of the GCM are all present in an RCM but at finer scale. The scale of an RCM is measured in tens of kilometers, as opposed to hundreds of kilometers for most GCMs. In exchange for higher resolution, RCMs must be run for regions, rather than for the whole planet, as their name implies. This trade-off of resolution for geographic scope is required by the limits of computational time required to run the model. Model runs of more than a week on a supercomputer are usually prohibitively expensive because the model must compete for other uses of the specialized computing facility, such as weather forecasting. Because the climate of one region is connected to the climate of neighboring regions, RCMs cannot be

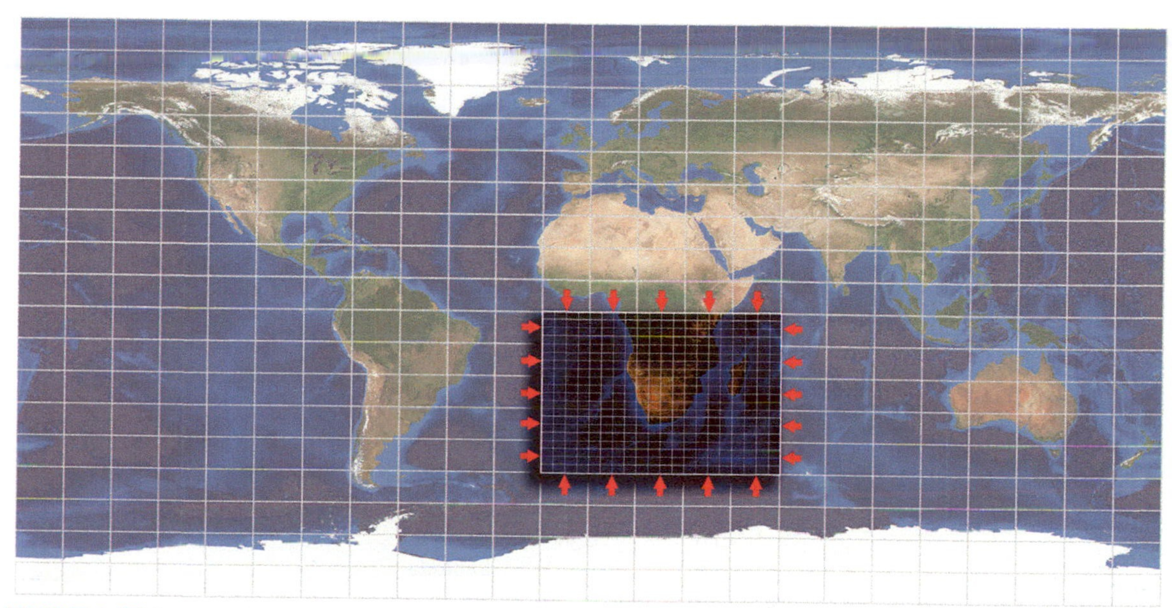

FIGURE 2.25 Regional climate model.
Regional climate models (RCMs) are run embedded in general circulation models (GCMs). They receive information at their boundary from the GCM. An RCM therefore cannot rectify errors in a GCM. It can, however, improve the simulation of regional change by resolving processes that cannot be captured at the resolution of a GCM. Here the domain of an RCM is illustrated embedded within the domain (global) of a GCM. Red arrows indicate transfer of information from the GCM to the RCM. *Source: Courtesy of NASA.*

run alone; they must be connected to other regions in some way. The most common way for an RCM to be connected to global climate is to embed an RCM into a GCM. The RCM then takes coarse-resolution GCM inputs at its edges and turns them into a finer-scale regional climate simulation.

The trade-off between spatial resolution (scale) and geographic scope (domain) of a model forces the use of adaptations of GCMs to address special problems. For instance, to study atmospheric phenomena in more detail, climate modelers will sometimes use a high-resolution atmospheric model but will couple it to a static ocean model to save computational demands.

NO MEAN TEMPERATURE

Global mean temperature is the political yardstick often used to measure climate change impacts and the success of international policy efforts. It is a simple and clear metric for these purposes, but it is the wrong metric for biological analyses. Biological impacts happen in specific places that all have their own unique climate characteristics important to species' survival—the global mean fuzzes all these meaningful regional variations into one number. For instance, the variation between islands, which are much cooler because their climates are dominated by cooler oceanic temperatures, and continental interiors, which are relatively much warmer, is completely obscured in global mean numbers (Figure 2.26). Global mean temperature is fine for international policy dialog, but biologists need to pay attention to regional on-the-ground variation.

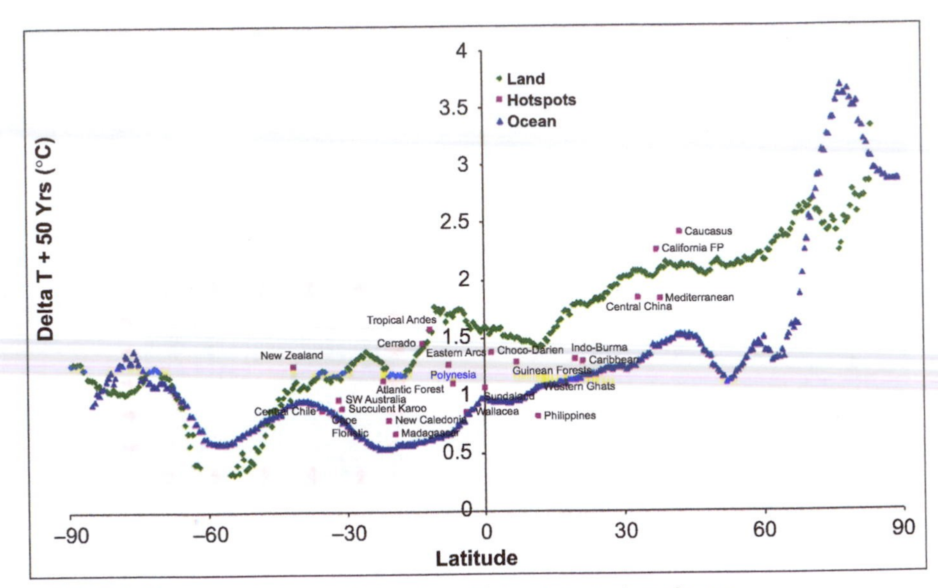

FIGURE 2.26 Global biodiversity hot spots and future temperature change.
Temperature versus latitude in a Hadley Center general circulation model (GCM) simulation for 2050 (A2 scenario). Blue circles indicate mean temperature change in the ocean at a latitude, green circles indicate mean temperature change over land for the latitude. Temperature change over oceans at midlatitudes is less than temperature change over land owing to the heating properties of water and continental interiors. Mean temperature change in the global biodiversity hot spots is indicated by red squares, for comparison. Note that island hot spots fall near the oceans line, while continental hotspots fall near the land line.
Source: Lovejoy and Hannah (2005). Reproduced with permission from Yale University Press.

GCMs are also used to establish the role of human emissions in climate change. For these assessments, GCM simulations are run for the recent past using only natural drivers of climate change and compared to observed warming trends (see Fig. 2.20). In general, GCMs are able to reproduce the full range of warming that has been observed since the last decades of the twentieth century only when human drivers of change ("human forcings") are included in the models. This is generally taken as strong evidence that human pollution is the cause of recently observed climate change.

Reconstructing past climates may be done with GCMs to either validate the models or investigate possible past conditions. GCMs may be tested by determining if they can reproduce past climates. Of course, past climate in these tests must be reconstructed from other sources, such as pollen records. Often, the past record is not robust enough to provide a very detailed test of GCMs, but GCMs can be tested to determine if they can reproduce the broad outlines of past climate, such as temperature changes over thousands of years. Because pollen and other records of past climate are fragmentary, GCMs can also be used to explore gaps in our understanding of past change. For example, GCMs have been used to try to explore the role of greenhouse gas forcing in past climates.

There are some types of past change, however, that GCMs do not represent well. Transitions between glacial and interglacial periods are not reproduced well by GCMs. This is probably because of positive feedbacks not captured well even by sophisticated current GCMs. Simpler models that simulate transitions between multiple states better reproduce glacial–interglacial transitions.

STATE MODELS FOR GLACIAL–INTERGLACIAL PERIODS

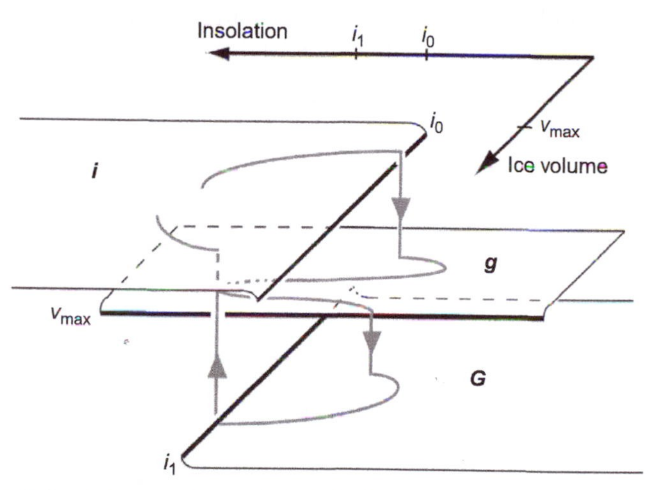

Schematic of simple three-state model of glaciation.
The diagram illustrates transitions between interglacial (i), mild glacial (g), and full glacial (G) climate regimes regulated by insolation and global ice volume. This model illustrates an i-G state transition when insolation drops below a critical threshold (io), with the depth and persistence of glacial states (g, G) determined by the existing ice volume as insolation changes. *Source: Palliard, 1998.*

Glacial-interglacial transitions are not well simulated in GCMs, but simpler "state" models reproduce some of the behavior of these transitions remarkably well. A state model represents glacial conditions as one state, with interglacial conditions as a second state, with transition coefficients between the two. These simple models are sometimes used to explore glacial–interglacial dynamics because GCMs, for all their complexity, cannot yet reliably simulate state-transition dynamics.

REGIONAL CLIMATE MODELS

GCMs provide some insight into climate changes on regional scales relevant to assessing impacts on biological systems, but the utility of GCMs in regional work is limited by their coarse scale. Most GCMs are run at scales of hundreds of kilometers, which means that a subcontinent may be represented by as few as five or six cells in the GCM. This means that many features that will be important in determining orographic rainfall and other important climate phenomena will not be resolved at the scale of the GCM.

For instance, in a GCM, all of the mountains of western North America will be represented as a single large "hump" that extends from the Sierra Nevadas of California to the Rocky Mountains (Figure 2.27).

To solve this resolution problem, RCMs are used. RCMs operate on the same principle as GCMs but with cells of considerably smaller dimensions—typically 10–80 km on a side. Because the dimension is squared to get the area of the cell, and cubed to get its volume, an RCM at 50 km has more than 100 times greater resolution than a GCM at 300 km.

The greater resolution of the RCM allows representation of mountains and other topographic features with greater fidelity. This in turn allows simulation of orographic rainfall, temperature variation with altitude, and other features lost in coarse-scale GCMs. The resolution of GCMs and RCMs is typically given as the

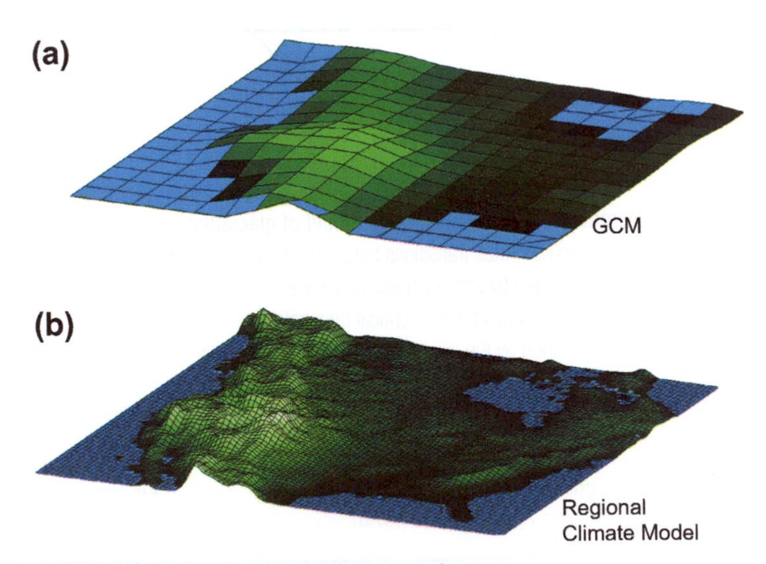

FIGURE 2.27 Regional climate model (RCM) resolution.
An RCM can resolve features such as mountain ranges that have important influences on climate. In this example from North America, all mountains from the Sierras of California to the Rocky Mountains are represented as a single hump at the horizontal resolution of a GCM, whereas they are better resolved at the resolution of the RCM. *Source: Lovejoy and Hannah (2005). Reproduced with permission from Yale University Press.*

length of one side of a grid cell, or horizontal resolution. The horizontal resolution of modern GCMs is typically 80–300 km, and that for RCMs is 10–80 km.

An RCM must be embedded in a GCM to function. At its edges, the RCM needs information about conditions in neighboring cells. For instance, an RCM cannot simulate orographic rainfall unless it knows the amount of moisture entering the region. These neighbor cell conditions, or "boundary conditions," are provided by the GCM in which the RCM is embedded (see Fig. 2.25).

The higher resolution of an RCM is appropriate for many regional impact assessment applications. At finer horizontal resolution, rainfall changes over many areas of a region may be resolved. Temperature changes, such as up-mountain slopes, can be resolved at scales relevant to cities, watersheds, and other planning units.

STATISTICAL DOWNSCALING

GCM projections are translated for regional impact assessment using either statistical or dynamic downscaling. Dynamic downscaling nests a fine-scale climate model (or RCM) within a GCM. Statistical downscaling uses observed relationships between large-scale climate phenomenon and local conditions to generate fine-grain projections from GCM output. For instance, rainfall at a site may be correlated with synoptic conditions such as regional atmospheric pressure fields. If such a relationship exists, it can be used to project future rainfall using pressure fields simulated by a GCM. Biologists doing regional impact assessment need to be aware of alternative downscaling methods because GCM simulations are too coarse to be useful in these applications.

For many biological applications, however, even 10–80 km is still very coarse. Movements of large animals in a landscape occur on scales of a few tens of kilometers. Movements of small animals may occur on scales of meters or kilometers. Plant dispersal events, particularly for structural species such as trees, may occur on scales as small as a few meters.

To address these relatively fine-scale phenomena in biological assessments, further reduction in scale may be achieved through smoothing and interpolation. In this process, present climate data are interpolated to a desired scale, such as 1 km. The difference between present and future GCM simulations is then added to the interpolated current climate data, yielding a future climate surface at the desired scale. The edges of GCM cells are smoothed to avoid "blocky" changes in the future surface. This process may be used to reduce either GCM or RCM (or statistically downscaled) data to fine scale. Much current biological impact assessment in species distribution and dispersal is now done with climate surfaces at a 1-km scale using this technique (available for download at sites such as Worldclim.org).

DOWNSCALING SIMPLIFIED

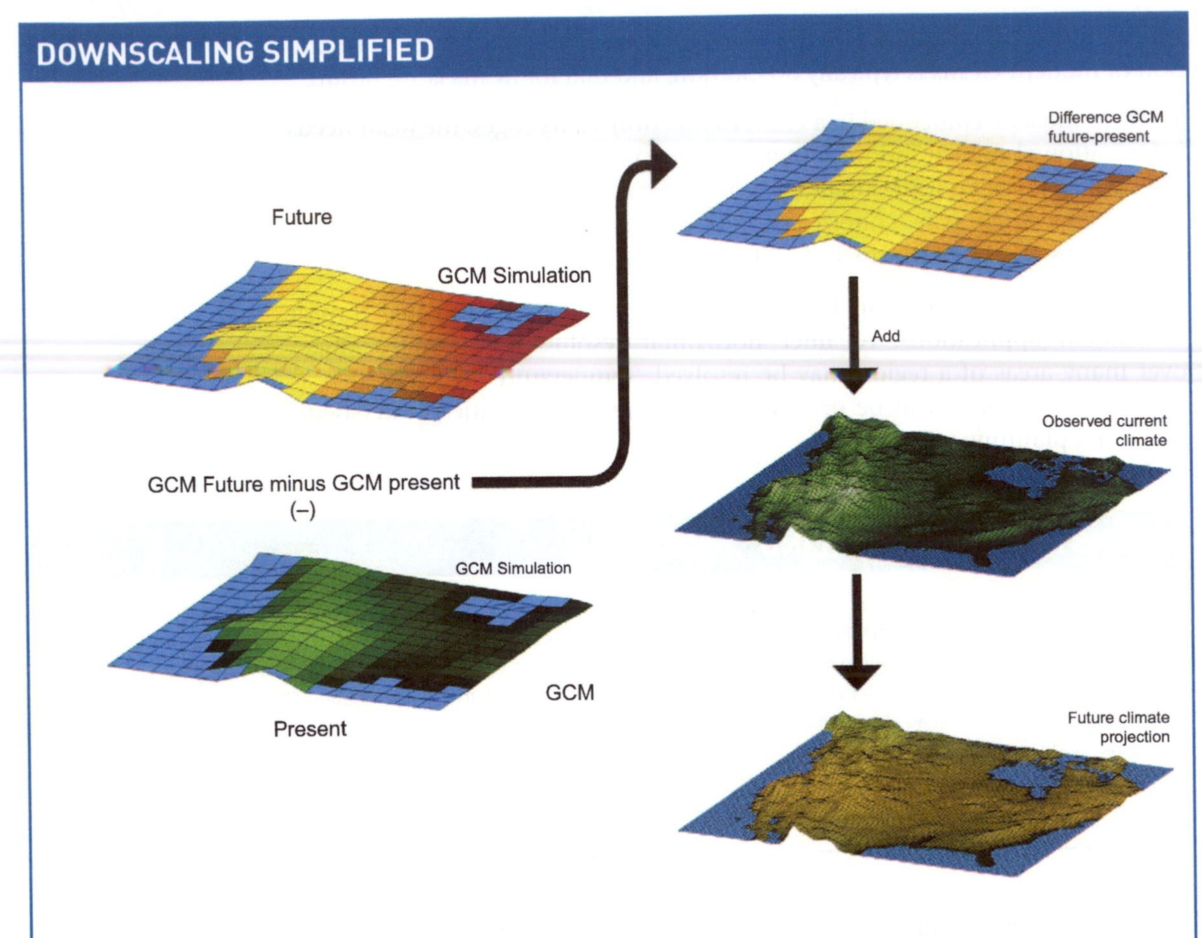

One approach to generating finer-scale regional climatologies from GCMs is simpler than either statistical downscaling or RCM approaches. The difference method subtracts the present value for a variable of interest, such as temperature, that a GCM projects from the future projected value. This difference is then added to current observed climate for that variable to obtain an estimate of possible future values. The difference method is used because GCMs do not faithfully reproduce present climate at fine scales, so comparing future GCM projections to observed climate may result in errors. The method takes the amount of change from the GCM but the spatial and temporal variability from observed (current) climate.

COMMONLY USED GCMs

Many GCMs exist, developed by universities, research centers, and national weather services. All are run on supercomputers or massively parallel computers. All use similar suites of physical equations but differ in the specifics of particular equations, complexity, and treatment of parameters.

SUPERCOMPUTERS AND MASSIVELY PARALLEL COMPUTERS

Source: Supercomputers. Courtesy of NOAA.

GCMs employ complex equations in a simulation of the entire globe, making them very computationally intensive. This means that an individual GCM simulation will take a long time to run on a conventional computer. To speed runs, supercomputers with large memory and processing capability are used. These are often the very same supercomputers that are used to run weather forecasts, although climate change research centers, such as Britain's Hadley Centre, have their own dedicated supercomputers. A less expensive alternative is to join many smaller workstations in parallel. Such systems are called massively parallel computing systems and have become an option for smaller labs and even some major international centers.

Most assessments of climate change use simulations from more than one GCM because no model simulates the future perfectly. Using more than one GCM therefore helps researchers explore the uncertainty in possible future climates. It is therefore important that several credible GCMs are available to choose from.

Among the best known and most widely used GCMs is probably that of the Hadley Centre in Britain. The Hadley Centre is a branch of the weather service (Meteorological Office or Met Office) in the United Kingdom. The Hadley Climate Model or HadCM is a relatively sophisticated model that includes active interaction between climate and land cover.

Other GCMs frequently used in impact assessments include the Community Climate Model produced by a consortium of universities (the climate research "community") led by the National Center for Atmospheric Research in Boulder, Colorado; the Canadian Climate Model produced by a research group at the University of Victoria using the supercomputer of the Canadian Weather Service; the GFDL model created by NASA's General Fluid Dynamics Laboratory; and the CSIRO GCM run by the Commonwealth Scientific and Industrial Research Organization in Australia. For most assessments, several of these models are used—often those that do particularly well at simulating current climate for the region of interest, though increasingly, ensembles of large numbers of GCMs (30+) are becoming best practice in impact assessments.

Assessments must also make assumptions about the magnitude of future greenhouse gas emissions. Humans are using increasingly more fossil fuel each year, resulting in increasing emissions of CO_2 and rising atmospheric CO_2 levels. How fast those emissions continue to grow will determine how fast and how much climate will change. No one knows with certainty how future energy use will unfold, so assumptions must be made in every GCM about how much CO_2 is released into the atmosphere over the time period being modeled.

EMISSIONS PATHWAYS

The Intergovernmental Panel on Climate Change (IPCC) has prepared a series of standard greenhouse gas concentrations for use in GCM simulations. These are known as Representative Concentration Pathways (RCP). Each RCP represents one possible trajectory for future greenhouse gas concentrations. Previous IPCC assessments have used emissions scenarios, with identifying codes such as A1 and B2. These are now outdated, though sometimes still an important source of information. The RCP series was constructed from a range of possible emissions curves, rather than from global storylines. The RCPs allow assessment of the implications of a range of possible pathways, including 'overshoot' scenarios in which global greenhouse gas concentrations exceed a policy target (e.g., 2° C global mean temperature change) and then return to the target level.

IPCC

The Intergovernmental Panel on Climate Change was formed in 1988 by the World Meteorological Organization to help promote a scientific and political consensus about the occurrence and impact of climate change. Leading climatologists and impact researchers conduct reviews of climate change research for the IPCC, which are then submitted to a vetting process open to all 192 United Nations member governments. Previous IPCC reports have been issued in 1990, 1995, 2001, and 2007. The current IPCC report is the fifth. The IPCC shared the Nobel Peace Prize with Al Gore in 2007.

Each RCP is identified by a number corresponding to the watts per square meter (W/m^2) forcing associated with the greenhouse gas concentrations in 2100 (Figure 2.28). For example, RCP6.0 is a pathway with a forcing of 6.0 W/m^2 in 2100. RCP2.6 is an 'overshoot' pathway, in which atmospheric CO_2 rises and then falls. RCP8.5 is most similar to current emissions trends.

GCM OUTPUTS

The equations of a GCM produce "weather"—daily rainfall, temperature, and winds values. However, most of these outputs are not saved in a GCM run. It is simply not practical to save such large amounts of data. Instead, summary statistics are saved, such as mean monthly rainfall and mean temperature. This allows a profile of the run to be saved without taking up huge amounts of storage space with unneeded data. The GCM output is often further simplified to produce a trace of global mean temperature increase (Figure 2.29).

However, GCM runs do not always save the data most relevant to biological analysis. Organisms may respond to extreme events, such as drought or severe

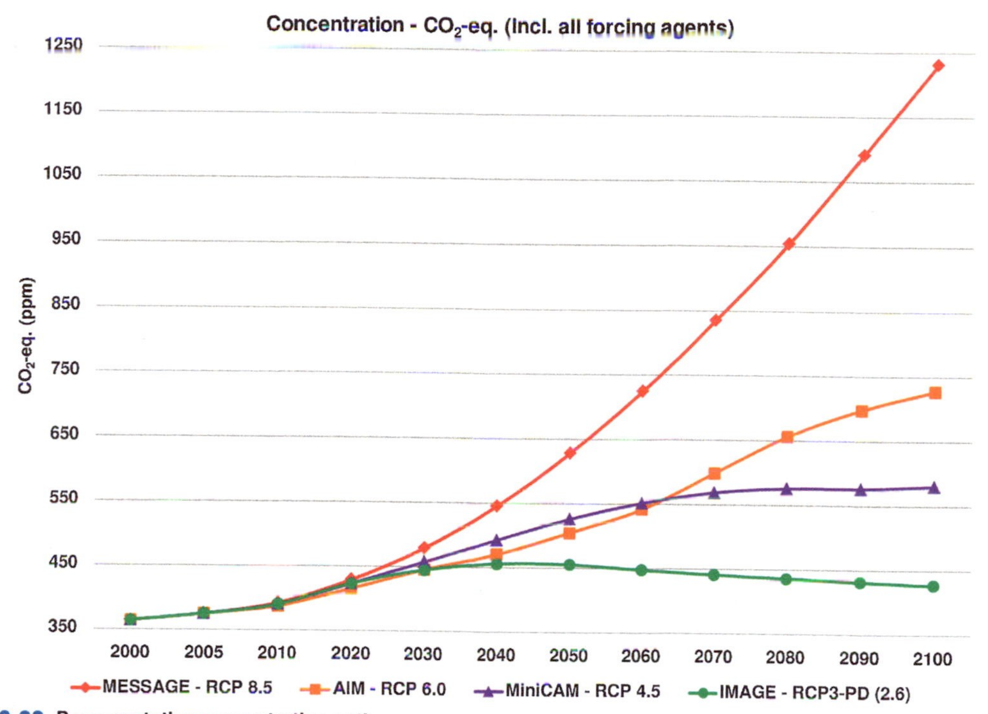

FIGURE 2.28 Representative concentration pathways.
The Intergovernmental Panel on Climate Change (IPCC) defines possible future buildup of greenhouse gases in the atmosphere as representative concentration pathways (RCP). RCPs represent atmospheric greenhouse gas concentrations resulting from possible differences in future emissions. Higher concentrations will result from higher emissions and lack of action to curtail emissions, whereas lower concentrations may result from lower economic growth or active efforts to reduce greenhouse gas emissions. RCP units are watts per square meter, corresponding to the radiative forcing of various concentrations in 2100. This figure shows the CO_2 equivalent of each RCP, the atmospheric concentration of CO_2 required to have a forcing equivalent to all greenhouse gases in the RCP (CO_2, methane, etc.). *Source: IPCC 2014.*

storms, that are not captured in mean monthly statistics. Biological studies may use typically archived statistics or work with climatologists to have more biologically meaningful outputs saved or extracted from GCM runs.

ASSESSMENT USING MULTIPLE GCMS

Assessments of climate change impacts may focus on individual disciplines, such as biology or agriculture, or be integrated multidisciplinary assessments. In either case, the use of multiple GCMs is recommended. Assessing possible outcomes against a variety of GCM projections can help capture uncertainty about possible futures. For instance, for many regions, projected change in rainfall varies from an increase in some GCMs to a decrease in others. Using several projections can help bracket these possible outcomes. In some circumstances, taking ensemble combinations of GCM projections can be more accurate than using them separately to bracket possibilities. Whatever approach is used, the use of several GCMs improves the credibility of impact analysis, whereas the use of too many GCMs results in a welter of findings that are difficult to sift through.

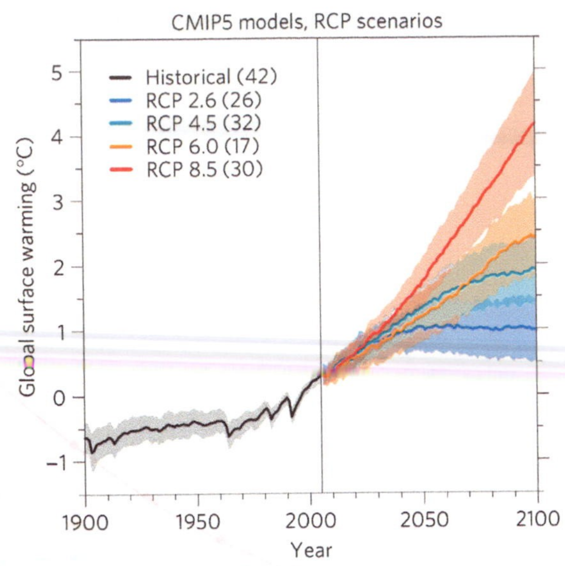

FIGURE 2.29 Global mean temperature estimated from general circulation models (GCMs). Using representative concentration pathways (RCP) scenarios, GCMs simulate global climate change. One summary statistic from these simulations is global mean temperature, shown here as it varies with RCP (heavy colored lines) and GCM (shading around colored lines). *Source: From Knutti and Sedlacek (2012). Reproduced with permission from Nature.*

BIOLOGICAL ASSESSMENTS WITH DOWNSCALED DATA

Assessments use GCMs to provide climate simulations against which biological changes can be judged. Such assessments can use either statistical or mechanistic models. Mechanistic models use equations describing biological processes such as photosynthesis to infer change in vegetation type or disturbance. Statistical models use statistical relationships or equations simulating biological processes such as photosynthesis to drive a specific quantitative result with GCM input. For instance, the vegetation type can be simulated by a dynamic global vegetation model, or habitat suitability for a species can be simulated with a species distribution model.

FURTHER READING

Intergovernmental Panel on Climate Change, 2013. Climate Change 2013: The Scientific Basis; Contribution of Working Group I to the Fourth Assessment Report of the Intergovernmental Panel on Climate Change. Cambridge University Press, Cambridge, UK.

Intergovernmental Panel on Climate Change, 2014. Climate Change 2014: Impacts, Adaptation and Vulnerability; Contribution of Working Group II to the Fourth Assessment Report of the Intergovernmental Panel on Climate Change. Cambridge University Press, Cambridge, UK.

Intergovernmental Panel on Climate Change, 2014. Climate Change 2014: Mitigation of Climate Change; Contribution of Working Group III to the Fourth Assessment Report of the Intergovernmental Panel on Climate Change. Cambridge University Press, Cambridge, UK.

2

SECTION

The Impacts of Human Induced Climate Change

CHAPTER 3

Species Range Shifts

Climate determines where plants and animals can live. Plants and animals all have an "address"—the combination of conditions in which they can survive and reproduce. This address is called their *niche*: by definition, organisms cannot survive outside of their niche.

Climate plays a major role in defining the niche of all species. Temperatures too hot or too cold, too much moisture or too little, all determine where plants can grow and where animals can survive. If you have tried to grow a palm tree in New York City or raise pears in Phoenix, you know that plants cannot flourish where the climate is wrong for them. Palm trees are killed by frequent New York frosts, and pears do not get enough chilling to set fruit in Phoenix. The niche space in which species can survive is determined in large part by suitable climatic conditions.

Species' ranges shift over time to track suitable climate. When climate changes in a location, some species may find themselves in suddenly hostile conditions. Others will find that previously unsuitable climates have changed in their favor. Individuals in unsuitable conditions will die or fail to reproduce, gradually disappearing from the location, whereas individuals near newly suitable habitat will gradually occupy areas in which they have not occurred previously.

GRINNELLIAN AND HUTCHINSONIAN NICHES

Joseph Grinnell first used the term niche in 1917. Charles Elton (1927) described species niche as something akin to an occupation: it was what the species did to survive, or its role in a biological community. Later, G. Evelyn Hutchinson expanded the concept to include the full range of environmental conditions that determine a species' fitness or survival. The Hutchinsonian niche is more analogous to a species' address. Climate change biologists are interested in the range of climatic conditions that determine a species' distribution and therefore more frequently employ the Hutchinsonian concept of niche.

When climate changes too rapidly, extinctions may occur. However, extinction is not always or even most often the end result of climate change. More often, species are able to track suitable climatic conditions, occupying new areas and leaving unsuitable locations as climate changes, a process sometimes termed "niche tracking."

Climate Change Biology. http://dx.doi.org/10.1016/B978-0-12-420218-4.00003-2

SPECIES RANGES

A species' range is the area in which it is found, including both its extent (extent of occurrence) and the locations within that extent that are actually occupied by the species (area of occupancy). Range is determined by the spatial distribution of individual populations. Abundance, which is often governed to some degree by climate, determines whether populations endure. When populations are lost on the range periphery, or when new populations appear that expand the range, a range shift occurs.

Range shifts can be driven by long-term changes in mean climate state, by short-term climatic extremes such as freezing, or by interactions with other species being driven by climate change. Many examples of each of these types of shifts have already been observed owing to climate change. Managing these movements is one of the great challenges for conservation in the twenty-first century.

This chapter focuses on range shifts of plants and animals in terrestrial, marine, and freshwater systems. These are among the most dramatic and best understood of the mounting biological impacts of human-driven climate change. The evidence is accumulating quickly and is far from complete—there are certainly major changes yet to be documented. However, the overall body of evidence clearly shows that range shifts are occurring in many species in response to global warming and changes in other climate variables.

FIRST SIGN OF CHANGE: CORAL BLEACHING

Perhaps the most severe and wide-ranging impact of climate change on biological systems is coral bleaching. Corals have microscopic algae, zooxanthellae, that inhabit their cells in a symbiotic relationship. The algae photosynthesize and pass nutrients to the coral host and the coral provides a physical structure that protects the algae and keeps them in adequate sunlight for photosynthesis.

ZOOXANTHELLAE—THE OTHER HALF OF CORALS

Corals harbor microscopic algae called zooxanthellae within their tissues. Zooxanthellae provide products of photosynthesis to the coral, and the coral in turn provides a physical reef structure that keeps the zooxanthellae near the surface, where light for photosynthesis is abundant. When this symbiotic relationship breaks down owing to high water temperature, corals expel their zooxanthellae, causing them to appear white or "bleached."

However, when corals are exposed to high water temperatures, they expel their algae. Without the photosynthetic pigments in the zooxanthellae, the corals lose their color and all that is visible is their calcium carbonate skeleton, which appears white. The coral thus appears "bleached" (Figures 3.1 and 3.2). Coral bleaching was undescribed in the scientific literature 50 years ago, yet it is so common and widespread today that almost all coral reefs in the world have been affected at one time or another.

FIGURE 3.1 Bleached coral.
El Niño events in 1982–1983 and 1997–1998 bleached corals in reefs throughout the world. Bleaching is an increasingly common phenomenon even in non-El Niño years. This coral head in St. Croix bleached in 1995. *Source: Courtesy U.S. National Oceanic and Atmospheric Administration.*

FIGURE 3.2 1997–1998: A deadly year for corals.
The right panel shows corals bleached in the El Niño event of 1997–1998. The left panels show a single coral head pre- and postbleaching: (a) prebleaching, (b) bleached coral head, (c) partially recovered coral head, and (d) fully recovered postbleaching. *Left Source: Manzello et al. (2007); Right Source: Courtesy U.S. National Oceanic and Atmospheric Administration.*

The intensification of coral bleaching is due to human-induced warming of the oceans. Corals live in the shallow surface waters of the ocean, which have warmed first and most quickly in response to atmospheric warming due to the greenhouse effect. As the atmosphere has warmed, some of that heat has been transferred to the surface of the ocean, resulting in warmed global mean ocean temperatures near the surface, or sea surface temperature (SST).

SEA SURFACE TEMPERATURE

SST drives many climatic phenomenon. SST is important because oceans comprise two-thirds of the Earth's surface and what happens at the air–water interface influences much of what happens near the habitable surface of the planet. SST is important in strengthening hurricanes and in determining the height of tropical cloud formations, among other phenomena. From disturbance regimes to cloud forest limits, change in SST is biologically relevant.

When these higher sea surface baseline temperatures are combined with warming in El Niño events, temperatures rise high enough for bleaching to happen. Bleaching occurs when SST rises more than 1 or 2 °C above normal summer maximal temperatures for periods longer than three to five weeks. Thus, both temperature and duration of exposure are important determinants of whether bleaching occurs and its severity once it happens. Because water temperatures vary between regions, the threshold temperature for bleaching is also different from region to region. It may also vary seasonally in the same region.

Once corals bleach, they often die. Recovery is possible but varies strongly depending on the severity of the bleaching event and conditions immediately after the event. Corals already weakened by other factors, such as pollution, sedimentation, or disturbance by tourism, are less likely to recover. For example, coral cover in the Great Barrier Reef in Australia declined 50% from 1985 to 2012 due to the combined effects of coral bleaching, cyclone damage and predation by crown-of-thorns starfish (*Acanthaster planci*).

Mortality due to bleaching may be severe enough to wipe out entire species over large areas. It is therefore a strong driver of range shifts in corals. For example, in the central lagoon of Belize, staghorn coral (*Acropora cervicornis*) was the dominant species until the 1980s, when it was wiped out by a combination of disease and rising water temperatures. The scroll-like coral *Agaricia tenuifolia* took over as the dominant coral, only to be wiped out in the high water temperatures of the 1998 El Niño event. These massive mortalities were the worst in at least 3000 years, resulting in range changes over large areas of the Caribbean for staghorn and other corals.

Reefs throughout the world are being hit with coral bleaching so severely that the future distribution of all tropical coral reefs is in question (Figure 3.3).

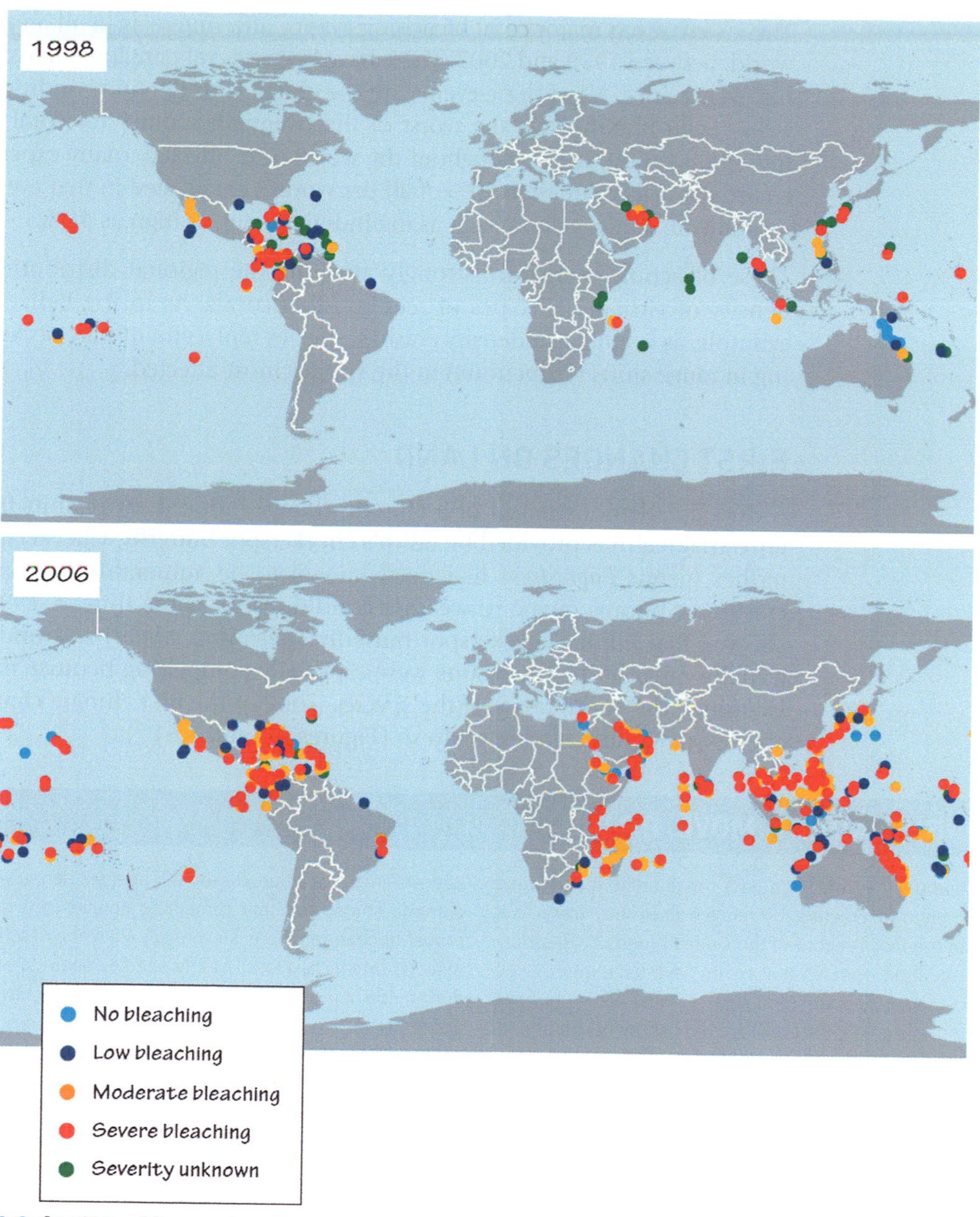

FIGURE 3.3 Coral bleaching events.
As global mean temperature rises, the frequency of events that exceed the bleaching threshold increases. The threshold varies in different regions. These global maps illustrate the severity of bleaching in the 1998 El Niño, which was the first major global bleaching event, and recorded bleaching in 2006. Every major coral reef region in the world has now suffered moderate to severe bleaching events. *Source: Marshall and Schuttenberg (2006).*

There were seven major coral bleaching events, affecting reefs in all areas of the world, between 1979 and 2002. There have been several panglobal mass bleaching events since. All of these events are associated with El Niño conditions. The 1997–1998 El Niño was the worst of the twentieth century for coral bleaching. In that event, reefs throughout the world were affected, many experiencing record damage. More than 10% of all the world's corals died in that event, with mortality in some regions, such as the Indian Ocean, as high as 46%.

These bleaching events are strongly affecting the regional distributions and density of virtually all types of reef-building corals. As in the Belize lagoon example, as population density crashes, species replacement may occur, resulting in range shifts (reductions) in the species most affected.

FIRST CHANGES ON LAND

A dramatic demonstration of a climate change-induced range shift from the terrestrial realm is provided by Edith's checkerspot butterfly. Checkerspot butterflies (genus *Euphydras*) had been known to be vulnerable to population crashes or booms owing to weather conditions for some time. A 1996 study showed that Edith's checkerspot butterfly (*Euphydras editha*) was undergoing a major range shift. The results were especially compelling because the study examined the entire range of the species—one of the first climate change and species' range shift studies to do so (Figures 3.4 and 3.5).

SPOTLIGHT: ADAPT, MOVE, OR DIE

Insects shed unique light onto past range shifts in response to climate change. Beetles currently known only from Asia are seen in the fossil record of the United Kingdom. Restricted-range endemic species that might have been interpreted as tightly evolved to local conditions are now known to have moved hundreds of kilometers or across continents on timescales of tens or hundreds of thousands of years. In many insects, affinity to climatic conditions seems to drive association with place rather than the other way around. Coope (2004) explores the implications of these findings for climate change biology and conservation. Coope suggests that a species faced with climate change has three options: adapt, move, or die. There is little evidence for extinction from the fossil insect record. An initial wave of extinction is seen at the onset of the ice ages, but once that spasm is past, few extinctions are associated with entry into, or emergence from, ice ages. One interpretation of this record is that the initial descent into glaciation eliminated species sensitive to major climatic shifts, and that remaining species are remarkably robust to change. There is even less evidence of adaptation in insects. One or two fossil beetle species seem to have arisen during the ice ages, but these seem not to have modern descendants, so they may represent adaptation or they may be anomalies. The final option, movement, is abundantly supported. Insect ranges moved long distances as the ice ages deepened and ebbed. Thus, for most of the world's species, moving in response to climate change seems to be a comfortable option. Whether it will remain so on a planet heavily altered by human action will determine the fate of millions of invertebrate species.

Source: Coope, G.R., 2004. Several million years of stability among insect species because of, or in spite of, ice age climatic instability? Philosophical Transactions of the Royal Society of London 359, 209–214.

Populations of Edith's checkerspot are found from Mexico to Canada, and populations in the south and in the lowlands were found to be disappearing faster than populations in the north and in the uplands. Continuation of this trend would lead to loss of lowland and southern range and an increase in range at upper elevations and poleward—exactly the pattern expected with climate change. Researchers were able to rule out competing causes for the shift, including habitat destruction, clearly indicating climate change as the cause of the range shift.

Since these early signs of range shifts, evidence has mounted for many species and many regions that climate change-caused range shifts are taking place.

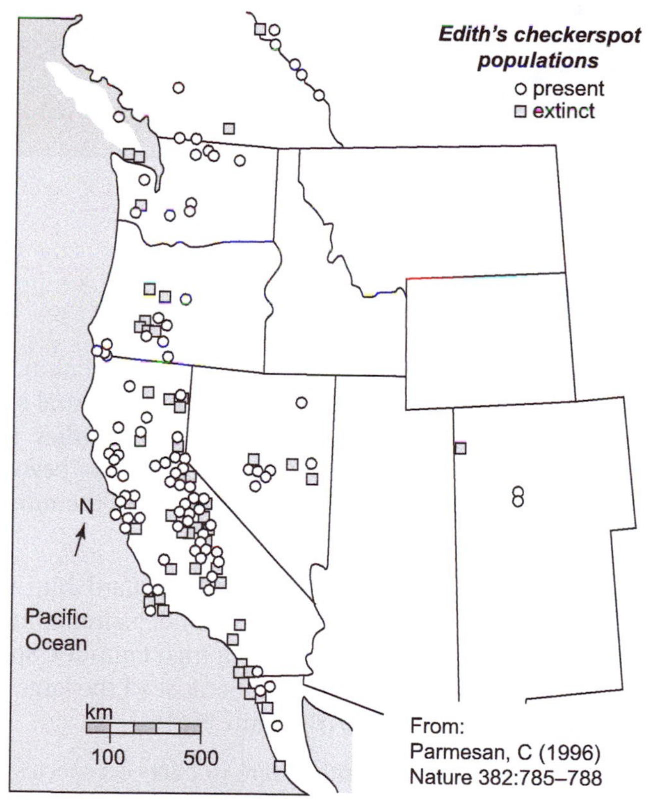

FIGURE 3.4 Edith's checkerspot butterfly range shift.
Southern populations of Edith's checkerspot butterfly are becoming extinct (shaded squares) more frequently than northern and montane populations, resulting in a northward and upslope range shift. *Source: Parmesan (1996). Reprinted with permission from Nature.*

FIGURE 3.5 Edith's checkerspot butterfly (*Euphydryas editha*).
Source: From http://www.nps.gov/pinn/naturescience/butterfly.htm.

MOUNTING EVIDENCE OF RANGE SHIFTS

Evidence of climate change-related range shifts and population changes has accumulated rapidly since the pioneering marine and terrestrial studies in the 1990s. Evidence of impact is particularly strong for butterflies and birds, as well as for species at high latitudes. However, changes have been recorded in many taxa and across latitudes. Here, we survey some of the more important results, moving from the poles to the equator.

Arctic and Antarctic species have been particularly hard hit, as would be expected from temperature records and climate model simulations indicating that climate change will be more pronounced at high latitudes. Species dependent on sea ice have been selectively affected because of the large losses in sea ice extent that have occurred since 1970 (Figure 3.6).

The Arctic is rapidly losing sea ice, with severe impacts on species that depend on sea ice for some portion of their life cycle, such as polar bears, walrus and spectacled eider (see Chapter 5). Large-scale range changes have been seen in the arctic fox, which is retreating northward as its more competitive cousin, the red fox, expands its range with warming.

ARCTIC FOX AND RED FOX RANGE CHANGES

Arctic fox (a) and red fox (b).
Source: From (a) Wikimedia Commons and (b) U.S. Fish and Wildlife Service.

The arctic fox (*Vulpes lagopus*) is declining owing to climate change, in a range shift apparently mediated by competition with the related red fox (*Vulpes vulpes*). Arctic foxes have light coloration and compete well in snowbound landscapes. As climate warms, snow cover decreases and the advantages of this coloration for avoiding predators are lost. In these areas, the darker red fox is more competitive and pushes out the arctic fox. Thus, the range of the arctic fox is moving northward because of climate change, but the proximate cause is competition with the red fox.

FIGURE 3.6 Penguins and climate change.
Emperor penguin (*Aptenodytes forsteri*) populations are declining in Antarctica with climate change.
Source: Photo courtesy of NOAA. Photographer: Giuseppe Zibordi.

In the Antarctic, penguin ranges have shrunk and populations declined as sea ice has reduced in area. Emperor penguin populations have undergone population declines as high as 50%, and Adelie penguins have declined as much as 70% in some locations. Populations farthest from the pole have been hardest hit, as expected, creating the conditions for a poleward range shift in these Antarctic species.

Decreases in Antarctic sea ice have led to declines in the abundance of algae that grows on the underside of the ice and resultant declines and range retractions in krill that feed on the algae (Figures 3.7 and 3.8). Krill support fish, birds, and mammals higher in the food chain, so follow-on changes in the abundance and ranges of these species are expected, many of which are already being observed.

Acidification further influences the base of the polar food chain. Many plankton near the base of polar food chains form calcium carbonate shells. They do this in some of the coldest, and therefore least saturated, waters on the planet. Initially, warming of these waters may increase calcium carbonate saturation, pushing food web interactions in one direction. Later, direct acidification effects may provide pushback as increasingly acid polar waters become once again less hospitable to calcium carbonate-secreting organisms.

SPOTLIGHT: HIDDEN ADAPTATION

Although the "move" part of the adapt, move, or die trilogy is dominant, adaptation has occasionally been recorded (Thomas et al., 2001a). At the range margins of British insects on the move north in response to warming, long-winged variants are more common (Figures 3.9 and 3.10). This makes evolutionary sense because long-winged forms are better able to disperse to newly suitable climatic space than are short-winged forms, which often have very poor flight ability. However, where does this adaptation come from? Apparently, the long-winged trait is recessive. This allows it to persist in the gene pool with no ill effects because it is not expressed, despite being maladaptive when climates are stable and favorable. When climate changes, there is heavy selection for the ability to move to a suitable climate, and the recessive trait is expressed and selected for. These hidden traits are literally invisible until needed. When needed, they play a key role in climate adaptation.

Source: Thomas, C.D., Bodsworth, E.J., Wilson, R.J., Simmons, A.D., Davies, Z.G., Musche, M., et al., 2001a. Ecological and evolutionary processes at expanding range margins. Nature 411, 577–581.

In temperate climates, studies involving large numbers of species indicate broad biological response to climate change. In Europe, a study of 35 butterfly species found that 63% had undergone northward range shifts, as would be expected with global warming, and only 3% had shifted southward. In the species tracking climate, the range shifts were large—between 35 and 240 km. Similarly, in 59 species of birds in Great Britain, a mean northward range shift of nearly 20 km was observed over 20 years.

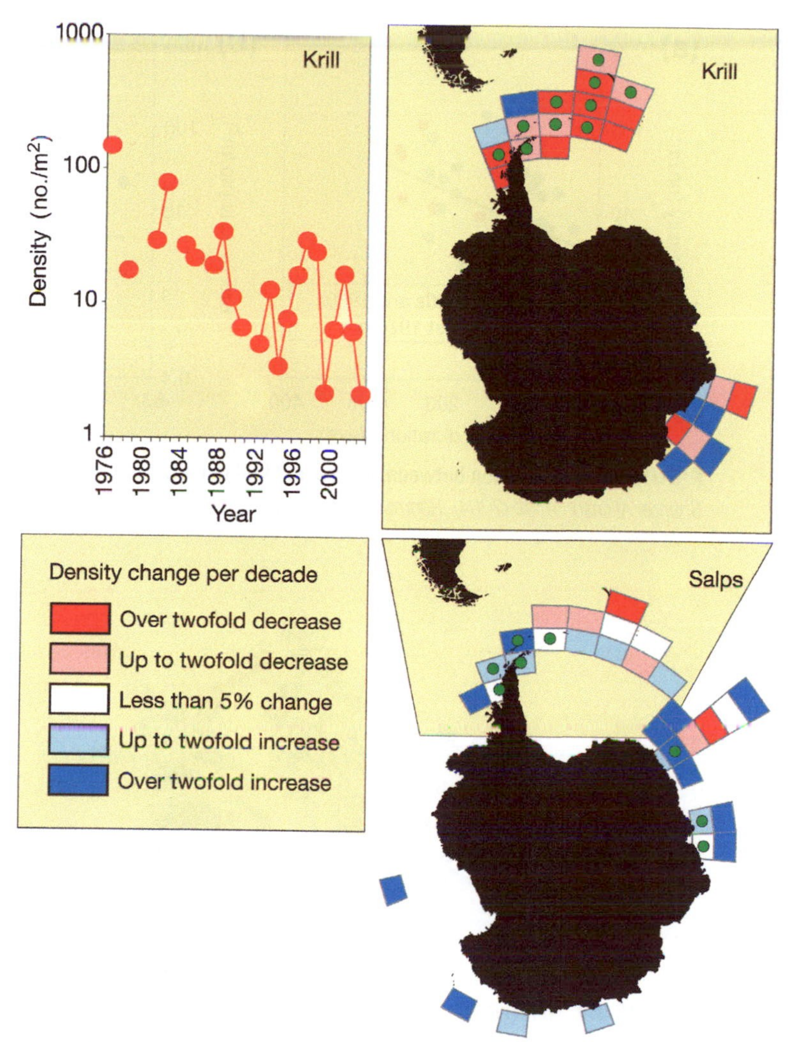

FIGURE 3.7 Shifting krill in Southern Oceans.

Krill abundance is decreasing in areas bordering Antarctica, whereas salp densities are increasing. Krill depend on ice algae for summer population growth. Declining sea ice due to climate change reduces algal density and depresses krill populations. Salps increase in their place. The maps show the change in krill (top) and salp (bottom) abundance. These changes have had profound impacts on food webs in the southern oceans. *Source: Atkinson et al. (2004). Reproduced with permission from Nature.*

Dragonflies in Great Britain have expanded northward; 23 of 24 well-documented species have shown a northward shift, with a mean shift of 88 km. A total of 77 lichen species have expanded their ranges northward into The Netherlands. Alpine plants have been moving upslope in Swiss mountains. In the United States, the pika (*Ochotona princeps*), a small montane mammal, is disappearing from lowland sites. These results are all for relatively well-known species for which

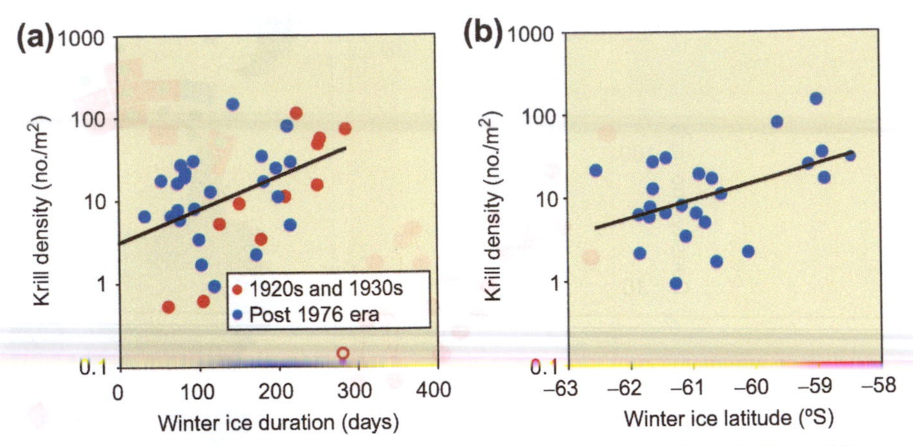

FIGURE 3.8 Correlation between sea ice and krill density from region shown in Figure 3.7.
Source: Atkinson et al. (2004). Reproduced with permission from Nature.

FIGURE 3.9 The silver-spotted skipper (*Hesperia comma*) has expanded its range threefold in Britain since 1982.
Source: Pimm (2001). Reproduced with permission from Nature.

good historical records exist. Butterflies and dragonflies are extensively collected by amateurs and professionals alike, whereas an avid bird-watching community generates exceptional amateur and professional sighting data, especially in Great Britain. Many other, less well-known species are likely to be shifting ranges in the temperate zone (Figures 3.9 and 3.10), and the number of documented cases is rising steadily, including invasions of nonnative species (Figure 3.11).

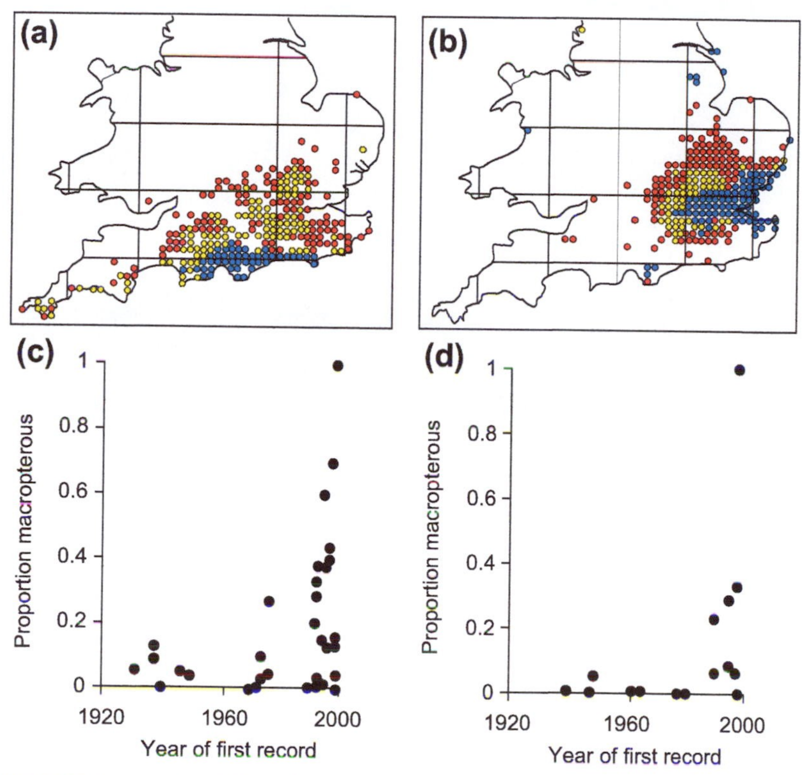

FIGURE 3.10 Large-winged forms increase in expanding range margins.
Insects have longer wings in expanding range margins. The conehead bush cricket, *Conocephalus discolor* (a), and Roesel's bush cricket, *Metrioptera Roeselii* (b), have undergone recent range expansions due to climate change. Blue circles indicate historical range, with yellow and red circles denoting progressive expansion in these species. Long-winged forms are more common in populations on the range margin (c and d). Locations with fewer years since first observation are recently colonized areas on the range margin. *Source: Thomas et al. (2001a). Reproduced with permission from Nature.*

Tree lines have shifted poleward and upslope in a wide variety of settings and regions, mostly northern, including eastern Canada, Russia, and Sweden. However, the tree-line picture is mixed, with tree lines in many areas showing no effect. Few tree lines seem to be retreating, leaving a net signature consistent with the expected trend due to warming.

THE DISAPPEARING PIKA: CLIMATE AND PHYSIOLOGY

Pika (*Ochotona princeps*) in typical rocky habitats.

Pika, small high mountain residents, are disappearing because of climate change. The physiology of the pika renders it susceptible to direct death from warming when its alpine habitats reach high temperatures, and elevated temperature inhibits foraging as a result. By 2004, pika had disappeared from 7 of 25 sites in the western United States, most at lower, warmer elevations.

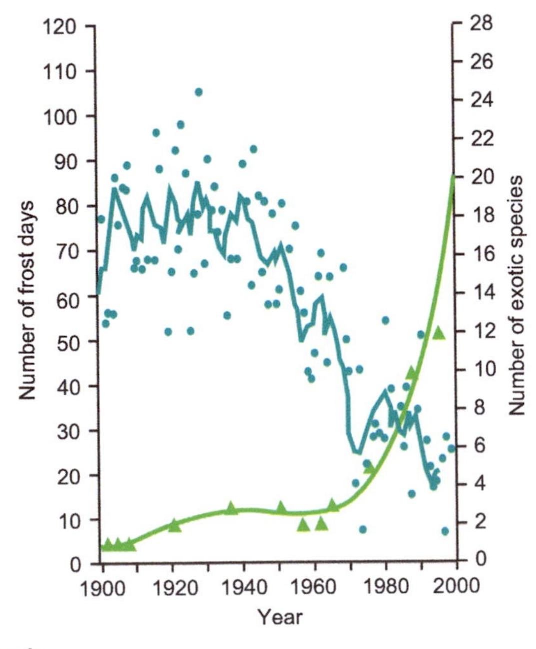

FIGURE 3.11 Climate-linked invasion.
Invasion of exotic plant species is correlated with reduced drought days in Switzerland. Removal of climatic stress may remove important constraints on the spread of these species. *Source: Walther et al. (2002). Reproduced with permission from Nature.*

Upslope migration of tree line is mediated by a number of factors, which in many instances has made it difficult to demonstrate a tree-line effect of climate change alone. Recent increased growth at tree line has been observed in the southwestern United States but not Alaska. Warming with increased rainfall may lead to a pronounced increase in growth, whereas warming and drying may not lead to enhanced growth at tree line. In the Sierra Nevada mountains of California, significant changes in mortality, growth form, and population density of trees have all been observed without changes in tree line.

In southern Africa, a range shift of the quiver tree (*Aloe dichotoma*) has been demonstrated across its entire range from South Africa into Namibia (Figure 3.12). This is the only instance of a well-documented range shift across the entirety of a species' range for plants. Quiver tree populations are declining in the north and at lower elevations, with the only northern stronghold remaining at high elevation.

Range shifts in the tropics are less well documented, partly because change in the tropics has been less pronounced than at high latitudes so far and partly because species in the tropics are less well known. Nonetheless, there is evidence of range shifts from several areas of the tropics. In the Monteverde cloud forest of Costa Rica, toucans and other birds are moving upslope in sync with warming. Mountaintop amphibians at Monteverde have declined or disappeared. Butterflies from North Africa are expanding their ranges into southern Europe.

SPOTLIGHT: COMMUNITY CHANGE

Vegetation associations change with climate. Current examples of these changes are emerging. We know from the paleo record that many combinations of species observed in present landscapes may not have existed in the past, and that species not currently found together may have coexisted. This ephemeral nature of species associations corresponds to Gleason's 1917 definition of community, in which species are found together by accident of their individual climatic tolerances, and contrasts with the view of communities as coherent entities, originally posited by Clements in the early 1900s. If this view is correct, we should find examples of communities dissociating and reassembling owing to human-induced climate change. And we do.

In the U.S. Southwest, drought is causing the dieback of pinyon pine (*Pinus edulis*), leaving a characteristic community, the pinyon–juniper woodland, without one of its signature components. Breshears et al. (2005) document the decline of pinyon over a large swath of four western states and attribute it to "climate change-type drought." The pinyon–juniper community in many areas of this region is no more. A new community, characterized by juniper with other subcomponents, is taking its place. As climate change continues, other vegetation associations throughout the world will be torn apart and reassembled.

The key to the pinyon dieback seems to be warming. A similar drought in the 1950s resulted in much less tree mortality. That drought was much cooler than the current drought, however. The combination of warm temperatures and low rainfall seems to push pinyon over a threshold of water stress from which it cannot recover. Because climate change projections call for both more drought and warmer temperatures, the future may hold many more of these climate change-type droughts.

Source: Breshears, D.D., Cobb, N.S., Rich, P.M., Price, K.P., Allen, C.D., Balice, R.G., et al., 2005. Regional vegetation die-off in response to global-change-type drought. Proceedings of the National Academy of Sciences 102, 15144–15148.

FIGURE 3.12 **The quiver tree, *Aloe dichotoma*.**
A. dichotoma is a bellwether of climate change. Populations are declining in the north and at lower elevations, with the only northern stronghold remaining at high elevation (Foden et al., 2007). *Source: Photo from Wikimedia commons.*

PATTERNS WITHIN THE PATTERNS

There have been so many documented cases of the poleward and upslope range shifts expected with climate change that it is possible to search for specific patterns within these movements. Can biological responses be used to prove that the global climate is warming? Two studies published in 2003 support this idea. Both studies examined a large data set of biological evidence compiled for the IPCC. One examined the data on a species-by-species basis, and the other examined individual research studies. Both concluded that the climate change signal exhibited by range shifts was so strong that it could result only from a change in global climate. Other causes, such as a bias toward publishing results that showed a change, were ruled out.

In the species-by-species study, the overwhelming majority of species showed the poleward and upslope shifts expected with warming. In 1700 species studied, poleward range shifts averaged 6 km per decade. A total of 279 of the species showed responses that tracked climate change—poleward shift during warming periods and shift away from the poles in cooling periods—but a net poleward shift. This gives strong indication of climate causality.

THE TOUCAN AND THE QUETZAL

Keel-billed toucan (*Ramphastos sulfuratus*).
Source: From Wikimedia Commons.

Keel-billed toucans have been moving upslope with warming. In the forests of Central America, these birds prey in part on young quetzals. The toucan's beak is shaped to reach into nesting hollows and pluck out quetzal chicks. As a result, as the toucan moves upslope, the lower elevation edge of the quetzal population is receding. The quetzal range is being affected by climate not directly but, rather, mediated by the toucan movement.

SPOTLIGHT: THE BELOWGROUND CONNECTION

Ninety percent of soil microbes are found around plant roots. Some of these microbes, especially mycorrhizae, make nutrients such as phosphorus more available, which plays critical roles in plant survival and fecundity. But how do these belowground elements move when climate changes? Limited evidence indicates that mixed stands of trees and belowground fauna facilitate migration. However, many belowground elements, such as fungi, seem to spread slowly. Range shifts in plants may be limited where soil fauna move more slowly than plant propagules. Perry et al. explored this topic for the first time in 1990. It remains a relevant and understudied element in plant range responses.

Source: Perry, D.A., Borchers, J.G., Borchers, S.L., Amaranthus, M.P., 1990. Species migrations and ecosystem stability during climate change: the belowground connection. Conservation Biology 4, 266–274.

The second paper examined 143 research studies, many of them the same as in the species-by-species analysis, and found that most studies reported shifts in the direction expected by climate change. The variety of organisms in the studies—from trees to grasses and mammals to mollusks—indicated the breadth and depth of the climate change effect in the natural world. These patterns within patterns indicate not only that the natural world is responding to climate change but also that natural responses are one of the lines of evidence that can be used to show that global climate is warming.

EXTINCTIONS

When range shifts become too severe for species to keep up with changing climatic conditions, extinctions can occur. The first climate-linked extinction to be documented was the disappearance of the golden toad (*Bufo periglenes*) from the Monteverde rain forest of Costa Rica. This formerly abundant species would come together in huge numbers to mate. During these mating aggregations, the golden toad was easily observed. At other times of the year, it was a more elusive, if brightly colored, resident of the cloud forest. In 1987, the mating aggregation failed to materialize and the golden toad was never seen again. Just that quickly, from one year to the next, the entire species vanished.

The cause of the golden toad disappearance has been linked to climate since the first scientific paper was written about it. The year of its disappearance followed a series of years of increasing dryness in the cloud forest, and the total population collapse of 1987 followed the driest year on record (Figure 3.13).

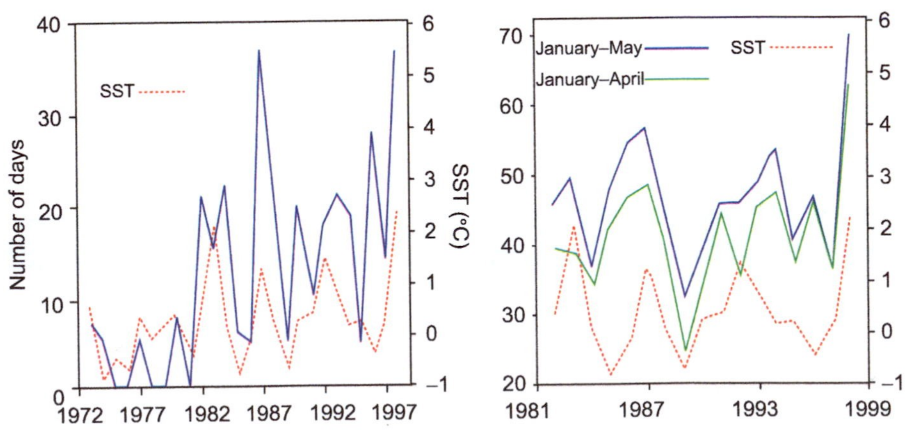

FIGURE 3.13 Drying trends in Monteverde cloud forest, Costa Rica.
Number of dry days per year in Monteverde and departure of nearby sea surface temperature from long-term average. Note the long-term increase in the number of dry days and the peak in 1987, which is the year of the disappearance of the golden toad. *Source: Pounds et al. (1999). Reproduced with permission from Nature.*

The mechanism of its climate alteration is the lifting of cloud bases within the cloud forest belt as SSTs rise, an effect that has been observed in both paleo data and general circulation model simulations.

SPOTLIGHT: CLIMATE CHANGE PULLED THE TRIGGER

Harlequin frogs of the genus *Atelopus* are going extinct at an alarmingly rapid rate in South America. Their disappearance is part of a global phenomenon of amphibian extinctions that affects hundreds of frog and toad species worldwide. Both disease and climate change have been implicated in the extinctions.

Alan Pounds of the Monteverde Research Station has examined *Atelopus* declines with a group of researchers from Central and South America and concluded that it is a combination of climate and disease that is killing the frogs (Pounds et al., 2006). The disease is a chytrid fungal infection that attacks the frog's skin. The chytrid fungus may be a normal resident of the skin that the frogs control by staying in conditions that are cooler or sunnier than the fungus needs to thrive. With climate change, temperatures warm, making it more difficult for the frogs to find cool conditions unfavorable to the fungus. At the same time, cloudy conditions are more common, so opportunities to limit the fungus by sun exposure are reduced.

Pounds and co-workers found that most *Atelopus* species were disappearing at elevations at which the combination of cloudiness and warmer temperature would be most favorable to the fungus. Furthermore, there was a high correlation between the years in which species were last seen and unusually warm years. Changing temperature and cloudiness seem to converge to create conditions that favor the fungus so strongly that it erupts, wiping out entire populations and species. Pounds has stated, "The fungus is the bullet that killed the frogs, but climate change pulled the trigger."

Source: Pounds, J.A., Bustamante, M.R., Coloma, L.A., Consuegra, J.A., Fogden, M.P.L., Foster, P.N., et al., 2006. Widespread amphibian extinctions from epidemic disease driven by global warming. Nature 439, 161.

Beyond Monteverde, toads and frogs have become extinct across Central and South America, all linked to climate change. These extinctions are all in the genus *Atelopus*, which is the genus that includes the harlequin frog. The harlequin frog was one of the species that disappeared locally at Monteverde along with the golden toad. Seventy-four species in the genus have disappeared, 80% of which were last seen just prior to an unusually warm year.

All of the extinct species are likely to have succumbed to a fungal disease caused by the chytrid fungus, *Batrachochytrium dendrobatitis*. The chytrid fungus naturally occurs in low levels in the skin of *Atelopus* species. The species seem to undergo major population crashes due to chytridiomycosis when the fungus gets out of control—a situation that corresponds strongly to warm years. Warm years correlated with dry spells and high minimum temperatures that seemed to create ideal conditions for the growth and deadliness of the fungus.

SPOTLIGHT: TROPICAL SENSITIVITIES

The tropics are sensitive to climate change. Deutsch et al. (2008) show that tropical insects have narrower fitness tolerances to temperature. These and many other tropical organisms are likely to be more sensitive to change because they are evolved to thrive in a relatively narrow temperature range. Tropical insects are also at or near their thermal optimum and thus have less scope to respond to warming than do temperate insects, which are typically below their optimum. Thus, although the magnitude of climate change may be highest at high latitudes and near the poles, sensitivity may be greatest in the tropics.

The impact of climate change is the product of the magnitude of change and sensitivity. Expressed from the standpoint of the species, exposure to change multiplied by sensitivity equals vulnerability. Vulnerability can be reduced by conservation actions to help in adaptation, which results in the ultimate impact on the species. However, for the most part, impact will scale with sensitivity and exposure. Physical change may occur first in high latitudes where warming is greatest, but the greatest biological damage may occur in the tropics, where sensitivity is greatest.

Source: Deutsch, C.A., Tewksbury, J.J., Huey, R.B., Sheldon, K.S., Ghalambor, C.K., Haak, D.C., et al., 2008. Impacts of climatewarming on terrestrial ectotherms across latitude. Proceedings of the National Academy of Sciences 105, 6668–6672.

In the case of the golden toad and *Atelopus* species, small range size, sensitivity, and range reduction all come together to cause extinction. All of the extinct species are narrowly distributed in mountaintop cloud forests. All have sensitivity to climate change because of the chytrid fungus. As climate change imposes shifts in the potential range of these species, they have nowhere to go because they are all mountaintop species. Thus, the classic combination of conditions for climate change extinction exists for them and has apparently already played out for these lost species.

Understanding the contributors to extinction risk helps conservationists design effective responses that can help save species. Based on our understanding of the extinctions so far, appropriate conservation efforts can focus on surviving montane frogs and toads. Some may be saved by artificial manipulation of habitat, such as spray mists for cooling, whereas others may require captive breeding until the dynamics of climate and disease are better understood. Because the species are relatively abundant during low-risk periods, taking a few pairs from healthy populations for captive propagation can help provide insurance against extinction due to population crashes in the wild population, without significantly depleting the numbers of wild individuals.

FRESHWATER CHANGES

Arctic lakes are yielding the first signs of freshwater changes consistent with those expected from past climates and theory. Range shifts in lake-dwelling species may take place over tens of meters rather than over tens or hundreds of kilometers. The first expected shifts are in the abundance of warm-water species and species that thrive along the shore. Thawing shorelines create new habitat and extended habitat availability in lakes that are frozen most of the year. As ice cover declines, species have more time to colonize shoreline habitats, and species typical of these habitats increase in number.

A study of 55 arctic lakes revealed recent shifts in diatom populations toward species most common in shorelines and warm open waters (Figure 3.14). Diversity in these lakes is therefore increasing, and the increase is greatest at high latitudes, where the ice-related effects are strongest. Species typical of mossy shorelines have increased as shorelines thawed enough for modest growth to occur, whereas species that inhabit open waters increased once ice had retreated.

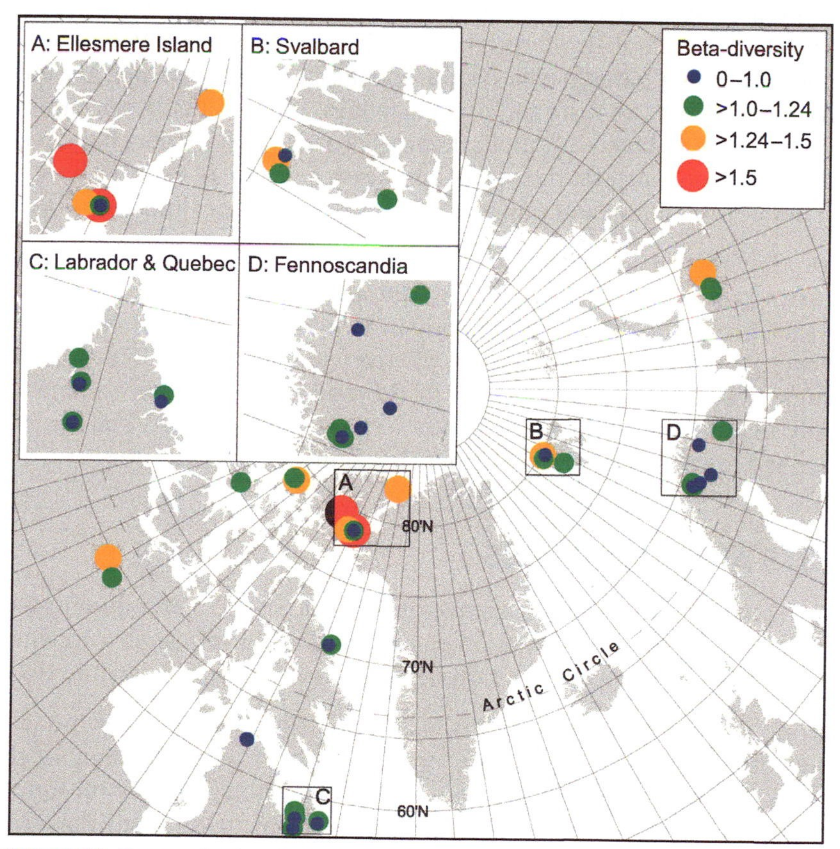

FIGURE 3.14 Changes in diversity in Arctic lakes.
Diatom diversity has increased in arctic lakes owing to warming during the second half of the twentieth century. An increase in diversity in sampled lakes is indicated for the arctic and in more detail for four regions with especially rich records in the insets. *Source: Smol et al. (2005) © National Academy of Sciences U.S.A.*

Lack of long-term monitoring hampers detailed detection of change in these and other lakes. Analysis of sediments allowed comparison of current diatom richness and abundance to levels from 1850 to the present. As change accelerates, detection using historical monitoring records will become increasingly feasible. Many lakes in other areas of the world are likely to be already experiencing the impacts of climate change, but it will be years, perhaps decades, before these changes are as well documented as the changes in arctic lakes.

Cold-adapted fish in many parts of the world are beginning to experience mortality events that have resulted or may ultimately result in range shifts. The sockeye salmon *(Oncorhynchus nerka)* of the Fraser River in British Columbia have undergone mass mortality when water temperatures have spiked. Early migrants are most heavily affected because of higher summer water temperatures. Later migrants are less affected, producing possible directional evolutionary pressure. The mechanism of the die offs is related to thermal physiology. When the sockeye overheat, their basal metabolism accelerates, leaving less room for metabolism directed at swimming and other activities. When water temperatures get too warm, the combined metabolic demand of activity plus higher basal metabolism proves too much for their heart and death results. Large dieoffs have been seen in Oregon salmon and other cold-adapted species for similar reasons.

Arctic char *(Salvilinus alpinus)* populations are declining across the species' southern range limit in the United Kingdom, during warming over the past 40 years. In a particularly detailed study in Switerland, changes in brown trout *(Salmo trutta)* were measured across 87 river sections from 1978–2001 were documented. Brown trout declined in the warmest river sections, and were mostly maintained in the coldest river sections, resulting in a shift in abundance and range toward cooler, higher elevation streams and rivers.

PESTS AND PATHOGENS

Range shifts are not limited to native or endemic species. Introduced species, pests, and pathogens may also undergo range changes in response to climate, often with dramatic impacts on ecosystems and species of conservation concern.

Upslope movement of malaria has been noted in many areas of the world, along with changes in distribution of other human diseases. Many of these changes are due to changing ranges of disease vectors, such as mosquitoes. Mosquitoes are cold-blooded (ectothermic), which makes them sensitive to climate. Mosquitoes simply cease to be able to thrive and reproduce above certain elevations. The exact limiting elevation depends on the species of mosquito and on local conditions and climate. When climate warms, other factors remaining equal, the mosquito vectors of human diseases will expand their range upslope.

At the same time, the growth of disease agents, such as the protozoans in the genus *Plasmodium* in the mosquito, is directly affected by temperature because the mosquito's body temperature is the same as the surrounding air temperature. As the air warms, the mosquito's body temperature increases, speeding the development of the protozoan larvae residing in the mosquito.

Malaria becomes an increasing problem in warming climates both because its mosquito vector moves up in elevation and because the larval *Plasmodium* matures more rapidly in warmer temperatures. An expanded and intensified disease belt results (Figure 3.15). Similar factors favor expansion of some other human diseases owing to climate change, especially those with cold-blooded insect vectors such as mosquitoes or ticks.

Like human diseases, animal disease incidence and range may expand with warming. Avian malaria and other animal diseases can have major impacts on animal populations. These diseases may also increase in warming climates. In species already affected by other factors, these changing disease incidences may have major consequences. As with the golden toad and *Atelopus* species, disease may mediate range shifts in other species.

Diseases of plants and plant pests are shifting up in elevation and poleward. The pine moth (*Thaumetopoea pityocampa*) has extended its range into France from Spain. Movements of pine blister rust and other fungal-mediated diseases are of major concern because these may be limited by cold temperatures and therefore move up in elevation and poleward in response to warming.

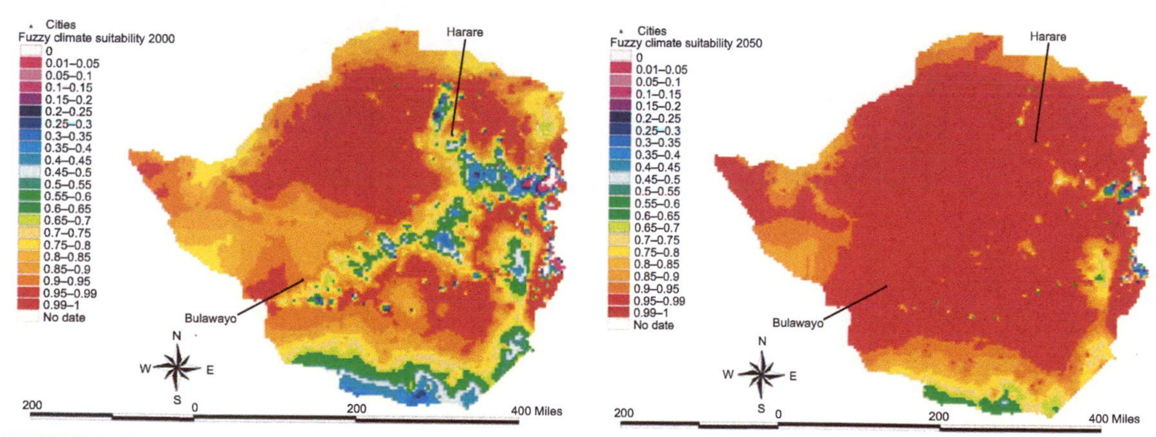

FIGURE 3.15 Expanding malaria zone.

Malaria is currently rare in the highlands of Zimbabwe (left panel). Malaria parasites mature up to 10 days more rapidly under projected temperature increases. This allows the disease to persist in formerly inhospitable areas. The right panel shows the projected spread of malaria into the Zimbabwe highlands by 2050 due to this effect. Orange and red colors denote suitable conditions for malaria transmission, and blue-green colors areas with poor conditions for transmission. *Source: Patz and Olson (2006) © National Academy of Sciences U.S.A.*

AVIAN MALARIA

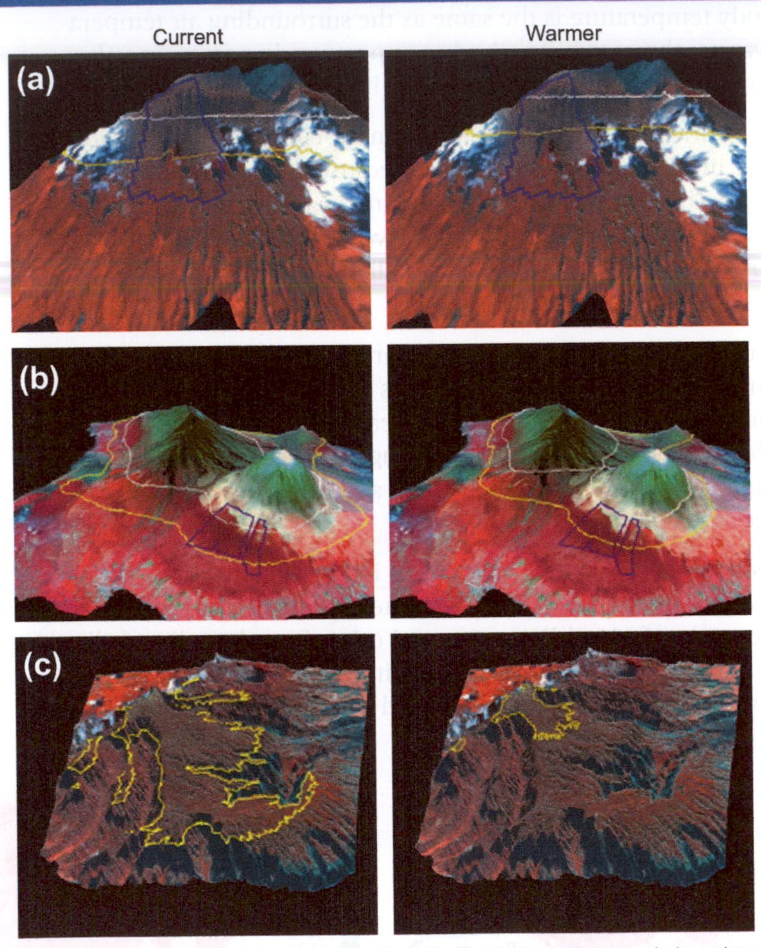

Projected changes in 17 (yellow) and 13 °C (white) isotherms that limit the distribution of avian malaria under current and 2 °C warming conditions. Changes are shown for Hanawi Reserve (blue boundary) on the island of Maui (a), Hakalau Refuge (blue boundary) on Hawaii (b), and the Alakai swamp region on the island of Kauai (c).
Source: Benning et al. (2002) © National Academy of Sciences U.S.A.

Like human malaria, avian malaria is moving upslope with climate change. The thermal tolerances of mosquito vectors of malaria are limited by cold temperatures, especially in tropical highlands. In areas such as Hawaii, where this disease has already decimated native bird populations, this upslope movement threatens some of the last upper elevation refuges of native species.

In the aquatic realms, the oyster parasite *Perkinsus marinus* has extended its northern range in the United States from the mid-Atlantic states to Maine. A kidney disease in trout has moved upslope in Sweden, resulting in declining trout stocks at lower elevations. Invasive species such as the zebra mussel (*Dreissena polymorpha*) may experience northward range expansion as temperatures become milder in the far north.

FURTHER READING

Grottoli, A.G., Warner, M.E., Levas, S.J., Aschaffenburg, M.D., Schoepf, V., McGinley, M., Matsui, Y., 2014. The cumulative impact of annual coral bleaching can turn some coral species winners into losers. Global change biology.

Hughes, L., 2000. Biological consequences of global warming: is the signal already apparent? Trends in Ecology and Evolution 15, 56–61.

Manzello, D.P., Berkelmans, R., Hendee, J.C., 2007. Coral bleaching indices and thresholds for the Florida reef tract, Bahamas, and St. Croix, US Virgin Islands. Marine Pollution Bulletin, 54 (12), 1923–1931.

Parmesan, C., Yohe, G., 2003. A globally coherent fingerprint of climate change impacts across natural systems. Nature 421, 37–42.

CHAPTER 4

Phenology: Changes in Timing of Biological Events Due to Climate Change

The cherry blossoms are out earlier this year than ever before. This statement has been true many times in Washington, DC, during the past two decades, and it is likely to be true this year and most years thereafter. Significantly, it is as true in Washington State as it is in Washington, DC. In fact, it is true in Japan, Europe, and other locations where cherries are grown. The cherry trees are blooming earlier because climate is getting warmer and winter is ending earlier.

Budburst and flowering are part of what biologists call phenology—the timing of biological events. Phenology literally means the study of timing, but biologists often use it as a synonym for the process of timing itself. The blooming of a tree is part of a species' phenology because it happens regularly, with timing that can be predicted, based on cues that are understood. Other examples of phenology are spring arrival and fall departure of migratory species, spring nesting in birds, length of growing season, and maturation of crops.

Budburst is determined in most plants by climatic cues. Temperature and temperature patterns, such as chilling followed by warming, set off a series of plant hormonal responses that lead to flowering. These climate cues act on tissues capable of growth and differentiation such as buds that are all derived from meristematic tissue. Meristematic tissue exists at the growing tip of a plant, in the apical meristem. In many plants, bits of meristematic tissue are also left behind in the branches and trunk as the plant grows, awaiting signals to begin growth or differentiation into flowers. Growth in these tissues is usually suppressed by hormones released by the apical meristem. In the case of flower buds, growth and differentiation into flowers is triggered by hormones released when temperature, sequences of temperature, or other conditions are appropriate.

Flowering is triggered by temperature, and so it makes sense that as conditions warm, flowering will occur earlier. Thresholds of temperature should be reached earlier in the spring, triggering the plant hormonal responses that lead to budburst and flowering. Growing degree days is an important temperature measure that often correlates well with changes in phenology. Growing degree days is the number of days with temperatures over a defined threshold.

Climate Change Biology. http://dx.doi.org/10.1016/B978-0-12-420218-4.00004-4

BUDBURST: THE PHYSIOLOGY OF PLANT RESPONSES TO TEMPERATURE

Budburst stages in North American fruit trees. From left to right: swollen buds, apricot, early April; apricot flower, late April; cherry buds, early April; budburst, cherry, late April. *Source: Courtesy of Jon Clements and the University of Massachusetts.*

Buds in many species form during one growing season and then overwinter, developing during the next growing season. In their overwintering form, the buds are protected by bud scales that protect them from desiccation in cold, dry winter air. When spring arrives, warm temperatures provide a cue, which triggers physiological changes that cause these scales to part, allowing the leaf or flower to begin to emerge. This temperature-sensitive phenomenon is budburst.

SPOTLIGHT: SIGNALING CHANGE

A pioneering review of climate change effects on species and ecosystems appeared in *Trends in Ecology and Evolution in February* in 2000 (Hughes, 2000). Lesley Hughes of Macquarie University cataloged the impacts of climate change on the natural world that had been observed to that point. Startlingly, even though at that time climate change was still being debated, dozens of biologists were already documenting its effects. Her review found evidence of changes in physiology, growth, phenology, distribution, and abundance. These findings have been confirmed and updated several times since (Intergovernmental Panel on Climate Change, 2007; Parmesan and Yohe, 2003; Root et al., 2003; Walther et al., 2002). However, Hughes provided the first and definitive call that the biological signal of climate change was apparent and growing.

Source: Hughes, L., 2000. Biological consequences of global warming: is the signal already apparent? Trends in Ecology and Evolution 15, 56–61.

The expected acceleration of flowering has been observed in plants throughout the world. Of course, because there is regional variation in climate change, the observed changes are not all of the same magnitude, but many recorded observations of flowering are occurring earlier in the season.

The general pattern of change in these phenological events is advance in spring and delay in fall, characteristic of a warming climate. The overwhelming majority of taxa that have been studied are showing these responses, in plants and vertebrates, in terrestrial, marine, and freshwater systems (Table 4.1).

Table 4.1 Summary of Data on Phenological and Distributional Changes of Wild Species[a]

Taxon	Total No. of Species (or Species Groups)	Spatial Scale			Timescale (Range, Years)	Change in Direction Predicted (*n*)	Change Opposite to Prediction (*n*)	Stable (*n*)	No Prediction (*n*)
		L	R	C					
Phenological Changes									
Woody plants	*n* ¼ 38 sp		2	1	35–132	30	1	7	—
Herbaceous plants	*n* ¼ 38 sp	1	1		63–132	12	—	26	—
Mixed plants	*n* ¼ 385 sp	1			46	279	46	60	—
Birds	*n* ¼ 168 sp	2	3	1	21–132	78	14	76	—
Insects	*n* ¼ 35 sp		1		23	13	—	22	—
Amphibians	*n* ¼ 12 sp	2			16–99	9	—	3	—
Fish	*n* ¼ 2 sp		1		132	2	—	—	—
Distribution/Abundance Changes									
Tree lines	*n* ¼ 4 sp þ 5 grps	2	1		70–1000	3 sp þ 5 grps	—	1	—
Herbs and shrubs	66 sp, 15 detailed		3		28–80	13	2	—	—
Lichens	4 biogeographic grps (*n* ¼ 329 sp)	1			22	43	9	113	164
Birds	*n* ¼ 3 sp		1		50	3	—	—	—
	N sp (*n* ¼ 46 sp)		2		20–36	13	15	18	
	S sp (*n* ¼ 73 sp)		2		20–36	36	16	21	6
	Low elevation (91 sp)	1			20	71	11	9	—
	High elevation (96 sp)	1			20	37	27	32	—
Mammals	*n* ¼ 2 sp		1		52	2	—	—	—
Insects	*n* ¼ 36 sp		1	1	98–137	23	2	10	1
	N boundaries (*n* ¼ 52 sp)		1		98	34	1	17	—
	S boundaries (*n* ¼ 40 sp)		1		98	10	2	28	—
Reptiles and amphibians	*n* ¼ 7 sp	1			17	6	—	1	—

Continued...

Table 4.1 Summary of Data on Phenological and Distributional Changes of Wild Species[a] *Continued*

Taxon	Total No. of Species (or Species Groups)	Spatial Scale			Timescale (Range, Years)	Change in Direction Predicted (*n*)	Change Opposite to Prediction (*n*)	Stable (*n*)	No Prediction (*n*)
		L	R	C					
Fish	4 biogeographic grps (*n* ¼ 83 sp)	1			—	2 grps	—	1 grp	1 grp
	N sp (1 sp)		1		70	1	—	—	—
	S sp (1 sp)		1		70	1	—	—	—
Marine inverte-brates	N sp (21)	1	1		66–70	19	2	—	1 sp not classified
	S sp (21)	1	1		66–70	20	1	—	
	Cosmopolitan sp (*n* ¼ 28 sp)	1			66	—	—	—	28
Marine zoo-plankton	Cold water (10 sp)		1		70	10	—	—	8 sp not classified
	Warm water (14 sp)		1		70	14	—	—	
	6 biogeographic grps (36 sp)		1	1	39	6 grps	—	—	—

N, species with generally northerly distributions (boreal/arctic); S, species with generally southerly distributions (temperate); L, local; R, regional (a substantial part of a species distribution; usually along a single range edge); C, continental (most or the whole of a species distribution). No prediction indicates that a change may have been detected, but the change was orthogonal to global warming predictions, was confounded by nonclimatic factors, or there is insufficient theoretical basis for predicting how species or system would change with climate change.

[a]*Some studies partially controlled for nonclimatic human influences (e.g., land-use change). Studies that were highly confounded with nonclimatic factors were excluded.*

Source: Root et al. (2003).

ARRIVAL OF SPRING

The earlier arrival of spring is perhaps the best documented change during the past two centuries because of the strong interest people have in the end of winter in the colder climates of Europe, Asia, and North America. In these regions, spring means the beginning of the planting season, the advent of ice-free conditions on rivers and lakes that permits navigation, and, in the past, the beginning of the end of a period of food scarcity. As a result, northern cultures have a long history of recording the first signs of spring—budburst on trees, the arrival of first migrants, the breakup of ice, and other signs of impending warmth. Many of these records are of biological phenomena and of sufficient accuracy to serve as a historical baseline against which the effects of warming may be judged.

Some of the best records come from Europe and involve changes in plant species that have been domesticated as food crops and are sensitive to climate. Wine grapes are quite sensitive to climate and harvest conditions are critical to wine quality, so meticulous records of wine grape phenology have been kept over centuries in wine-growing regions of Europe. In western Europe, harvest time in 2003 was the earliest harvest in a 500-year record. A strong correlation in this data exists between harvest date and summer temperature, with April–August temperatures explaining 84% of the harvest time variation.

Another long record is that for cherry blossoms in Japan. This six-century-long record indicates earlier blooming during the past 200years. Timing begins to advance at approximately the same time that human fossil fuel use intensifies, becoming statistically significant after 1900. The rate of acceleration in advancement increases after 1950, tracking the acceleration in the use of fossil fuels in the latter half of the previous century. By the end of the record, timing of the Japanese cherry bloom was several days earlier than at the start of the record in 1400.

Other plants show similar patterns, with earlier blooming having been recorded for lilac (*Syringa vulgaris*) and honeysuckle (*Lonicera tartarica*) in the western United States. Table 4.2 summarizes these and other illustrative studies of earlier timing of spring events in species from a wide range of taxa and geographic settings.

SPOTLIGHT: TIMING MATTERS

Pied flycatchers (*Ficedula hypoleuca*) in Europe have declined in abundance by 90% because prey availability is getting earlier with warming and chick hatching has not shifted sufficiently earlier to keep pace (Both et al., 2006). Caterpillar peak abundance is becoming earlier with climate change, but flycatcher hatching is not keeping pace in most populations. In flycatcher populations in which hatching is occurring earlier, populations show little decline, whereas populations with weak advances in hatching show strong declines. These results indicate that numerous species may go through bottlenecks because of timing mismatches, where the phenology of prey changes faster than phenology in the predator.

Source: Both, C., Bouwhuis, S., Lessells, C.M., Visser, M.E., 2006. Climate change and population declines in a long-distance migratory bird. Nature 441, 81–83.

Table 4.2 Studies Showing Earlier Arrival of Spring

Location	Period	Species/Indicator	Observed Changes (Days/Decade)	References
Western USA	1957–1994	Lilac, honeysuckle (F)	21.5 (lilac), 3.5 (honeysuckle)	Cayan et al. (2001)
Northeastern USA	1965–2001	Lilac (F, LU)	23.4 (F), 22.6 (U)	Wolfe et al. (2005)
	1959–1993	Lilac (F)	21.7	Schwartz and Reiter (2000)
Washington, DC	1970–1999	100 Plant species (F)	20.8	Abu-Asab et al. (2001)
Germany	1951–2000	10 Spring phases (F, LU)	21.6	Menzel et al. (2003)
Switzerland	1951–1998	9 Spring phases (F, LU)	22.3	Defila and Clot (2001)
South-central England	1954–2000	385 Species (F)	24.5 days in 1990s	Fitter and Fitter (2002)
Europe (Int. Phenological Gardens)	1959–1996	Different spring phases (F, LU)	22.1	Menzel and Fabian (1999), Menzel (2000), Chmielewski and Rotzer (2001)
	1969–1998		22.7	
21 European countries	1971–2000	F, LU of various plants	22.5	Menzel et al. (2006)
Japan	1953–2000	*Gingko biloba* (LU)	20.9	Matsumoto et al. (2003)
Northern Europe	1982–2004	NDVI	21.5	Delbart et al. (2006)
United Kingdom	1976–1998	Butterfly appearance	22.8–23.2	Roy and Sparks (2000)
Europe, North America	Past 30–60 years	Spring migration of bird species	21.3–24.4	Crick et al. (1997), Crick and Sparks, (1999), Dunn and Winkler (1999), Inouye et al. (2000), Bairlein and Winkel (2001), Lehikoinen et al. (2004)
North America (US, MA)	1932–1993	Spring arrival, 52 bird species	10.8–29.6[a]	Butler (2003)
North America (US, IL)	1976–2002	Arrival, 8 warbler species	12.4–28.6	Strode (2003)
England (Oxfordshire)	1971–2000	Long-distance migration, 20 species	10.4–26.7	Cotton (2003)
North America (US, MA)	1970–2002	Spring arrival, 16 bird species	22.6–210.0	Ledneva et al. (2004)
Sweden (often by)	1971–2002	Spring arrival, 36 bird species	22.1–23.0	Stervander et al. (2005)

Table 4.2 Studies Showing Earlier Arrival of Spring *Continued*

Location	Period	Species/Indicator	Observed Changes (Days/Decade)	References
Europe	1980–2002	Egg-laying, 1 species	21.7–24.6	Both et al. (2004)
Australia	1970–1999	11 Migratory birds	9 Species earlier arrival	Green and Pickering (2002)
Australia	1984–2003	2 Spring migratory birds	1 Species earlier arrival	Chambers et al. (2005)

F, flowering; LU, leaf unfolding; 2, advance; 1, delay.
[a]Indicates mean of significant trends only.

Earlier spring may be reflected in multiple species at single locations. Two studies in the United States indicate earlier arrival of spring in multiple species. The famed ecologist Aldo Leopold recorded spring events for 55 species during the 1930 and 1940s. When these species were resampled in the 1990s, 33% showed earlier spring timing, whereas that of almost all of the rest remained the same (the timing for one became later). A study of spring vocalization in six frog species from the early 1900s was compared with similar measurements taken almost 100 years later, revealing an advancement of almost 2 weeks during the course of the century.

In birds and butterflies, spring migrants are arriving earlier, and the date of first flight for species in diapause is advancing. California birds are arriving earlier but not departing significantly later. Migrant birds of the North Sea have arrived several days earlier every decade since 1960. Tree swallows in North America are laying eggs earlier in response to warming. In butterflies, advances in emergence and first arrival of spring migrants have been identified in 26 species in the United Kingdom. A separate study found 17 butterfly migrants arriving earlier in Spain. Seventeen species of butterfly in California (72% of those studied) have advanced the date of their first flight in spring.

Birds are nesting and laying eggs earlier as well. An analysis of more than 70,000 nest records in Britain revealed an 8-day advance in first egg-laying in 20 species from the 1970s to mid-1990s. Mexican jays (*Aphelocoma ultramarina*) are nesting earlier in Arizona, as are tree swallows (*Tachycineta bicolor*) throughout the United States and Canada and pied flycatchers (*Ficedula hypoleuca*) in Europe. Brunnich's guillemot (*Uria lomvia*) is breeding earlier around Hudson Bay in response to decreasing sea ice resulting from warming.

FRESHWATER SYSTEMS

Ice breakup in lakes and rivers has been watched carefully by people living in cold climates, allowing us to observe progressive earlier breakup associated with climate warming (Figure 4.1). Ice breakup is important to people because it indicates that waterways blocked by ice will soon be navigable, allowing transport of goods, hunting, recreation, and other activities of economic and social importance. Its occurrence was therefore recorded historically by a number of industries and hobbyists. These historical records may be compared among themselves where continuous or with modern observations to determine trends in ice breakup. The vast majority of such records indicate that physical changes are occurring earlier in lakes and streams.

These physical changes have important biological consequences because they determine the growth and stratification conditions that affect the entire freshwater food web. Algae at the base of the food web bloom earlier in arctic lakes as a result. Stratification changes in more temperate lakes will affect growing season and competitive relationships for species in different strata (Figure 4.1).

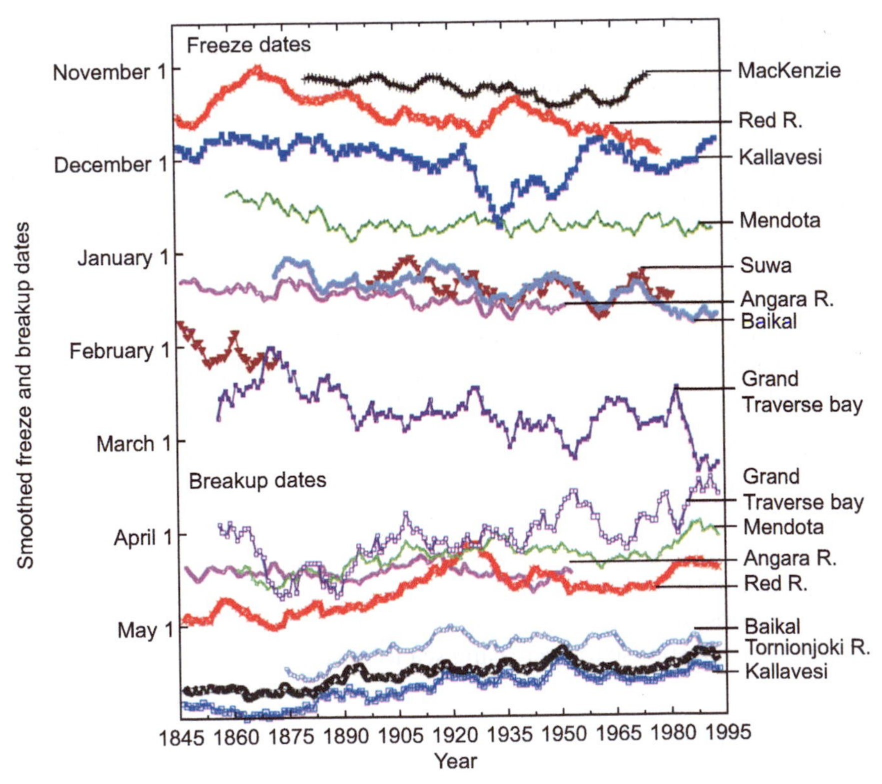

FIGURE 4.1 Accelerating ice breakup and delayed ice formation.
Ice breakup and formation is shown for major rivers and lakes of North America. Ice breakup is occurring significantly earlier, whereas ice formation is significantly delayed. *Source: From Magnuson et al. (2000). Reproduced with permission from AAAS.*

SPRING AHEAD, FALL BEHIND

The growing season is getting longer in the Northern Hemisphere. The longer season results because spring is occurring earlier and fall arriving later as the planet warms.

Evidence from direct measurement of CO_2, satellites, and weather stations indicates a longer growing northern season. Changes in Southern Hemisphere growing season are more difficult to detect because of the limited land at temperate latitudes in the south. Changes in tropical growing season, driven by rainfall changes rather than temperature change, are occurring as well.

The CO_2 "sawtooth" in the Mauna Loa CO_2 measurement record is increasing, meaning that more CO_2 is taken up in Northern Hemisphere vegetation each year. The Northern Hemisphere dominates the sawtooth response because there are large landmasses at temperate latitudes in the Northern Hemisphere and very few in the Southern Hemisphere. Each season, as Northern Hemisphere plants enter spring earlier and begin to grow, the uptake of carbon to build new shoots and leaves measurably decreases the amount of carbon in the atmosphere. Each fall, the loss and decay of deciduous leaves puts a similar amount of carbon back into the atmosphere. This cycle is clearly visible in the measurements of CO_2 in the Mauna Loa record. It is the amplitude of this sawtooth that has been increasing. The growing season may be getting longer in the Southern Hemisphere as well, but it would not be recorded because the much greater Northern Hemisphere landmass dominates the effect.

Growing season assessments from satellite images verify the lengthening season. The Normalized Difference Vegetation Index (NDVI) is a "greeness" index that can be used to assess when plants are actively growing. For example, NDVI is used for early detection of crop failure and famine in Africa and other areas of the world. Similarly, NDVI can be used to detect the onset and length of growing season in temperate climates. NDVI increased between 1981 and 1991, with the areas of strongest change between 45° and 70° North latitude.

Weather station records and test plots support the NDVI and Mauna Loa studies. The European Phenological Gardens are a network of sites that record phenological information. Data from this network show a lengthening of growing season in Europe by almost 11 days from 1959 to 1993. Dates of first and last frost in Europe are getting later and earlier, respectively. The European growing season is estimated to have lengthened by 1.1–4.9 days per decade, depending on location, based on these weather station data.

TROPICAL FOREST PHENOLOGY

Phenology in tropical forests is dominated by rainfall-related events. Because there is little temperature seasonality in tropical forests, many phenological events commonly observed in temperate climates (e.g., arrival of spring and first frost) are not present. What is present is some variation in rainfall, and some, but not all, tropical forest plants and trees respond to this seasonality of precipitation.

Flowering and fruiting in tropical forests is often governed by drought or rainfall intensity. A common pattern is for trees to flower during annual drought periods, with fruit appearing later, when rains have returned (Figure 4.2). This pattern is much more common in forests that experience a pronounced dry period.

Some tropical forest plants have highly specialized phenology. They may flower for only one or two days, following specific events such as the first heavy rains after drought. In some cases, this response may be highly synchronized, with hundreds of trees blooming throughout a forest at the same time, and persisting for only one or two days. These patterns may pose unique evolutionary challenges for pollinators and frugivores.

The fruit produced in these dry-period timed sequences plays a major role as food for forest animals. Because many trees flower and fruit at the same time, there is a large annual peak in fruit abundance. This means that rain forest animals in these settings see a peak in food and then a prolonged period of low food availability. Significantly, in years when the dry season is unusually wet, flowering and fruiting may fail entirely. When this happens, animal food resources dwindle and starvation may result.

Controls other than a dry season on tropical forest phenology are less pronounced. Studies in Asia have shown few responses to temperature variables, except at the northern limits of the tropics. Similarly, at individual and community levels, there is no strong phenological signal in areas without a pronounced dry period. There are some multiyear cycles, but these have not been linked strongly to temperature or other climatic cues. In contrast, wherever there is a pronounced dry season in Asian forests, pronounced phenology of flowering and fruiting is found, with subsequent ecological implications for animals.

The seasonality of some tropical forests may interact with forest fragmentation in ways that produce strong ecological effects. Forest edges next to deforested areas may dry relative to similar regions with more complete forest cover. This may accentuate the dry season, affecting phenology. At the same time, the amount of forest area available for animals to forage across is reduced in areas with heavy deforestation, reducing the ability of seed- or fruit-eating species to buffer shortages by moving about the landscape. The combination of more pronounced feast and famine created by accentuated phenologic responses and the reduced ability to smooth feast and famine by moving may create serious stresses for tropical rain forest birds and mammals.

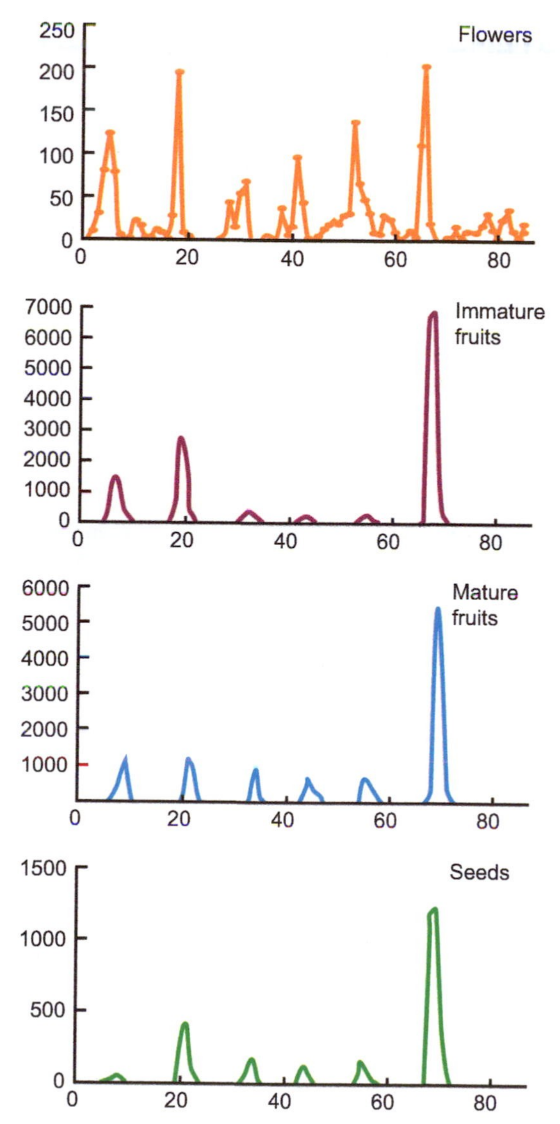

FIGURE 4.2 Phenology of flowers, seeds, and fruits in a tropical forest tree.
The production of flowers, seeds, and fruits in a single tropical tree from a 7-year record. Time is indicated in months from January 1987. Flowering, fruiting, and seed production are all cued to the end of the dry season, peaking in September in most years. Note the unusually high production in September 1992, followed by failure of fruit and seed production the following year (approximately month 80—September 1993). *Source: From Condit (1998). With kind permission from Springer Science Business Media.*

MARINE SYSTEMS

Water warms earlier and ice breaks up earlier in the spring in the oceans, causing important phenological changes in marine systems. Ice breakup in spring affects processes as diverse as polar bear hunting in the Arctic and the growth of algae that are at the base of Antarctic food webs. Many phenological events in the marine realm have implications for ecosystem structure and process.

Among the most important of these changes are alterations in the base of the marine food web—changes in plankton abundance and distribution in time and space. These changes have repercussions all the way up the marine food chain. Decreases in krill may mean declines in whale populations. Food webs with cod as a major element have been replaced by webs in which shrimp or crabs play dominant roles.

PLANKTON: THE BASE OF THE MARINE FOOD WEB

Plankton are drifting creatures, many of them microscopic, that inhabit the ocean's surface waters. The term "plankton" derives from the Greek *planktos*, which means to drift or wander. Phytoplankton are microscopic plants that depend on photosynthesis to survive. Their blooms can contain billions of cells. Large blooms are visible from space. Zooplankton are herbivores that feed on phytoplankton, and they range in size from microscopic organisms to large jellyfish. This tiny, complex food web generates large amounts of biomass, supplying a range of active swimming macro-organisms from small fish to giant baleen whales.

Just as plants bloom earlier on land, marine phytoplankton are blooming earlier as water temperature warms. Zooplankton that feed on the phytoplankton in turn undergo changes in timing of peak abundance. Zooplankton in the German bight go through seasonal cycles up to 11 weeks earlier during warm years. Changes in the biomass of copepods in the Pacific go through seasonal timing that is tightly linked to temperature, with the variation between cool and warm years meaning differences in peak abundances varying between May and late July.

However, the changes in phenology may not be tightly linked between groups. In some areas, phytoplankton blooms and peak abundance are occurring earlier, but the abundance shifts of the zooplankton that feed on them are not advancing as rapidly. In the North Atlantic, the diatom *Rhizosolenia alata* and dinoflagellate *Ceratium tripos* are peaking 33 and 27 days earlier, respectively, whereas the copepod that feeds on them, *Calanus finmarchicus*, is peaking only 11 days earlier. The reasons for these uncoupled responses are not entirely clear, but they have important implications for the structure and interrelationships of marine communities. For example, species that feed on both phytoplankton and zooplankton may be favored, whereas those that feed only on zooplankton may find reduced biomass of preferred prey at key times of the year. Whether uncoupled responses are occurring in all oceans is not currently known.

Differential responses are seen based on life-history traits. Plankton that are planktonic forms of nonplanktonic adults seem to be affected more by warming than are organisms that are permanent members of the planktonic community. Thus, planktonic larvae of sea urchin or jellyfish have been more affected by warming than have species that spend their entire lives as plankton.

MECHANISMS: TEMPERATURE AND PHOTOPERIOD

The timing of plankton blooms is determined in many systems by either temperature or photoperiod. Photoperiod is the duration of the daily light–dark cycle. Species typically respond to temperature directly with increased growth or use temperature as a cue for the initiation of a directed physiological change such as going into diapause. Diapause is the state of suspended growth and activity, typically during cold winter months, although some species exhibit summer diapause. Diapause may be initiated and terminated through either temperature or photoperiod cues. Photoperiod cues are triggered by increases or decreases in day length.

Changes in temperature may cause changes in species responding to thermal cues but not in species responding to photoperiod. For example, some freshwater plankton species overwinter in diapause in the mud at lake bottoms. Warming in relatively shallow lakes reaches the resting forms of these species and may accelerate emergence and growth. The copepod *Thermocyclops oithonoides* rests overwinter as copepodids in lake mud in the Mugglesee and other lakes in Europe. Shallow lakes such as the Mugglesee do not stratify in spring, so warming reaches bottom water, warming the mud in which the copepodids overwinter. This causes earlier development and emergence of the *T. oithonoides* copepodids, probably through direct thermal effects on metabolism. Earlier emergence is followed by more rapid growth in subadults and earlier peaks in abundance. Other freshwater species in the same lake may complete additional generations during each summer due to the physiological acceleration from warmer water temperatures.

In contrast, species that cue to photoperiod for emergence from diapause would see no such effect of warming. In marine plankton, groups of diatoms with many species that cue to photoperiod show little response to recent warming trends. Diatom records from the Continuous Plankton Recorder (CPR) surveys showed low correlation of the seasonal cycle of abundance with warming from 1960 to 2000, whereas other plankton groups showed marked correlations. Dinoflagellates and the larvae of jellyfish and urchins showed strong correlation between seasonal timing and temperature, with copepods showing a somewhat weaker but still pronounced linkage of timing and warmer water. The main difference between the dinoflagellates, copepods, larval plankton, and diatoms is that the diatoms have many species that cue to photoperiod. Whereas photoperiod responses may rapidly evolve, at least in the short term the diatom response to warming is muted.

Trophic mismatches among entire trophic groups may result. In the North Atlantic, diatoms that form the base of the food chain were not advancing in seasonal timing, whereas the timing in dinoflagellates and copepods that feed on them was advancing in response to warmer water temperatures. The results may be rearrangements of food webs, as has been seen in fisheries (e.g., cod) in which bottom-up effects of warming on the base of the food chain interact with top-down food web changes from fishing.

CONTINUOUS PLANKTON RECORDERS

In 1926, a new invention changed scientists' views of plankton. Alister Hardy invented the CPR, a simple device that was towed behind a boat and recorded variation in plankton over time. Previously, plankton were collected in simple dragnets and were assumed to be uniformly distributed in the water column. Hardy's CPR showed that plankton distribution was instead strongly patchy. A major international research program is now devoted to understanding the diversity and spatial distribution of plankton, which will greatly enhance understanding of the effects of climate change on the base of the marine food web.

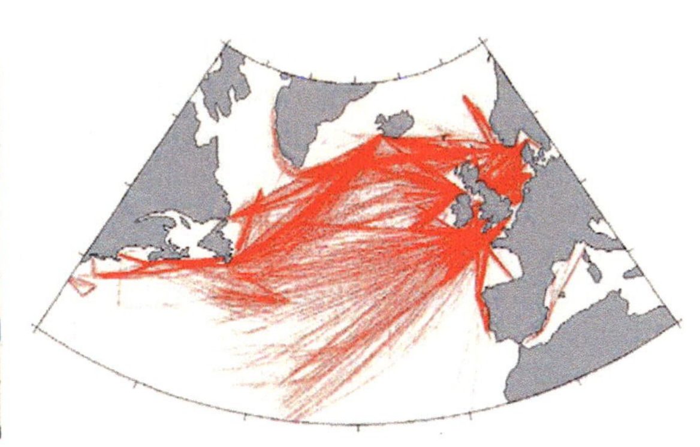

Continuous Plankton Recorder and Results. *Source: From Hays, G.C., Richardson, A.J., Robinson, C., 2005. Climate change and marine plankton. Trends in Ecology & Evolution 20, 337–344.*

LIFE-CYCLES OF INSECT HERBIVORES

Many insect herbivores have complex life histories that may be accelerated by warming. For some species, this can mean completing additional life-cycles in a single growing season. The completion of an additional generation can result in large increases in abundance due to warming, turning forest pests into agents of mass destruction.

Insect life histories may be cued to environmental conditions; their speed may be determined by environmental conditions, especially temperature; or they may move to take advantage of changing environmental conditions. Like their arthropod cousins in the seas, insects may go through diapause or have life-cycles with several distinct stages. Each stage may be influenced by the physiological effects of warming, as may the transition between stages.

Phenology of insect life-cycles may be tightly linked to phenology in their food plants. Butterflies and other insects may be highly dependent on specific plants to complete their larval stage. Growth of the preferred plant that is too early or too late may result in large population crashes in the associated insect. For instance, a late-summer frost in 1992 resulted in the death of host plants (plantain; *Plantago* spp.) for Edith's checkerspot butterfly in the Sierra Nevada mountains of California, leading to the extinction of all populations that inhabited open clearings.

Where tight phenological matches with host plants are not limiting, warming may lead to rapid population growth in insects that feed on plants. Insects that can complete more than one life-cycle per year are known as multivoltine. Early emergence in multivoltine insects may lead to an additional generation in the course of a summer, resulting in rapid population expansion. Milder winters may result in increased overwinter survival and summer population growth in univoltine species. The resulting population increases can change insect herbivory into a catastrophic disturbance, resulting in the defoliation or death of host plants over large areas.

Outbreaks of the mountain pine beetle *Dendroctonus ponerosae* in western North America have resulted in the death of more than 100 million lodgepole pines (*Pinus contorta*). In British Columbia alone, more than 80 million trees have been lost across an area in excess of 450,000 ha. The beetle is killed by winter temperatures below −25 °C. Successive winters without killing temperatures resulted in population growth in mountain pine beetles in the 1980s and again from 1997 onward. Warmer winters and earlier springs meant that bark beetles could complete multiple life-cycles in a single growing season, resulting in population explosions (Figure 4.3). The 1980s outbreak was checked by the return of cold winter temperatures. The 1997 outbreak is ongoing. In Colorado, the outbreak peaked between 2005 and 2008. The beetle is a natural occupant of healthy forests, but its numbers are kept in check by a diversity of tree species and ages. Fire suppression and logging have resulted in large areas of even-aged, mature trees susceptible to beetle attack, whereas warm winters have promoted population growth sufficient for an outbreak causing widespread devastation. The consequences of this phenology-related outbreak are explored in the next chapter.

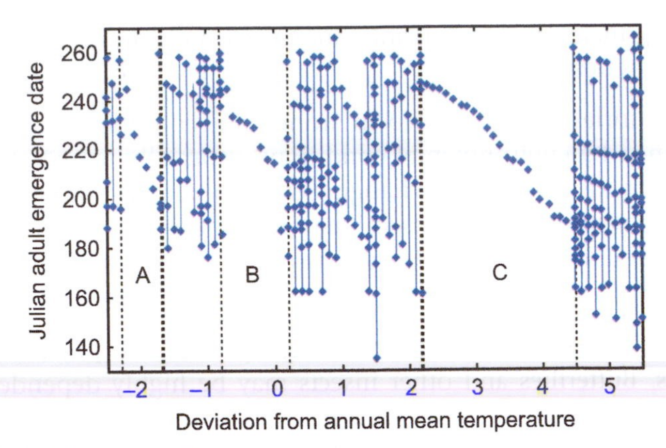

FIGURE 4.3 Modeled bark beetle emergence.
Pine bark beetles emerge in synchrony in defined temperature bands. The regions of the graph marked A, B and C are characterized by damaging synchronous emergence. This model output shows the simulated emergence synchrony associated with a slightly over 2 degree C increase in phloem temperature—a threshold that was crossed in much of western North America between 1998 and 2006. *Source: Logan et al. (2003). Reproduced with permission from the Ecological Society of America.*

SPOTLIGHT: GHOST FORESTS

Entomologists predicted that climate change would trigger an outbreak of bark beetles in western North America based on the projections of a computer model in 1999. By the time their research was published, the outbreak was already under way. Logan and Powell (2001) predicted that warming would trigger mountain pine beetle outbreaks in high-elevation, five-needle pines such as whitebark pine (*Pinus albicaudalis*) and limber pine (*Pinus flexilis*). Temperatures in the upper elevation habitats of these pines are generally too cold for bark beetles to reproduce more than once in a growing season. As temperatures warm, the phenology of the beetles changes: multiple breeding cycles become possible, allowing a population explosion (outbreak) to occur. Such an outbreak took place during the 1930s "dust bowl" years, when temperatures in the western United States were unusually high. Logan and Powell saw evidence of the 1930s outbreak in groves of dead trees still standing. These "ghost trees," killed in the 1930s, have not decayed due to the cold, dry conditions of the upper elevations. The model results, coupled with evidence of past outbreaks, led Logan and Powell to predict future movement of the beetle upslope and from the western to the eastern United States. They also predicted that beetle infestation might jump the high-latitude, high-elevation barriers along the continental divide, allowing beetles to infest eastern jackpine forests that had previously been beetle free. An outbreak in whitebark pine occurred in 2001 and has been progressing according to prediction since. Millions of acres of lodgepole pine have been lost in western Canada, suggesting that the prediction of a jump to eastern jackpine forests may be valid as well. These rapid confirmations of modeling predictions indicate that models may in some cases have great utility for foretelling and avoiding impacts.

Source: Logan, J.A., Powell, J., 2001. Ghost forests, global warming, and the mountain pine beetle (Coleoptera: Scotytidea). American Entomologist 47, 160–173.

The interaction of warming, insect life-cycles, and management regimes has had major consequences for forest mortality driven by insect outbreaks in North America. Similar factors may lead to mass mortality of trees or other plant host in other regions. In the near term, return to cold winters may occasionally occur and limit mountain pine beetle damage. In the long term, continued warming will ensure the persistence of larger populations of mountain pine beetle and other insect pests, meaning that control through management actions such as reduced fire suppression and increased diversity of forests will be increasingly important in determining damage to forests.

TIMING MISMATCHES BETWEEN SPECIES

Different species respond to warming at different rates, leading to possible mismatches in timing between species. The trophic mismatches in marine systems described previously are one example. Species interactions may be more sensitive when they involve changes in multiple climate cues. Predator–prey, herbivore–food plant, pollinator–plant, and other species–species interactions are other types of interactions that may be affected. Some of these species–species mismatches are already being observed. However, the ultimate effects of such mismatches are known for very few, if any, interactions.

For example, in The Netherlands, leaf emergence in trees is advancing, resulting in earlier peak insect abundances. Blue tit populations are responding to this change, laying eggs earlier. The effect of tree phenology is being transmitted up the food web, ultimately changing nesting behavior in blue tits (Figures 4.4 and 4.5).

FIGURE 4.4 Blue tit (*Cyanistes caeruleus*) resting on a branch.
Source: From Wikimedia Commons.

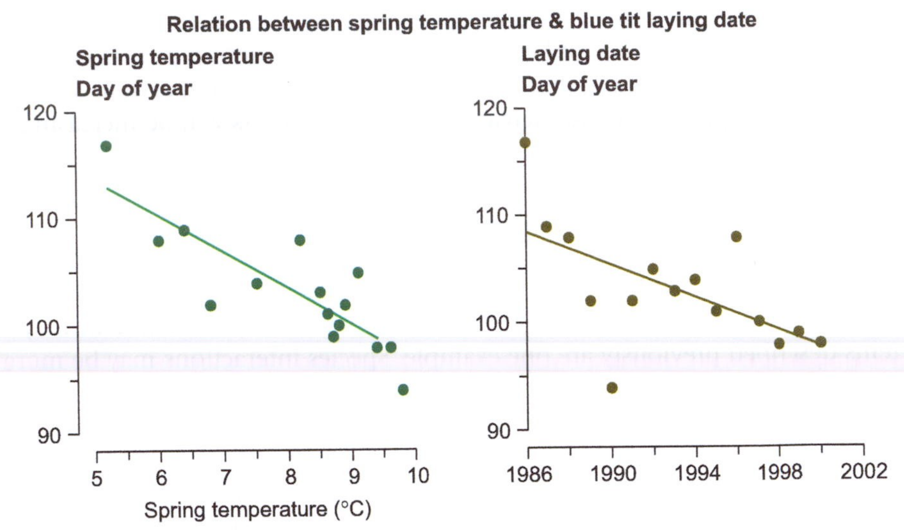

FIGURE 4.5 **Blue tit egg laying is earlier in warmer years, and progressive warming is resulting in an advance of more than 10 days in less than two decades.**
Source: Courtesy of Environmental Data Compendium.

When species involved in interactions respond to different environmental cues, mismatches are more likely to result. For example, oak budburst is controlled by temperature, whereas the emergence of winter moth (*Operophthera brumata*), which feeds on the oak, depends on freezing. The adult moth emerges in late fall, mates, and lays eggs in the oak canopy. Caterpillars emerge at the time of budburst and feed on buds. If the caterpillars emerge too early, there are no buds to feed on; if they emerge too late, they must eat less palatable and nutritious leaves. In England, warming has not been accompanied by reduction in freezing, so the moths are emerging after budburst. This timing mismatch is causing winter moths to feed on less desirable vegetation, with consequences for development and population size.

SPOTLIGHT: TIMING IS EVERYTHING

Caribou populations in Greenland are declining due to warming. The cause is a mismatch between caribou food requirements and plant availability. Caribou time their spring migration to coincide with emergence of food plants essential for lactation. Without adequate nutrition, female milk production decreases and calf survival declines. If the flush of spring growth occurs before caribou arrive, calf survival suffers. Prior to detailed study of warming, biologists believed that caribou would balance their nutrition by moving to areas of the landscape in which food was available.

However, Post et al. (2008) showed that both experimental and observed warming decreased the spatial variability in plant phenology: Plants tended to become most nutritious more synchronously as climate warmed. This decrease in spatial variability eroded the caribou's ability to maintain nutritious food by moving around the landscape. As a result, the condition of mothers declined, impacting calf survival. The observed caribou declines had been explained: It was the consequence of higher calf mortality as plant food availability became too clustered in time.

Source: Post, E., Pedersen, C., Wilmers, C.C., Forchhammer, M.C., 2008. Warming, plant phenology and the spatial dimension of trophic mismatch for large herbivores. Proceedings of the Royal Society B: Biological Sciences 275, 2005–2013.

FIGURE 4.6 Tracking penguins from space.
Changes in ice cover and temperature are affecting phenology of penguin food sources and the location and size of penguin colonies. Scientists are tracking changes in penguin colony locations from space using the guano marks they leave on the ice. In this photo, a colony and its guano mark show up distinctly against the white background of snow and ice. *Source: Courtesy of the British Antarctic Survey.*

Most mismatches that have been documented involve predator–prey interactions (Figure 4.6), including herbivore–plant relationships. For example, results from the Rocky Mountain Biological Laboratory show that a 1.4-degree rise in temperature has resulted in marmots (*Marmota flaviventris*) emerging earlier from hibernation. In the same period (1975–1999) snowmelt and plant flowering remained unchanged, resulting in a 23-day mismatch between emergence of marmots and their food plants.

Mismatches in animals may result from combinations of climate and non-climate phenologies, just as in climate–photoperiod mismatches in marine systems. The pied flycatcher, *F. hypoleuca*, is laying eggs earlier in Europe in response to warming but not early enough to match peak abundance of insect food, which is also advancing due to warming. The flycatchers are unable to lay early enough to match peak prey abundance because their migration is cued by day length. Thus, in a single species, warmth-cued laying is responding to climate change, whereas photoperiod-cued initiation of migration is not responding to warming. As a result, the flycatchers are partly keeping pace with earlier peak insect abundance (by laying eggs earlier) but are unable to fully match the advance in prey because their arrival from migration sets a limit on how early they can lay. Chicks hatching earlier emerge when caterpillars are abundant and so will have more food prey, and may be expected to survive better, than chicks hatching later. There is selective pressure for earlier laying, which may eventually lead to an evolutionary change in the photoperiod-cued timing of migration.

Physiological responses therefore respond more quickly to climate change, whereas photoperiod-cued responses require evolutionary change to adjust. Both are likely to occur in populations where there is mismatch between predator and prey phenologies. Whether evolutionary changes in photoperiod-mediated responses can take place quickly enough to match the rapid pace of human-induced climate change remains to be determined. However, species with photoperiod-mediated responses linked to climate have presumably adapted to climate change many times in the past. A larger question is whether population sizes are large enough after human hunting and habitat reduction for species to survive the population bottlenecks that will drive evolutionary response.

FURTHER READING

Both, C., Bouwhuis, S., Lessells, C.M., Visser, M.E., 2006. Climate change and population declines in a long-distance migratory bird. Nature 441, 81–83.

Clark, J.S., Salk, C., Melillo, J., & Mohan, J. 2014. Tree phenology responses to winter chilling, spring warming, at north and south range limits. Functional Ecology.

Cleland, E.E., Chuine, I., Menzel, A., Mooney, H.A., & Schwartz, M.D. 2007. Shifting plant phenology in response to global change. Trends in Ecology & Evolution 22 (7), 357–365.

Korner, C., Basler, D., 2010. Phenology under global warming. Science 327, 1461–1462.

Root, T., Price, J.T., Hall, K.R., Schneider, S.H., Rosenzweig, C., Pounds, J.A., 2003. Fingerprints of global warming on wild animals and plants. Nature 421, 57–60.

Post, E., Pedersen, C., Wilmers, C.C., Forchhammer, M.C., 2008. Warming, plant phenology and the spatial dimension of trophic mismatch for large herbivores. Proceeding of Royal Soceity B: Biology Science 275, 2005–2013.

CHAPTER 5

Ecosystem Change

Impacts of climate change on any individual species can cascade through biological systems, affecting other species, many of which may be responding to climate change themselves. These complex ecosystem impacts are among the most important in climate change biology because they impact the services these ecosystems deliver for human well-being and livelihoods.

Ecosystem services include provisioning services such as production of food and fiber, cultural services such as recreation and spiritual value, and supporting services such as nutrient or carbon cycling. Each type of service provides billions of dollars of benefits each year, and each may be threatened by climate change. Provisioning services such as supply of freshwater can be dramatically affected by rainfall changes. Wild-caught fish are susceptible to major abundance changes that correlate to climate shifts. Cultural services such as recreation may become difficult or impossible when events such as large fires or insect outbreaks affect large areas of forest. Supporting services such as biomass production and nutrient cycling are intimately tied to climate and human additions of carbon to the atmosphere.

The complexity of ecosystem change makes it difficult to detect and quantify. Nonetheless, pioneering studies are beginning to shed light on at least some of the ecosystem alterations already under way due to climate change. Modeling cannot always capture the richness of ecosystem interactions, but perspectives on areas of possible future concern are emerging. This chapter explores climate change impacts on ecosystems, observed and predicted.

TROPICAL ECOSYSTEM CHANGES

Tropical forests are warm-adapted systems that are generally more sensitive to drought than to warming. Drought sensitivity influences the distribution of tropical forest trees even in areas with relatively high rainfall. Because drought and length of dry season are intimately connected to climate, climate change is most likely to influence tropical forest diversity through these factors.

Climate Change Biology. http://dx.doi.org/10.1016/B978-0-12-420218-4.00005-6

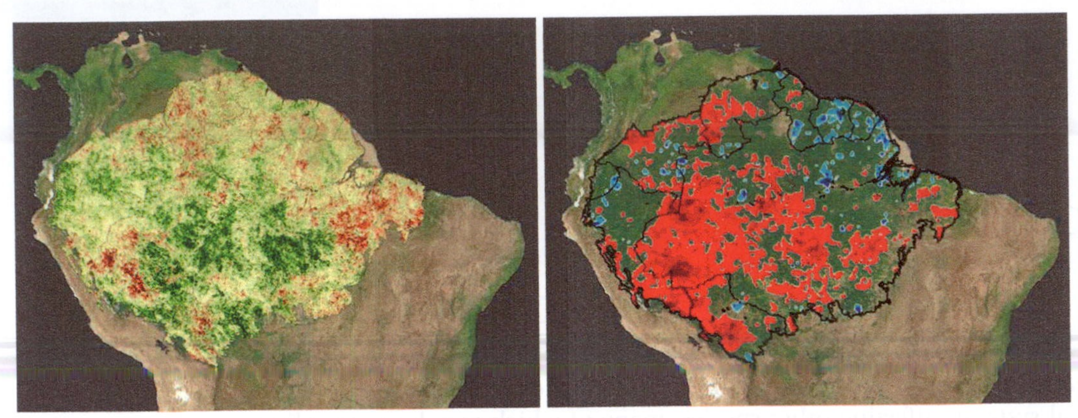

FIGURE 5.1 Drought in the Amazon, 2005.
The Amazon was affected by a major drought in 2005. Rainfall deficit (red in right panel) indicates the areas most severely affected. Primary productivity (left) increased (green) despite the drought in many areas. The mechanism for this unexpected result is the subject of continuing research. *Source: Kamel Didan, University of Arizona Terrestrial Biophysics and Remote Sensing Lab.*

The most dramatic of these effects is the occurrence of megadrought in the Amazon (Figure 5.1). Much of the Amazon experienced major drought in 2005 and again in 2010. In 2005, rainfall was lower than normal across 37% of the Amazon. The 2010 drought was even more extensive, with over half (57%) of the Amazon having lower than normal rainfall. These droughts have favored burning and resulted in large amounts of carbon being released to the atmosphere. The 2005 drought was a one-in-one hundred event, so the Amazon experienced two 100-year droughts in 5 years. The CO_2 emissions from each of the droughts has been estimated as larger than the annual CO_2 emissions of the United States.

The Amazon droughts may be reinforced because the Amazon has internal recycling of moisture, and raise concerns for other tropical forests under climate change. Climate change is expected to alter the hydrologic cycle, causing increased rainfall in some areas and increased drought in others. Because tropical forests are very sensitive to drought, areas like the Amazon that experience increased drought will be at high risk.

Changes in tropical forests due to drought have been documented at Barro Colorado Island (BCI) in Panama (Figure 5.2), providing some insights into possible ecological repercussions of tropical forest drought. BCI is the site of a long-term research effort in which a 50-ha plot is censused repeatedly for every plant with a stem thicker than a pencil (10 mm). This intensively studied area provides some of the first confirmation of the expected effects of climate change on tropical forests. At BCI, forest composition has changed as trees have died in response to drought. Drought at BCI has evolved since the 1970s, coinciding with the marked human-induced climate changes.

Prior to 1966, BCI experienced dry seasons with less than 100 mm of rain approximately every 6 years; whereas after 1966, one year in three was

FIGURE 5.2 Tropical forest test bed.
Barro Colorado Island (BCI) has been described as the premier test bed for tropical ecology. The forest canopy at BCI (right) often shows synchronous flowering in one or more species. A satellite photo (left) shows how rising lake waters during construction of the Panama Canal isolated the island. *Source: Left photo by NASA; right photo by Christian Ziegler, Wikimedia Commons.*

characterized by such droughts. The overall drying trend is accentuated in El Niño years, the strongest of which occurred in 1983. In that year, dry season precipitation was only 3 mm and temperatures were 2 °C above normal.

Of all mature and young trees (trees 200 mm dbh) in the BCI plot, 4.3% were killed by the 1983 drought. Seventy percent of tree species had higher mortality during the drought than in the subsequent 5 years (1985–1990). These high death rates have not affected forest structure—the ratio of canopy to understory trees remains the same—but are changing other aspects of the forest in important ways.

HISTORY OF BCI

BCI was formed during the construction of the Panama Canal and has been run continuously as a tropical forest reserve since 1923, and as a research station by the Smithsonian Institution since 1946. The Panama Canal uses gravity-fed freshwater to fill its locks. A new lake, Lake Gatun, was created in the construction of the canal to supply this freshwater. Barro Colorado was isolated when Lake Gatun filled, isolating it from surrounding forest and disturbance.

Since the 1980s, a group of botanists, including Peter Ashton of Harvard University and Richard Condit of the Smithsonian Institution, have been maintaining permanent forest plots on BCI. All of the plants growing in these 50-ha plots are censused and identified. Changes in time are recorded through re-censusing.

The effects of drought and other climate change manifestations are now being detected through these permanent, repeatedly resampled plots, making them critical reference points in the quest for knowledge about global change.

The impact of the drought had a major effect on forest composition. Moisture-demanding species became less common, whereas species with higher drought tolerance quickly filled space vacated. A total of 33 of 37 moisture-demanding species declined in abundance from 1982 to 1995. Some moisture-demanding tree species, such as *Poulsenia armata*, underwent population crashes of 50%. Small trees and shrubs were hardest hit, perhaps because of their less extensive root systems. Seventeen of 18 small-stature trees declined during and after the drought. The tree fern *Cnemidaria petiolata* disappeared entirely.

The effects indicate that major changes in tropical forests are occurring due to climate change, largely undocumented. Very few tropical forests have monitoring in place that would detect these changes. The BCI plot, a global network of closely studied sites, and remote sensing (Figure 5.2) provide insight into the vast majority of forests that are unmonitored. The BCI results suggest that significant changes may be under way where climate change is leading to regional drying. This is confirmed by data from at least one cloud forest site with a similarly intensive monitoring record.

CLOUD FORESTS

Tropical cloud forests are expected to be impacted by climate change through the lifting cloud base hypothesis. This theory, supported by modeling results, suggests that as sea surface temperatures warm with climate change, moisture formation will occur at higher elevations. This means that cloud bases will rise, moving the point at which montane forests intersect clouds upslope. The net effect will be a shift in suitable climatic conditions for cloud forests toward the tops of mountains.

Biological changes in cloud forests are being observed that fit the lifting cloud base theory. The most dramatic and well-documented changes have occurred at the Monteverde cloud forest in Costa Rica (Figure 5.3). Changes in other cloud forests are occurring as well.

At Montverde, the site of the disappearance of the golden toad (see Chapter 3), many range shifts have been linked to drying driven by rising cloud bases. Range shifts have occurred in Monteverde in reptiles, amphibians, and birds in a wide variety of species.

All of these changes have been linked to drying trends associated with rising sea surface temperatures in the region. As sea surface temperature has increased in the region, the number of dry days per year has mounted, precipitation has

FIGURE 5.3 **Cloud forest, Monteverde, Costa Rica.**
Source: Photo by John J. Messo, NBII, USGS.

SPOTLIGHT: LIFTING CLOUD BASE EFFECT

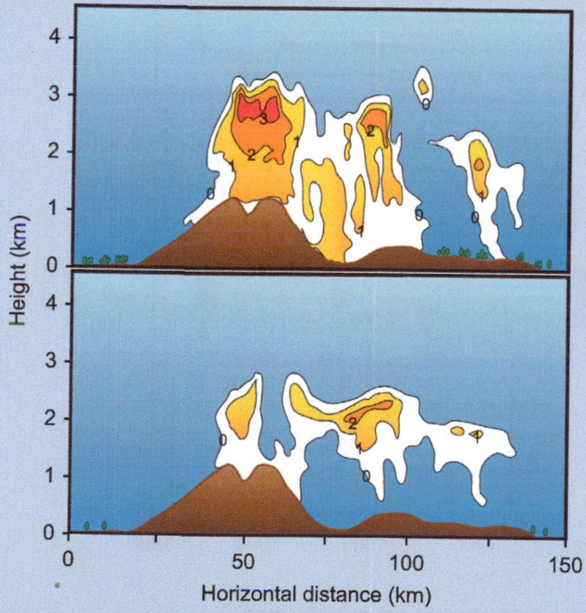

Tropical cloud forests form where clouds intersect mountain slopes (top). Under climate change or lowland land clearing, lowered relative humidity at altitude means clouds will form higher (bottom), reducing the area of intersection with mountains and decreasing the extent of cloud forest, possibly causing the loss of some of the many endemic species found there. In this schematic, increasing relative humidity and cloud condensation are indicated by shades of orange.
Source: Lawton et al. (2001).

Cloud bases are expected to lift in the tropics due to climate change, which may have serious present and future consequences for tropical cloud forests (Still et al., 1999). Cloud base lifts in warmer conditions because cloud formation depends on relative humidity—warmer air requires more moisture for cloud formation to occur. Thus, while warmer sea surface temperature increases evaporation and the moisture content of air near the surface, at elevation this increase in moisture is not enough to offset warming. Decreased relative humidity and cloud formation at higher altitudes result. Model results suggest that cloud bases were lower at the peak of the last glacial period because of cooling and will be higher in the future due to anthropogenic warming. Cloud base under doubled CO_2 simulations shows a rising cloud base, which may already have contributed to amphibian decline by promoting chytrid fungal disease, and it may result in drying and water stress in the future. An alternative hypothesis suggests that land clearing may reduce moisture recycling, resulting in the increase in dry days and increase in height of cloud base already observed in Costa Rica (Lawton et al., 2001). Both of these effects may come into play, indicating a double threat to tropical cloud forests.

Source: Lawton, R.O., Nair, U.S., Pielke, R.A., Welch, R.M., 2001. Climatic impact of tropical lowland deforestation on nearby montane cloud forests. Science 294, 584–587; Still, C.J., Foster, P.N., Schneider, S.H., 1999. Simulating the effects of climate change on tropical montane cloud forests. Nature 398, 608–610.

decreased, and streamflows have fallen. Perhaps the most important of these trends has been the increase in dry days. Like tropical trees, tropical amphibians are especially sensitive to dry periods.

Forty percent (20 of 50 species) of frogs and toads present at Monteverde in the early 1980s disappeared in a spectacular population crash in 1987 (Figure 5.4), which was a particularly dry year, with more than 50 dry days. Surviving frogs show strong declines and continuing fluctuations in synch with dry periods.

Birds typical of low elevations have shifted upslope during the drying period. Of particular note is the keel-billed toucan, which has expanded upslope at the expense of the resplendent quetzal. The toucan preys on the young of the quetzal, so the upslope shift in the quetzal appears to be triggered by predation and competition from the toucan. Other lower elevation species have moved upslope as well. At one plot at 1540 m in elevation, 15 previously absent lower elevation species have established breeding populations since 1976.

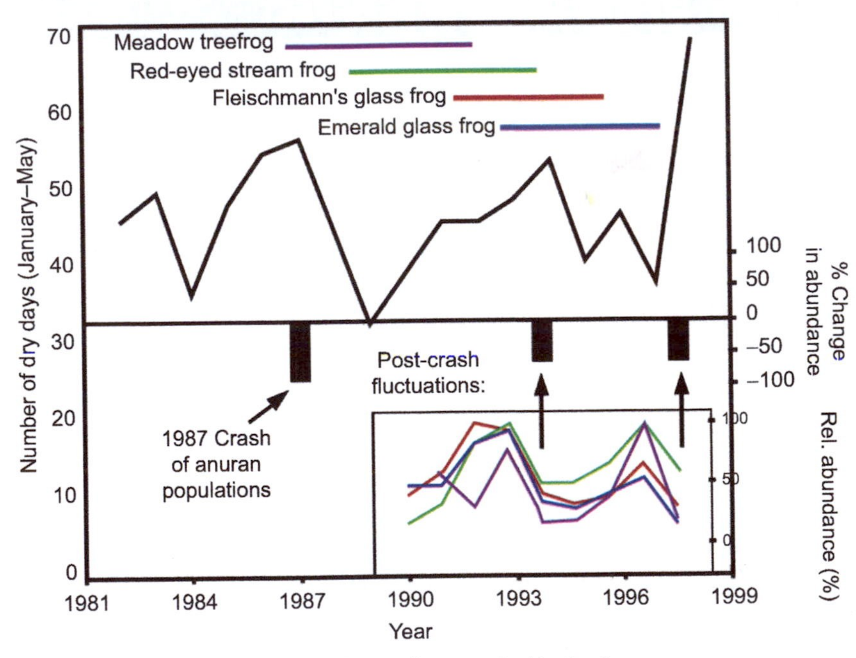

FIGURE 5.4 Monteverde population fluctuations synched to dry days.
Twenty species of frogs and toads disappeared from the Monteverde cloud forest in Costa Rica (first black bar) after an unusually long run of dry days (solid line). The golden toad (*Bufo periglenes*) was locally endemic, so its disappearance represented a global extinction, perhaps the first extinction linked to climate change. Subsequent long dry spells have caused other frog population crashes since 1987 (inset). Increasing frequency of dry spells in cloud forest is linked to climate change through the lifting cloud base effect. Dry periods appear to favor pathogenic growth of the fungus that is the ultimate cause of death in affected frogs. *Source: Pounds et al. (1999). Reproduced with permission from* Nature.

The changes at Monteverde fit the classic pattern of range shifts expected with warming—shifts upslope and the disappearance of upper elevation species—but they are also much more complex than this. Species interactions are changing, and population changes are being driven by multiple factors. In this case, the shifts and population losses are caused by drying and warming rather than by warming alone. Other causes, such as habitat loss, have less power to explain the observed changes. For instance, deforestation in the lowlands near Monteverde dates from several decades before the declines were observed. The changes at Monteverde and BCI clearly establish the importance of precipitation change, and especially change in dry season, in determining changes in tropical systems.

TEMPERATE ECOSYSTEM CHANGE

As discussed in Chapter 4, climate-driven changes in insect phenology are resulting in the deaths of hundreds of millions of pine trees in North America. The ecosystem changes that follow this devastation are profound. In many areas, more than 80% of all mature trees are killed, resulting in massive changes in light penetration, carbon storage, and canopy cover. Not all of the ecosystem repercussions of these changes are fully understood, but as the destruction progresses, insights into the ensuing ecological changes are progressing. The ecology of lodgepole pine ecosystems both facilitates the beetle attack and results in landscape-scale change where beetle outbreaks occur (Figures 5.5 and 5.6).

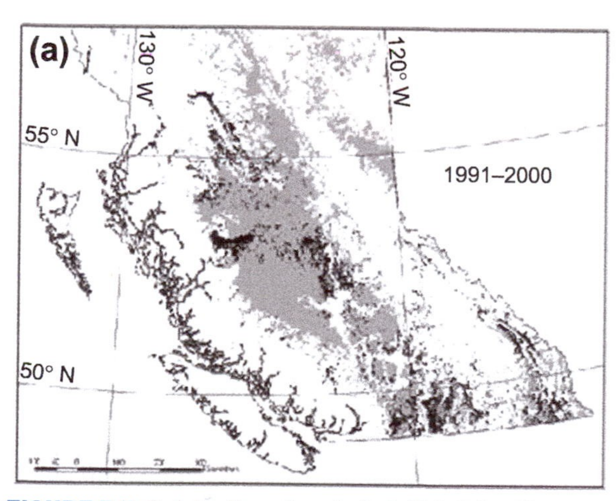

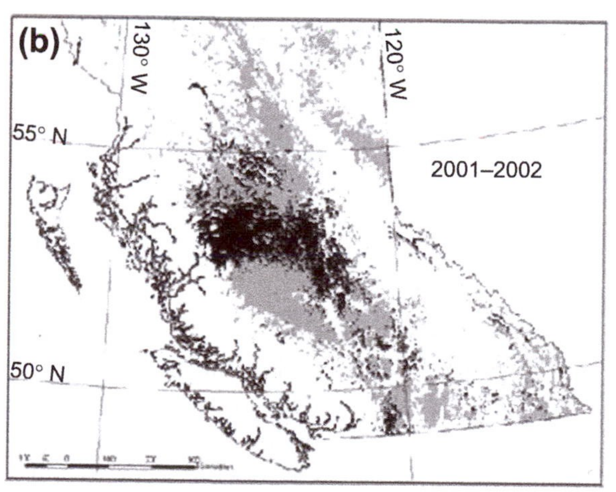

FIGURE 5.5 Bark beetle outbreaks in British Columbia.
Bark beetle outbreaks spread rapidly through British Columbia in the 1990s (a) and early 2000s (b). In 2001, peak increase in outbreak area occurred. *Source: Copyright Her Majesty the Queen in Right of Canada, Canadian Forest Service, as originally published in* Nature.

FIGURE 5.6

Bark beetle-killed trees (a) and bark beetle damage in tree limb (b). *Source: (a) Copyright University Corporation for Atmospheric Research. Photo by Carlye Calvin. (b) Deborah Bell, Smithsonian National Museum of Natural History.*

Lodgepole pine form dense, even aged stands as part of their fire ecology. Lodgepole pine seeds are encased in dense cones, thick with pitch, that only open when burned. The cones remain on the trees or on the ground until a fire releases the seeds. When a stand is burned, all of the trees are killed, but a huge release of seed is triggered. This reproductive strategy allows lodgepole pine to dominate large areas in which few other tree species exist and virtually all trees are the same age (Figure 5.7).

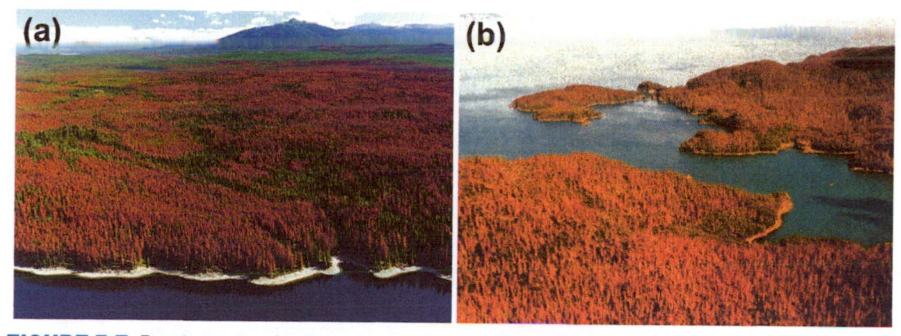

FIGURE 5.7 Dead stands of lodgepole pine in British Columbia.
Source: Reproduced with permission from the Ecological Society of America.

The outbreak has been facilitated by modern fire suppression techniques. Prior to human arrival, burning of lodgepole pine occurred randomly, resulting in a patchwork of fire scars, time-since-fire histories, and age stands of lodgepole. After approximately 1920, effective firefighting changed this pattern. Fires no longer raged out of control; the random mosaic of past burns was gradually turned into large areas of even-aged stands interrupted by areas that had been logged. The existence of even-aged stands facilitated outbreak in the beetle because there were no barriers to dispersal to disrupt synchronization. Beetle populations, once synchronized, could move from tree to tree over short distances, allowing synchronization to be maintained. The result has been mass destruction of trees.

The mountain pine beetle has always been native in western forests and absent from eastern forests. It now appears that warming will change this dynamic, allowing the beetle to undergo a range shift, spreading eastward from the gigantic British Columbia outbreak (Figure 5.8).

Warming has allowed the beetle to extend its range northwards in British Columbia, breaching the Continental Divide, the last effective barrier between the beetle and eastern pine plantations. In the past, the grasslands of the Midwest and the cool, high elevations of the northern Continental Divide had kept the beetle from spreading from west to east. With the British Columbia outbreak has come northward and upslope range expansion that has taken the beetle up and over the Continental Divide.

Eastern forests of jack pine may now be vulnerable to mountain pine beetle outbreaks. If the beetle is able to establish and move through jack pine, it is likely to extend its range across Canada and into the forests of the eastern seaboard. Because temperatures in the south will be favorable to beetle outbreaks, the range may eventually extend into the great loblolly pine regions of the U.S. Southeast, decimating stands of large commercial and biological importance.

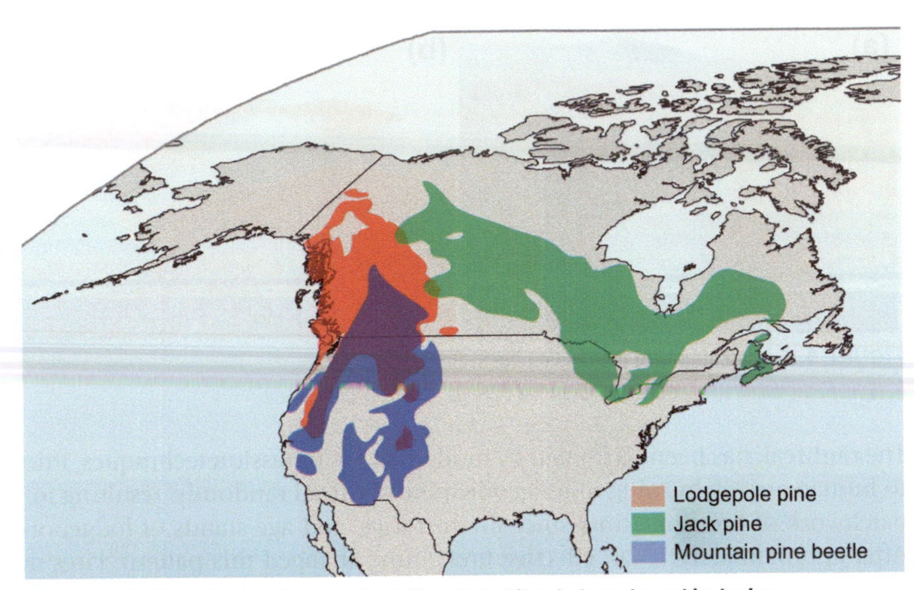

FIGURE 5.8 Map of current and potential beetle habitat, lodgepole and jack pine.
Eastern jack pine forests have been isolated from bark beetle habitat by the continental divide and grasslands of the Midwest. Extension of beetle habitat upslope with warming is crossing the continental divide and skirting grasslands through the continuous forests to the north in Canada. *Source: Logan et al. (2003). Reproduced with permission from the Ecological Society of America.*

SPOTLIGHT: FIRE WEATHER

Forest fire activity in the western United States is increasing, and this increase is linked to changes in climate. Westerling et al. (2006) found that the frequency of western U.S. wildfires correlated with warming and snowfall variables. Fire frequency tracked temperature between 1970 and 2005. Snowmelt may be the mechanism that mediates this effect of temperature. Large fire frequency correlates with years of early snowmelt. Warming apparently accelerates snowmelt, which results in drier, more flammable forests later in the year. The large fires that are resulting open large areas of the western U.S. landscape to ecological succession. This may accelerate species range shifts in response to climate change by freeing newly suitable climate from competition from established species.

Source: Westerling, A.L., Hidalgo, H.G., Cayan, D.R., Swetnam, T.W., 2006. Warming and earlier spring increase western U.S. forest wildfire activity. Science 313, 940–943.

A similar warming-related die-off has been observed in pinyon pine (*Pinus edulis*) in Arizona and New Mexico. This dieback is also driven by bark beetle infestation. Tree death of pinyon is associated with a large-scale drought in the region from 2001 to 2004. The drought, one of the largest on record for the region, is associated with higher than normal temperatures. More than 90% of mature trees have been killed in some areas. The net effect of this dieback has been to replace pinyon-dominated woodland with juniper (*Juniperus monosperma*) thicket over much of the Southwest. Species

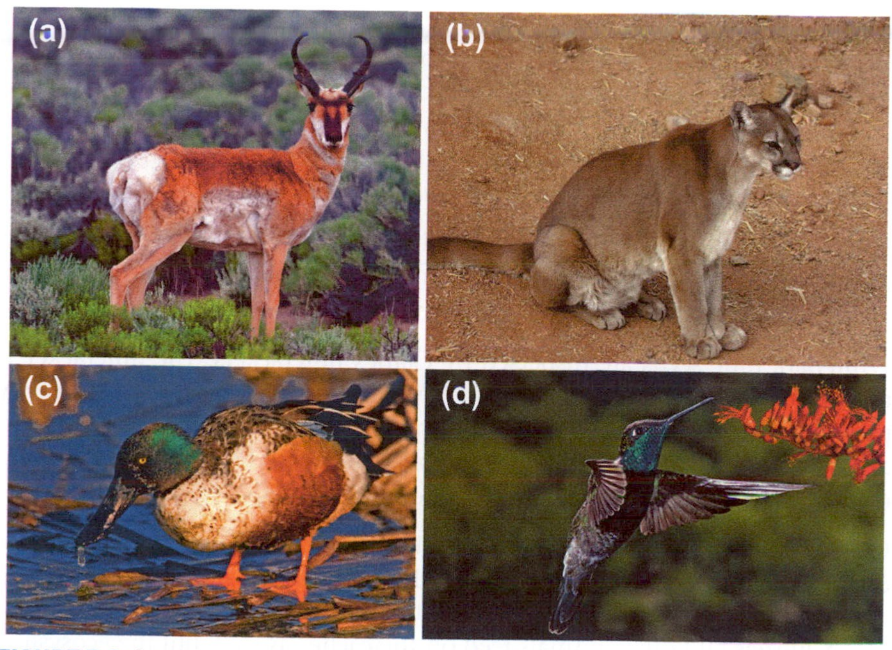

FIGURE 5.9 Species associated with pinyon woodland.
(a) Pronghorn (*Antilocarpa americana*); (b) Cougar (*Puma concolor*); (c) Shoveler (*Anas clypeata*); and (d) Magnificent hummingbird (*Eugenes fulgens*). *Source: From Wikimedia Commons.*

dependent on pinyon are experiencing loss of habitat over large areas (Figure 5.9).

HIGH MOUNTAIN ECOSYSTEMS

High-elevation five-needle pines such as whitebark pine (*Pinus albicaulis*) are being affected by bark beetle outbreaks almost as severe as those seen in lodgepole pine. Five-needle pines grow at high elevations previously too cold to sustain beetle outbreaks, but now outbreaks are occurring in Idaho, Montana, and Wyoming.

FIVE-NEEDLE PINES

Pine species growing at high elevation or under water stress often have shorter, more abundant needles. Pines with five needles per bundle that are found especially at higher elevations include whitebark pine (*P. albicaulis*) and limber pine (*Pinus flexilus*). These high-elevation pines enjoy long, cold seasons that limit damage from bark beetles. Recently, however, warming is bringing beetle outbreak damage to these pines.

Whitebark pine is particularly affected by the new high-elevation outbreak. Like other high-elevation pines, whitebark pine was previously protected from mountain pine beetle infestation by winter freezes that killed beetle larvae and by a summer warm season that was too short to allow beetles to complete a full life-cycle in a single year. A previous precedent for the whitebark outbreaks occurred in the 1930s. Outbreaks during the 1930s accompanied a run of unusually warm years, similar to the string of warm years currently being experienced in the region. The trees killed in the 1930s did not decay in the cold, high-elevation climate typical of the past, so their skeletons have remained in the landscape as "ghost forests."

The ecological consequences of the whitebark pine infestation affect food webs in Yellowstone National Park and surrounding conservation areas, providing a good example of the ecosystem and conservation consequences of this outbreak.

Whitebark pine seeds are dispersed by Clark's nutcrackers. The nutcracker is the only animal with the ability to get within the pine's dense and heavily pitch-laden cones. Nutcrackers stash caches of seeds, from which recruitment of new pines takes place. Whitebark pine are often found in clusters, composed of individuals all descended from a single cache of seeds.

Grizzly bears feed on the seed caches. This is a critical resource for bear reproduction. Recruitment of bear cubs into the population is lower and the condition of cubs is poorer in years in which whitebark pine cone production is low. Squirrels and other animals feed on the caches as well, but the ecosystem consequences are greatest for bears.

BEARS, PINES, AND CLIMATE

Bears are omnivores that eat a remarkable variety of things. Young grizzly bears feed heavily on the seeds from whitebark pine cones at certain times.

As climate change increases beetle-kill in whitebark pine, this critical food source may be reduced, jeopardizing the health and growth of young grizzlies.

Whitebark pine trap snow at upper elevations, which impacts hydrologic regimes of all downslope ecosystems. The pine's needles reduce wind velocity, holding snow in "windrows" behind the trees. The shade of the canopies reduces sublimation (evaporation from ice phase) of snow, retaining more water for spring melt.

The loss of whitebark pine over large areas will mean less runoff and less late runoff for downstream systems. In areas of whitebark pine death, lower streamflows are being observed, coupled with lower recruitment of trout. These impacts reinforce those of warming, which is reducing streamflow and trout recruitment directly.

Together, then, the impacts of whitebark pine death due to beetle infestations are affecting both terrestrial and freshwater ecosystems over areas much broader than the high-elevation range of the trees. Fewer and less fit grizzly cubs mean changes in bear population dynamics that are rippling to the many species on which bears feed. Lower flow and less late flow are reducing trout recruitment and populations, significantly altering stream ecology. Most of the long-term endpoints of these alterations have yet to be fully studied, but the cascade effects of the loss of whitebark pine are clear.

The future of whitebark pine is not entirely dark. The pine can recruit after beetle outbreaks. Beetles do not attack immature trees. It appears that beetles prefer trees older than 80 years, whereas whitebark pines begin reproducing at approximately 60 years of age. This 20-year window of cone production prior to beetle infestation may provide a buffer that will allow population recovery. The ancient whitebark pine may slowly disappear from western North American landscapes, never to be seen again. However, generations of young whitebark pine may spring up in their place, reproducing and being killed by beetles in cycles that will partially restore mountain ecology in the region. A major remaining question is whether future generations of Clark's nutcrackers can relearn exploitation of the whitebark cone after 50–60 years of shifting to other foods following the initial whitebark population crash.

GLACIER AND SNOWPACK-DEPENDENT ECOSYSTEMS

Loss of snow and ice is driving change in terrestrial and freshwater systems from the tropics to the poles—change that will accelerate throughout this century. Snow and ice comprise the major parts of the cryosphere—that part of the Earth that is covered by snow or ice, including permafrost. The cryosphere is changing rapidly due to global warming. Glaciers are melting in all areas of the world, and snowpack is disappearing earlier in spring in temperate areas. We have dealt with loss of ice cover at the poles in other sections; here, we focus on biological consequences of temperate and tropical changes in the cryosphere.

Tropical glaciers are disappearing throughout the world. Virtually all tropical glaciers are experiencing warming temperatures that are causing them to lose more mass to melting than they accumulate from snowpack each year. In some regions, this effect may be heightened by changes in surrounding land use, which can reduce precipitation (snowfall).

On Mount Kenya in eastern Africa, of 18 glaciers present in 1900, only 11 remained by 1986. During the same period, 75% of the glacier cover of the mountain was lost (from 1.6 to 0.4 km^2). The Qori Kalis glacier in Peru, the main outlet glacier of the biggest tropical icecap in the world, is retreating at a rate of more than 1 km per decade. The physical consequences of melting are exposure of land previously covered by ice, flow increases in rivers fed by glaciers in the near term, and large losses in river flow in the longer term once glaciers have disappeared (Figures 5.10 and 5.11).

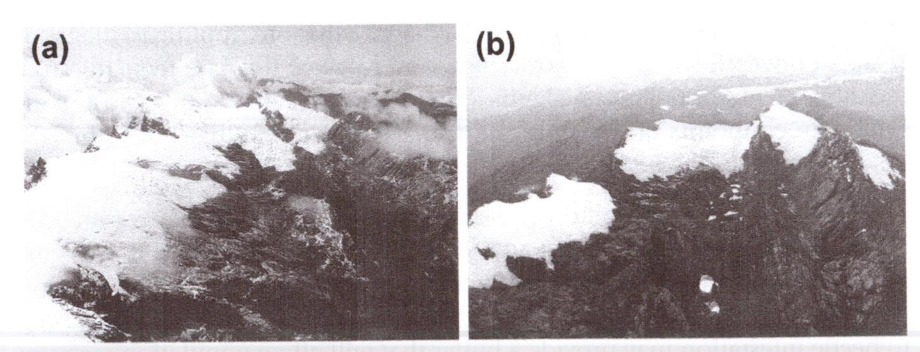

FIGURE 5.10 **Loss of tropical glaciers.**
Photos of Puncak Jaya glacier in Papua New Guinea from 1936 (a) and 1972 (b). Tropical vegetation is moving into areas formerly covered by the ice of this glacier. *Source: From Wikimedia Commons.*

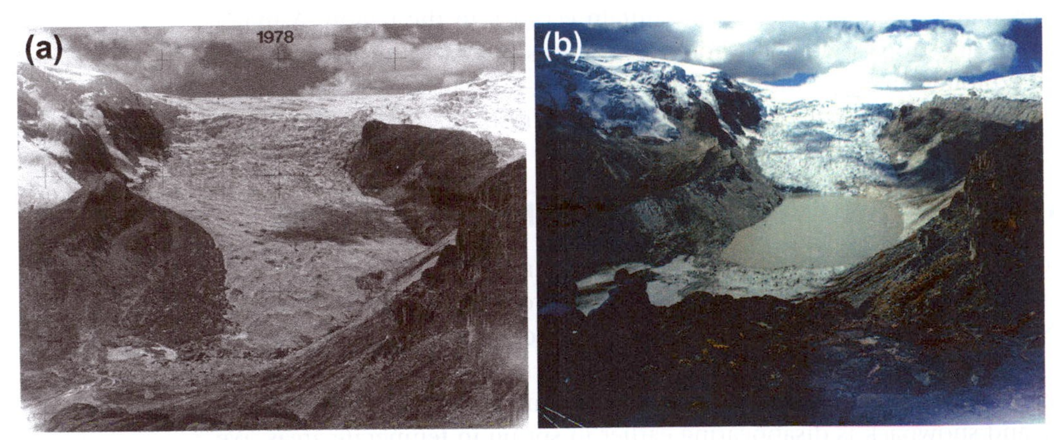

FIGURE 5.11 **Qori Kalis Glacier, Peru.**
The Qori Kalis glacier is the most significant ice outlet from the Quelccaya ice cap on the Cordillera Vilcanota in southeast Peru. Its extent in 1978 (a) was much larger than in 2004 (b). In only 5 years, this glacier has retreated more than half a kilometer. New ecosystems are developing in the freshwater ponds left behind by the glacier. *Source: Photos courtesy Lonnie Thompson, Ohio State University.*

The biological consequences of river flow changes result from both loss of flow and warming water temperatures. As glaciers lose all mass, coldwater flows are dramatically reduced. Lower water levels result in reduced habitat for freshwater species in glacial streams and rivers. As important, reduced flow results in warmer water temperatures because the reduced flow volume will warm more rapidly. This temperature effect, which is compounded by global warming, reduces habitat for coldwater species.

Loss of snowpack may have even greater impacts in temperate environments. Glaciers are melting in temperate regions as diverse as the Alps, Glacier National Park, and the Himalayas. However, in all these regions and throughout the temperate zone, snowpack is decreasing and melting earlier as well. Much more streamflow is dependent on snowmelt than on glacier outflow in these temperate ecosystems.

SPOTLIGHT: MOVING UP

Tropical glaciers are melting. Biological response to that change is not far behind. Seimon et al. (2007) found that life is tracking glacial retreat in the Tropical Andes. Ponds formed by glacial retreat during the past 80 years now have three species of frogs and toads inhabiting them that have moved up from lower elevations. Because these anurans are near the top of the pond food chain, their presence indicates that complete ecosystems have assembled anew in these habitats in the few decades since they have been formed. Chytrid fungal disease has followed these species, indicating that the interplay of disease and climate will be complex in these new habitats.

Source: Seimon, T.A., Seimon, A., Daszak, P., Halloy, S.R.P., Schloegel, L.M., Aguilar, C.A., et al., 2007. Upward range extension of Andean anurans and chytridiomycosis to extreme elevations in response to tropical deglaciation. Global Change Biology 13, 288–299.

Loss of snowmelt and earlier melt results in less streamflow in critical summer dry periods. Snow serves as a water storage device, gradually releasing water as melt progresses through the spring and summer. Especially in dry regions, this delayed release of stored precipitation provides a much-needed buffer against drought. As snowpack decreases and melts earlier, drought deficit in freshwater systems and riparian vegetation becomes more pronounced (Figure 5.12).

Freshwater species such as salmon and trout may be particularly severely affected. Warming will reduce coldwater upper elevation habitats for these species at the same time that loss of snowpack will reduce coldwater habitat driven by snowmelt. The net effect may be loss of high- and mid-elevation coldwater habitats, with no possible source of replacement.

Loss of snowpack can impact plants and trees in the snowpack zone as well as riverine and riparian vegetation. For instance, ponderosa pine (*Pinus ponderosa*) have

FIGURE 5.12 **Sockeye salmon (*Oncorhynchus nerka*) is an anadromous species sensitive to climate change in both its freshwater and its marine life stages.**
Source: From Wikimedia Commons.

moved upslope several 100 m in the past century in the Sierra Nevada mountains of California (Figure 5.13). The mechanism driving this shift appears to be retreating snowpack. Young ponderosa pine depend on meltwater from snowpack to survive late spring and summer drought. When snowpack disappears, recruitment of ponderosa pine is inhibited. Where mature trees are cleared for timber, the lower range limit of ponderosa pine shifts upslope, following the retreating snowpack.

POLAR AND MARINE SYSTEMS

Since 1978, ice cover in the Arctic has declined by nearly 20%. As sea ice retreats, annual ice will be found over deeper waters, farther from the continental shelf. These changes are affecting many species in the region, especially marine mammals such as polar bears (*Ursus maritimus*) and ringed seals (*Pusa hispida*).

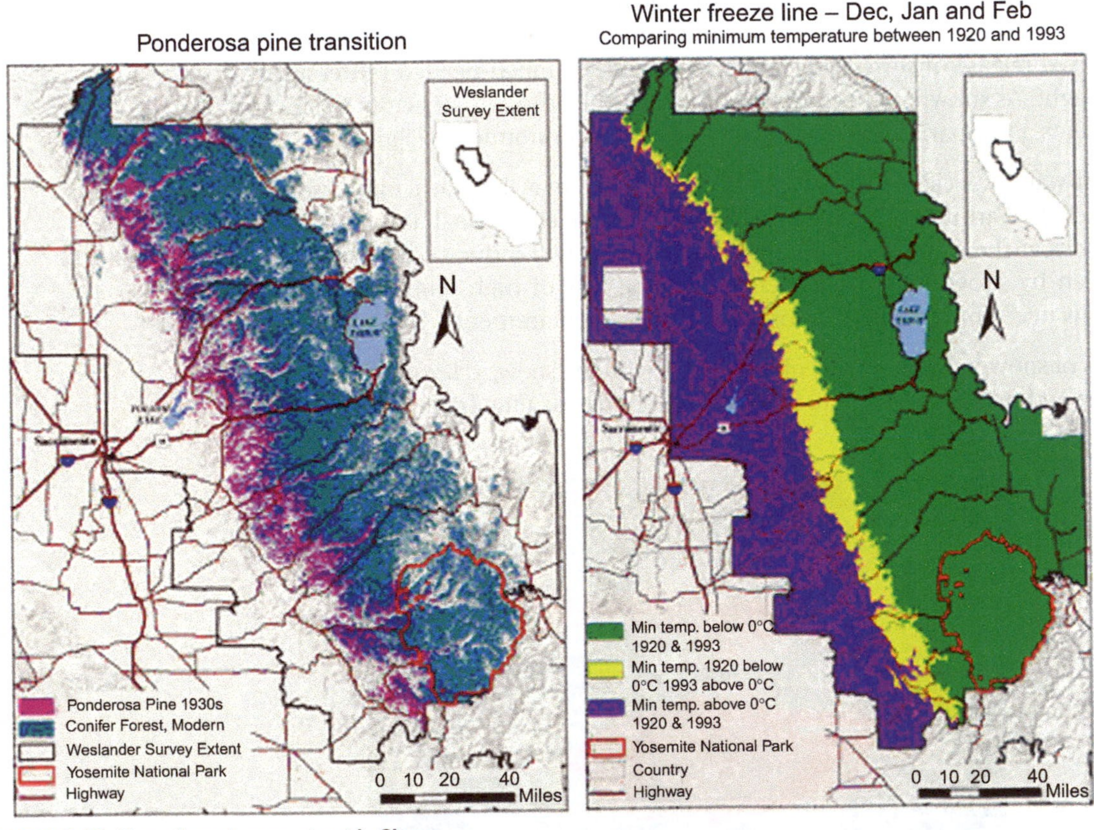

FIGURE 5.13 Map of ponderosa retreat in Sierras.
Ponderosa pine range has been reduced in the Sierra Nevada mountains of California since 1930. Upslope movement of montane hardwoods (dominated by *Quercus* sp.) has been replacing the lower range margin of ponderosa pine (left) while temperature has been increasing in the region (right). Upslope loss in ponderosa pine is detected by comparing vegetation surveys from the 1930s (Wieslander VTM survey) to modern vegetation maps (Thorne et al., 2008). The area of retreat in freezeline (yellow, right) closely corresponds to the area of pine loss (red-purple, left). *Source: Courtesy of James H. Thorne.*

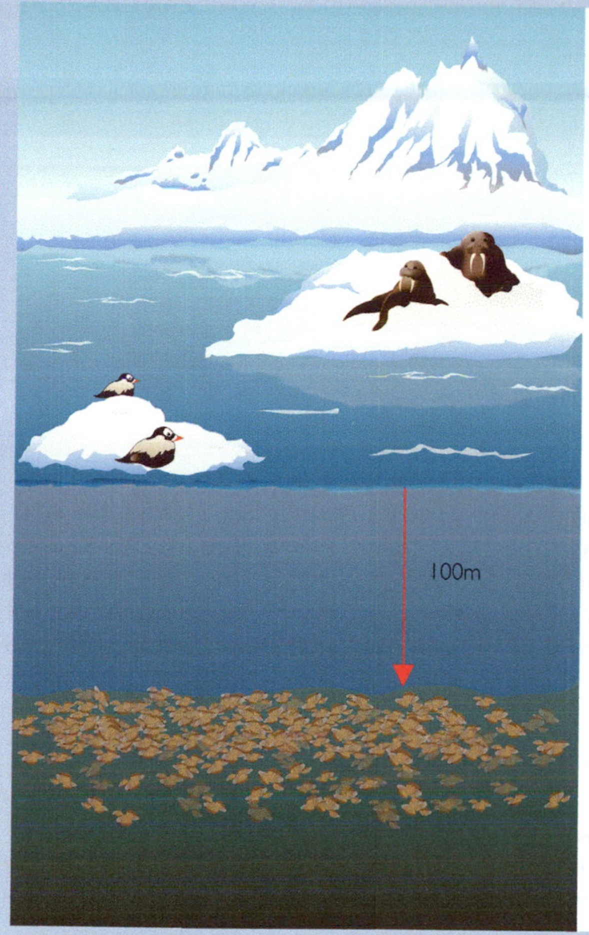

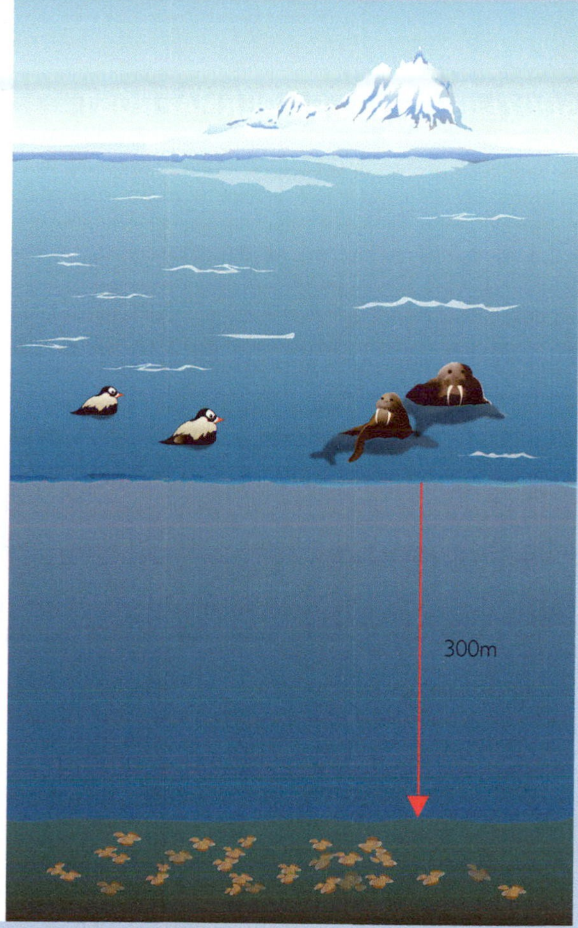

Walrus and spectacled eider rest on sea ice and dive deep to feed on bottom fauna (left). With warming, sea ice melts over the continental shelf, forcing walrus and eider into deeper water (right). Resting in cold water consumes more energy then resting in air, even on ice. So diving deeper, to less abundant food prey, and losing more heat while resting on the surface place species in negative energy balance, resulting in increased mortality and decreasing populations.

Sea ice is rapidly retreating in the Bering Sea between the United States and Russia. The consequences for species that use sea ice are profound (Grebmeier et al., 2006). Spectacled eider populations have declined more than 90% in the past three decades. Walrus populations are threatened as well. Both of these species feed on benthic organisms that are being dramatically affected by changes in ocean temperatures near the sea floor.

Clams, polychaete worms, and brittle stars are changing in abundance in this area. Because clams are much more nutritious for walrus and diving ducks such as eider, these changes at the base of the food web are responsible for large population changes in consumer species.

Here, the food chain is short, with benthic organisms feeding on detritus and top carnivores feeding on clams and other benthic invertebrates. Changes in water temperature have led to thick-shelled, less nutritious clams becoming more abundant than thin-shelled clams and to polychaetes and brittle stars (both low in nutrition) replacing clams.

A sill of cold water just south of Saint Lawrence island has kept groundfish out of the Bering system, but this cold water sill is now breaking down. With this barrier removed, groundfish will enter the system, competing with walrus and diving ducks for food. More important, trawling fisheries are likely to follow the fish stocks, resulting in destruction of bottom habitat. International protection is urgently needed to prevent further population declines of walrus and spectacled eider.

Source: Grebmeier, J.M., Overland, J.E., Moore, S.E., Farley, E.V., Carmack, E.C., Cooper, L.W., et al., 2006. A major ecosystem shift in the northern Bering Sea. Science 311, 1461–1464.

SPOTLIGHT: NOT SO PERMAFROST

Permafrost is melting in the low Arctic due to warming temperatures. Melted areas heave and slump as the ice crystals in the soil turn to liquid. Heaving and slumping in turn expose soil, which warms and becomes more favorable for plant establishment. Green alder (*Alnus viridis*) in particular becomes more abundant in slumps. This increased vegetation can lead to more absorption of solar energy, warming, and more slumping. Slumping can therefore magnify the effects of temperature on plant communities in the Arctic (Lantz et al., 2009).

Source: Lantz, T.C., Kokelj, S.V., Gergel, S.E., Henryz, G.H.R., 2009. Relative impacts of disturbance and temperature: persistent changes in microenvironment and vegetation in retrogressive thaw slumps. Global Change Biology 15, 1664–1675.

Polar bears hunt on ice through the winter and then fast through the summer as the ice edge retreats from land. Bears will forage over wide areas hunting seals, returning to land or multiyear ice at the start of summer ice breakup to begin their fast. Females must store enough fat during the hunting season to survive the 4-month open-water fasting period and successfully reproduce. Females with body mass less than 200 kg do not reproduce.

Warming has changed the amount of ice and snow protecting seal lairs, resulting in increased predation by polar bears. Unusually mild conditions on Baffin Island in 1979 resulted in seal lairs being covered by soft snow rather than ice. Without a protective layer of ice, seal pups were much more liable to detection by polar bears. Predation on seal lairs was three times higher that season as a result (Figure 5.14).

Later, more profound changes in sea ice have forced polar bears to return to land earlier, effectively shortening the hunting season. Based on current rates of decline of 5 kg/year, most females in western Hudson Bay may soon be unable to give birth. Air surveys have recorded mass bear drownings resulting when bears have attempted to prolong the hunting season by staying on the ice after breakup and swimming long distances back to land.

Polar bears exist in 14 populations in five countries (the United States, Russia, Canada, Norway, and Greenland (Denmark)), each population with unique ecological characteristics. All populations hunt on annual ice, rather than multiyear ice, because seals are better able to maintain breathing holes in thinner ice. When annual ice breaks up, southern populations move to land, whereas northern populations move to multiyear ice. Changing sea ice conditions (Figure 5.15) therefore affect these populations differently.

Northern populations may be the last holdouts against warming, as might be expected. In the far north, thinning ice may temporarily lead to better hunting

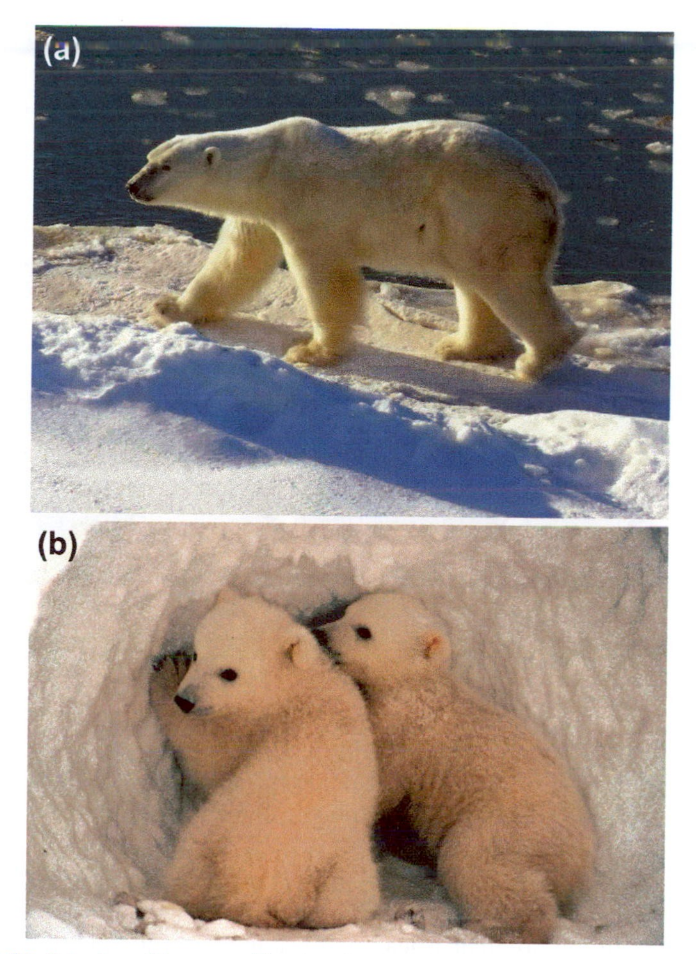

FIGURE 5.14 Polar bear (*Ursus maritimus*) and cubs in ice den.
Source: From Wikimedia Commons.

conditions for bears. As ice thins and retreats, these populations may see annual ice disappear and multiyear ice convert to annual ice.

Earlier retreat of sea ice is forcing bears into conflict with humans. Returning to land before their fat stores are replenished, some bears turn to human garbage or raids on food supplies to supplement their diet. This has led to increased human–bear conflicts that often result in the death of bears.

For these and other reasons, the International Union for Conservation of Nature (IUCN) has listed the polar bear as vulnerable to extinction. The IUCN polar bear specialist group estimates that bear populations could decline 30% in as little as 35 years, representing the loss of between 7000 and 8500 bears in a global population of less than 25,000. This group has estimated that polar bears may

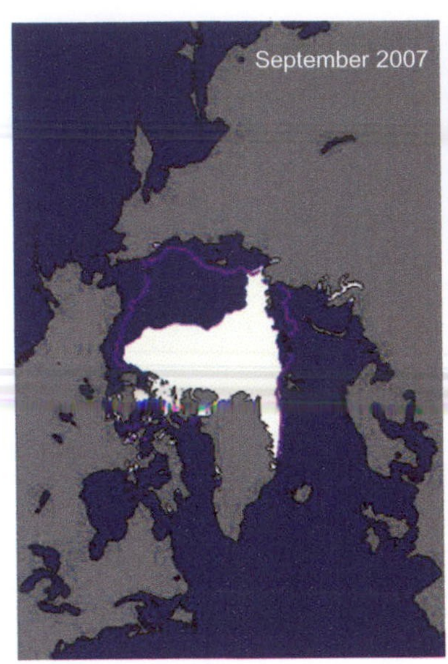

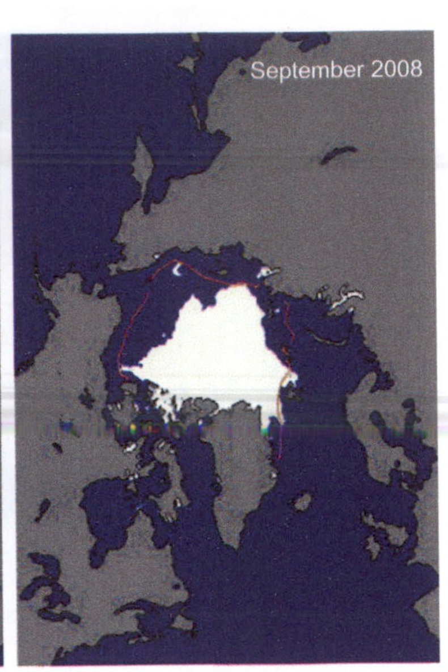

FIGURE 5.15 Retreating Arctic sea ice.
As sea ice extent decreases in the Arctic, ice retreats away from the continental shelf, requiring polar bears to return to land earlier in the year and diving species such as walrus and eider to dive deeper to obtain food. *Source: Images courtesy of the National Snow and Ice Data Center, University of Colorado, Boulder.*

disappear from most of their current range by 2100. The U.S. has listed polar bears as threatened, a decision that will be explored in detail in Chapter 15.

POLAR FOOD WEBS: CHANGES IN THE SOUTHERN OCEAN

Decreasing ice in the Antarctic is leading to changes in the base of the southern marine food chain. Krill that are prey for penguins, whales, and a host of other creatures are declining. Krill (*Euphausia superba*) are being replaced by salps (*Salpa thompsonii*) in the southwest Atlantic. This replacement represents a wholesale switch in the composition of planktonic grazers in these oceans (Figure 5.16).

As a result, marine mammal, penguin, and seabird populations are declining. On Bird Island, South Georgia, populations of fur seals, black-browed albatrosses, gentoo penguins, and macaroni penguins have shown declines of 20–50%. Between 1980 and 2000, the macaroni penguin population on the island declined from just over 1200 to less than 500. In other areas of the Antarctic, Emperor penguin populations have declined by 50%.

The mechanism for the declines in the populations of predators such as penguins seems to be climate-mediated decline in krill. Krill feed on plankton that grows on the underside of sea ice, so declining ice area means less primary productivity on which krill can feed. Ice is correlated with krill

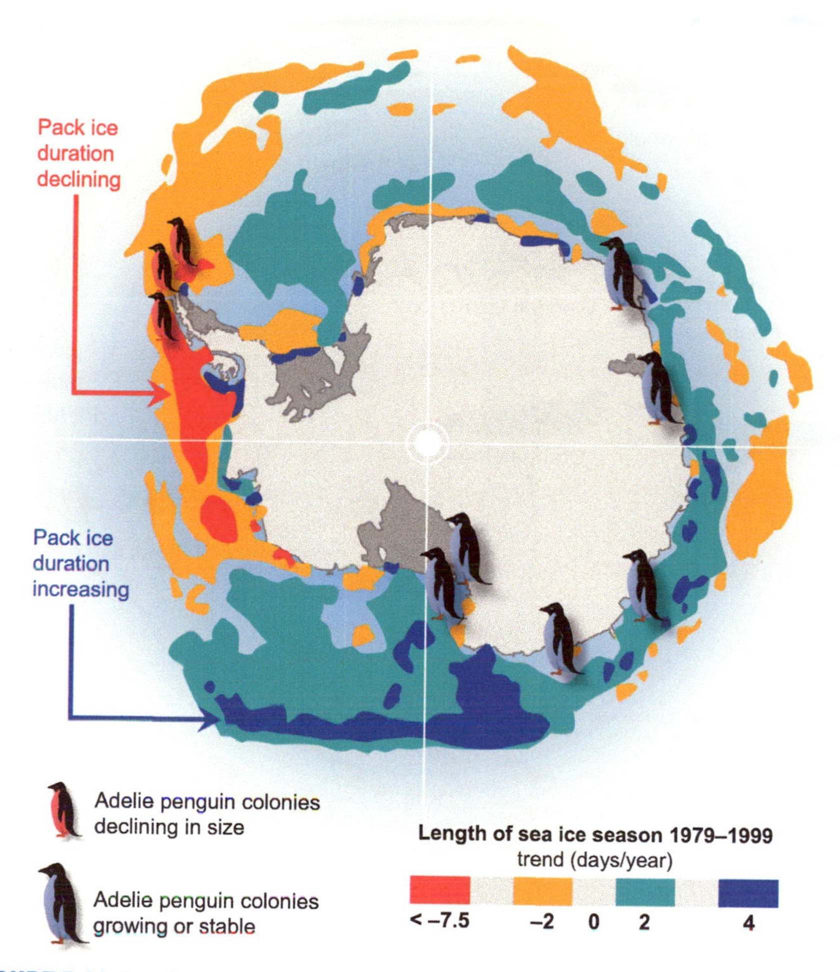

FIGURE 5.16 Pack ice changes and declining penguin populations in the Antarctic. Sea ice changes in the Antarctic are less straightforward than the continual declines in the Arctic. In some areas Antarctic pack ice is lasting longer, while in other places it is declining in duration. Associated with these changes, are changes in penguin populations driven by changes in food availability as plankton habitat is altered by the changes in sea ice. Decreases in pack ice duration are being driven by warming, while increases in pack ice duration are being driven by changes in winds (which may also be driven by climate change) *Source: Atkinson et al. (2004).*

overwinter survival in their larval stage, so decreasing ice may also reduce recruitment into the adult population (Atkinson et al., 2004). Both of these effects may contribute to the large declines in krill that have been observed since 1926. The crash in krill, in turn, has knocked the bottom out of the Antarctic food web, resulting in population declines in many predators (Figure 5.17).

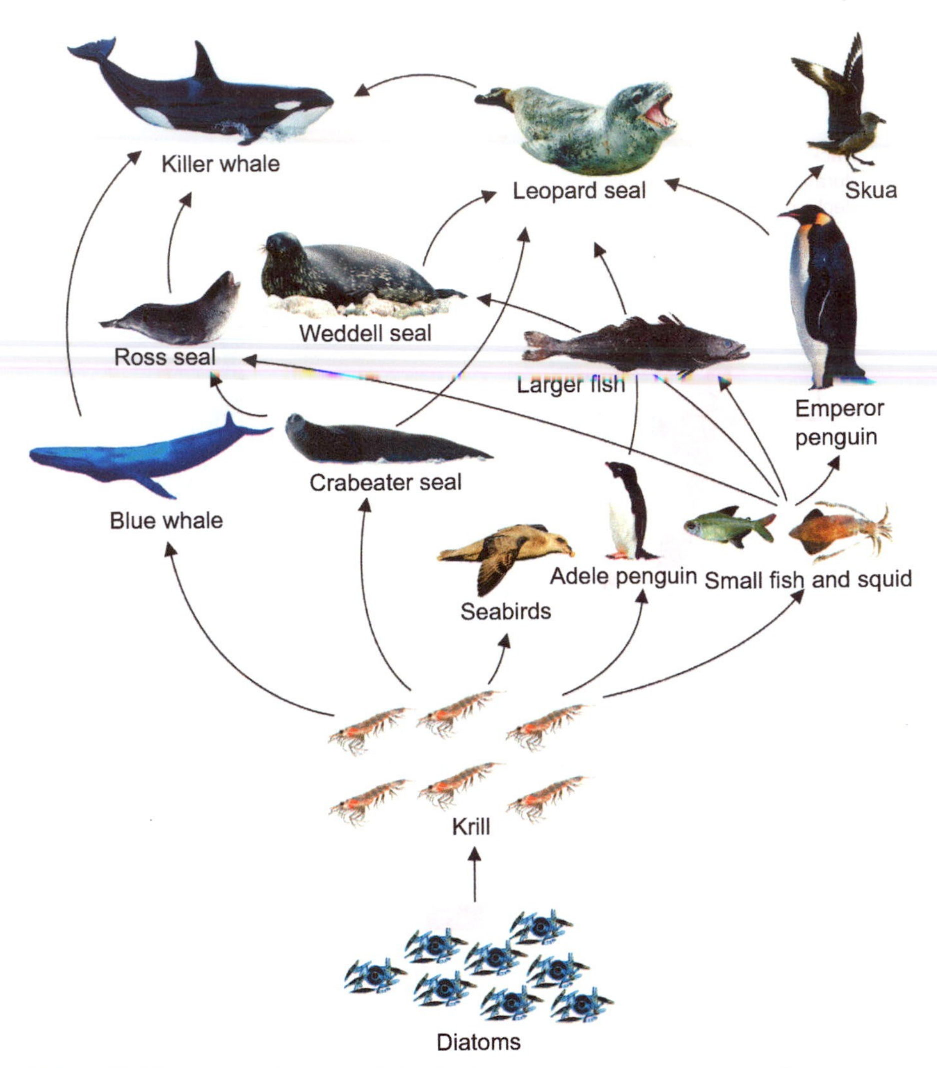

FIGURE 5.17 **Example of an antarctic food web.**
Diatoms dependent on sea ice support a diverse food web, including great whales that feed directly on plankton and several food chains that have diatoms at their base.

TROPICAL MARINE SYSTEMS

Coral bleaching due to climate change (see Chapter 3) results in wholesale ecosystem changes for coral communities. Corals are ecosystem engineers that literally shape the physical framework in which species interactions take place. When corals are removed, dramatic shifts in species composition may occur. This is seen in the past record, where coral reefs have been replaced by algal communities or other types of reefs under different temperature and ocean

FIGURE 5.18 Coral reef catch in Raja Ampat Indonesia, (the Indonesia side of New Guinea). Catches of coral-associated marine life and fish may decline following reef bleaching episodes. *Photo courtesy of Conservation International © Conservation International/photo by Sterling Zumbrunn.*

chemistry conditions. It is also seen less dramatically in modern coral communities after a bleaching event.

When corals die after a bleaching event, they may be replaced by algae or other noncorals. This results in changes throughout the food web, as species that depend on or feed on corals, such as parrotfish, are replaced by species that feed on algae. For instance, the genus *Acropora* alone harbors 20 species that are obligate symbionts, and more than 50 species of crustacean have been found associated with a single coral species (*Pocillopora damicornis*). If bleaching damage is local, the transformation may result in a mixed assemblage of coral-associated species and algae-associated species. If bleaching is widespread, replacement with algae-associated species may be complete.

Human uses of coral reefs (Figure 5.18), especially fishing and tourism, are impacted by these changes. Much of the attraction of coral reef diving—colorful corals and associated fish—is lost in algae-dominated systems. The loss in tourism revenue associated with bleaching may therefore be substantial. Long-term or short-term declines in tourism may result. Where reefs have to be closed to tourism to promote recovery from bleaching, short-term loss of revenue results. Where recovery fails, the attractiveness of the site declines, resulting in losses of visitation and revenue.

Conversely, fishing harvest seems less affected by bleaching. The few available studies indicate that coral-associated fish are replaced relatively rapidly by algae-associated fish, resulting in only a short-term decline in biomass (or harvest). Total replacement of coral by algae may not result in a major reduction in fish take. The economic impact of bleaching on fisheries may be driven heavily by the relative market demand and prices of coral-associated and algae-associated species. Where coral-associated species are highly prized in fisheries or more valuable, bleaching may cause major economic or social hardship even if total take rebounds.

PELAGIC MARINE SYSTEMS

Pelagic systems are also heavily influenced by climate. Evidence from throughout the world suggests that fish abundances are highly correlated with climate indexes such as the Atmospheric Circulation Index (ACI), North Atlantic Oscillation, Pacific Decadal Oscillation, and El Niño Southern Oscillation. Herring, cod, salmon, and pollack are among the species for which effects have been documented (Figure 5.19).

Fisheries records provide insight into the effects of climate on marine systems. Fisheries catch data provide information on both range limits and abundance of pelagic species. Many fisheries show climate-linked variations in range, abundance, or both.

For instance, the horse mackerel (*Trachurus trachurus*) catch in the North Sea increases dramatically during years of warm water temperatures. The warm water allows the phytoplankton and zooplankton on which the mackerel feed to expand their range northward (Figure 5.20). The expanded area and abundance of plankton results in a booming mackerel population and catch.

Climate responses are complicated by existing overexploitation. Coastal fisheries supply food to hundreds of millions of people, but exploitation during the past several centuries has caused major ecosystem changes and decline in fish stocks. Industrial fishing is pushing overexploitation into the last remote and lightly populated areas of the world. Even deep-water, long-lived fish such as the Patagonian toothfish (marketed as "Chilean sea bass") and orange roughy, are being fished to commercial extinction.

Climate change has resulted in large-scale ecological rearrangements that have been recorded in fisheries. For instance, increasing meltwater from Greenland and Northern Canada has caused wholesale ecological upheaval in the Georges Bank ecosystem and fishery (Figure 5.21). Fresh water from melting sea ice increased stratification in shelf waters off Canada and Europe. When the dense warm waters of the Gulf Stream mixed this light, cold fresh water, they became less likely to sink. Phytoplankton stayed closer to the surface and photosynthesized more, changing the base of the food web. The rise in phytoplankton caused an autumn spike in zooplankton, ultimately increasing stocks of commercially valuable zooplankton feeders such as herring. These changes combined with overfishing of cod (*Gadus Morhua*) to cause major changes in

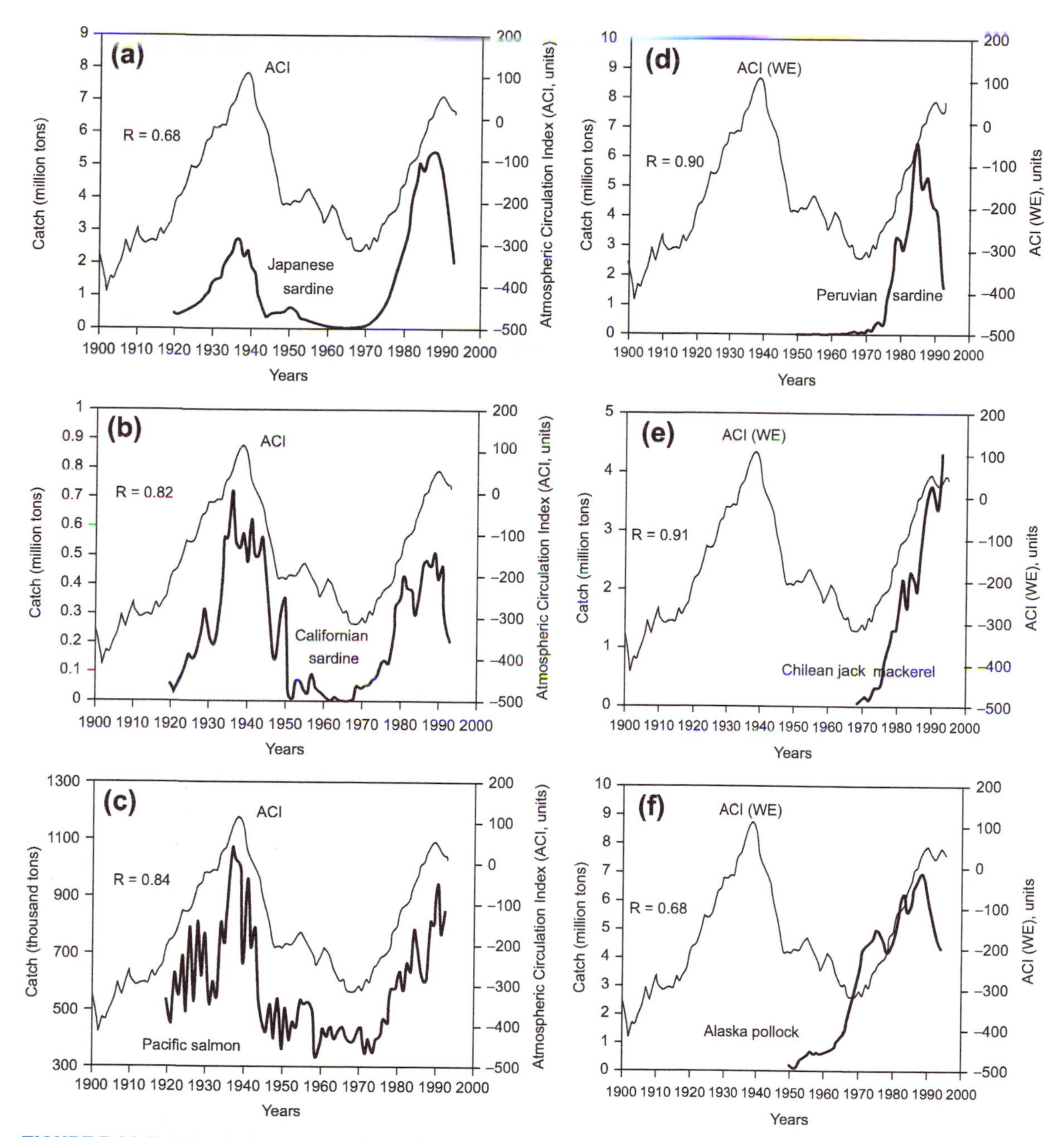

FIGURE 5.19 Variation in fisheries correlated with climate indices.

Fish catches correlate strongly with an index of atmospheric circlulation patterns, the Atmospheric Circulation Index (ACI). The ACI reflects changes in atmospheric flows that influence ocean currents and sea surface and subsurface temperatures. *Source: Klyashtorin (2001). Reproduced with permission from the Food and Agricultural Organisation of United Nations.*

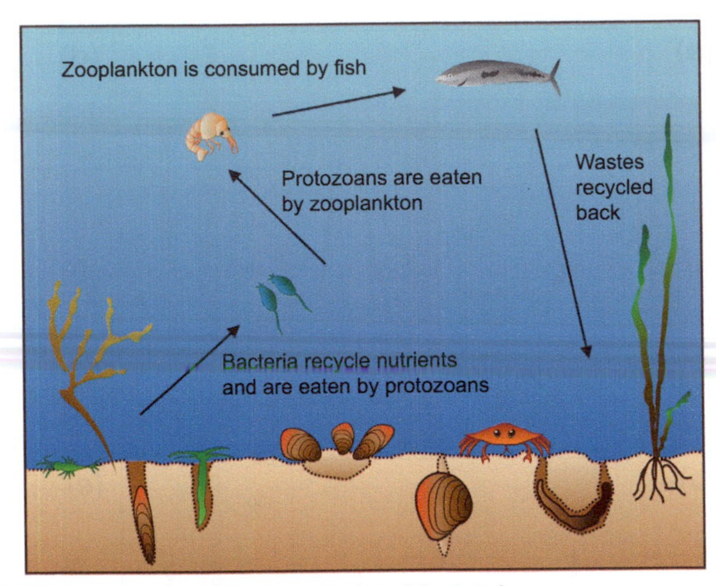

FIGURE 5.20 **Horse mackerel (*Trachurus trachurus*) food chain.**
Source: Reproduced with the permission of Her Majesty the Queen in Right of Canada, 2010.

the fishery of Georges Bank and surrounding areas. A previously abundant cod fishery collapsed and couldn't recover. Snow crab and shrimp increased in the place of cod. The top-down influence of the fishery accelerated a massive change in the ecosystem driven by climate change from the bottom-up.

The combination of climatic change and fishing pressure results in simultaneous top-down and bottom-up changes in fisheries that may be difficult to tease apart.

Similarly, southern ocean fish that depend on krill may collapse as krill decline, whereas industrial fishing is hitting these remote stocks for the first time recently. In both cases, climate may hasten declines in the fishery resulting from overfishing in ways that may push fisheries to collapse too quickly for fisheries management agencies to foresee or prevent.

OCEAN ACIDIFICATION

Direct chemical change of the oceans is occurring as a result of the massive amounts of human-produced CO_2 being released into the atmosphere. CO_2 dissolves in water to form carbonic acid. CO_2 is in equilibrium with carbonic acid in seawater, with most remaining as dissolved CO_2. The CO_2 that is converted to carbonic acid releases hydrogen ions into seawater, raising its pH. Already the hydrogen ion concentration in seawater has been increased by approximately 30% (0.1 pH unit). Further acidification is inevitable as CO_2 already in the atmosphere and that currently being added continue to dissolve in the oceans.

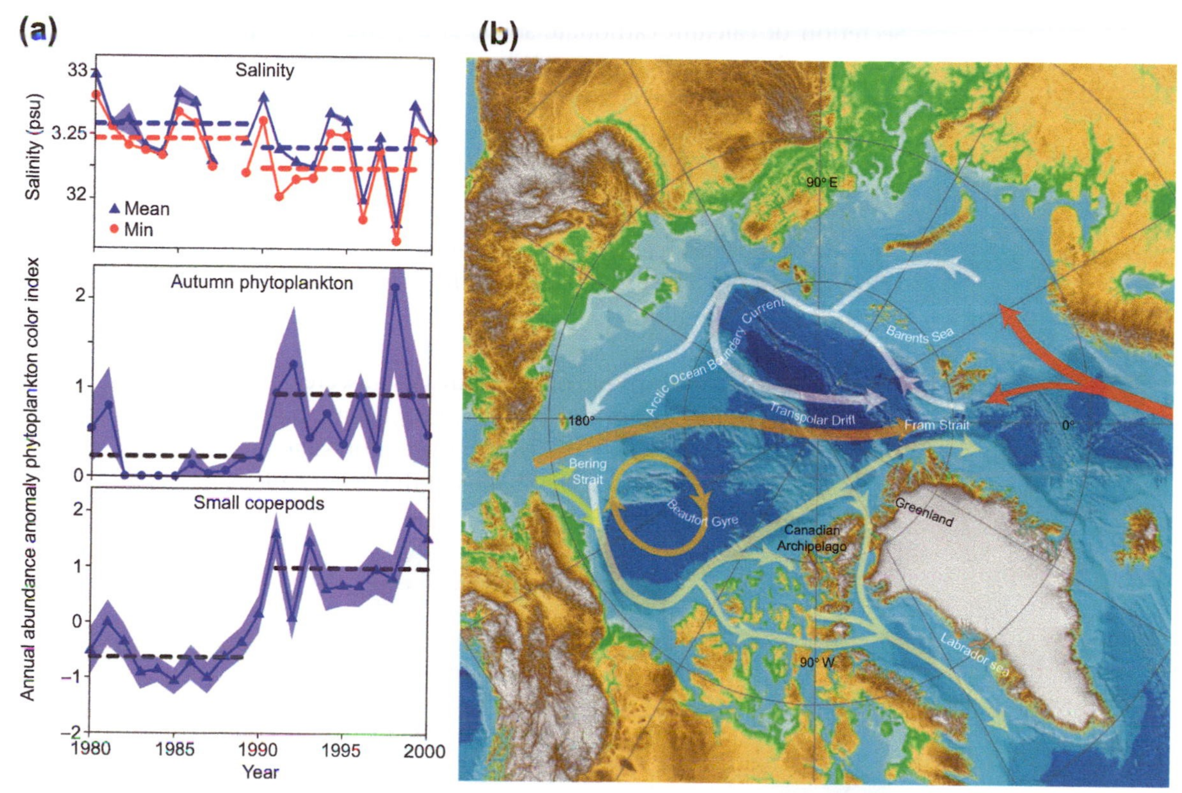

FIGURE 5.21 A regime shift in the Gulf of Maine and Georges Bank occurred in 1990 in response to climate change. Salinity dropped, resulting in increases in phytoplankton and zooplankton (copepod) abundance. The cause of the salinity change was large-scale reorganization of ocean circulation in the Arctic (map). In the late 1980s, warm saline Atlantic water entered the Barents Sea. This reduced the size of the Beaufort Gyre and caused increased flow of low-salinity water out the Canadian archipelago west of Greenland. When this water reached Georges Bank (around 1990), it triggered the ecological regime shift in plankton (left panels). *Source: Greene and Pershing (2007). Reproduced with permission from AAAS.*

CALCITE AND ARAGONITE

Calcite and aragonite are two forms of calcium carbonate, essentially two alternative crystalline structures, which may be favored by specific depositional or biological processes. For many marine organisms that secrete calcium carbonate shells, tests, and skeletons, the particular form of calcium carbonate secreted depends on the biochemical makeup of the species. Corals and clams, for instance, both secrete calcium carbonate, but most modern corals secrete their skeletons in the form of aragonite, whereas most bivalves secrete calcite shells.

The biological effect of ocean acidification is expected to affect a wide range of organisms. All organisms that secrete calcium carbonate shells are affected by the pH and saturation state of seawater. Other organisms may have physiological processes that are sensitive to changes in seawater pH.

Secretion of calcium carbonate shells is dependent on pH because it affects the saturation state of seawater. When seawater is saturated or supersaturated, calcium carbonate shells can be formed. When pH is low, high hydrogen ion levels cause seawater to become undersaturated, which makes secretion of calcium carbonate shells difficult. If the pH (and saturation state) is low enough, erosion of shells that have already been formed may occur.

Organisms may secrete calcium carbonate shells in two forms—calcite and aragonite (Figure 5.22). There is no difference in the chemical composition of calcite and aragonite: they are both calcium carbonate. Their crystal structure varies, however, and they are produced by different enzymes and by different organisms. Modern tropical corals, for instance, secrete their calcium carbonate shells in the form of aragonite, whereas many clams and other organisms secrete calcite. Each form of calcite has its own saturation state in seawater, influencing the ability of organisms to secrete shells of that form (calcite or aragonite).

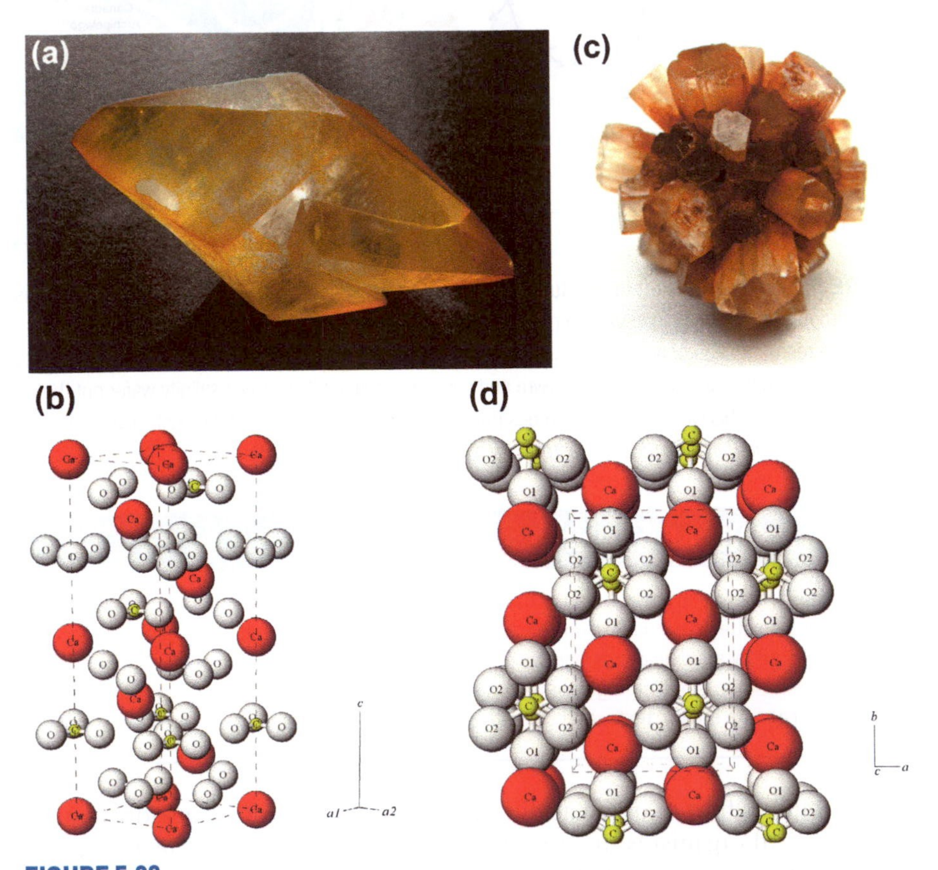

FIGURE 5.22

External appearance and crystalline form of calcite (a, b) and aragonite (c, d). *Source: (a) and (c) from Wikimedia Commons; (b) and (d) from National Institute of Advanced Science and Technology.*

ECOSYSTEM FEEDBACKS TO CLIMATE SYSTEM

Some ecosystem effects of climate change produce new ecosystem properties that feed back to the climate system. Feedback to climate usually occurs where the balance of greenhouse gases produced by an ecosystem is altered by climate. CO_2 from plant respiration and decomposition and methane from decomposition are the two gases most often involved. Climate feedback may also occur where change in vegetation changes the albedo (reflectance of land) or the aerodynamic roughness of the surface.

The carbon balance of ecosystems is particularly important because plants produce CO_2. Plant respiration tends to increase with increasing temperature, releasing more CO_2 with warming. However, plant growth also increases with temperature, making the balance between carbon fixed into plant tissues and that respired important in determining the net effect of warming.

The interplay of increased growth, increased respiration, and increased decomposition will determine the net impact of a warming ecosystem on greenhouse gas release. For instance, thawing of arctic soils facilitates plant growth and respiration at the same time that it allows increased release of CO_2 and methane from decomposition. Because arctic soils experience a summer growing season but almost no decomposition, they accumulate large amounts of frozen organic matter. As these soils thaw, the decomposition of the stored organic matter produces methane, increased plant growth sequesters carbon, and enhanced respiration releases CO_2. The net effect is a positive warming feedback due to the large amount of stored organic material and still-limited growing season (Figure 5.23).

Large-scale ecosystem feedback to climate may also occur with changes in the large tropical forests of the Amazon or Central Africa. These extensive forests

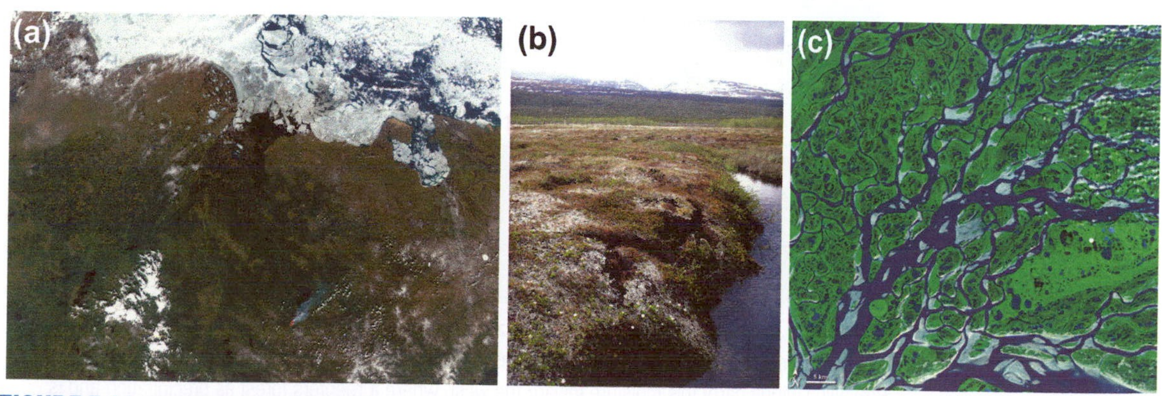

FIGURE 5.23 Thawing permafrost soils.

As permafrost thaws (a), it expands (b), rupturing the surface (c). This can cause damage to vegetation or structures, opens up new habitat, and impacts nutrient cycling. *Source: (a) and (c) Courtesy of NASA/GSFC/MITI/ERSDAC/JAROS and the U.S./Japan ASTER Science Team; (b) from Wikimedia Commons.*

harbor enough carbon to impact global climate if released. Some studies indicate that feedbacks between climate, fire, and vegetation are already altering the Amazon ecosystem.

Feedback scenarios involving the Amazon are of particular concern because of its complex internal system of moisture recycling. In the Amazon, evapotranspiration in one part of the basin provides rainfall in the next. This moisture recycling from vegetation to the atmosphere cascades across the Amazon to the Andes. Should the Amazon dry sufficiently to eliminate forest in some areas, the reduction in moisture recycling would lead to forest drying and loss in adjacent areas. A positive feedback loop could be established that would result in loss of forest cover over large areas. In turn, the CO_2 released from decaying vegetation if this happens is of large enough magnitude to affect global climate (Figure 5.24).

A second high-latitude example of climate feedback from ecosystem change comes from boreal forests. Tundra is low stature and light in color and thus has high albedo (reflectance), whereas conifer forests are high stature with dark foliage, resulting in low reflectance. Conifer forests are expanding northward

FIGURE 5.24 Moisture recycling in Amazon.
Moisture transpired by trees in the Amazon basin enters the atmosphere, contributing to cloud formation. Prevailing winds carry this moisture toward the west, where it reenters forest as precipitation and is transpired again. This process continues until air masses are blocked by the vertical rise of the Andes. This transpiration and precipitation cycle is important in maintaining forest cover in the Amazon in times of climate change. *Source: Figure courtesy of Ocean O'Graphics UCSB.*

due to warming in approximately the past 50 years and are expected to replace tundra over large areas as the planet continues to warm. The decrease in snow cover on these lands, coupled with forest expansion, will make these high-latitude lands darker and more absorbant of solar energy, which will further warm the planet. Particularly in high latitudes, changes in vegetation cover, decomposition, and albedo may drive climate changes of global importance. These potentially dramatic changes already under way give great urgency to the search for answers about the future. Interestingly, one of our best sources of information about the future is the past. The Earth has gone through many past climate changes—large and small, fast and slow—and the biological responses to these changes provide many clues about the changes we are currently seeing and those yet to come. Exploring the lessons of these past changes is the subject of the next section.

FURTHER READING

Greene, C.H., Pershing, A.J., 2007. Climate drives sea change. Science 315, 1084–1085.

Hoegh-Guldberg, O., Bruno, J.F., 2010. The impact of climate change on the world's marine ecosystems. Science 328, 1523–1528.

King, K.J., Cary, G.J., Bradstock, R.A., & Marsden–Smedley, J.B., 2013. Contrasting fire responses to climate and management: insights from two Australian ecosystems. Global change biology 19 (4), 1223–1235.

Logan, J.A., Powell, J.A., 2001. Ghost forests, global warming, and the mountain pine beetle (Coleoptera: Scotytidea). American Entomologist 47, 160–173.

Moline, M.A., Claustre, H., Frazer, T.K., Schofield, O., Vernet, M., 2004. Alteration of the food web along the Antarctic Peninsula in response to a regional warming trend. Global Change Biology 10, 1973–1980.

Lessons from the Past

CHAPTER 6

Past Terrestrial Response

The lesson of the past is that when climate changes, species move. For instance, as the world emerged from the last ice age, temperatures rose dramatically across North America and Europe. Areas that had been frozen under solid ice—outside the niche of virtually any organism—suddenly thawed and supported rich assemblages of life. Less dramatic changes in the tropics led to different climates replacing one another in the space of a few 1000 years. Many other climate shifts have occurred throughout the Earth's history, changing the average climate of the planet and also the living conditions of every place on the planet.

The record of the past is dominated by evidence of species' ranges moving in response to climate change. In temperate and polar regions, this has included species moving long distances to occupy lands as glaciers retreated; or retracting large distances in the face of advancing ice. In the tropics, the changes are more subtle but still dramatic in their rearrangement of ecosystems.

Every species moves in a unique way in response to climate. Combinations of species change as a result—the vegetation communities we see today are temporary—different species have occurred together in the past, and different combinations will be found in the future.

Paleoecology is a rich discipline and cannot be fully summarized here. This chapter focuses on vegetation shifts that give insight into the scope of vegetation change in geologic time, and explores in more detail our understanding of response to emergence from the last ice age, as this evolving thinking is critical to understanding possible biotic response to large human-induced climate change.

SCOPE OF CHANGE

Our record of climate change and biological responses is best for the past 500,000 years. The resolution of the climate record has rapidly improved for this period, revealing many rapid changes that occur in decades, centuries, or millennia. Our understanding of deeper time is much less detailed. It is

Climate Change Biology. http://dx.doi.org/10.1016/B978-0-12-420218-4.00006-8

likely that many biologically important rapid climate changes also occurred in the earlier ice ages and beyond. This picture is beginning to emerge with high-resolution pollen records from the Andes and other tropical regions. But outside of these areas, sketching out climate and biological change comes in broad brushstrokes, missing the intricate details. Even this coarse picture yields many clues essential to our understanding of the biological influence of climate.

In deep time, continents move, changing their position relative to the great tropical and temperate climate belts. It is therefore essential to understand the general features of continental movements to get a full picture of the climatic changes in deep time.

THE EARTH MOVES

The movement of the continents is the most important concept in modern geology. Continents ride on large crustal plates that shift position on time scales of millions of years. Plates are constantly being formed at volcanically active mid-ocean ridges and constantly being destroyed by subduction as their leading edges are driven back into the crust where they meet another plate.

More than 200 million years ago, the continents were all connected in a supercontinent called Pangaea (Figure 6.1). What are currently northern South America and Africa and southern North America and Europe were all equatorial at this time. The continents came together to form Pangaea after being widely scattered.

Approximately 190 million years ago, Pangaea began to break up into Gondwanaland and Laurasia. Northern Africa and South America remained equatorial, whereas North America and Europe moved poleward. All the southern continents (South America, Africa, Antarctica, and Australia) and India were connected in Gondwanaland at this time. This explains the distribution of plant and animal families that are related but are currently found in widely separated regions, such as the ratite birds in South America, Africa, Australia, and New Zealand.

SPOTLIGHT: STOMATA WHAT?

Plants use stomata to exchange CO_2 with the atmosphere. When more CO_2 is present, plants need less stomata to get the job done. This simple fact has been used to get a proxy for atmospheric CO_2 over 300 million years (Retallack, 2001). Fossil leaves were examined under a microscope to count their stomata. Fewer stomata mean lower atmospheric CO_2 concentrations. The derived CO_2 concentration estimates agree with other proxies. They indicate close correlation of high CO_2 with global warm periods.

Source: Retallack, G.J., 2001. A 300-million-year record of atmospheric carbon dioxide from fossil plant cuticles. Nature 411, 287–290.

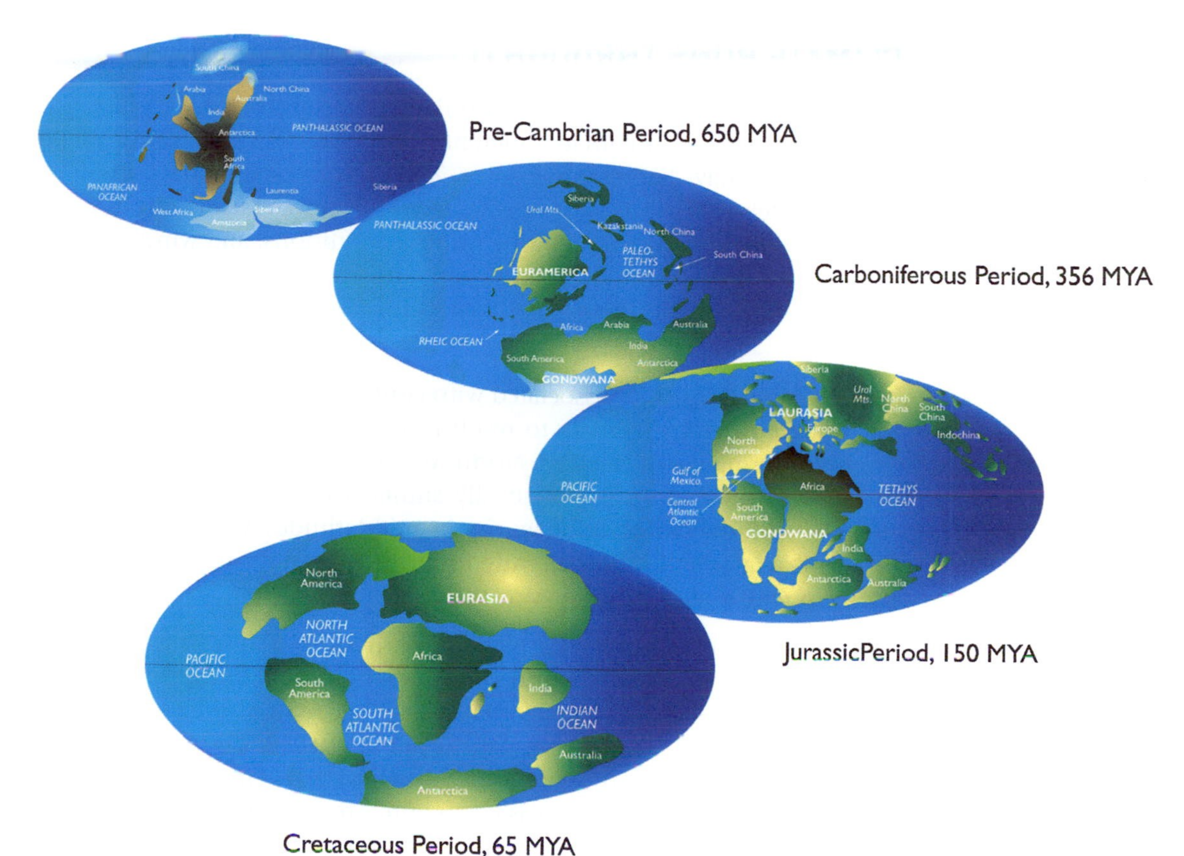

FIGURE 6.1

Pangaea (a), Laurasia and Gondwanaland (b). *Source: (a) Courtesy of Canadian Geographic. (b) From Wikimedia Commons.*

Gondwanaland broke up beginning approximately 150 million years ago. First, India rafted off toward Asia, eventually slamming into the plate there and causing the uplift of the Himalayas. Somewhat later, Africa and South America separated. The last to go, Australia, broke off from Antarctica 50–60 million years ago, allowing the circumpolar current to be established around Antarctica approximately 47 million years ago.

Since the Gondwanaland breakup, major mountain building and the closing of the isthmus of Panama have had global and regional climatic consequences. The uplift of the Himalayas, beginning about 40 million years ago, resulted in global cooling associated with effects on the carbon cycle (see Chapter 19). The closure of the land bridge at Panama separated the Atlantic and Pacific oceans, with accompanying changes in ocean circulation that affected precipitation and temperature patterns in multiple regions. The effect of the uplift of the Andes had major impacts on South American climate and biogeography.

CLIMATE RUNS THROUGH IT

As the continents moved, they crossed the world's great climatic belts, resulting in long periods of climate different from those that exist on those same landmasses today. For example, over 200 million years, India moved from approximately 30° south to approximately 15° north, starting out in temperate vegetation, moving through true tropical conditions, and winding up with its current tropical/subtropical biota and climate. North America and Europe moved from tropical climates and vegetation to temperate during the same period.

These climate transitions associated with continental movement superimpose on changes in global climate to result in complex biological variations. For example, during warm climates, North America had warmth-loving evergreen broadleaf forests and was climatically similar to the subtropics today, even though its continental position was in the mid-latitudes. We know this because fossils of relatives of trees currently found in the subtropics have also been found in temperate North America.

Warm evergreen broadleaf forests are of particular interest because they currently harbor most of the world's biodiversity. These forests are similar to tropical forests of today, but they were found in the subtropics and temperate latitudes in the past. The one constant about the forests is that they reflect very warm conditions. Thus, they are sometimes known as "megathermal forests." Here, however, we refer to them as warm rain forests, and we refer to their drier cousins as warm seasonal forests.

Approximately 90–100 million years ago, warm rain forests and warm seasonal forests dominated all continents except Antarctica (Figure 6.2). Even Antarctica had some warm rain forest, although it was dominated by temperate forests. North America, South America, Africa, and Asia were all covered in warm rain forest, except in the northernmost extremes.

Between 90 and 60 million years ago, the warm rain forests contracted as climate cooled. By 50 million years ago, climate was again very warm (the Eocene climatic optimum), and warm rain forests occurred on all continents except Antarctica (Figure 6.3). This represented the greatest extent of the warm rain forests since the evolution of most modern plants, with warm modern rain forest taxa extending into what is now Alaska and across Australia.

These extensive warm rain forests of the early Eocene provided connections for the exchange of plant species across the northern continents, between South America and Australia via Antarctica and dispersal between North America and South America across a relatively small expanse of water. These unprecedented exchanges would help mold the biological affinities of modern floras. The intermittent connections between North America and Europe via Greenland

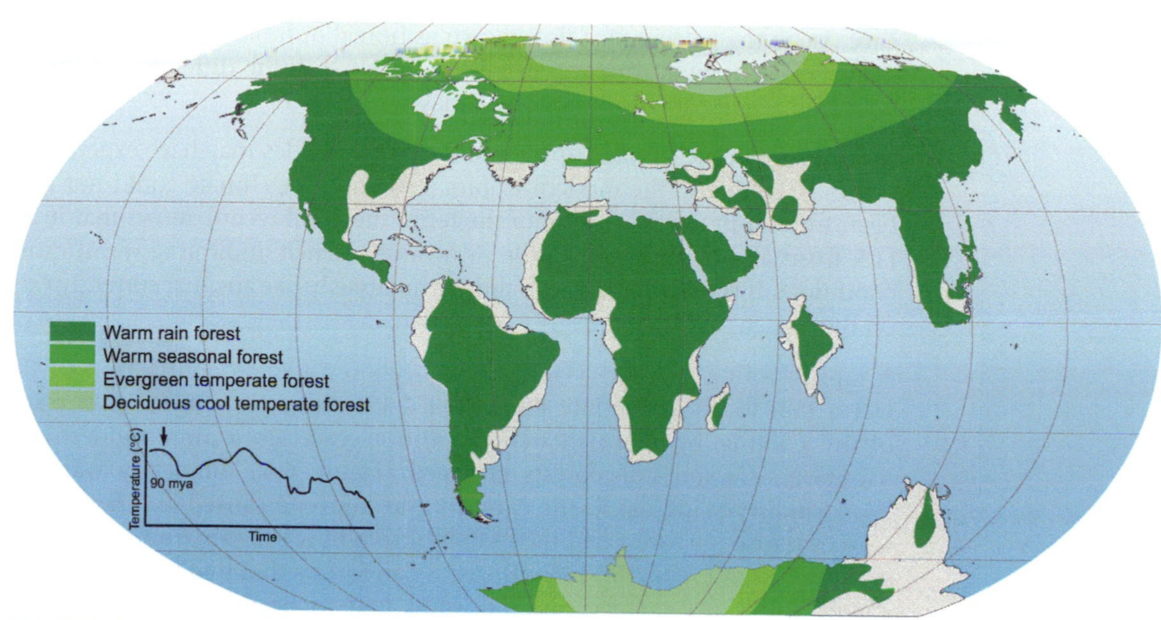

FIGURE 6.2 Distribution of forest types, 90 million years ago.
Warm rain forest existed at high latitudes in the warm climate of this period. Global mean temperature was much warmer at this time than at present (see inset). *Source: Adapted from Bush et al. (2007).*

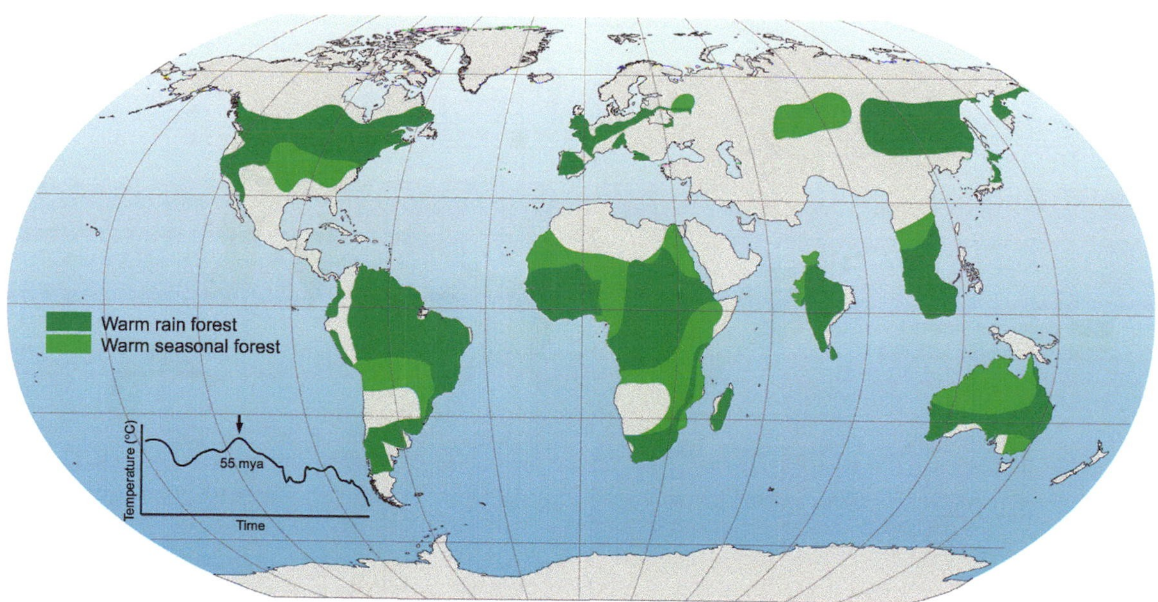

FIGURE 6.3 Distribution of warm rain forest and warm seasonal forest 55 million years ago, near the early Eocene Climatic Optimum (inset).
Source: Adapted from Bush et al. (2007).

gave rise to the paleoarctic (also known as boreal or holarctic) flora. The divisions between South America and Africa resulted in the differences between neotropical and paleotropical floras today.

During the Eocene climatic optimum, the mid-latitudes may have experienced a climate similar to the modern tropics. Although the climate could not have been tropical in the strict sense of the term, the fossil record shows that it did lack frosts and that there was little or no water deficit, indicating warm moist conditions. It is likely that it was a climate not exactly like any that currently exist on the planet.

A rapid cooling and the onset of ice sheets resulted in contraction of the warm rain forest in the late Eocene (45–35 million years ago). As a result of this cooling, by approximately 30 million years ago a more modern distribution of warm rain forests emerged (Figure 6.4). Planetary warming 25–15 million years ago in the early to mid-Miocene caused another expansion of warm rain forest and the redistribution of warm rain forest and warm seasonal forest (Figure 6.5). Southeast Asian forests that had been warm seasonal 30 million years ago became warm rain forest during those periods. Near-modern distribution of forest types emerged as the planet descended into the ice ages of the Pleistocene (Figure 6.6).

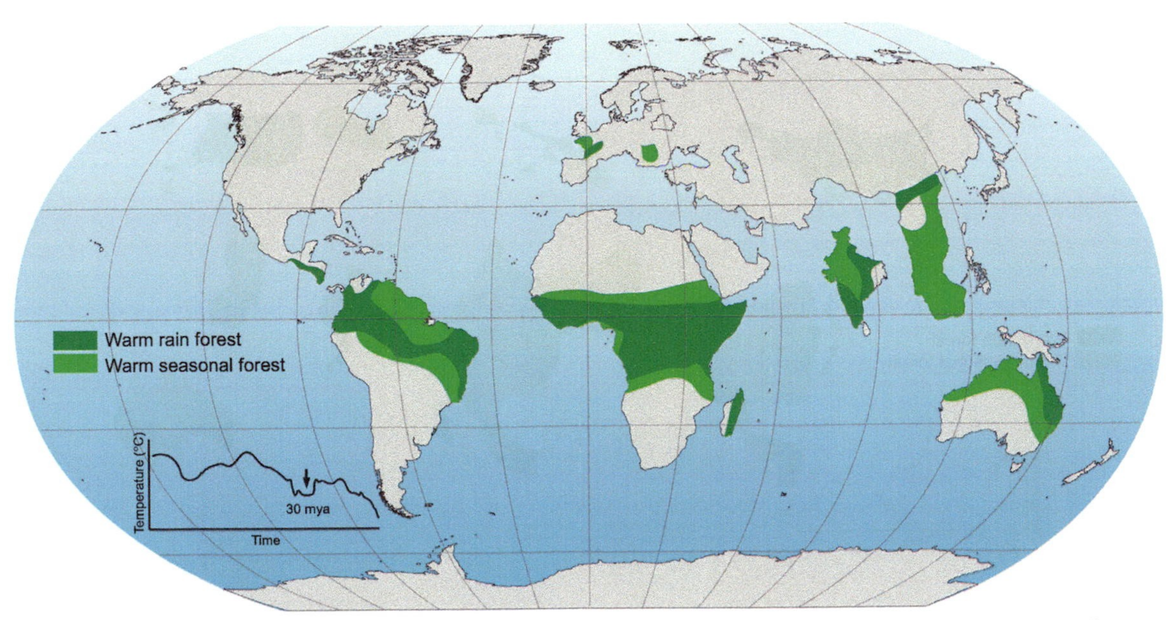

FIGURE 6.4 Distribution of warm rain forest and warm seasonal forest 30 million years ago, just after the global cooling associated with the formation of Antarctic ice sheets (inset).
Source: Adapted from Bush et al. (2007).

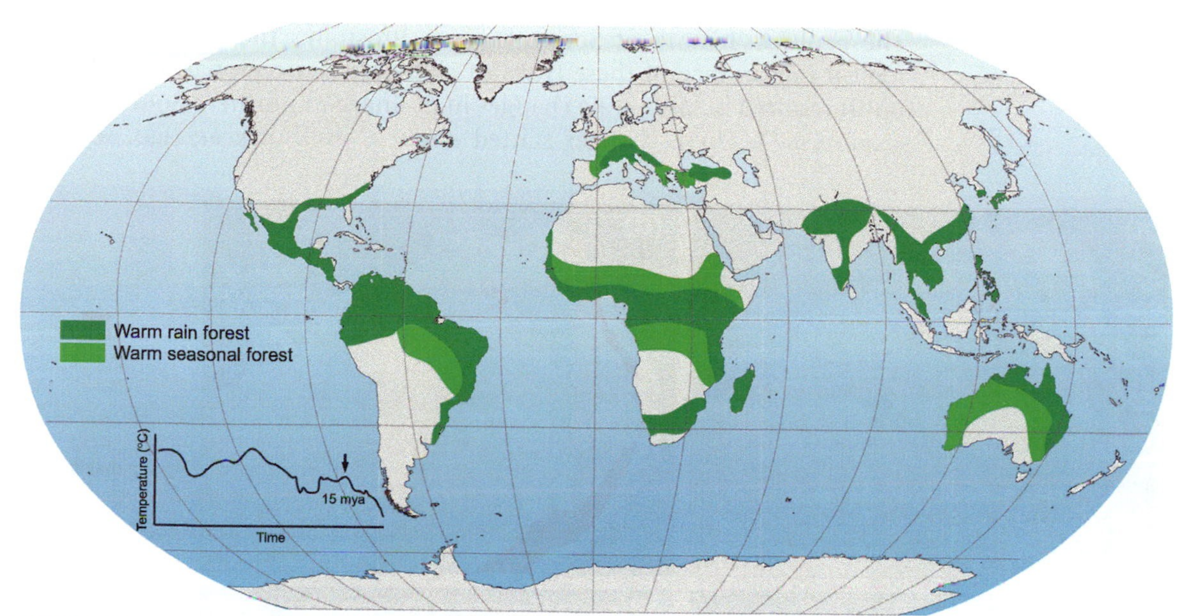

FIGURE 6.5 Distribution of warm rain forest and warm seasonal forest 15 million years ago.
Source: Adapted from Bush et al. (2007).

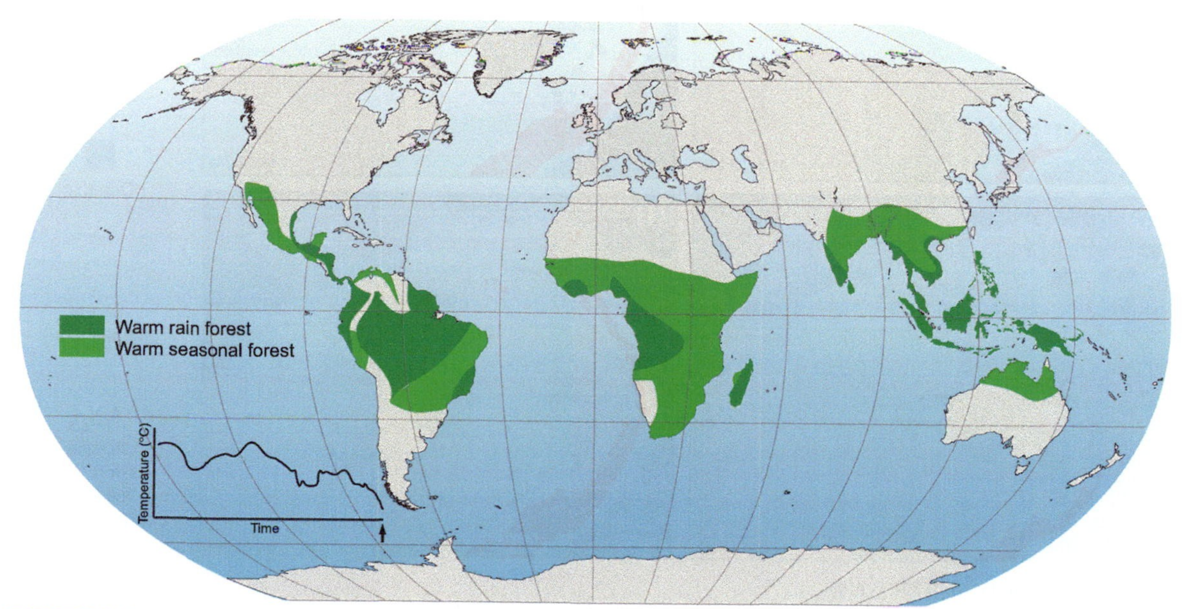

FIGURE 6.6 Present distribution of warm rain forest and warm seasonal forest.
Source: Adapted from Bush et al. (2007).

The geologic history of South America illustrates the climatic and biological importance of these types of changes on a regional scale. Andean uplift resulted in large-scale changes in drainage in northern South America (Figure 6.7). These changes started about 65 million years ago, with the

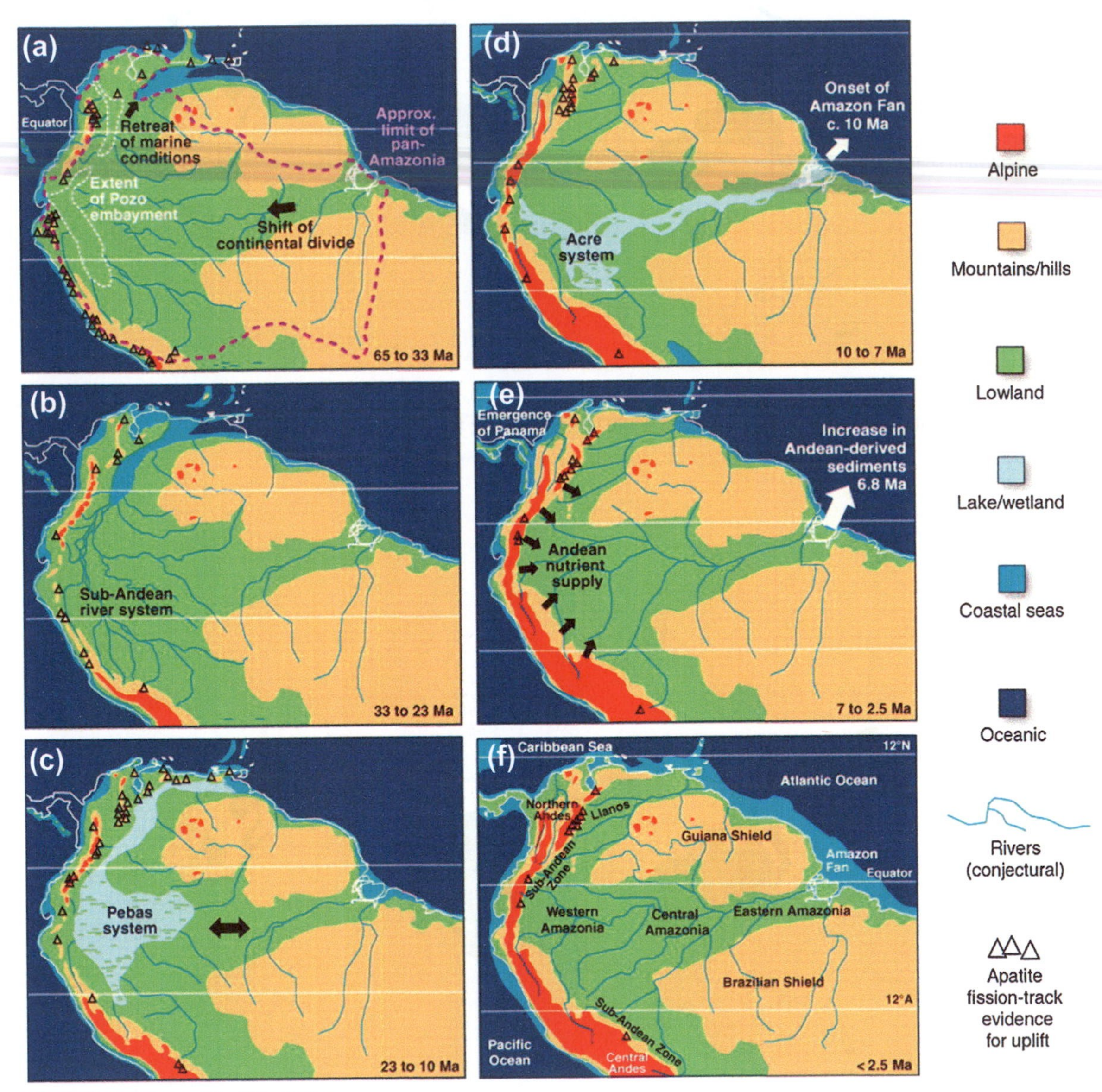

FIGURE 6.7 The uplift of the Andes.

Effects of the uplift of the Andes are shown on river courses, wetlands, and lowland forest extent from 65 million years ago to the beginning of the Pleistocene 2 million years ago. Note the large wetland (B, C) prior to the rise of the northern Andes and the shift in drainage toward the Amazon (B, C, D). *Source: Hoorn et al. (2010). Reprinted with permission from AAAS.*

establishment of large wetlands about 30 million years ago. The wetlands persisted in western South America until about 10 million years ago, when ongoing Andean uplift eventually formed the modern course of the Amazon. The rise of the central and northern Andes created orographic rainfall and steep temperature gradients. Soils were deposited in western Amazonia from the erosion caused by rain in the new uplands. Biological data shows that the uplift of the northern Andes has lead to biological radiations in many groups, such as lupines. Plant diversity increased 10–15% as the vast wetland habitats were replaced by forest. Modern tree diversity is highest in areas of relatively richer soils deposited in the western Amazon as the Andes rose.

Understanding of changes in the distribution of global vegetation during the alternations between glacials and interglacials in the Pleistocene is still evolving, and it requires a closer look at this complex period.

FAST AND FAR: THE RECORD OF THE ICE AGES

Ice ages have gripped the Earth for much of the past 2 million years, the geological period known as the Pleistocene. During this time, predominantly icehouse, glacial conditions have been interrupted by brief interglacial periods many times, including five times in the past 500,000 years. The planet is currently in one of these warm, interglacial periods. In fact, all of human civilization has developed in the past 10,000 years of unusually warm, stable climate. To better understand the biological significance of glacials and interglacials, we need to review the record of the ice ages, assessing the size and speed of temperature changes.

The ice ages culminate a period of global cooling in which first the Antarctic and then the Arctic and Northern Hemisphere experienced significant ice buildup. Antarctic ice sheets began to form approximately 40 million years ago. This ice was intermittent—it built up and disappeared repeatedly. The record in the Arctic is more obscure, with strong evidence of intermittent ice beginning 10 million years ago and fragmentary evidence showing Northern Hemisphere ice buildup as far back as 40 million years. Thus, northern and southern ice buildup may have initiated at approximately the same time, 40 million years ago, but the evidence from the south is more robust.

Following a long period of intermittent ice, permanent ice at the poles developed approximately 10 million years ago in the Antarctic and at the start of the Pleistocene, 2 million years ago, in the Northern Hemisphere. The establishment of large continental ice sheets marked the beginning of the ice ages and the Pleistocene. These large ice masses were interrupted only occasionally during the past 2 million years. Transition to the current warm interglacial period began approximately 20,000 years ago, with full warm conditions occurring approximately 10,000 years ago.

Continental ice sheets were more than 1 km thick during the last glacial period. They extended as far south as the current location of New York city in North America and as far south as Copenhagen in Europe. The melting and retreat of these massive ice sheets took place in little more than 10,000 years in the transition to the current interglacial. Similarly rapid melting characterized the onset of past interglacials. Descent back into glacial conditions is more gradual, occurring over many tens of thousands of years.

The biological response to these cycles reflects rapid change over long distances. Plant species occupied new habitats exposed by the retreating ice very quickly. Full assemblages of plants and animals were typically in place within a few thousand years of the ice retreat, sometimes representing huge range expansions. Hippos existed in what is now the Thames River in the heart of London during the last interglacial. How these rapid biological responses could occur is one of the ongoing debates in paleoecology.

SPOTLIGHT: INDIVIDUALISM AND COMMUNITIES

Plants and animals respond individualistically to climate change, pursuing climatic conditions that meet their own specific set of tolerances. Graham and Grimm (1990) were among the first wave of researchers to recognize the implications of this lesson from paleoecology for future biological response to human-induced climate change. Using examples of beech, hemlock, and small mammals, these authors argued that past climate change had produced novel associations of species and that the same might be expected in the future. This suggested that attempts to conserve communities, or even define conservation outcomes in terms of community, might be difficult. This view has gained increasing support, both from research results and from conservation planning theory, during the intervening decades. The view of Frederic Clements that communities were interdependent entities has largely been replaced by that of Henry Gleason, who argued that communities were really just passing associations of species that rather ephemerally shared a common set of climatic conditions. The term "association" is favored by many climate change biologists because it indicates a less permanent relationship among species than does "community." Communities seem destined to be torn apart and reassembled by future climate change.

Source: Graham, R.W., Grimm, E.C., 1990. Effects of global climate change on the patterns of terrestrial biological communities. Trends in Ecology and Evolution 5, 289–292.

ICE RACING IN NORTH AMERICA AND EUROPE

Ice sheets retreated hundreds of kilometers in the Northern Hemisphere at the end of the last ice age, and plants and animals followed. Species tracked the retreating ice not in coherent communities but at differing paces, each species finding its own speed (Figure 6.8).

HOW AN ICE AGE STARTS

Ice ages start with cool summers. Combinations of solar forcings that lead to cool summers in the Northern Hemisphere allow ice to be retained through the warm season and continental ice sheets to form in North America and Europe. A similar dynamic for cool southern summers does not exist because there is little landmass to hold ice in South America or Africa at high latitudes. In the late 1800s, scientists believed that cold winters led to ice ages. Milutin Milankovitch, a Serbian geophysicist and engineer, recognized that cool summers were the key to ice buildup. Cycles in solar forcing—Milankovitch cycles—bear his name in recognition of his contribution to understanding their role in the ice ages.

The result was a constant shifting of species associations. Species found together at the height of the glaciation might not be found together at the advancing edge of vegetation occupying the habitats newly vacated by the ice. Several species might be found together in one region, only to have one or more of the group be absent in another region. The mix of species and where they were found together shifted as the continents continued to warm and the ice retreat progressed.

One remarkable facet of the advance was that at its end, some species with poor dispersal attributes were found a long distance from their glacial strongholds. By the late 1800s, biologists noticed that this posed a problem. There was no evidence for oaks in England during the height of the glacial, and the best evidence available indicated that they had not occurred farther north than the Alps. This meant that they had migrated hundreds of kilometers in less than 20,000 years. Assuming acorns fell, at most, a few meters from the parent tree, and that oaks take 30–50 years to reach reproductive age, there simply was not enough time since the last glacial to account for oaks in England. Yet they were there, and they were abundant.

Biologists therefore proposed that rare, long-distance dispersal events must account for the long-distance range shifts seen in trees in both Europe and North America. This supposition was similar to that of the dispersalists, such as Darwin, who argued that disjunct distributions in groups such as the ratite birds (ostrich and emu) must be explained by unusual long-distance dispersal. Another group of biologists at that time proposed land bridges between continents. Continental drift finally explained these distributions without resorting to either freak dispersal or disappearing land

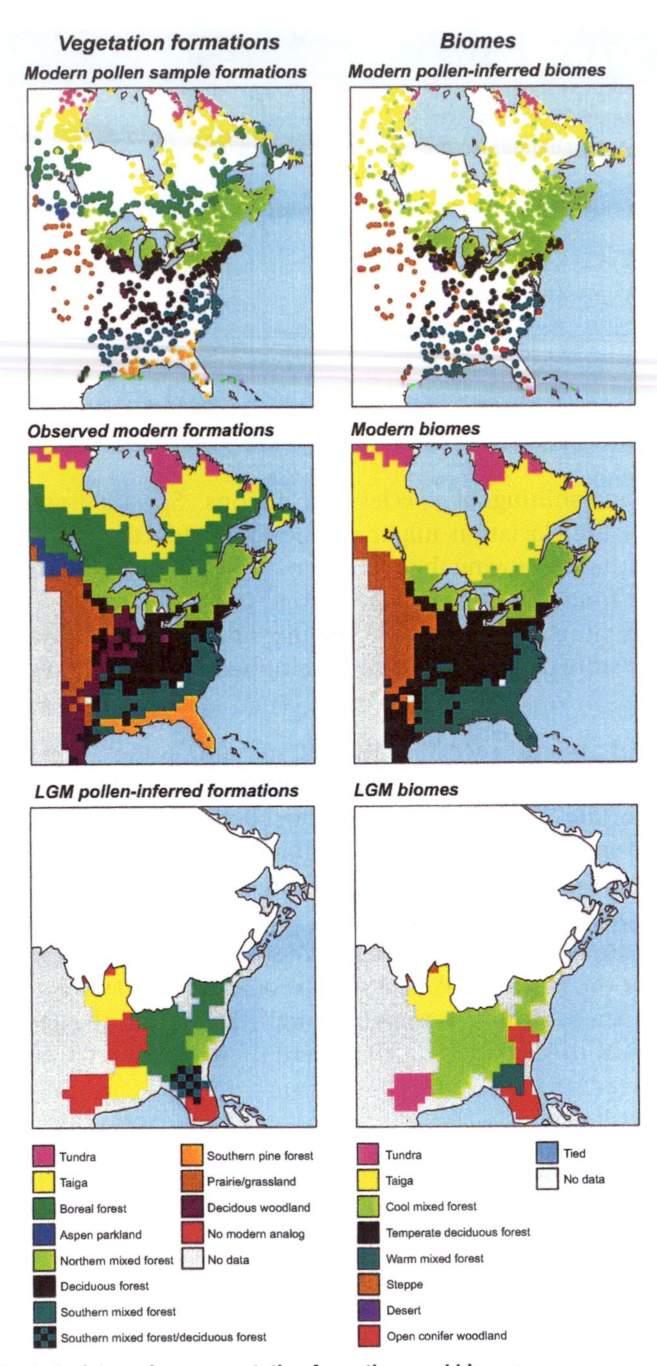

FIGURE 6.8 Present and last glacial maximum vegetation formations and biomes.
Pollen analysis can be used to assign vegetation formations and biomes for past climates. The top panel shows vegetation formations and biomes inferred from analysis of current pollen rain. The middle panel shows the actual current vegetation formations and biomes. The bottom panel shows vegetation formations and biomes for the last glacial maximum, inferred from pollen data. *Source: Jackson et al. (2000).*

bridges. However, no such resolution has been found for Reid's paradox: biologists still debate how poorly dispersed tree species could move so far in so few generations.

SPOTLIGHT: REID'S PARADOX

Once glaciers receded after the last ice age, oak trees would have had to move approximately 1000 km to their present location. Clement Reid pointed out in 1899 that this would take approximately 1 million years if oaks took decades to reach reproductive maturity and acorns fell meters from the tree. However, it has taken less than 15,000 years for oaks to do this. Clark et al. (1998) termed this "Reid's paradox." Initial resolution of this apparent paradox focused on long-distance dispersal. Typical acorns do fall meters from the tree, but the leading edge of a range shift might be driven by atypical acorns that happen to disperse much farther than the norm. These long-distance dispersals made up the "tails" of a probability curve for dispersal distance. Models indicated that speed of range shift was very sensitive to the number of propagules in these probability tails. "Fat" tails gave very rapid range advances, whereas "thin" tails slowed range shifts to a crawl.

By the 1990s, however, a second explanation had emerged. Perhaps small pockets of vegetation, hanging on in suitable microclimates, were important in rapid recolonization as the glaciers receded (McGlone, 1995). First proposed for Southern Hemisphere sites where glaciation had been less severe, this mechanism is gaining increasing acceptance even for Northern Hemisphere sites. It is likely that both long-distance dispersal and micropockets of vegetation are important in enabling plant migration during rapid climate change. Which is more important may only be resolved as human-induced climate change actually unfolds.

Source: Clark, J.S., Fastie, C., Hurtt, G., Jackson, S.T., Johnson, C., King, G.A., et al., 1998. Reid's paradox of rapid plant migration—dispersal theory and interpretation of paleoecological records. BioScience 48, 13–24; and McGlone, M.S., 1995. The responses of New Zealand forest diversity to quaternary climates. In: Huntley, B., Cramer, W., Morgan, A.V., Prentice, H.C., Allen, J.R.M. (Eds.), Past and Future Rapid Environmental Changes: The Spatial and Evolutionary Responses of Terrestrial Biota. Springer-Verlag, Berlin, pp. 73–80.

OUT OF LAND: THE SOUTHERN TEMPERATE RESPONSE

In the Southern Hemisphere, ice and biodiversity followed a much different course, largely because the lay of the land. The small amount of land at high latitude in the Southern Hemisphere makes it impossible for very cold continental winters to become established. Therefore, instead of kilometer-thick ice sheets, the south had extensive montane glaciers interspersed with large areas of non-ice habitats. Much of the high-latitude southern lands are mountainous, so the areas not covered by ice supported a variety of vegetation. Cold south-facing slopes might have supported alpine vegetation, whereas warmer slopes supported forests or mixed vegetation.

The small area of land at high latitudes in the south also means that there was no poleward land for vegetation to occupy as climate warmed. There could not be wavelike poleward advances because Africa, South America, and New Zealand all taper to southern tips. A plant trying to go poleward in these southern areas would run into ocean.

SPOTLIGHT: RAPID CHANGE

In the 1990s, new ice cores from Greenland changed the way researchers viewed climate change in the North Atlantic and the globe (Dansgaard et al., 1993). The Greenland Ice Core Project (GRIP) record resolved fluctuations in climate on an annual scale. Individual layers in the ice core were laid down in a single snowfall season. The picture that emerged was one of continuous change. Large and rapid changes in temperature (as indicated by an ^{18}O proxy) were common throughout the 250,000-year record, both in cool periods (ice ages) and in warm interglacials. Many of the changes corresponded to "flickers" of 1–3 °C in a matter of decades. This picture of rapid change challenged previous notions that held changes of 1000 years to be "rapid." It challenged climate change biologists to propose mechanisms for vegetation response that could operate on annual or decadal time-scales. The biological mechanisms are still being elucidated, but it is clear that rapid large climate change has been faced by species (at least in the North Atlantic) many times in the past. This provides some hope that those same mechanisms may be able to foster plant responses to rapid, large human-induced climate change this century.

Source: Dansgaard, W., Johnsen, S.J., Clausen, H.B., Dahl-Jensen, D., Gundestrup, N.S., Hammer, C.U., et al., 1993. Evidence for general instability of past climate from a 250-kyr ice-core record. Nature 364, 218–220.

As a result, the picture of postglacial vegetation dynamics in the Southern Hemisphere was not of large-scale cross-continental marches but rather of local rearrangement and expansion from pockets of favorable microclimate. As the montane topography of the south warmed, patches of forest on warm north-facing slopes quickly expanded, resulting in very little lag between warming and forest occupation of even the most poleward locations. When there were rapid climate reversals, dominant vegetation shifted over large areas extremely rapidly. Such large, wholesale shifts are unlikely to be explained by long-distance dispersal. Expansion from patches seemed a much better fit to the Southern Hemisphere picture.

NORTH FACING AND SOUTH FACING: WHICH IS WARM?

The temperature difference between north- facing and south-facing slopes can be several degrees at temperate latitudes. South-facing slopes are warmer in the Northern Hemisphere, whereas north-facing slopes are warmer in the Southern Hemisphere.

NORTH MEETS SOUTH

Evidence of rapid shifts from patches of favorable habitat in the Southern Hemisphere influenced biologists' views of Northern Hemisphere responses. If expansion from patches dominated southern responses, could they also have played a role in northern responses to deglaciation? To answer this question, we have to understand the methods of paleoecology.

SPOTLIGHT: MICROREFUGIA EMERGE

At about the same time as McGlone was working out postglacial change in New Zealand (see Spotlight box on page 144), Vera Markgraf of the University of Colorado produced records of very rapid switches in vegetation dominance in Patagonia. These changes in response to climate flickers were so rapid that only microrefugia, pockets of vegetation holding on in microclimates until favorable climate returned, could seem to explain them (Markgraf and Kenny, 1995). McGlone and Markgraf's ideas lead to a Southern Hemisphere view that is now influencing thinking about Northern Hemisphere postglacial processes, despite the massive ice differences between the two settings (large continental ice sheets in the north and none in the south). Fossil evidence indicates that forest may have persisted near even the large continental ice sheets of North America and Europe (Willis and van Andel, 2004). In the future, climate may not return to present reference levels, so microrefugia may be a thing of the past.

Source: Markgraf, V., Kenny, R., 1995. Character of rapid vegetation and climate change during the late-glacial in southernmost South America. In: Huntley, B., Cramer, W., Morgan, A.V., Prentice, H.C., Allen, J.R.M. (Eds.), Past and Future Rapid Environmental Changes: The Spatial and Evolutionary Responses of Terrestrial Biota. Springer-Verlag, Berlin, pp. 81–102; and Willis, K.J., van Andel, T.H., 2004. Trees or no trees? The environments of central and eastern Europe during the Last Glaciation. Quaternary Science Reviews 23, 2369–2387.

The postglacial advance of species, especially trees, had been worked out for the Northern Hemisphere in the 1960 and 1970s. The paleoecologists who mapped out the advance of trees used pollen trapped in lake sediments as evidence. A location under ice obviously supported no trees. As the ice melted, lakes formed. When trees arrived at the shore of the lake, their pollen fell into the lake and was preserved in the muddy sediment at the bottom. Paleoecologists could dig up this mud, carefully preserving the layering using a hollow drilling core (Figure 6.9). They then counted pollen in each layer, and the layer in which the pollen for tree species rapidly rose they pegged as the arrival time of that species. By coring hundreds of lakes, the timing of ice retreat and tree arrival could be worked out for large areas and many species.

However, there was a catch. Some tree pollen was often found in the lake sediments in layers laid down long before the main burst of tree pollen arrived. Northern Hemisphere paleoecologists assumed that this "trace" pollen was the result of long-distance wind-borne pollen settling in the lake. Their hypothesis was that trees moved north in a front, and that some pollen produced by the main population of trees blew a long way to the north. This was a perfectly reasonable assumption, but the evidence from the Southern Hemisphere suggested another possible explanation: What if that trace pollen came from small patches of forest near the ice sheet? Instead of being blown a long distance from a large population of trees, the trace pollen might have come a short distance from a small patch population. If this were true, the maps of tree species marching northward masked the real mechanism of vegetation change—expansion from micropockets.

Thus, two possible explanations exist for the speed of Northern Hemisphere tree migration after the last glacial: Either some seeds underwent unusual long-distance dispersal, or patches of forest existed close to the ice sheets that served as points of expansion (Figure 6.10).

FIGURE 6.9 Drilling a sediment core.
Paleoecologist Nicole Sublette lowers a coring device to a lake floor in the Peruvian Andes with help from Eric Mosblech, a teacher volunteer participating in a field expedition as a means to improve high school curriculum development. The sediment in this core was dated based on annual variations in deposition, and the pollen in each strata counted under a microscope to assess the vegetation surrounding the lake at the time the sediment layer was deposited. *Source: Courtesy of Mark Bush, Eric Mosblech, and Nicole Sublette.*

Which of the two explanations is correct has not been fully settled, but evidence of micropockets is growing. First, climatologists have calculated that the huge continental ice sheets were large enough to influence local climate. A kilometer-thick ice sheet would actually be high enough to partially block the jet stream that normally brings cold arctic air into central North America or Europe. Conditions near the ice sheet might then be warmer than one might expect near so much ice, perhaps even warm enough to support forest. Evidence for trees close to the edge of the ice sheet has been discovered in many locations, indicating that the southern perspective may provide insights for the north.

SPOTLIGHT: ISLAND HOPPING

Rapid vegetation shifts in response to climate change may be mediated by micropockets of vegetation, long-distance dispersal, or both. Although evidence favoring micropockets is mounting, indications that long-distance dispersal has occurred are also accumulating. In the Svalbard archipelago of the Arctic, genetic evidence shows that multiple colonizations have occurred from different sources during the past 20,000 years (Alsos et al., 2007). Alsos et al. used amplified fragment-length polymorphism to determine genetic fingerprints of source colonists to Svalbard. Their results indicate many colonization events but also at least one species that may have survived the full glacial at Svalbard.

Source: Alsos, I.G., Eidesen, P.B., Ehrich, D., Skrede, I., Westergaard, K., Jacobsen, G.H., et al., 2007. Frequent long-distance plant colonization in the changing Arctic. Science 316, 1606–1609.

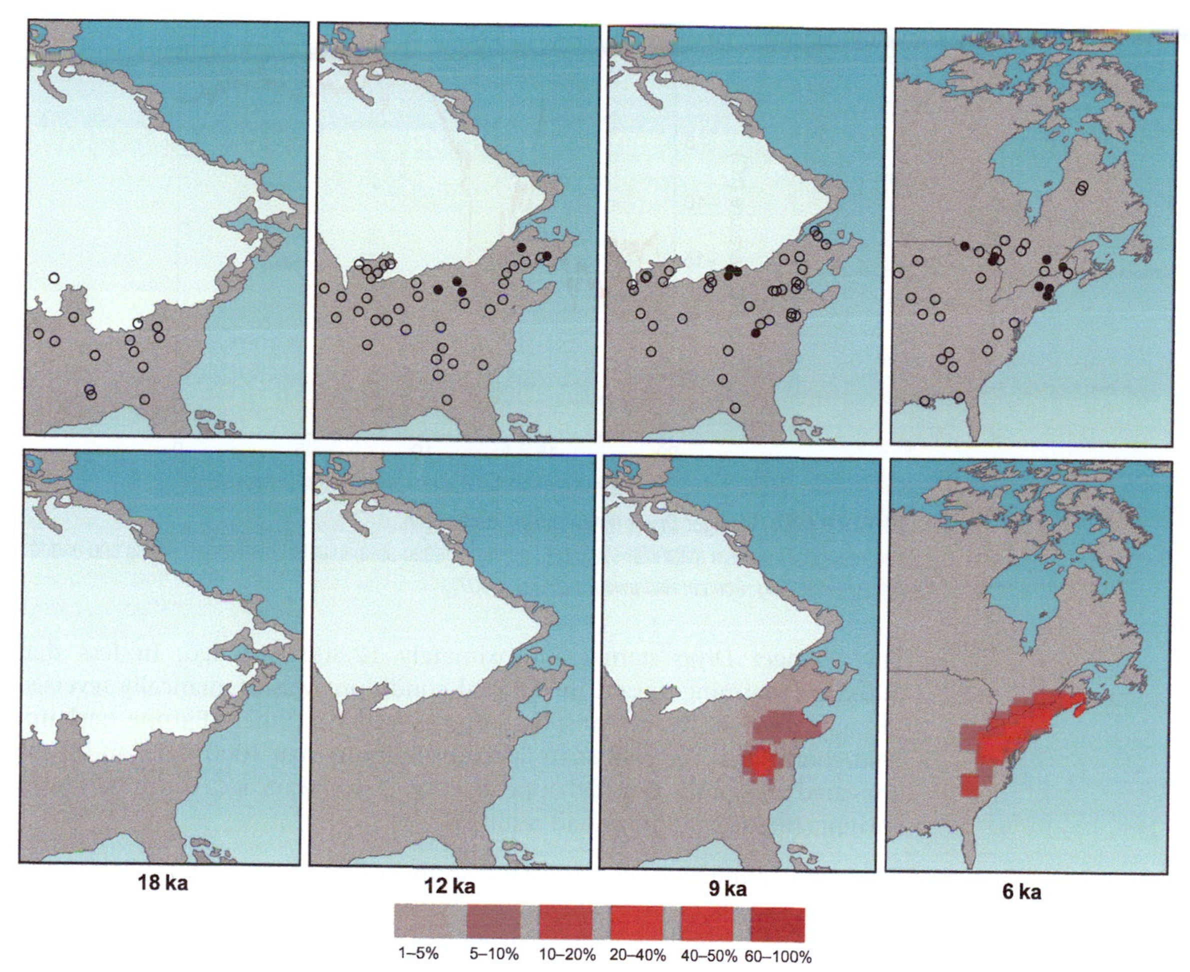

FIGURE 6.10 Macrofossil and pollen records near the Laurentide Ice Sheet.
Trace pollen amounts interpreted as wind-blown input from distant forests in many studies may actually represent pollen micropockets near the ice sheet. Macrofossil records indicate the presence of forest trees near the ice sheet even in times for which little pollen has been recorded from these locations. *Source: Jackson et al. (1997).*

RAPID CHANGE: THE YOUNGER *DRYAS*

Rapid climate changes have marked the record frequently in the transition from the last interglacial to the present. One of the most marked of these changes is a rapid cold snap known as the Younger *Dryas*. This cold snap happened as the northern continents were well on their way to warming and losing their ice cover. The warming was interrupted by as sudden return to cold conditions. Nearly as rapidly, the cold snap ended and warming continued (Figure 6.11).

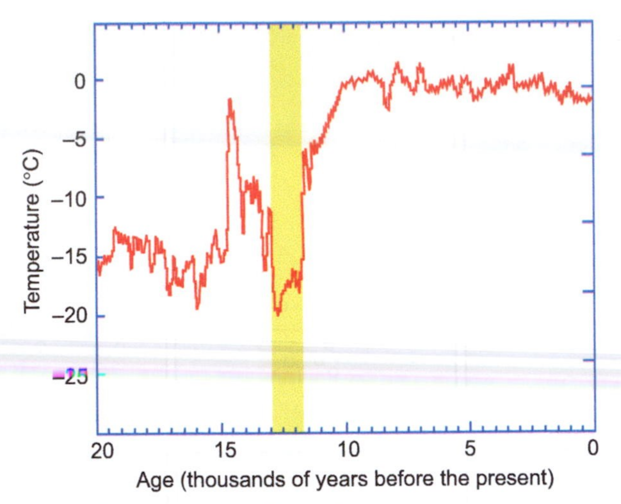

FIGURE 6.11 **Younger *Dryas* temperature fluctuation.**
The Younger *Dryas* is a millennial-scale cool period (shaded area) initiated by a rapid cooling and ended by rapid warming. *Source: Redrawn from Alley (2007).*

The Younger *Dryas* started approximately 12,800 years ago. In less than 500 years, warming toward interglacial conditions was dramatically reversed, plunging the Northern Hemisphere, and especially northern Europe, back into cold conditions. The cold snap lasted little more than 1000 years, and it disappeared as rapidly as it had appeared. By 11,500 years ago, warming toward current climate conditions had resumed.

The temperature changes during the Younger *Dryas* were dramatic. In Greenland, the swings at the opening and closing of the cold snap have been estimated to be more than 10 °C in a century. In Britain, a warming of 10–15 °C has been estimated for the end of the cold snap. Temperature swings in other areas of the world were less, but there is evidence that the Younger *Dryas* was a global phenomenon.

This event was first noted in European sediments, in which a tundra plant, *Dryas octopetala*, was seen twice in the fossil record (Figure 6.12). The first instance was in a previous brief swing back toward glacial conditions, in which this cold-loving plant flourished (the Older *Dryas*). The second time was in a band representing approximately 1300 years of sediment (the Younger *Dryas*). It disappeared again from the record in Europe as climate warmed, remaining absent until the present. Because *D. octopetala* is a cold-hardy plant found in tundra or alpine environments, its absence indicated warming, and its brief reappearance indicated a cold snap. More modern temperature reconstructions have verified this interpretation.

The end of the Younger *Dryas* is closely followed by the beginning of the Holocene geologic period. The Holocene is the past 11,000 years of relatively warm,

FIGURE 6.12 ***Dryas Octopetala*, known commonly as Mountain Avens or White *Dryas*.**
Source: From Wikimedia Commons.

stable climate. The beginning of this warm stable period coincides with the end of the Younger *Dryas*.

Several extinctions and large distribution shifts in many species have been noted in response to the Younger *Dryas*. The most famous of the distribution shifts, of course, is the range change in *D. octopetala* that gave the cold snap its name. Extinctions, such as the disappearance of the spruce *Picea critchfieldii* from North America, have been noted but were not extensive, which is remarkable given the severity of the change. The fact that the vast majority of species, even in hardest-hit Europe, survived the Younger *Dryas* means that they were able to withstand both a very rapid cooling and a very rapid warming. This may give some clues regarding the importance of patches in helping vegetation survive and reestablish from unfavorable climatic conditions. It may also provide insight into the temporal dynamics of climate-driven extinctions. Perhaps it takes many generations for a species to finally be driven to extinction by rapid climate change. The Younger *Dryas* was so short that species may have been able to recover before chance drove all populations to extinction.

The cause of the Younger *Dryas* is thought to be shutdown of the North Atlantic thermohaline circulation due to an outpouring of glacial meltwater from North America (see Chapter 2). As climate warmed and continental ice sheets melted, large quantities of freshwater accumulated in what are now the Great Lakes of the United States and Canada. Ice blocked drainage of these lakes for many years. Finally, the blockage was breached, and huge amounts of freshwater gushed out the St. Lawrence seaway and into the North Atlantic. There, the freshwater diluted the salinity of the Gulf Stream, preventing it from sinking. This disrupted the thermohaline circulation and shut down the Gulf Stream. The heat pumped from the equator by the Gulf Stream suddenly ceased to arrive in the North Atlantic, resulting in a rapid cooling. When the freshwater pulse stopped, the thermohaline circulation turned back on and the region rapidly resumed warming.

TROPICAL RESPONSES

Tropical responses to deglaciation are much more subtle than temperate responses, but major changes are still evident. Large-scale, cross-continental revegetation following retreating ice is absent in the tropics simply because continental ice sheets did not exist there. However, smaller scale revegetation following the retreat of montane glaciers is evident, and there is strong evidence for vegetation type and composition shifts on tropical mountains. Retraction of lowland tropical forest at the peak of glaciation seems to have occurred, although not to the extent once proposed (see box on page 155).

DANSGAARD-OESCHGER EVENTS

FIGURE 6.13 Greenland ice core.
Source: Courtesy of NASA/JSC.

One type of very rapid climate change is Dansgaard–Oeschger (D–O) events. These are very rapid warming events in the North Atlantic that take place on timescales of 10–20 years, followed by a slower cooling over half a century or more. These warm snaps and cool rebounds were first noticed in the GRIP and Greenland Ice Sheet Project ice cores drilled in the early 1990s (Figure 6.13). They are corroborated by glacial debris in sediments. As icebergs drift south in the cool phase of a D–O cycle, they melt and deposit debris that can be detected in ocean sediment sampling. Both the ice record and sediment analysis suggest that D–O events are relatively large temperature excursions that occur on decadal or finer time-scales. The biological challenges and signals from such very rapid change are one of the important sources of insight that may tell us how species will react to very rapid human-induced warming.

There is strong evidence that tropical vegetation shifted during the last ice age, reflecting globally cooler temperatures. Most of this evidence is based on pollen records recovered from lake beds with sufficient stratification for accurate dating. The abundance and length of such records are limited, however. We have many fewer lake cores for the tropics than for the temperate zone and, like the temperate record, few reliable records date back further than 100,000 years. Part of the reason for fewer records in the tropics may be less intensive study and less road access to remote areas, but the major reason is simply that there are fewer tropical lakes. This is partly because there were no continental glaciers to leave pothole lakes as they retreated.

As temperatures warmed toward the current interglacial climate, tropical forests expanded and moved upslope. Because there are more records from montane lakes in the tropics than from the lowlands, the evidence for upslope movement is most compelling.

Pollen records from all tropical regions reflect a forest line much lower at the height of the last glacial than the present forest line, which then moved upward to track warming temperature (Figure 6.14).

Different species mixes moved upslope as forest expanded upwards. Although some forest classifications are somewhat arbitrary because species move individualistically, in general, lowland forest, mid-elevation forest, and upper montane forest all moved upslope. These forest types, usually identified by keying in on characteristic genera, reflect individualistic species movements, not monolithic shifts in forest communities of constant composition.

Emerging high-resolution records reveal very rapid vegetation variation along with general upslope movements. These rapid vegetation shifts are similar to those observed in records from temperate and high-latitude sites. They correspond to D–O events and match ice-core records of climate flickers.

For example, in the tropical Andes, a lake core spanning 120,000 years reveals extensive rearrangement of genera as climate warmed from the last glacial to the present. Significantly, comparison of this and other records in the region reveals common patterns of rapid vegetation response to climate flickers. The Andes pollen records match Greenland ice core and Meditteranean sea surface temperature records, indicating global patterns of synchronous change. This contrasts to other Andean sites where noise obscures patterns (Figure 6.15).

In tropical island systems, rapid fluctuation between dominant forest species have been recorded. Some of these fluctuations seem to be initiated by climate flickers such as the Younger Dryas. Others follow these climate-triggered events but do not seem to correlate with climate events. It is probable that these later tree dominance changes are the result of internal forest dynamics, indicating that a climate flicker can set off vegetation change that may resonate with biological dynamics for several centuries.

SPOTLIGHT: PLEISTOCENE REFUGIA: A THEORY DISPROVED

In the late 1960s, a petrogeologist and avid bird watcher, Jurgen Haffer, published an influential theory of Amazonian speciation. Haffer proposed that the Amazon was cooler and drier during glacial periods, resulting in forest contracting into patches (refugia) surrounded by grassland. The repeated process of isolation and reconnection as ice ages came and went in the Pleistocene was proposed to account for biogeographic patterns and species richness in the Amazon. However, major points of the theory could not be substantiated. Limited paleoecological evidence suggested that the forest might have still been continuous over large areas in glacial climates. Genetic evidence indicated that splits in Amazonian lineages were much older than the proposed ice age drivers. The "Pleistocene refugia" theory is now widely seen as discredited.

Source: Hoorn, C., Wesselingh, F.P., Ter Steege, H., Bermudez, M.A., Mora, A., Sevink, J., Antonelli, A., 2010. Amazonia through time: andean uplift, climate change, landscape evolution, and biodiversity. Science 330 (6006), 927–931.

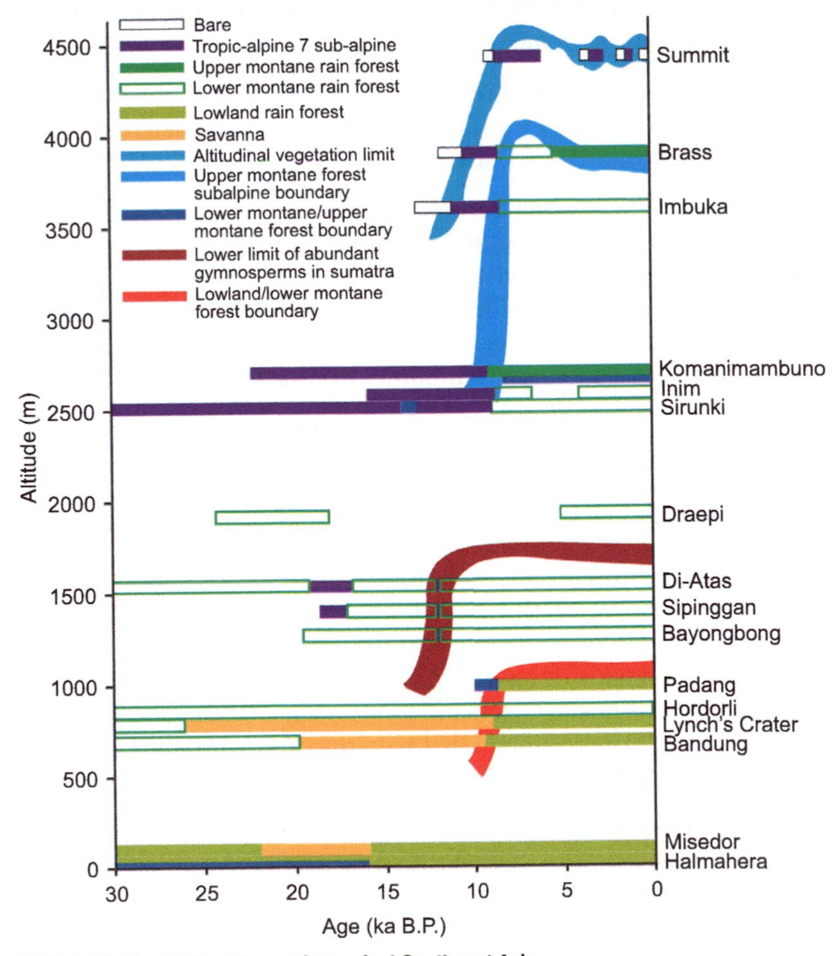

FIGURE 6.14 Vegetation change in tropical Southeast Asia.
Vegetation types (horizontal bars) and limits (vertical bands) inferred from pollen analyses for sites in Southeast Asia. *Source: Flenley (1998).*

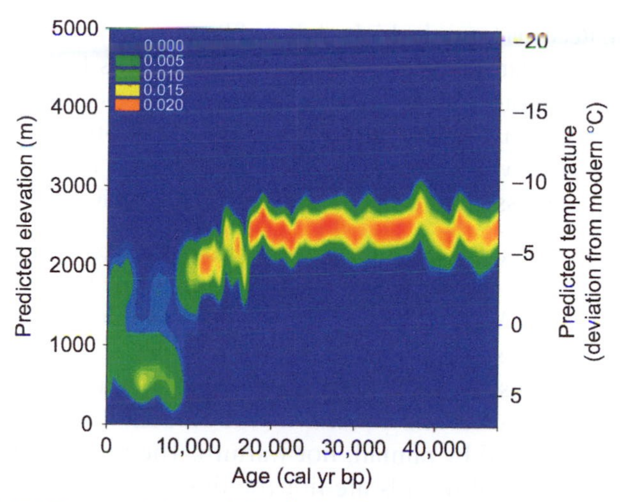

FIGURE 6.15 48,000 years of change in the tropical Andes.
Paleotemperature and paleoelevation inferred from pollen spectra from Lago Consuelo, Peru. Color indicates the probability that a sample came from that elevation or temperature (blue, low; orange, high). Note that time runs right to left in this figure. *Source: Bush et al. (2004). Reproduced with permission from AAAS.*

MILANKOVITCH FORCING IN THE BIOLOGICAL RECORD

Major climatic cycles are driven by changes in the amount of energy from the sun reaching the Earth, as discussed in Chapter 2. Seemingly small changes in incoming solar energy due to variations in the Earth's orbit can result in large climatic consequences. Ice buildup reflects sunlight, setting up a positive feedback loop that can result in accelerated cooling or ice melt. Changes in solar energy input feed these reinforcing positive-feedback mechanisms. These changes are reflected in long-term paleoecological records. The 23,000-, 41,000-, and 100,000-year cycles of precession, obliquity, and eccentricity are seen to dominate at different times in biological records, indicating that climatic variation is important in driving vegetation change over millennia.

Biological records extending far enough into the past to reflect cycles of tens or hundreds of thousands of years are not abundant. However, increasing numbers of tropical lake cores offer pollen records going back hundreds of thousands of years, and some marine records offer sediments to preserve the shells of planktonic species or other proxies over a similar length of time.

For example, a long lake core from Lake Titicaca on the border of Peru and Bolivia provides a record of vegetation changes across 370,000 years. This record reflects mostly precessional forcing (cycles of 23,000 years and multiples). However, within the dominant precessional forcing, shorter-term events are common. Records from the high plains of Bogotá and adjacent areas reflect obliquity (41,000 year) forcings. Paleoecological records from other areas of South America and other areas of the world are in agreement with the Titicaca record, confirming that within the past 100,000 years precessional forcing is important in driving vegetation changes. In the Amazon, for instance, this precessional forcing is associated with changes in precipitation.

However, it is important to note that these orbital forcings are not in sync in all areas of the world or even in different areas of the same region. For example, sites in the northern Andes reflect precessional forcing in sync with solar insulation in July, whereas sites in the southern Andes reflect precessional forcing in sync with solar insolation in December. This is because the rainy seasons are 6 months apart across the equator in the Andes. Thus, the local precessional forcing is caused by rainfall, not warmth. The precessional cycles seen in these paleoecological records are not the direct result of changes in solar insulation; rather, they are driven by global climatic changes that in turn affect local climate.

LESSONS OF PAST CHANGE

The lessons of past change from the terrestrial realm include some broad principles that are applicable in marine and freshwater realms as well. These include the individualistic nature of species response to climate change and the resultant splitting up and reassembly of communities. Associated with these effects are the emergence of novel assemblages of species and the disappearance of others. The terrestrial realm shows clear marks of change due to Milankovitch cycles, which it holds in common with water realms.

Past climatic change holds the key to understanding vegetation change across both spatial and temporal scales. The affinity of northern floras in North America and Eurasia and in southern floras and faunas from South America to Australia results from geographic and climatic connections dating back 50 million to 100 million years. Response to more recent, rapid climate changes indicates that vegetation has been able to respond even to very rapid climate flickers of 100–1000 years. Whether this response capacity will be fast enough to keep pace with human-induced climate change, especially in landscapes heavily altered by human activity, remains to be seen.

FURTHER READING

Bush, M., 2002. Distributional change and conservation on the Andean flank: a palaeoecological perspective. Global Ecology and Biogeography 11, 475–484.

Clark, J.S., Fastie, C., Hurtt, G., Jackson, S.T., Johnson, C., King, G.A., et al., 1998. Reid's paradox of rapid plant migration—dispersal theory and interpretation of paleoecological records. BioScience 48, 13–24.

Davis, M.B., Shaw, R.G., Etterson, J.R., 2005. Evolutionary responses to changing climate. Ecology 86, 1704–1714.

Hooghiemstra, H., 2013. Pollen-based 17-kyr forest dynamics and climate change from the Western Cordillera of Colombia; no-analogue associations and temporarily lost biomes. Review of Palaeobotany and Palynology, 194, 38–49.

CHAPTER 7

Past Marine Ecosystem Changes

Past analogues to future climatic and chemical conditions in the oceans offer one of our most important sources of insights into the possible biological effects of human-induced change. The vastness of the oceans, their huge contribution to primary productivity, and their critical role in the climate system make these past insights particularly valuable.

Multifaceted marine responses to climate change are known from the paleoecological records. Marine communities respond to climate change with individualistic range changes, changes in composition, and changes in body size. Sea level changes accompanying glaciation have an especially important effect in the marine realm because they affect available habitat for organisms in tropical, temperate, and arctic regions. Atmospheric CO_2 dissolves in seawater, changing its acidity, which affects the growth of calcium carbonate-secreting organisms such as corals and clams. Climate change can alter major circulation patterns in the oceans, affecting local conditions, nutrient transport, and long-distance dispersal of planktonic larvae. Temperature, sea level, circulation, and acidification are the major drivers of marine response to climate change.

Much of deep-time marine paleoecology focuses on major extinction events. These analyses are the focus of Chapter 9. This chapter summarizes other insights from marine paleoecology that illuminate possible response to future climate change.

EFFECTS OF TEMPERATURE CHANGE

Temperature change has a profound effect on the distribution of marine organisms, from tropical corals to arctic clams, and from pelagic plankton to the deepest benthos. Precipitation change, so important on land, has little effect in the oceans. Temperature dominates climatic constraints on species' ranges in the two-thirds of the planet covered by water.

Records dating back several million years indicate range changes in response to temperature. For instance, warmer climate in the Miocene, 15–17 million years ago, was accompanied by range extensions of mollusks, with many species that

Climate Change Biology. http://dx.doi.org/10.1016/B978-0-12-420218-4.00007-X

are currently found in the tropics and subtropics extending far to the north in the Pacific—some reaching as far as Alaska. Similar changes were observed in plankton. When climate cooled again after the mid-Miocene global temperature peak, these species' ranges retreated southward.

Such range shifts lead to changes in community composition—sometimes transitory, sometimes more long-lasting. For example, Antarctic communities of 4 million years ago were much different from Antarctic communities today. When global cooling set in approximately 3.5 million years ago, crabs, sharks, and many predatory fish were lost from Antarctic communities. As a result, these communities are currently dominated by invertebrates even at higher trophic levels. Brittle stars, basket stars (ophiuroides), and feather stars (crinoids) are abundant (Figure 7.1). This makes present cold Antarctic waters resemble deep-sea communities, whereas past warmer Antarctic marine communities would have been more similar to current temperate food chains.

In past benthic communities, cold-water species have found refuge in deep waters during times of warming, recolonizing shallower waters when temperatures drop. Benthic organisms on the North Atlantic continental shelf were severalfold more diverse during warm interglacial periods than during glacial conditions due to these temperature-related shifts. Nutrient shifts accompanying temperature changes may have been an important mediator of these changes.

For the Pleistocene (last 2 million years), fossil data are available for many types of species, with strong indications of temperature responses. The responses associated with warming are not uniform. For example, in contrast to the

FIGURE 7.1 Four-million-year-old Antarctic shift.
A crinoid, typical of current Antarctic marine communities. The circumpolar current formed around Antarctica 4 million years ago, resulting in major ecosystem changes. Sharks and bony fishes were lost, resulting in proliferation of crinoids and ophiuroids (brittle stars). *Source: Photo courtesy of Jeff Jeffords.*

North Atlantic benthic response, plankton in the Mediterranean were more diverse under glacial conditions, becoming less rich as waters warmed. Some responses to climate change may have been simple range shifts in response to temperature, while others, such as the Mediterranean example, may have been mediated by changes in nutrient availability and freshwater inputs.

Latitudinal shifts are evident as well—poleward with warming and toward the equator with cooling. In one particularly well-studied example off the coast of Southern California, bivalve ranges changed significantly with temperature change. Species that existed between Santa Barbara and Ensenada during a warmer interglacial 125,000 years ago are currently confined to tropical or subtropical waters.

SPOTLIGHT: CHILL OUT

Foraminifera ("forams") are used as proxy for past sea surface temperature (Figure 7.2). They are microscopic organisms that inhabit the open ocean. These protists preferentially incorporate ^{18}O into their calcium carbonate shells or "tests." The ^{18}O isotopic composition of foram's calcite shells varies with temperature. The ^{18}O content of the oceans varies primarily with the amount of water trapped in large ice sheets. Expressed as parts per 1000 (ppt or ‰), the ratio of ^{18}O to ^{16}O tells paleoclimatologists the temperature regime under which the shell was formed. The technique is well-established but not free from controversy. Until 2001, paleoclimatologists believed that tropical oceans warmed less than high-latitude oceans. Even in very warm periods, such as the Paleocene, tropical oceans showed less warming than did higher-latitude oceans. Pearson et al. (2001) showed that this "cool tropics paradox" was an artifact of partial dissolving of foram tests in tropical oceans. When Pearson et al. chose well-preserved forams, the cool tropics disappeared and foram results indicated that the tropics warmed in line with higher latitudes during global warm periods over the past 100 million years.

Source: Pearson, P.N., Ditchfield, P.W., Singano, J., Harcourt-Brown, K.G., Nicholas, C.J., Olsson, R.K., et al., 2001. Warm tropical sea surface temperatures in the Late Cretaceous and Eocene epochs. Nature 414, 470.

This same region provides indications that marine fauna may have responded to climate "flickers." Some species currently found only north of Santa Barbara are found in the same fossil beds with species that are currently found only south of Ensenada. The best explanation for this may be that the warm-water species established farther north during the warm interglacial, which was then interrupted by a millennial-scale cold snap that brought in the more northern, cold-water species. This change was so sudden that the shells became preserved in the same fossil layer (Figure 7.3).

Body size appears to be important with regard to marine range changes. Larger-bodied organisms are more likely to undergo range expansion as climate changes. In California bivalves, species whose ranges shifted during the Pleistocene had larger body sizes than average. Whether this bias extends to other marine taxa is not clear. It has important implications, however, because large-bodied bivalves have also been shown to be more invasive when

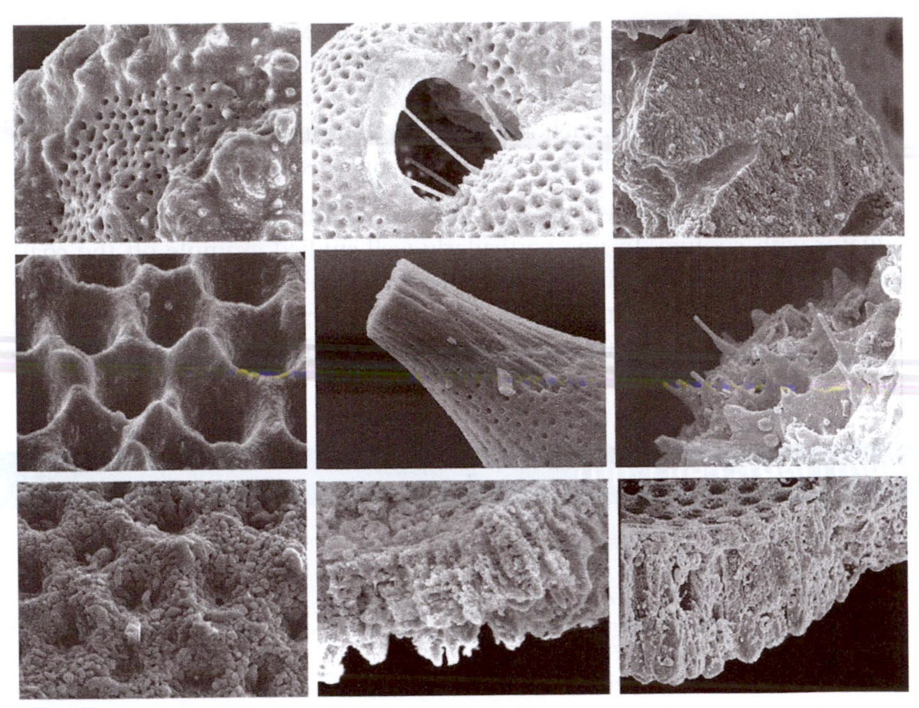

FIGURE 7.2 Electron micrograph of foraminifera tests.
Source: Pearson et al. (2001). Reproduced with permission from Nature.

introduced outside their native range. These factors may provide clues to which physiological and life history characteristics may make species most vulnerable to future temperature variations.

EFFECTS OF SEA-LEVEL CHANGE

Global warming fuels sea level change in two ways. First, water expands as it warms, so higher mean global temperature will result in a rise in sea level due to thermal expansion. A far greater effect may result from the melting of land ice. Ice in Antarctica and Greenland alone, if totally melted, would raise world sea level by dozens of meters—enough to flood approximately 1 million square kilometers of coastal lowlands (Figure 7.4). The melting of all ice in Greenland and Antarctica would raise sea level approximately 75 m. It is important to note that melting of sea ice does not have any effect on sea level because sea ice already displaces water. Therefore, almost all of Greenland ice melt contributes to sea level rise, whereas the melting of sea ice from around Antarctica contributes nothing to sea level rise. Melting of the sea ice around Antarctica would have profound food web implications but would not be a major contributor to global sea level rise.

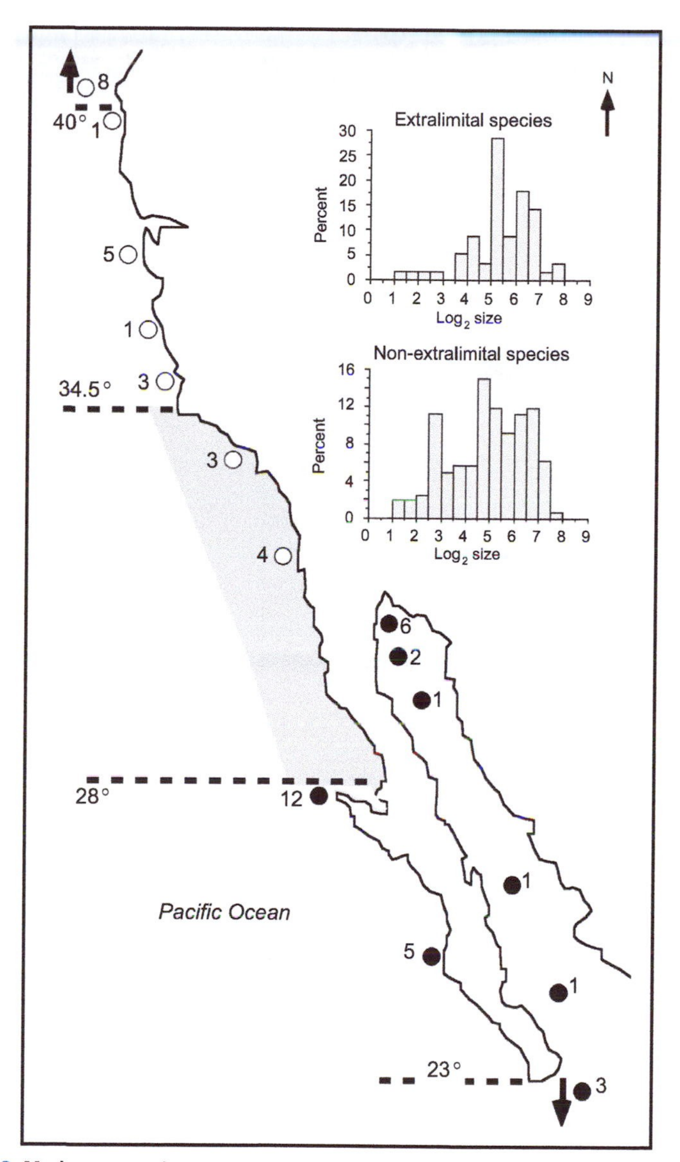

FIGURE 7.3 Marine range changes.
At the Last Glacial Maximum (LGM), 12 species were found together along the Southern California coast that now are now nearly absent from the area. This plot shows the present southern and northern range limits (points) of species that were co-occurring in the shaded area at the LGM. *Source: Roy et al. (2001).*

Past warming and cooling has raised and lowered sea levels repeatedly, giving some indication of possible physical and biological effects. These fluctuations have been particularly strong during the glaciations and deglaciations of the Pleistocene ice ages (past 2 million years). Each time massive land ice has

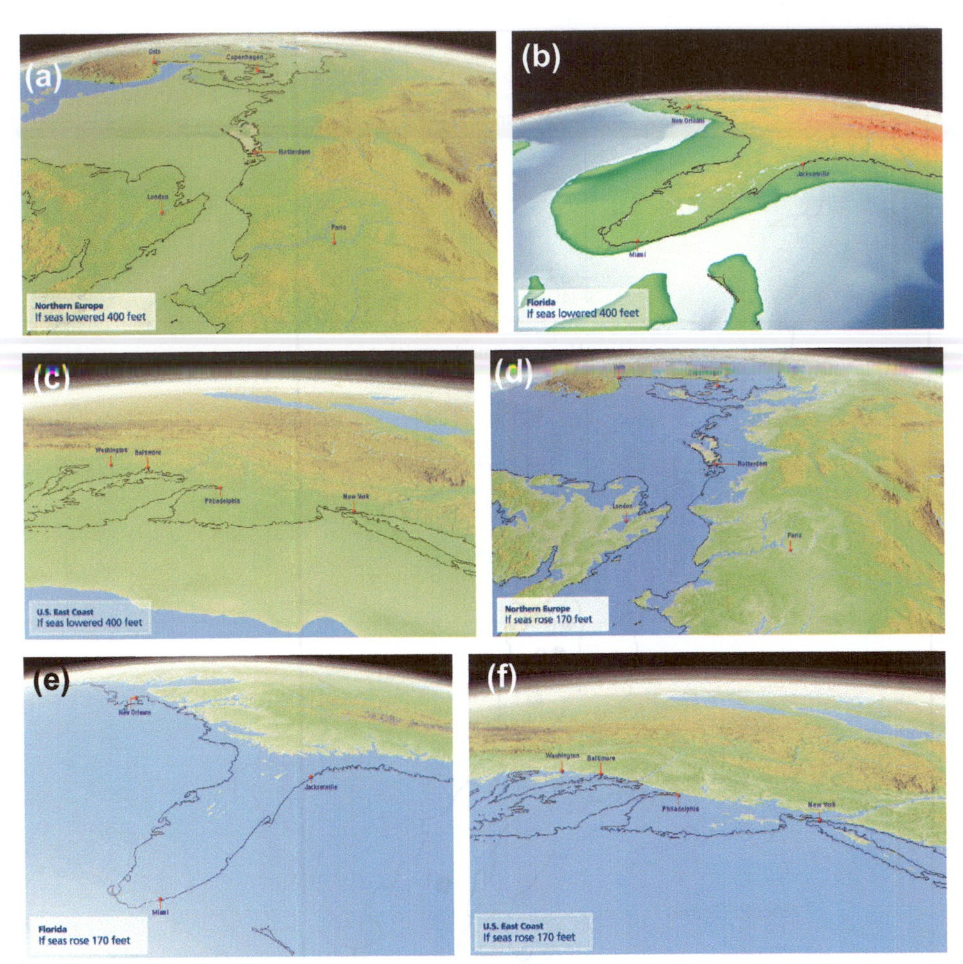

FIGURE 7.4 **Current coastline superimposed on coastline at LGM and coastline with Greenland/Antarctica melted for Northern Europe (a and d), Florida (b and e) and Southeast Asia (c and f).**
Source: Courtesy of William F. Haxby, Lamont-Doherty Earth Observatory of Columbia University.

formed in a glaciation, the removal of water from the oceans has resulted in a major sea level drop. Each time the ice sheets have melted, the water returning to the oceans has raised sea level.

Records of marine organism range shifts in response to sea level change exist from as far back as 55 million years ago. Off the east coast of North America, foraminiferal communities changed repeatedly during the Eocene and Pliocene, with species entering or disappearing from habitat as sea level shifted. The species composition following these shifts varied greatly, even at similar sea levels. However, all species were drawn from a common pool, with substitutions responsible for the shifts in community composition with sea level.

Sea level change has affected species distributions in the tropics as well. In the Pacific islands, distributions of reef species changed with timing that coincided with solar cycles over several million years. Inner reef specialists were stranded and became locally extinct on many islands when sea level dropped. When sea level rose, these inner reef species recolonized the newly submerged habitat. Outer reef species persisted through both low and high sea level stands.

SEA-LEVEL EFFECTS OF LAND ICE AND SEA ICE

When land ice melts, it contributes to sea level rise, but melting of sea ice does not. This is because sea ice already displaces water; its melting does not change net sea level, just as melting ice in a glass does not cause the glass to overflow. Land ice, on the other hand, does not displace water. When land ice melts, the meltwater eventually reaches the sea, increasing the amount of water in the oceans. When huge amounts of land ice melt, such as the melting of continental ice sheets of an ice age, sea level is substantially increased—by approximately 120 m in the case of the melting that occurred from the Last Glacial Maximum to the present.

Pleistocene glacial–interglacial cycles have driven repeated changes in sea level, which are recorded in many records of biological response. Particularly good records exist for clams, oysters, and other bivalves. Bivalves in tropical Pacific islands changed distribution repeatedly with changes in sea level, resulting in changes in community composition. Effects such as these have also been observed in bivalves, gastropods, and other species in Fiji, Aldabra, off the coast of Kenya, and elsewhere.

CHANGES IN OCEAN CIRCULATION

Multiple changes in ocean circulation may result from climate change. The earth's oceans play a major role in determining how changes in climate occur across regions. The oceans absorb heat and CO_2, transport them, and, more slowly, mix them into deep waters (Figure 7.5). Ocean circulation plays an important role in all of these functions.

Two major types of ocean circulation changes stand out as having had exceptional influence on both climate and biodiversity in the past. These are changes in thermohaline circulation and changes in teleconnections.

Shutdown of thermohaline circulation may result from the input of large pulses of freshwater in the North Atlantic, as discussed in Chapter 2. Light freshwater dilutes the heavier saline water of the Gulf Stream, preventing it from sinking, thus breaking the "conveyor belt" effect of thermohaline circulation. When this happens, sudden temperature shifts occur on land and in the sea. The effects on terrestrial biodiversity of these sudden cold snaps were discussed in

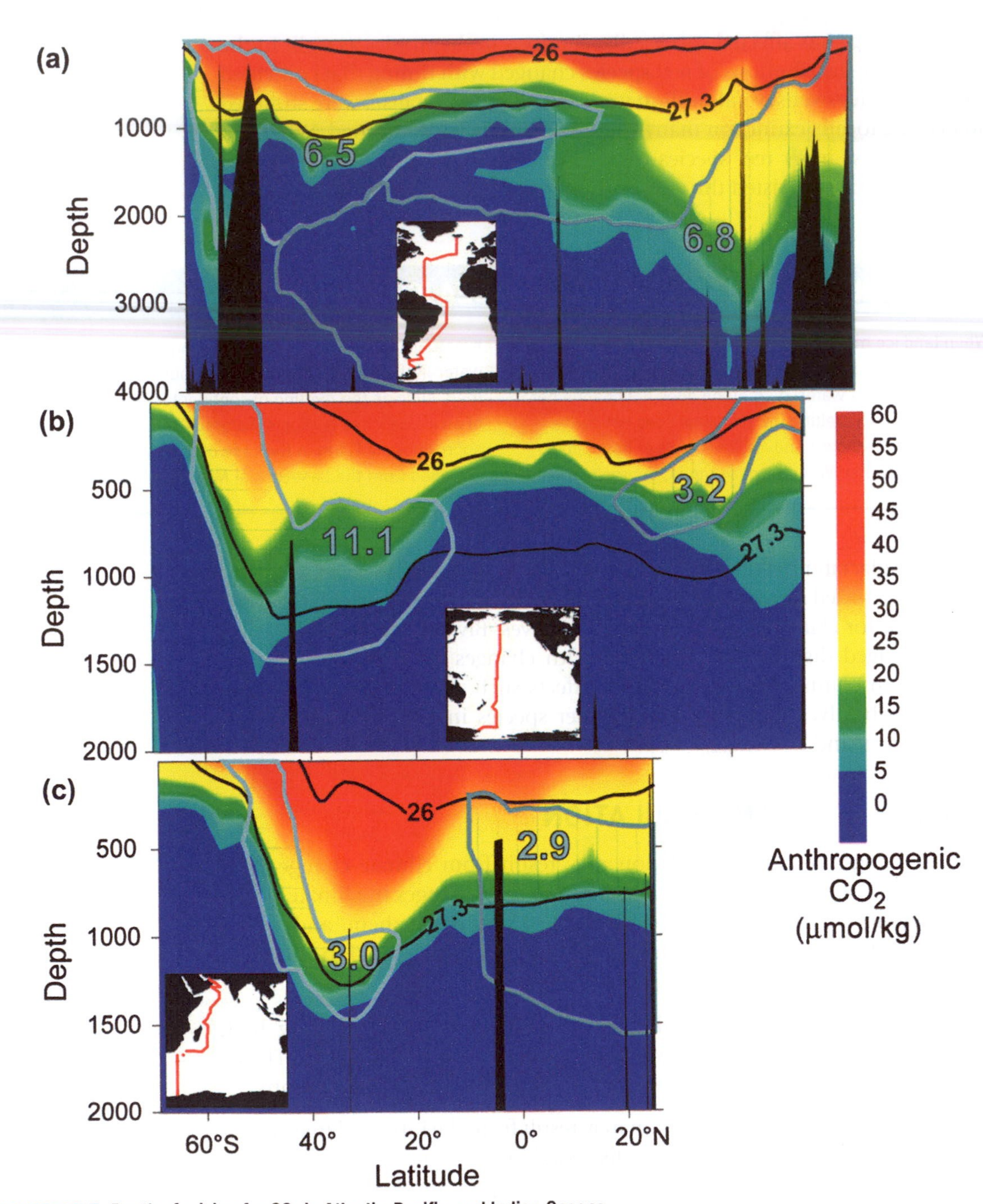

FIGURE 7.5 Depth of mixing for CO_2 in Atlantic, Pacific, and Indian Oceans.

The three color panels (a-c) indicate depth of mixing along the transects mapped in the insets. a) Atlantic; b) Pacific and c) Western Indian Ocean. CO_2 from human emissions mixes into surface waters more rapidly than deep waters. CO_2 is poorly mixed in bottom waters in all three oceans, but is highest in North Atlantic waters, where it is carried downward in thermohaline circulation. *Source: Sabine et al. (2004).*

Chapter 6. Our understanding of marine effects of thermohaline shutdown is less detailed, but it is clear that profound changes in marine life result.

Biological ties to the thermohaline circulation are strong. Nutrients accumulate in deep waters, making the upwelling zones of the circulation the "breadbasket" for the base of many marine food chains. Nutrient upwelling results in high productivity of phytoplankton, which in turn results in high diversity and abundance of other levels of the food chain and rich fisheries.

Deep water formation resulting from the sinking of heavy, cold saline waters provides oxygen to the depths. When thermohaline circulation shuts down, affected deep waters become oxygen deprived, which can lead to the death of many organisms. We will see in Chapter 9 that oxygen deprivation in deep waters may be a cause of major extinction events in the past.

The second major circulation change is alteration in global teleconnections such as El Niño. There are several of these variable atmospheric–oceanic circulation patterns, including El Niño/Southern Oscillation (ENSO), the North Atlantic Oscillation (NAO), and the Atmospheric Circulation Index (ACI). All include some change in ocean circulation, coupled with changes in atmospheric circulation. For instance, El Niño episodes are associated with decreased upwelling of deep nutrient-rich waters and changed air circulation patterns over the Pacific (Figure 7.6).

TELECONNECTIONS

Some weather patterns occur on such a large scale that they have repercussions throughout the world. These effects, where change in one area of the planet is consistently associated with changes in other regions or other continents, are known as teleconnections. When El Niño occurs in South America, the changes in air masses over the Pacific are so large that they cause ripple effects throughout the world. Failure of the rains in southern Africa, for instance, is associated with El Niño conditions. These teleconnections are important in climate change because change in one area of the world may result in associated changes in many other areas.

Marine systems, especially fisheries, respond rapidly to changes in these circulation indices. Changes in circulation indices, including ENSO, NAO, and ACI, have been linked to changes in fish stocks for species including Pacific herring, Atlantic cod, sardines, anchovies, Pacific salmon, horse mackerel, Chilean jack mackerel, and Alaska pollock. Past response to changes in circulation were clearly widespread, but are poorly resolved in the fossil record.

CHANGES IN OCEAN CHEMISTRY

Major changes in ocean chemistry are recorded in the fossil record, particularly in the shells and skeletons of marine organisms. One of the most critical ocean chemistry changes recorded in these shells is that of ocean CO_2 and

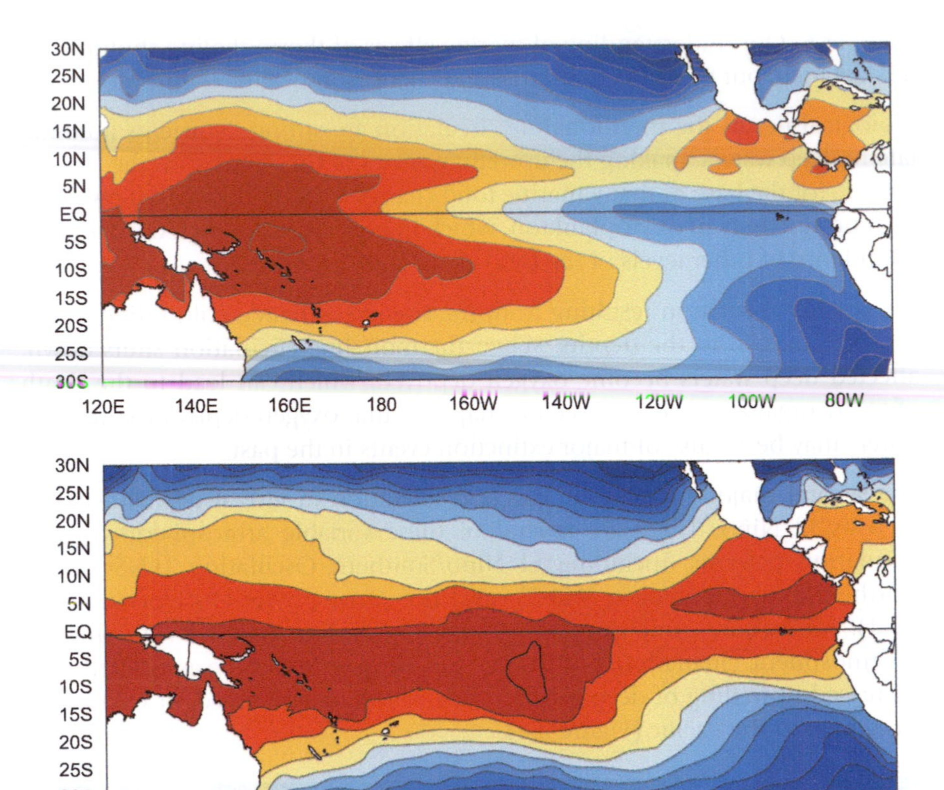

FIGURE 7.6 Sea surface temperature response to El Niño.
Normal conditions (top) and El Niño conditions (bottom). In El Niño conditions, warm water pools more widely in the tropical Pacific and water temperatures are higher in the eastern tropical Pacific, blocking upwelling along South America and in the Galapagos. *Source: Courtesy of Mark Bush. From* Ecology of a Changing Planet, *3rd edition. Benjamin Cummings.*

pH levels. As we have seen in previous chapters, ocean CO_2 and pH affect the ability of organisms to form hard calcium carbonate structures that can be preserved in the fossil record. Tropical corals are major ecosystem engineers that form calcium carbonate reefs. Benthic calcium carbonate-secreting organisms important in the fossil record include clams, mussels, and tube worms. Among pelagic organisms, foraminifera are perhaps the most important taxa leaving calcium carbonate fossil remains (Figures 7.7 and 7.8).

Remains of these shells may provide important fossil evidence of past sea conditions. Corals and foraminifera are among the leading indicators of past climate and paleoecology. These and other calcium carbonate-secreting organisms are important windows into the past because of the hard structures they leave behind.

Foraminifera are deposited in layers deep enough to be recovered in drilling cores of the ocean floor. The individual shells in these deposits can then be

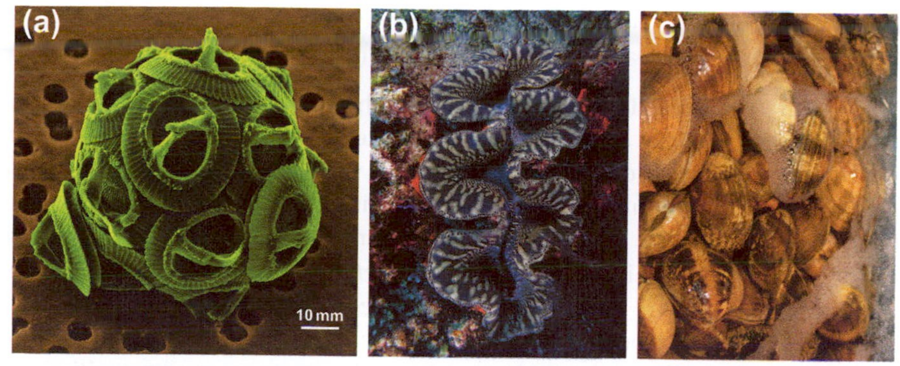

FIGURE 7.7 **Calcite-secreting organisms a) Coccolithophore; b) tridacnid clam; c) manila clam (*Venerupis philippinarum*).**
Source: From Wikimedia Commons.

FIGURE 7.8
Aragonite-secreting organisms Scleractinian corals (a-b) and a pteropod (*Limacina helicina*). Coral reefs of aragonite provide structure which is the base of many tropical marine foodwebs. Pteropods are important food sources for salmon, mackeral and cod.. *Source: Courtesy U.S. National Oceanic and Atmospheric Administration (NOAA).*

used to identify the foraminifera occupying ancient oceans at a location, often to the genus or sometimes species level. By comparing these ancient ocean denizens to the climatic affinities of modern foraminifera or by analysis of the isotopic composition of the shells (known as tests), past ocean temperatures and conditions may be reconstructed.

Coral Reef Distribution

Modern coral reefs are composed of scleractinian corals. In the past, however, other types of reef-building organisms dominated under differing temperature and CO_2 regimes. One hundred million years ago, when CO_2 levels were approximately double present levels, corals were a minor component of tropical reefs, which were dominated by bivalves and other noncoral organisms. This is presumably because the elevated CO_2 changed the pH and saturation state of seawater, favoring organisms that secreted their shells in the calcite variant of calcium carbonate. Modern corals, which secrete their skeletons in the aragonite form of calcium carbonate, were much less dominant in the tropics at this time.

The fossil record indicates that coral reef growth is not determined by temperature alone, despite the importance of coral bleaching in response to warmer water temperatures. Viewed over millions of years, the correlation between global mean temperatures and coral reef dominance is poor. Coral reefs have been dominant in the ice age conditions of the past 2 million years, but in other cool conditions in deep time, coral reefs were less dominant. Factors other than temperature seem to control coral reef latitudinal distributions. The latitudinal range of modern corals has extended further south along Australia within the past million years at some sites, whereas species composition, at least at one site in New Guinea, has remained remarkably constant.

Thus, although *growth* rates in corals correlate strongly to temperature and are widely used to infer past climates, the *distribution* of corals is not a reliable indicator of past climate. This paleoecological evidence is supported by observations of factors affecting growth in modern corals. Growth of scleractinian corals is sensitive to multiple factors, including salinity and turbidity in addition to temperature and calcium carbonate saturation state (Figure 7.9). Photosynthetic productivity drops rapidly with high salinity, just as it does with

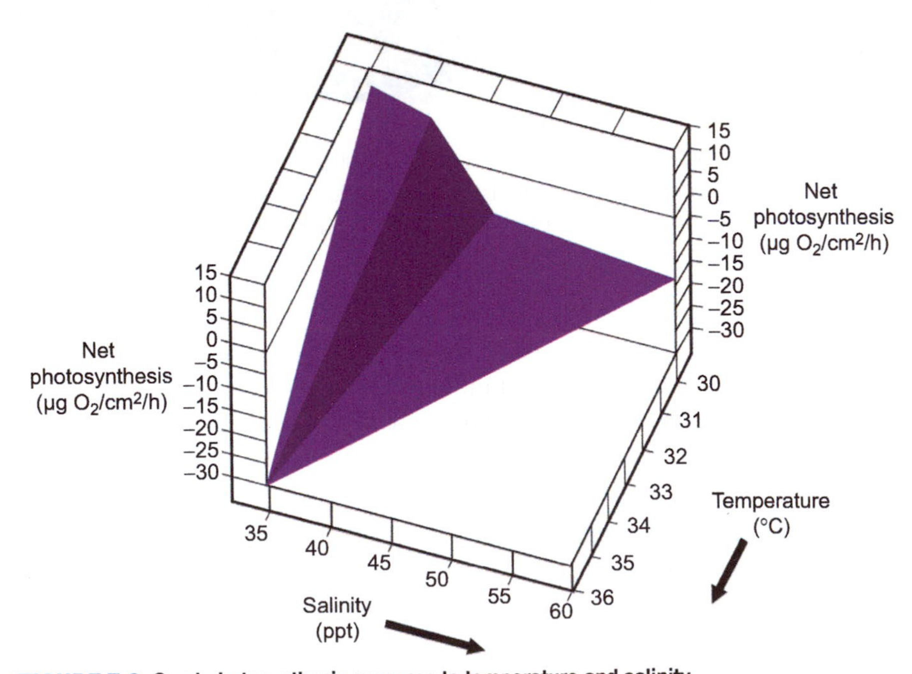

FIGURE 7.9 Coral photosynthesis response to temperature and salinity.
Temperature and salinity exert strong controls over growth and distribution of species. In this example, the photosynthetic response of corals to both temperature and salinity are shown. *Source: Roessig et al. (2004). With kind permission from Springer Business Media.*

high temperature. Turbidity reduces light input and decreases photosynthesis in corals. The pH and calcium carbonate (aragonite) saturation state of seawater determine the rate of secretion and the ability to secrete the coral's skeleton.

Salinity can change on regional scales in geologic time due to alterations in ocean mixing or thermohaline circulation. Salinity varies between regions due to differences in evaporation (higher near the equator) and upwelling. There may also be local variation in salinity due to local upwellings, freshwater inputs, evaporation, and other processes.

Turbidity is primarily a local phenomenon, but it can also vary over geologic timescale due to such factors as change in sea level and continental erosion. High turbidity occurs around river mouths where large sediment loads are released to the

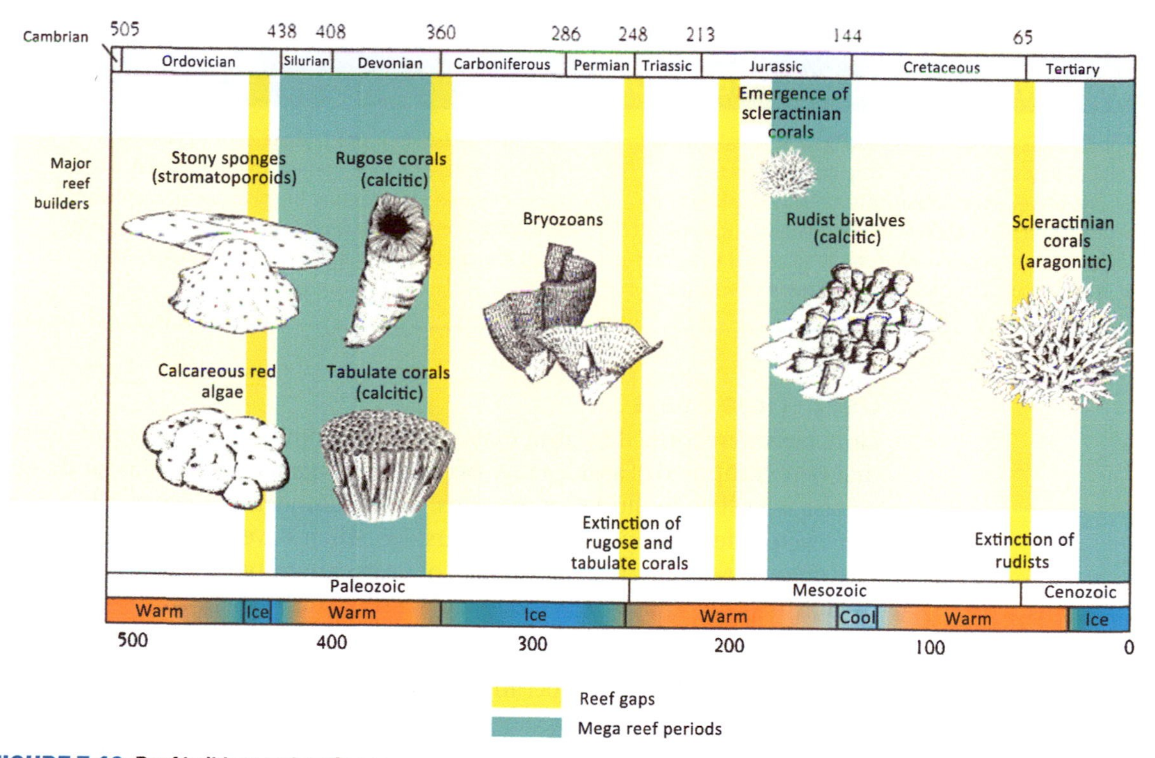

FIGURE 7.10 Reef builders and reef gaps.

Reef builders have varied over the past 500 million years. Dominant reef builders have only been modern scleractinian corals for the past 50 million years. Prior dominant reef builders included stony sponges, calcareous algae, bryozoans and other types of corals (rugose, tabulate). Some of these reef builders, such as calcareous algae are important (but not dominant) in building modern reefs. Others, such as rugose corals have become extinct, opening the way for the emergence of scleractinian corals 220 million years ago. The major coral extinctions often coincide with 'reef gaps', five periods in which there is no evidence of major reef building anywhere in the world. The reef gaps are times for which there is no fossil record of reefs—no fossil reef has ever been dated to these periods. In contrast, 'mega-reef' periods have left extensive fossil reef deposits in many parts of the planet. As discussed in Chapter 9, the reef gaps correspond to major extinction episodes in other species. *Source: Stanley and Fautin (2001).*

sea (Figure 7.10). It may also be high where wave action stirs bottom sediments. More regional but less severe changes in turbidity may result from upwelling or unusual events such as volcanic eruptions. Turbidity reduces the amount of light available for photosynthesis by zooxantheleae, thereby reducing coral growth.

These factors, along with sea surface temperatures and acidity, have helped determine the global distribution of coral reefs in the past. In many periods, coral reefs have existed in the tropics, but at other times coralline sponges and microbe-built reefs have hugged the equator, with coral reefs outside of this band and into the subtropics and with reefs built by bryozoans at higher latitudes. Algal reefs have dominated during some cool climates. Modern scleractinian corals are not the only reef builder, underscoring the unique and changing nature of biological relationships to climate.

WHAT WOULD AN ALGAL REEF LOOK LIKE?

Not all tropical reefs have been built by corals in the past. Millions of years ago, tropical reefs were built by calcite-secreting organisms such as bivalves or by algae. Coral reefs bleached by climate change are often replaced by algae where mortality is high. If this process continues, could the world's coral reefs be replaced by algal reefs? If so, the suite of coral-associated species would change greatly. The few studies of this phenomenon suggest that the biomass of reef fishes might be comparable when algae replace corals but that the species composition is greatly different.

Deep-Time Reefs

Early reefs were formed by algal mats, but by 500 million years ago, reefs existed that were composed of a wide range of calcifying organisms—tabulate and rugate corals, as well as mollusks, crustaceans, and echinoderms (Figure 7.10). These reefs were extensive by 250 million years ago. Their development was relatively continuous over hundreds of millions of years, interrupted only by "reef gaps," periods in which no reefs occur in the fossil record, which correspond to the major mass extinctions that have affected life on earth (yellow bars in Fig. 7.10).

Rugose and tabulate corals became extinct 250 million years ago, and scleractinian corals, more similar to modern corals, emerged. The end of the more ancient coral lineages has been linked to changes in ocean chemistry, particularly alterations in CO_2.

The scleractinian corals proliferated, building extensive reefs in the tropics. These reefs were formed by corals, calcareous algae, and calcifying sponges. Their distribution and abundance peaked approximately 220 million years ago, then declined in extinction episodes coupled with persistent gradual decline. These colonial, reef-building forms existed at the same time as solitary scleractinian corals, which live in colder and darker waters. Modern scleractinian corals include both warm-water, reef-building corals and solitary, cold- and deep-water corals. The abundance and distribution of both forms are highly likely to have been influenced by CO_2 levels in deep time, but the record of past CO_2 levels is not sufficiently resolved to allow mechanisms or timing to be definitively identified.

Implications for the Future of Tropical Reefs

Scleractinian coral reefs have existed for the past 220 million years. Before that, reefs much different from those of today were the rule. This indicates that tropical systems may be variable, with several stable states relative to climate. Which of these states is actually assumed under a particular climate is likely to be dependent on the climate history (warming and cooling) over both short (tens of thousands of years) and long (millions of years) time-frames.

Human interference in such a complex system is likely to have unpredictable results. Addition of large amounts of CO_2 to the atmosphere is changing temperature and acidity of seawater on very short time-scales. How these changes will balance or interact, and the final endpoints of that process, is difficult to assess. Additional human changes will include development of near-shore land and continuing fishing pressure. These factors will influence turbidity and food-chain structure and function.

Based on the past dominance of noncoral reefs under different climatic conditions, it seems possible that the sum of human influences will tip tropical reefs into a new state, much different from current reef systems that are so productive for fisheries and tourism.

Marine impacts of climate change manifest themselves differently in different regions and different taxa. Some regions and taxa are dominated by temperature-driven species substitutions, whereas others experience wholesale shifts in functional type and trophic structure due to ocean chemistry. Managing all aspects of marine response to climate change will be an immense challenge for the future—one that can be informed by these lessons of the past.

FURTHER READING

Pandolfi, J.M., 1999. Response of Pleistocene coral reefs to environmental change over long temporal scales. American Zoologist 39, 113–130.

Roy, K., Jablonski, D., Valentine, J.W., 2001. Climate change, species range limits and body size in marine bivalves. Ecology Letters 4, 366–370.

Schneider, R., 2014. Climate science: Sea levels from ancient seashells. Nature 508 (7497), 465–466.

CHAPTER 8

Past Freshwater Changes

Freshwater systems are among the most diverse and highly impacted natural systems in the world. Freshwaters harbor approximately 30,000 described species, including a staggering 41% of all fish species. The provision of water for human consumption and use constitutes a huge set of ecosystem services provided by rivers and lakes, one that is largely oversubscribed. Extinction rates are therefore very high in freshwater systems as dams, diversions for human uses, and pollution take their toll on these habitats. Extinction rates among freshwater taxa may be three or four times those of their terrestrial counterparts in the near future. For these reasons, the impacts of climate change on freshwater systems have immense consequences for biodiversity and human endeavor alike.

The freshwater realm is relatively tiny. Surface freshwater systems composed of rivers, lakes, streams, and wetlands account for less than 1/10th of 1% of the surface of the Earth. These limited systems are distributed unevenly across the face of the planet, accentuating this scarcity in many regions. As much as one-fourth of total freshwater supply is already used by humans, and pollution renders another significant fraction unusable and unsuitable as natural habitat.

Important biological differences mark freshwater systems. The base of freshwater food chains is often occupied by microorganisms, in contrast to terrestrial systems, in which primary production is most often centered in macroplants. These biological differences affect our understanding of past changes because some freshwater organisms are abundant and well preserved, whereas others preserve relatively poorly compared to terrestrial organisms.

Streams, rivers, and lakes are surface features whose existence and physical characteristics are very vulnerable to climate change. These freshwater features are ephemeral in geologic time: They often last less than 1 million years and seldom more than 10 million years. Their very existence is dependent on the balance of precipitation and evaporation, making climate a fundamental determinant of their longevity. The changing nature of freshwater systems leaves indelible imprints on the biota that inhabit lakes, rivers, and streams.

Climate Change Biology. http://dx.doi.org/10.1016/B978-0-12-420218-4.00008-1

Lakes and rivers are points and lines in a landscape, reducing the dimensions for change in biotic systems. Species ranges may migrate linearly in streams and rivers, or vertically in lakes, but large-scale, continuous range shifts in many directions are not possible in these systems.

Our understanding of past change in freshwater systems is limited by a fossil record dominated by lakes and microorganisms. The highly limited geographic scope of lake records combines with good temporal resolution and specialized taxonomic resolution favoring microorganisms to provide challenging and widely spaced "snapshots" of past conditions. Some aspects of past changes in lake systems are known in surprising detail, whereas others must be mostly inferred, and the record of stream and river changes is exceedingly limited. We then must use present conditions and responses to climate, coupled with this fragmentary fossil record, to reconstruct some understanding of past freshwater responses to climate change.

LAKES AS WINDOWS TO PAST CLIMATE

Among freshwater systems, lakes have a special place because they provide information about climate change as well as about its past biological effects. Lakes are one of the best sources of information about past climates. Lakes preserve abundant fossil microorganisms, many of which may be used to infer past climates. Larger organisms preserve more poorly in lake-bottom conditions, but special events yield treasure troves of macrofossils, known as laggerstatten ("mother lode") (Figure 8.1). Rivers and streams, in contrast, seldom harbor conditions that yield reliably datable fossils because annual depositional strata are rare and because periods of bed cutting destroy incipient fossils.

Lake sediments, especially in long-lived basins, provide important information about past climate. Lake lowstands and highstands may be indicated by geochemical and biological changes in sediments, such as increased precipitation of minerals in drying conditions or increased biological traces in times of high water and high productivity. These indicators are meaningful not only for freshwater systems in which the record is deposited but also for terrestrial and sometimes marine systems in the surrounding area. These implications may be strongest for the watershed of a particular lake, but there may also be implications for a broader region if, for example, lake levels reflect broad-scale regional changes in precipitation.

One of the most noteworthy findings of these lake records is the occurrence of what have been called "mega-droughts." Mega-droughts are droughts lasting from decades to centuries that bring prolonged harsh conditions to large areas. They have happened in all regions of the world, repeatedly, during the past

FIGURE 8.1 Fossils of a laggerstatten.
Laggerstatten are concentrations of well-preserved fossils that provide unique macrofossil evidence of past species assemblages. *Source: Matt Friedman - Specimen is held in the Field Museum, in Chicago.*

2000 years. Lake Naivasha in Kenya, Lake Titicaca in Bolivia, and Mono Lake in California are well-known examples of lake systems that show pronounced lowstands over the past two millennia (Figure 8.2). The lake records are borne out by other lines of evidence in some cases. For instance, lowstands in Moon Lake in the northern Great Plains correlate with evidence of active dune fields in the Great Plains.

On longer timescales, lakes provide insights into regional climatic changes, teleconnections, rapid climate change, and drivers of climate cycles. Isotopic and lake level indices are important in most of these investigations, whereas unique features such as pronounced stratigraphy (varves) are crucial in a few.

Regional climate changes may be quite complicated and variable within the context of global change. Global changes in mean temperature are often inferred from ice cores from Greenland or the Antarctic (Vostok). Regional temperature and especially precipitation changes are discernable from lake records and follow more complex patterns. For example, in a topographically complex area such as the Andes, lake records in one location (Lake Titicaca on the altiplano) may indicate precipitation effects opposite to those tens or hundreds of kilometers away (on the Andean flank). Lakes in North America and Europe provide records of change as glaciers have advanced and retreated. Lake level records from the western United States indicate that glacial ice sheets may have deflected the jet stream southward, resulting in

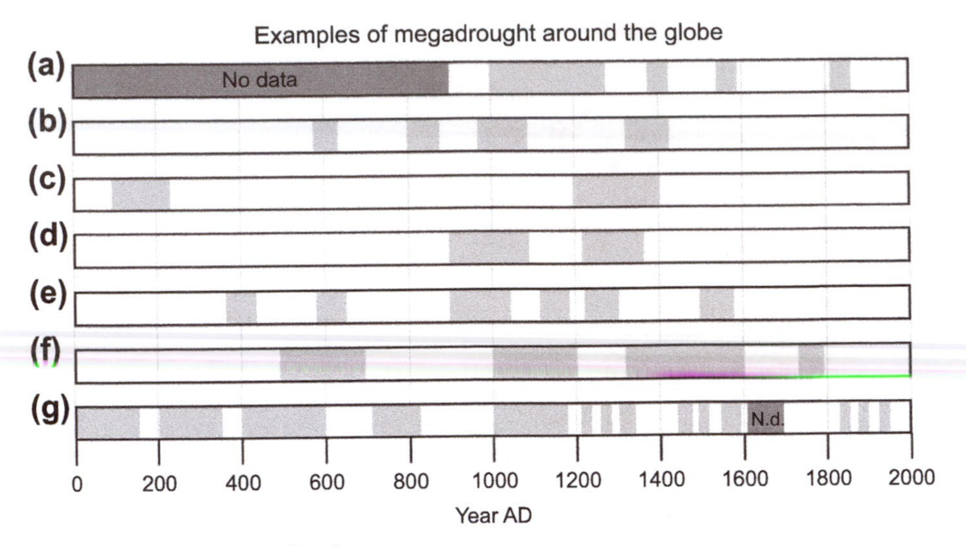

FIGURE 8.2 Mega-droughts over two millennia.
Mega-droughts are droughts that last for more than a century. Numerous mega-droughts have been documented from lake sediments in many regions of the world. Note that time runs from right to left in this figure. (a) Lake Naivasha, Kenya; (b) Lake Punta Laguna, Mexico; (c) Lake Titicaca, Bolivia/Peru; (d) Lake lowstands, Califorina, USA; (e) Tree-rings, S. Nevada, USA; (f) Active dunes, Great plains, USA; (g) Moon Lake, N. Dakota, USA. *Source: Overpeck et al. (2005). Reproduced with permission from Yale University Press.*

increased precipitation and higher lake levels. The jet stream currently intersects North America in southern Canada, but it may have been pushed as far south as California during the Last Glacial Maximum, as reflected in higher lake stands during that time.

Paleo-lake stands can yield information about phenomena such as El Niño events by reflecting precipitation changes in regional patterns typical of the teleconnection. For example, the rapid cooling of the Younger Dryas (12,800–11,500 years ago) affected climates in the North Atlantic, North America, Africa, and possibly other continents. Lake records provide a major portion of this evidence, with changes in level, isotopic indications of cooling, and changes in depositional regimes all indicating rapid cooling and rainfall effects. Records in North America and Europe indicate strong cooling at the time of the Younger Dryas, which is corroborated by Greenland ice core records. In Africa, the Younger Dryas is marked by lake drying, probably representing a mega-drought during the cooling spell. Both lake and ice core records are in agreement that by 11,000 years ago, the cooling spell had reversed in all regions and warming toward current interglacial conditions resumed.

SPOTLIGHT: FLOOD FORECAST

As more energy is retained in the climate system, more extreme events may result. For freshwater systems, large storms and flooding are relevant extremes to consider. Because the current climate is the warmest in the past 100,000 years or more, the paleorecord of flooding is incomplete and difficult to calibrate to climate change. On the other hand, historical records show increasing flooding, and models suggest this trend will continue.

Milly et al. (2002) examined historical floods in large river basins and modeled future flood regimes. They found that in basins larger than 200,000 km^2, the frequency of floods exceeding 100-year levels has risen since 1950. Modeling indicated that this trend is likely to continue and intensify in the future. Among five climate-change scenarios, the rate of great floods (100-year floods) was two to eight times higher than that during the historical period of observations.

For biological systems, such an increase may cause shifts in areas dominated by seasonally flooded forest and change river channel dynamics. Scouring from large floods can change habitats from river edge to river bottom. Furthermore, human responses to flooding, such as levees and dams, may have severe effects on migratory fish and shallow-water habitats.

Source: Milly, P.C.D., Wetherald, R.T., Dunne, K.A., Delworth, T.L., 2002. Increasing risk of great floods in a changing climate. Nature 415, 514–517.

Rapid climate changes, such as the warm reversal at the end of the Younger Dryas, are uniquely revealed by some lake records. Varves are especially important in studies of rapid climate change. Varves are annual layers created by biological or biogeochemical processes. For example, turnover in stratified lakes can create an annual pulse of biological detritus discernable as a varve. These annual markers provide a high-resolution "clock" of climate change. In the case of the Younger Dryas, varved lake records indicated that warming at the end of the cool spell was remarkably rapid—as much as 8–12 °C in a decade or less. Counting the number of varves associated with the transition at the end of the Younger Dryas provides a high-resolution estimate of its duration (Figure 8.3). Lake and ice core records are essential to our emerging understanding that many climate changes have been extremely rapid. Lake cores alone are extensive enough in geographic scope to indicate the regional extent of these rapid changes.

Lake records provide important confirmation of climate changes associated with Milankovitch cycles. Some sediment records support 100,000-year cyclic drying or temperature effects, whereas others in different time windows show stronger association with 23,000-year cycles associated with timing of tilt (precession). For instance, along a fault in Greece, the Ptolemais lake basin shows distinct alternating beds of carbonate (white) and organic lignite (dark), in distinct bands. These bands show a strong 21,000-year cycle, indicating solar forcing as the ultimate cause of the changes in deposition in the lake (Figure 8.4).

Thus, lakes contribute knowledge not only of their own physical and biological changes but also of changes in regional climate and vegetation, interregional

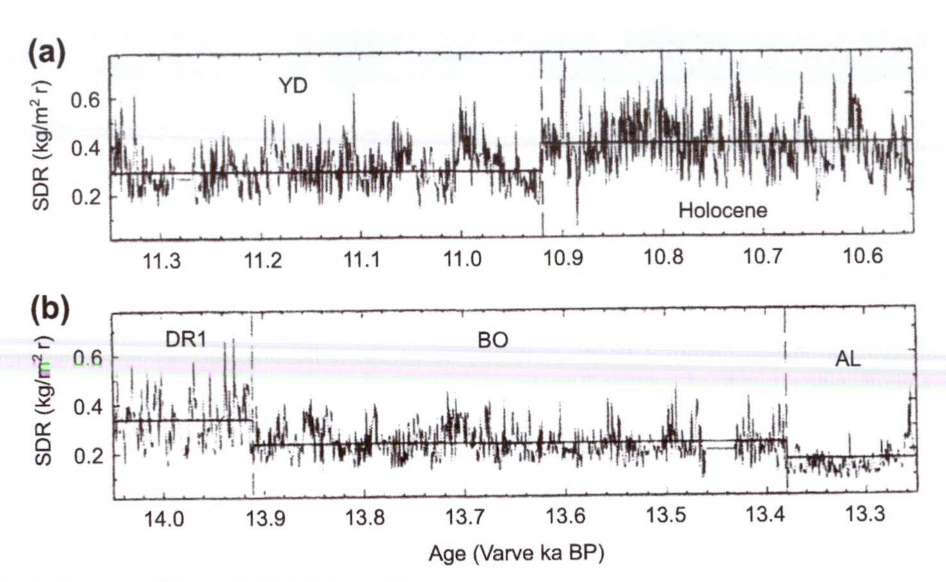

FIGURE 8.3 Varve record of the end of the Younger Dryas.
Varved (annually resolved) lake sediments make possible this highly resolved record of climate change around the Younger Dryas, Bolling, and Allerod events. *Source: Landmann et al. (1996).*

FIGURE 8.4 Milankovitch forcing in lignite deposits.
Interbedding driven by orbital cycles is clearly evident in these lake deposits in northern Greece. *Source: Cohen (2003).*

synchrony, and global magnitude, cycles, and speed of change. These remarkable records provide invaluable insights because their geographic coverage is much more extensive than glacial ice, the other key resource for closely timed records of climate change. At the same time, because the freshwater realm is tiny compared to overall land surface area, the insights from lake records may also seem tantalizingly incomplete. Further research, coring of remote lakes, and improved dating techniques will help resolve many current questions. Many others will remain, requiring new limnological techniques combined with other lines of evidence to piece together the intertwined climatic and biogeographic history of the planet.

TYPES OF FRESHWATER ALTERATION WITH CLIMATE

Freshwater systems are strongly affected by changes in climate in a number of ways. Knowledge of many of these types of changes is possible because they occur within short time-frames, often decades or centuries. Changes that have occurred in historic time must certainly also have occurred in response to longer-term climatic variations on scales of decades, centuries, and millennia. Changes that are important on both these short- and long-term scales include alterations in streamflow and temperature, watershed fragmentation and capture, lake drying and filling, and changes in lake thermal zonation and mixing.

Changes in streamflow result when precipitation and evapotranspiration vary. Climate change is expected to exacerbate both drought periods and intensive storm events. These apparently contradictory trends result because of an enhanced hydrologic cycle coupled with warming. More intense storm events result from heightened evaporation and transpiration in warmer temperatures. Once air masses have lost moisture in major storm events, however, warmer conditions may act to draw water from soils and extend droughts, leading to dryer and more drought-prone continental interiors. Streamflow will obviously increase with increased precipitation and decrease or cease in droughts. Timing of streamflow will also vary with changes in snowmelt and storm intensity. Snowmelt will generally be earlier as temperatures increase, resulting in early season peaks in flow followed by late-season flows that are reduced relative to current conditions. Intensification of storm events may result in increases in annual streamflow but also a compression of the annual flow volume into shorter bursts, resulting in intensified low-flow conditions, increased streambed cutting, and resulting habitat alterations.

Stream temperature follows air temperature, with shallower streams responding more quickly to changes in temperature. Short, shallow streams (first- and second-order streams) dominate stream and river systems, so the proportion of temperature-sensitive waterways is high. Streams and rivers with greater contributions from surface water relative to groundwater will also be more sensitive to changes in air temperature. Because many high-mountain systems contain unique, cold-adapted species such as trout, warming may have major consequences for biological systems. As discussed later, the orientation of streams and rivers plays a major role in determining species' ability to undergo range migration to adjust to rising temperatures.

Habitat fragmentation and climate change impacts may interact in freshwater systems. For instance, deforestation increases solar radiation input on freshwater systems, which can elevate temperatures in streams or small lakes in deforested environments. Increased solar incidents in these environments can also increase UV exposure in freshwater organisms, further increasing their vulnerability to thermal effects.

FIRST- AND SECOND-ORDER STREAMS

The first permanent stream in a river system is known as a first-order stream. When two first-order streams join, the result is a second-order stream, and so on. First- and second-order streams make up the majority of hydrological systems worldwide. For instance, in the United States, 85% of the total riverine system is composed of first- and second-order streams. Many of these are upper elevation freshwater systems that are most at risk from climate change.

Habitat change may be directly driven by climate. Lakes frequently dry out in periods of drought and fill in with sediment as they age. Climate change can accentuate periods of drying and, through reduced or increased vegetation, influence sedimentation rates. Extended periods of very long droughts are recorded for many regions of the world, each of which had profound effects on lake levels, salinities, and habitats. Lakes form and disappear in thousands or millions of years. The process of lake formation involves opening of fissures or lowlands with limited outlet through geologic processes, which is followed by sedimentation that alters depth, temperature, bottom substrates, and other features. Dry periods will speed these filling processes due to lack of vegetative cover, whereas wet periods will slow sedimentation and its attendant physical and biological changes.

Deep lakes have a defined temperature zonation that is affected by warming or cooling. Warmer water is less dense than cold water. Heat inputs to a lake increase the surface temperature, causing the warmer water to "float" above the colder water. Lakes become stratified when mixing due to wind cannot offset the buoyancy created by surface heating. When this happens, the lake develops two distinct zones—the mixed zone or epilimnion and the nonmixed zone or hypolimnion. The division between these two zones is known as the thermocline. The epilimnion is well oxygenated and biologically productive, receiving inputs of sunlight to allow photosynthesis. The hypolimnion is virtually excluded from interactions with the atmosphere during stratification. Because the hypolimnion is usually below the penetration depth of sunlight, primary production is limited and heterotrophic processes dominate. In certain conditions, oxygen may become depleted. In other conditions, sufficient oxygen may remain, and the hypolimnion can act as distinct cold-water habitat for certain organisms, the most obvious being cold-water fish species.

Stratification is strongly related to seasonal temperature fluctuations. Reduced temperature difference between the warm surface layer (epilimnion) and deep water (hypolimnion) reduces the stability of stratification and mixing can occur. Some lakes are fully mixed all the time, others are stratified part of the year until seasonal temperature fluctuations break down stability, whereas others are permanently stratified. Variations in climate that affect seasonality and temperature can alter the stratification regime of any given lake.

Two contrasting latitudinal gradients determine the stratification and stability of lakes. Mean surface water temperatures decrease with latitude, whereas seasonal temperature variations increase. Because the rate of decrease in water density increases at higher temperatures, smaller differences in temperature are required to create stratification in warm tropical waters versus lakes at higher latitudes. Therefore, only small changes in surface temperature in low-latitude lakes will lead to the breakdown of stratification. Colder, high-latitude lakes require a greater temperature gradient for stable stratification to develop, but because of colder temperatures and greater temperature variation at high latitudes, the magnitude of stability is decreased. Thus, lakes of intermediate latitude (30°–40°) are likely to develop the most stable stratification. This band of greatest stability changes in latitude as climate changes.

Some lakes stratify once a year, some stratify in both winter and summer, whereas others remain stratified all year long (Figure 8.5). These differences in stratification regime are climate-linked. Summer stratification results from warming of surface waters. Winter stratification is "upside down," with warmer water at the bottom. This results because water density has the unusual property of peaking at 4 °C: Both ice and near-freezing water are less dense than 4 °C water, so in lakes that freeze, 4 °C water sinks to the bottom, and cooler, almost frozen water and ice are above it. When ice breaks up or summer temperatures decline in fall, the conditions for stratification are removed, the lake becomes unstable, and the thermal layers mix.

Lakes that mix once a year are called *monomictic* and include both lakes that stratify only in winter (cold monomictic) and those that stratify only in summer (warm monomictic). Lakes that stratify in both summer and winter mix in the spring and fall and are called *dimictic*. *Meromictic* lakes are those that never completely mix and are most common in the tropics in lakes with deep waters, such as Lake Tanganyika. *Oligomictic* lakes are found in lowlands and the tropics and mix only irregularly, usually due to storm events or other extreme weather.

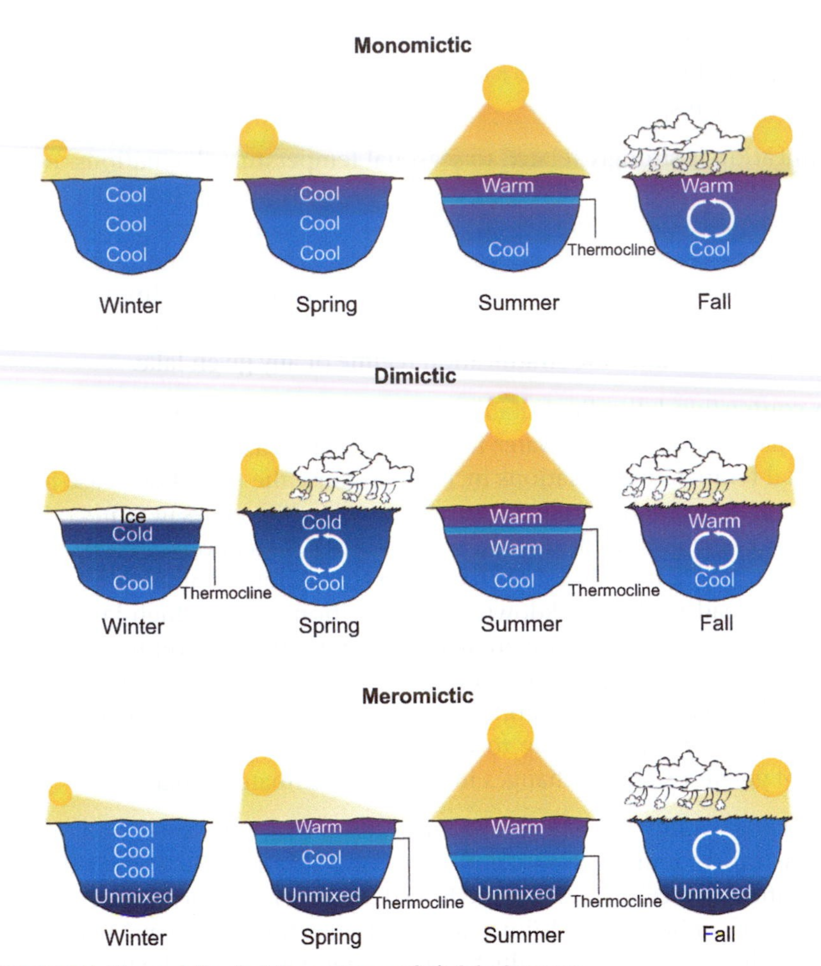

FIGURE 8.5 Monomictic, dimictic, and meromictic lake turnover.
Winds, ice, and temperature conditions determine lake turnover, resulting in lakes that turnover once (monomictic), usually in fall; or twice (dimictic)—once in fall and once in spring. Meromictic lakes mix incompletely, resulting in a deep unmixed layer.

Because freezing and summer warming drive stratification, past changes in climate have altered the mixing regimes of lakes at many times in the past. In periods of warming, cold monomictic lakes have become dimictic, dimictic lakes have become warm monomictic, and warm monomictic lakes have become meromictic. The reverse changes have taken place in periods of cooling. The appearance and extent of ice cover also varies with changes in climate. These shifts in mixing and stratification have major implications for food webs and lake biology, as discussed later in this book. The record of past freshwater systems then reflects multiple responses to climate change that may influence species and ecosystems.

FRESHWATER BIOTAS, HABITATS, AND FOOD CHAINS

Freshwater species are almost all cold-blooded, leaving them with little metabolic adaptability to climate change. Lake food chains are strongly affected by stratification, so past changes in stratification and mixing have had major effects on lake species and their interactions.

The base of freshwater food chains is often invisible. Whereas macroplants (macrophytes) are important in some freshwater settings, such as marine systems, planktonic forms dominate much of freshwater primary productivity. In turn, many freshwater herbivores are microscopic as well. Thus, in contrast to terrestrial systems, the base of the freshwater food chain is typically microscopic (Figure 8.6).

These microscopic photosynthesizers and zooplankton are often preserved in lake sediments, giving a good pint-sized look at past physical, chemical, and biological conditions. Layers of sediment accumulate in lake bottoms, sometimes with annual strata coinciding with spring sediment loads delivered by swollen rivers or by cycles in organic detritus due to seasonal mixing. These sediments may be dated, by radiocarbon techniques or by counting annual strata backward in time, providing a "clock" that can be used to calibrate the time of changes in biotic composition or climate.

Larger fossils of fish and other free-swimming freshwater denizens are rarer. However, unusual preservation events do occur, leaving laggerstatten of macrofossils. These tend to occur irregularly, and they provide snapshots into past

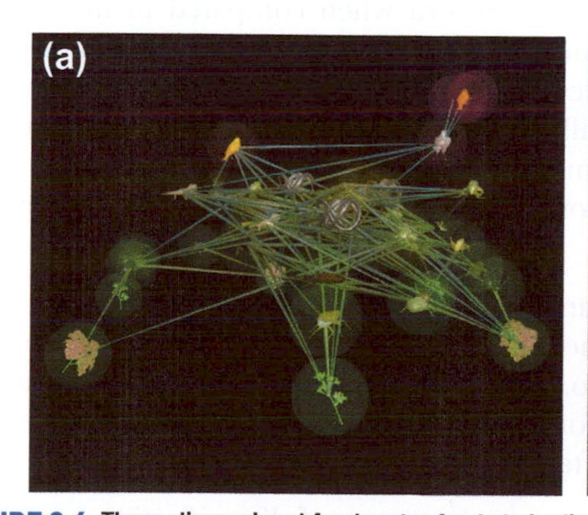

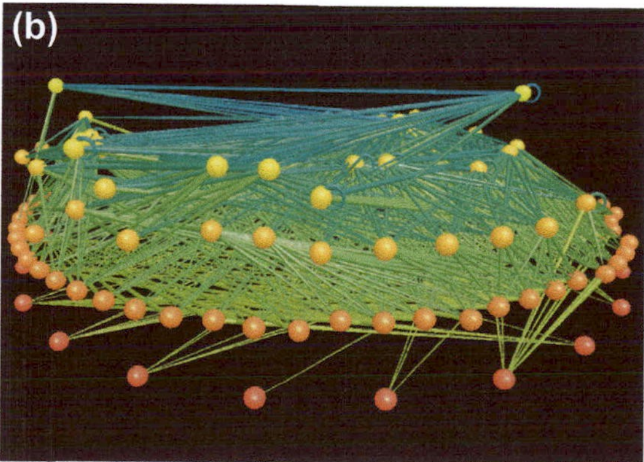

FIGURE 8.6 Three-dimensional freshwater food chain diagrams.
Dr Neo Martinez at the National Center for Ecological Analysis and Synthesis constructs three-dimensional food webs that capture the multiple trophic strategies of freshwater systems. Lower levels of the food chain are interconnected due to multiple midlevel feeders acting on primary producers. Upper levels of the food chain become progressively simpler, often converging on single top carnivores.
Source: Courtesy of Paulo C. Olivas.

biologies for a range of larger organisms. Many of these events can be timed by noting their relationship to annual stratification, but few occur regularly enough to give coherent views of the timing of changes.

In permanently stratified lakes, two largely separate food chains may develop—one above the thermocline in the epilimnion and one below it in the hypolimnion. In lakes that overturn, food chains typical of stratified lakes operate during stratification but interact when mixing occurs. This reassembly of food chains happens once each year in monomictic lakes and twice each year in dimictic lakes.

Changes in climate have affected these food chains with consequences for local extinction, genetic modifications, and diversity. Some of the complexity of these changes is captured in microfossil assemblages, whereas other parts are lost forever due to incomplete macrofossil preservation.

DEEP TIME: PACE OF EVOLUTION AND SPECIES ACCUMULATION

The number of species in freshwater systems has steadily risen during the past half billion years. Major extinction spasms discussed in the next chapter are seen in the freshwater record. At least four of the five major extinction events are evident in records of turnover and accumulation of freshwater families (Figures 8.7 and 8.8) An additional extinction spasm, unique to freshwater, appears at 400 million years ago.

Freshwater systems evolved later than terrestrial and marine biodiversity, accounting for the late start of the record when compared to the 600 million year record available for marine organisms. Phosphate limitation was a major reason for the delayed development of freshwater species. Phosphate in marine systems built up over millions of years of terrestrial erosion, so it ceased to be a limiting factor for marine species more than a billion years ago. Short-lived lakes, in contrast, did not persist long enough to build up phosphate through geologic processes.

Diversification of life in lakes and rivers awaited development of sufficient terrestrial plant life to deliver phosphate from decaying vegetation. At approximately half a billion years ago, vegetation on land was extensive enough to provide the needed nutrients for freshwater life to begin. The reliable freshwater fossil record begins soon after this time.

Once life began in freshwater, successive waves of diversification could take place. Early systems were probably dominated by plankton and fish, perhaps derived from marine systems that became enclosed and progressively less saline. Later, crustacean and mollusk diversity increased. By 350 million years ago, shrimplike organisms had invaded freshwater. Freshwater aquatic insects

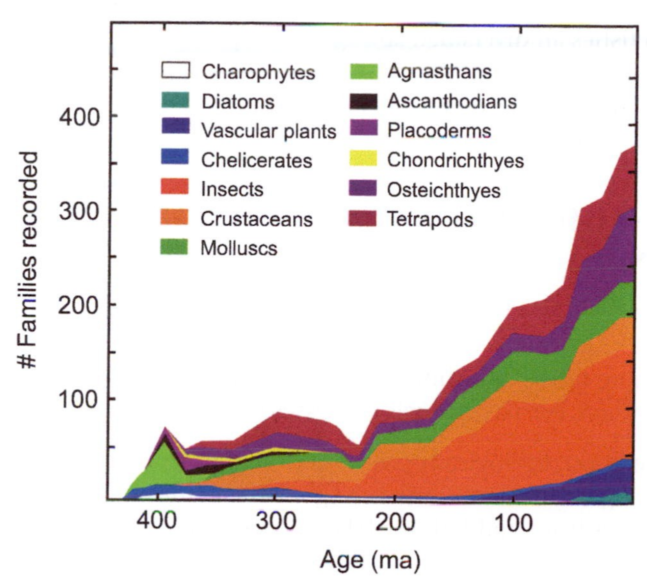

FIGURE 8.7 Freshwater family accumulation in deep time.
Source: Cohen (2003).

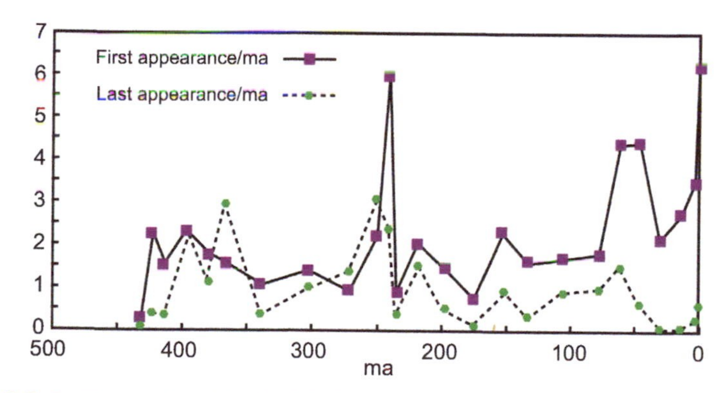

FIGURE 8.8 Freshwater family first and last occurrences.
Source: Cohen (2003).

evolved approximately 300 million years ago, with freshwater lineages derived from terrestrial insects.

Freshwater family diversity has been rising dramatically for the past 200 million years. A steady rise in diversity, with few plateaus and no dips, has taken place since the beginning of the Triassic period. Two geological factors contributed to this rise. First, the breakup of Pangaea gave rise to independent continental lineages, where before there had been pan-global species. Second, rifting associated with the breakup gave rise to deep, long-lived rift lakes in which speciation could flourish. Following Pangaea's breakup, aquatic insects, plants, and fishes all diversified strongly in response to the geological changes and in response to one another. The aquatic record is unique in that in shows no decline at the time of the K–T extinction event that was profound for terrestrial systems (see Chapter 9).

RECENT-TIME (TERTIARY AND PLEISTOCENE) RECORDS OF CHANGE

The processes evident in deep time continue today, but in the past 65 million years the evidence of these processes and the species that result is much more detailed. On these more recent and shorter time-scales, biological and climate processes are more evident, even as geologic processes remain prominent.

One of the most biologically important changes occurred with the strong global cooling event 34 million years ago, which eliminated many present-day tropical species from mid-latitude lakes. This end-Eocene cooling, discussed in Chapter 9 as the cause of major extinctions in terrestrial species, affected freshwater systems as well. Species that spanned the mid-latitudes and tropics prior to 35 million years ago were wiped out in their higher-latitude locations by this cooling. Antarctic ice formation began at this same time, so periodic freezing conditions probably caused this loss in mid-latitude freshwater species. Many of the clades affected are now found only in the tropics. This event therefore defined the divide between tropical and mid-latitude freshwater biology that is evident today.

The ice ages have had a major impact on lake ecology because retreating glaciers have left a large number of lakes and freshwater connections. Changes associated with the ice ages have affected freshwater systems far from the ice sheets (Figure 8.9). Fragmentation of freshwater habitats, resulting in speciation, has resulted from both sea level and temperature changes associated with glacial–interglacial cycles.

Coastal freshwater habitats have been successively fragmented and reunited during the glacial cycles. Sea level has risen repeatedly as ice sheets have melted during the past 2 million years of the Pleistocene. It has fallen when ice sheets have formed. Low-lying areas are flooded in the melting cycles and reexposed as sea level drops. In interglacial high sea level stands, lowland freshwater habitats are separated by barriers of ocean water, promoting allopatric speciation.

Temperature change acts in an analogous manner in highland freshwater systems, with interglacial warm periods resulting in habitat fragmentation. In glacial periods, global temperatures are cooler, resulting in expanded, connected highland cold-water habitats. In interglacials, these cold-water habitats shrink upslope, becoming isolated in mountaintop fragments. During periods of fragmentation, speciation can occur.

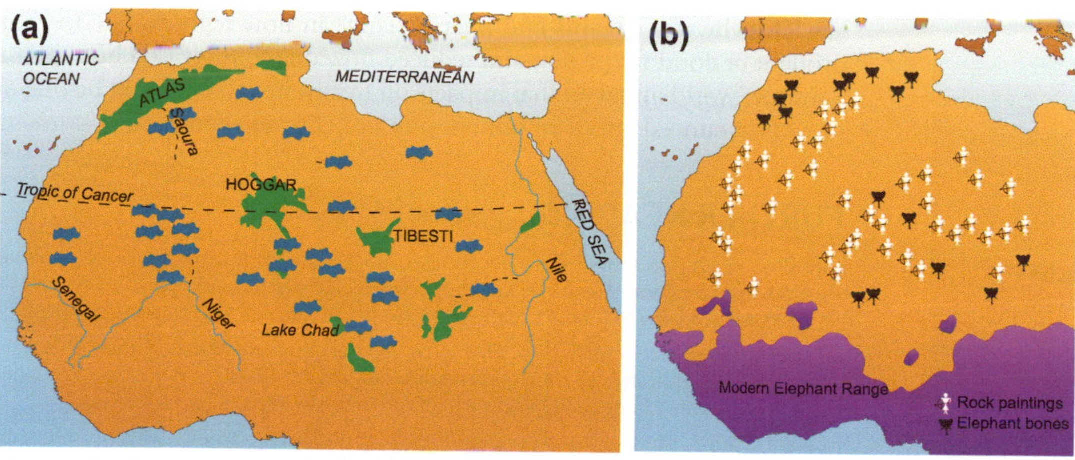

FIGURE 8.9 Fossil lake and human occupation sites in the Sahara.
Fossil lakes (left) in the Sahara indicate that the conditions were much moister there between 2500 and 10,000 years ago. Remains of human occupations and elephants (right) include rock art (symbols), finely worked tools, and signs of hunts. Modern elephant range is shown for comparison. *Source: Wilson et al. (2000).*

Because interglacials are short relative to glacial conditions, periods of fragmentation are short for both coastal and highland freshwater systems. This may limit the speciation associated with these cycles. Nonetheless, genetic patterns confirm that both lowland and highland speciation have occurred in freshwater systems due to glacial–interglacial cycles.

For example, in Brazil, phylogenies (genetic histories) of species of the genus *Characidium* reflect patterns consistent with both coastal (sea level) and highland (temperature) fragmentation. In this genus, the coastal species occurs in isolated patches separated by ocean but shows little genetic difference between patches. This is consistent with formerly continuous populations of a species that has been recently divided by rising sea levels as the Earth warmed into the current interglacial. Conversely, there are several highland species that are derived from a single ancestor but exist in isolated mountaintop patches. The pattern in these species is consistent with repeated fragmentation in interglacials leading to divergence and speciation.

FAST FORWARD

From driving habitat fragmentation in upper-elevation streams to changing the mixing patterns that determine food webs in lakes, climate change has altered freshwater systems and driven evolution in the past. In the future, heavy human use of these systems will dramatically alter freshwater–climate interactions. Among the early effects of future climate change are likely to be changes in montane fish habitats. In the long term, altered lake mixing, altered flow in glacial meltwater streams and rivers, change in flow regimes due to changed drought or flood frequency, and reduced flow due to increased human uses in a warmer world all mean that impacts on freshwater systems will be one of the most pronounced effects of human-induced climate change.

FURTHER READING

Axford, Y., Briner, J.P., Cooke, C.A., Francis, D.R., Michelutti, N., Miller, G.H., et al., 2009. Recent changes in a remote Arctic lake are unique within the past 200,000 years. Proceedings of the National Academy of Sciences of the United States of America 106, 18,443–18,446.

Buckup, P.A.M., Marcelo, R.S., 2005. Phylogeny and distribution of fishes of the *Characidium lauroi* group as indicators of climate change in southeastern Brazil. In: Lovejoy, T.E., Hannah, L. (Eds.), Climate Change and Biodiversity. Yale University Press, New Haven, CT.

Isaaks, D., 2014. Climate Aquatics Blog. http://www.fs.fed.us/rm/boise/AWAE/projects/stream_temp/stream_temperature_climate_aquatics_blog.html.

Schindler, D.W., 2009. Lakes as sentinels and integrators for the effects of climate change on watersheds, airsheds, and landscapes. Limnology and Oceanography 54, 2349–2358.

CHAPTER 9

Extinctions

There have been five major mass extinctions in the history of the Earth, and most are linked to climate in one way or another. Many scientists believe that we are in the midst of the sixth great extinction event, driven by destruction of natural habitat by people. Whether the sixth extinction event will be made worse by human-induced climate change is one of the major questions of climate change biology.

The first modern-day extinctions linked to climate change have now been recorded, so it is relevant to examine the record of the past and ask what role climate has played in past extinction events. Our knowledge of past extinctions is clouded by time, but major extinction events, involving the loss of large percentages of all living creatures, are clearly marked in the fossil record. The extinction of all dinosaurs is one such event, but there have been even greater extinction events deeper in Earth history. This chapter examines the timing of the great extinction events and then discusses potential causes and the role of climate change.

THE FIVE MAJOR MASS EXTINCTIONS

The five major mass extinctions marked in the fossil record occurred at irregular intervals, dating back almost half a billion years. The impact of these events was so profound that they often mark the boundaries of geologic time periods (Figures 9.1 and 9.2). The fossil layers older than the extinction contain plants and animals so different from the fossil layers after the extinction, and these differences are so widespread, that early geologists used them to mark off major periods in the Earth's history. This gives the extinction events some cumbersome names, but nongeologists can recall them by the life-forms that were lost and the time periods in which they occurred (Table 9.1).

The first mass extinction event happened approximately 440 million years ago (mya). At this time, life was concentrated in the seas and dominated by benthic marine organisms. The extinction event wiped out more than 100 families of marine life, including approximately half of all genera. Because it is difficult

Climate Change Biology. http://dx.doi.org/10.1016/B978-0-12-420218-4.00009-3

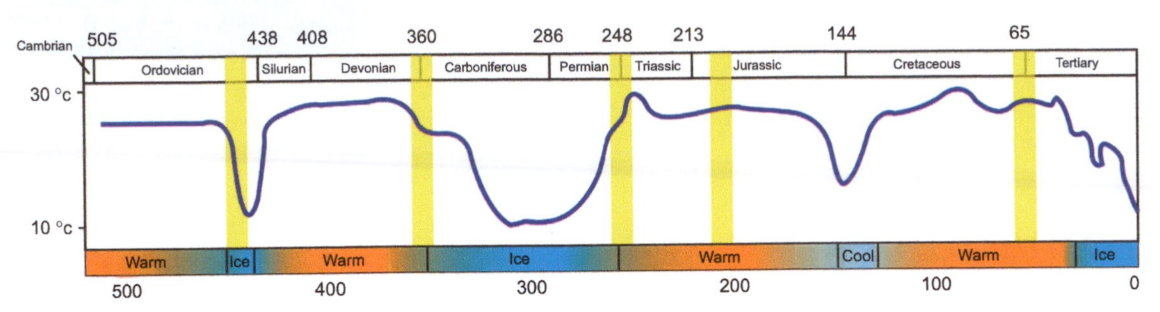

FIGURE 9.1 Timeline of extinction events.

Major extinctions are indicated by yellow bars. Global coral reef gaps accompany each of the 5 major past extinction events. Along with climate change, impacts and volcanic episodes are leading possible causes of major extinction spasms. *Source: Reproduced with permission from Christopher R. Scotese.*

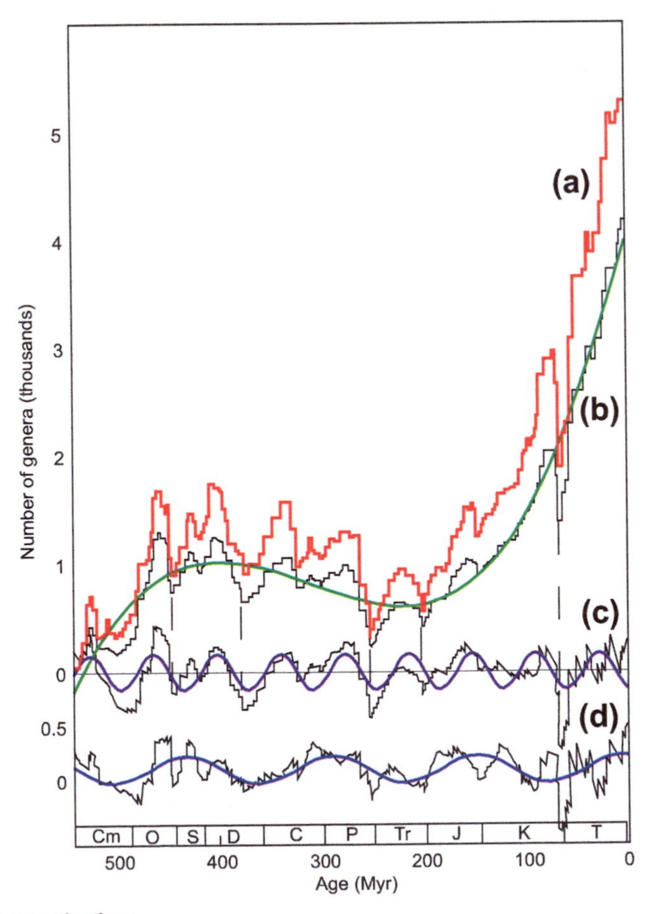

FIGURE 9.2 Major and minor extinctions.

Diversity of genera over 500 million years. (a) The red plot shows the numbers of known marine animal genera versus time. (b) The black plot shows the same data, with single-occurrence and poorly dated genera removed. The trend line (green) is a third-order polynomial fitted to the data. (c) Same as (b), with the trend subtracted and a 62-million year (Myr) sine wave superimposed. (d) The detrended data after subtraction of the 62-Myr cycle and with a 140-Myr sine wave superimposed. Dashed vertical lines indicate the times of the five major extinctions. *Source: Rohde and Muller (2005).*

Table 9.1 The Five Largest Mass Extinction Events

Time Frame (Millions of Years Ago)	Geologic Marker	Biological Impact	Possible Cause
440	Ordovician–Silurian	100 families of marine life extinct, including half of all genera	Rapid cooling
365	End-Devonian	20% of all families lost, mostly marine organisms—perhaps in several episodes	Removal of CO_2 from the atmosphere after the emergence of land plants
250	Permian–Triassic	Extinction of 90% of all species—land and marine	Massive volcanism (Siberian Traps), methane release
200	End-Triassic	Loss of large amphibians	Unclear
65	Cretaceous–Tertiary (K–T)	Extinction of dinosaurs and many marine species	Extraterrestrial impact(s)

to identify species in fossils this old, these genus-level counts are more reliable than species estimates. After the extinction, surviving lineages diversified and overall marine species diversity slowly recovered. This oldest of all known extinction events is called the Ordivician–Silurian event.

Approximately 100 million years later—365 million years ago—another huge extinction event took place. This event was smaller than the Ordovician–Silurian but still catastrophic. Land plants had begun to evolve at this time, and sharks and bony fishes appeared in the oceans. Extinctions were focused on marine organisms, particularly reef-building corals and other marine invertebrates. This event may have occurred in a series of events; its timing is still being worked out. Its geologic name is the end-Devonian event.

The worst mass extinction event occurred 250 million years ago (mya), wiping out 90% of all species (Figure 9.3). This was the third mass extinction, but it was the first to hit land species with major losses—more than half of all land species, mostly plants, were lost. This event marked the transition from the Permian to Triassic geologic periods and is called the Permian–Triassic or end-Permian extinction.

Only 50 million years later, approximately 200 mya, another extinction spasm hit, this one affecting large terrestrial animals as well as plants and marine species. The end-Triassic event paved the way for the evolution of the dinosaurs by wiping out many large animals, mostly amphibians.

The most famous of the mass extinctions occurred 65 million years ago. This fifth and most recent event was dramatic in its abruptness and the thoroughness with which it eliminated all dinosaurs from the face of the planet. Only the clade that evolved into modern birds would survive. Dinosaurs had

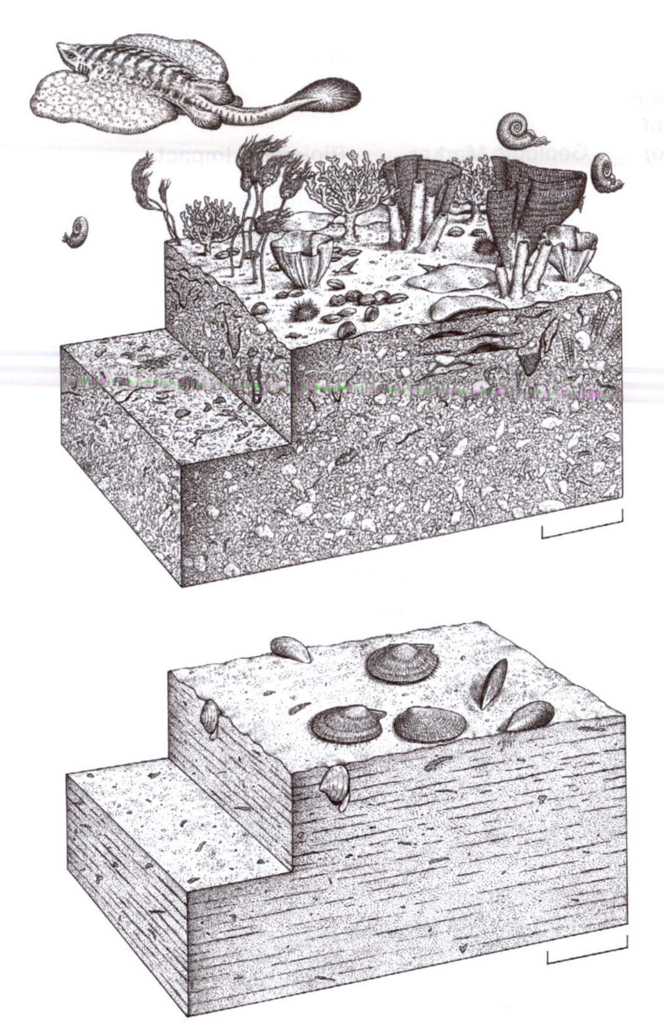

FIGURE 9.3 Marine benthic habitats before and after the Permian–Triassic extinction event. A marine fauna of 100 or more species is reduced to less than six species based on seabed reconstructions off south China. *Source: Benton and Twitchett (2003).*

emerged as the dominant life-form among large animals since the extinction of the dominant large amphibians 135 million years earlier. They in turn were suddenly wiped out in this event, which took place at the Cretaceous–Tertiary boundary (abbreviated K–T following European spelling).

Along with these "big five" mega-extinction events, there have been multiple marked minor extinctions (see Fig. 9.2). The lesser extinction events are much less discussed, but they are still very significant. Many of them are linked to climate change, especially cooling episodes.

CAUSES OF EXTINCTION EVENTS

Climate change has been implicated in all of the mass extinction events, as either a direct or a contributing cause. Where it is a direct cause, rapid temperature change and sea level rise or fall are often cited as driving extinctions. Where climate change is a contributing cause, these same factors drive extinctions but are set in motion by forces outside the climate system, such as asteroid impacts or volcanic eruptions.

Perhaps the best understood of the major extinction events is the most recent—the K–T boundary event that wiped out the dinosaurs 65 mya. This event was caused by the impact of an asteroid in what is now southern Mexico. The huge crater there suggests that an asteroid slammed into Earth just at the time the extinctions began. Debris from the impact shot up into space, some of it generating intense burning radiation as it reentered the atmosphere. Much of the debris remained in the atmosphere for decades or centuries, blocking out sunlight, cooling the planet, and altering living conditions so drastically that no dinosaur survived the change.

The oldest extinction event is the next best understood, and here it is climate change that is the cause. The Ordovician–Silurian extinction occurred 440 mya when the Earth's climate suddenly shifted from greenhouse to icehouse conditions. Gondwanaland passed over the pole, initiating a period of glaciation. Glaciation hugely altered ocean temperatures, wiping out many life-forms. Associated lowering of sea level eliminated many shallow seas and changed near-shore habitats, causing many other extinctions. The rebound of sea levels as glaciation retreated may have led to further loss of species.

The causes of the remaining three extinction events are more controversial. Asteroid impacts, huge volcanic eruptions, and climate change linked to massive methane releases have all been suggested to explain one or more of these "middle three" (end-Devonian, Permian–Triassic, and end-Triassic) extinctions.

The cause of the end-Devonian extinction (365 mya) is still debated by researchers, but it may be linked to the evolution of land plants. Terrestrial plants first flourished about this time, and one result of their emergence would have been the removal of a large amount of CO_2 from the atmosphere. Because CO_2 is the primary greenhouse gas, reduction in CO_2 would have cooled the planet. A major rapid cooling took place in step with the end-Devonian extinction, so climate change triggered by the emergence of land plants may have been the cause.

An undersea crater north of Australia and another off Antarctica suggest that asteroid impacts may have played a role in the end-Permian event 250 mya. Alternatively, massive lava outpourings in Siberia that occurred at this time may have altered the climate enough to cause extinctions. Rapid warming due to massive releases of methane from the ocean floor has also been suggested. A runaway greenhouse warming of 6 °C or more has been implied from a combination of volcanism and methane release. Computer models suggest that rapid warming at this time may have been sufficient to directly cause extinctions on land and to shut down ocean circulation, robbing ocean waters of enough oxygen to cause extinctions in marine organisms. Whether one or all of these events were the cause of the extinction remains controversial. What is clear is that climate is likely to have been involved in driving the greatest extinction in the history of the planet, whatever the primary cause.

Lava outpourings and impact craters are associated with the end-Triassic extinction 200 mya. However, these causes have not been conclusively linked to these extinctions, and more work is required to fully understand the mechanism underlying this event.

CLIMATE AS THE COMMON FACTOR IN MAJOR EXTINCTIONS

Climate change is the factor that turns regional events into global killers. Asteroid impacts and volcanic eruptions occur locally and regionally; they are never global events in themselves. What then turns them into drivers of global species loss? It is the effect these events have on global climate that transmits their impacts more broadly.

Both asteroid impacts and volcanic eruptions spew massive amounts of particulate matter into the atmosphere. These particulates intercept incoming sunlight, blocking solar warming of the Earth. Particulates from even a single volcanic event can have a measurable cooling effect on global climate. For instance, the eruption of Mount Pinatubo in the Philippines cooled global climate for years after the event (see Chapter 2). Multiple volcanic events or an asteroid impact can result in a major global cooling, which can be fatal to animals, especially large organisms or species that cannot thermoregulate and to plants sensitive to freezing. If the cooling is extreme, both plants and animals may have difficulty moving to a suitable climate quickly enough. Drops in sea level as water is locked up in ice can devastate continental shelf habitats and eliminate entire shallow seas, with major repercussions for marine life.

IMPACTS AND CLIMATE

Asteroid impacts have been implicated in several of the big five extinction events, with resulting alterations in climate as a major cause of the extinctions. In the first hours after the Chicxulub impact (and a possibly even larger impact off India at the same time), animal populations throughout the world were probably decimated by rapid warming and infrared radiation. A large asteroid impact ejects material into the outer reaches of the Earth's atmosphere. As these particles reenter the atmosphere, they heat up, just as the surface of a spaceship heats as it reenters the Earth's atmosphere. When there are huge numbers of particles heating on reentry, they generate enough infrared radiation to rapidly heat the surface of the planet.

THE BLOW THAT KILLED THE DINOSAURS

The extinction of the dinosaurs is one of the most heralded events in paleoecology. It was caused by an extraterrestrial impact but has lessons for climate change. The dinosaurs' fate was sealed when a large asteroid struck Earth near the Yucatan Peninsula in what is now Mexico. The ejecta from this impact resulted in intense radiation and large amounts of atmospheric ash. The radiation caused rapid warming across the globe, most intense near the impact site. The ash and debris in the atmosphere blocked sunlight, resulting in cooling after the initial global warming. These changes combined ensured the extinction of the dinosaurs.

Large animals would be killed by this heating, whereas smaller animals able to shelter beneath rocks or trees or in lakes would be more likely to survive. Mortality of plants would be massive. These patterns match postimpact evidence. Large species became extinct, whereas smaller species survived. Water-dwelling species such as turtles and crocodiles survived, whereas strictly terrestrial species vanished. There is evidence of postimpact carbon layers, probably laid down after massive vegetation dieback or fires.

After the initial onslaught of radiative heat, cooling would have set in. Smaller particulates from the impact would have stayed suspended in the atmosphere for months or years, blocking incoming sunlight. Cooler, lower light conditions would have prevailed on the surface, resulting in major changes in plant communities.

Because dinosaurs were cold-blooded, or ectothermic, they relied on incoming sunlight to warm their bodies to operating temperature. Large cold-blooded animals might simply shut down metabolically in a long period of global cooling. This would almost certainly be fatal to enough individuals to wipe out populations and entire species. Blocking of sunlight would result in a collapse of primary production and the death of many plants and undermine the foundation of the food chain. Larger organisms feeding higher on the food chain would not have enough prey to survive. Large herbivores might simply starve. Smaller organisms and omnivores able to feed on a wide range of foods would be more likely to survive the impact and the climatic turmoil that followed.

This model serves well for other asteroid impacts and volcanic events. Particulates spewed into the atmosphere by impacts or by volcanism cool the planet and block light. Changed climate and light conditions drive large-scale changes in vegetation, leading to extinctions in animals. Extinction events triggered by impacts and volcanism are ultimately caused by climatic effects. Climate change, in one form or another, is therefore implicated in all five major extinction episodes.

DOES CLIMATE CHANGE ALWAYS CAUSE EXTINCTION?

Although the largest extinctions are linked to climate change, not all climate change produces extinction. Major shifts in climate have often, but not always, come with large numbers of extinctions. Here, we explore the largest recorded climate shifts of the past billion years, comparing the levels of extinction associated with each.

Many climatic changes are associated with the big five major extinctions and with multiple minor extinction events obvious in the fossil record. Although smaller than the big five, the "minor" extinctions have caused noticeable drops in global biodiversity.

Even smaller extinction events associated with climate change give important clues about how climate can drive extinctions. Such events are especially well studied in the more recent fossil record of the past 100 million years. In this record, we are sometimes able to resolve extinction events associated with climate change that may not register on the global "Richter scale" of major extinctions. Finally, for the past 2 million years, the climate record and fossil record of the ice ages are more highly resolved, bringing us relatively detailed understanding of the events in this period. We therefore divide our discussion of the links between climate change and extinction into three parts: deep time (100 million to 1 billion years ago), the past 100 million years, and the ice ages (2 million years ago to present).

CLIMATE AND EXTINCTIONS IN DEEP TIME

"Snowball Earth" was the first major climate shift of the past billion years. Approximately 900 mya, the Earth plunged into an extreme icehouse period. All continents seem to have been glaciated at this time, and the oceans may have frozen over. Early life consisted of single-celled marine organisms at this time and was probably strongly affected. Our understanding of extinctions is not well established, however, because single-celled organisms do not preserve well in the fossil record. After the snowball Earth period, temperatures rose and multicelled organisms continued to evolve.

OXYGEN ISOTOPES: DETERMINING PAST TEMPERATURE

Oxygen is present in the biosphere in several isotopic forms, one of which holds the key to inferring past temperatures. ^{16}O is the most abundant form (99.8% of all oxygen), whereas ^{18}O is a rare form. The ratio of ^{16}O to ^{18}O indicates past sea surface temperature. Because ^{18}O is heavier than ^{16}O (by two neutrons), it evaporates less readily. Water vapor in the atmosphere and precipitation are therefore enriched in ^{16}O.

As ice sheets form from precipitation, they lock up more ^{16}O than ^{18}O, and the oceans become relatively enriched in ^{18}O. Ocean waters high in ^{18}O therefore represent cooler climates, whereas lower $^{18}O/^{16}O$ ratios indicate warmer, nonglacial climates. Because some marine organisms fix oxygen in their shells, fossil shells can be analyzed for $^{18}O/^{16}O$ ratio to determine past temperatures and past climate shifts.

The climate warmed after the snowball Earth period, leading to an extended greenhouse period that ended when icehouse conditions returned approximately 440 mya (see Fig. 9.1). The return to the icehouse resulted in a well-documented, massive extinction—the Ordivician–Silurian event. Life was concentrated in the seas at this time, and the swing in temperature and change in sea level associated with the formation of glaciers on land was devastating to marine life. Perhaps as many as half of all marine species disappeared at this time.

The climate warmed after the Ordovician–Silurian extinction and then declined back into icehouse conditions beginning approximately 365 mya. The first major cooling event occurred at approximately the time of the end-Devonian extinction spasm. Other factors have also been suggested as causes of the end-Devonian extinction, but rapid cooling is a likely cause in this extinction. As described previously, the emergence of land plants may have sucked CO_2 out of the atmosphere and plunged the Earth into icehouse conditions accompanied by an extinction spasm.

A return to warm greenhouse conditions 250 mya was associated with the end-Permian extinction event. Again, this climate transition is only one of several possible causes of the mass extinction, and it is unclear whether impacts or volcanism might have triggered the climate shift. Computer modeling has reconstructed a rapidly warming climate leading to shutdown of ocean circulation. The crash in ocean circulation robbed deep waters of oxygen according to the models, which would account for massive marine extinctions. The same models suggest raised temperatures over land that might account for terrestrial extinctions. This was the last major climate shift prior to the past 100 million years.

THE PAST 100 MILLION YEARS

Three major climate events can be identified in the past 100 million years (Figure 9.4). The first, the thermal maximum of 55 million years ago (the Paleocene–Eocene thermal maximum (PETM)), was an abrupt warming spike. The second was the onset of major land glaciation in Antarctica, and the third was the onset of the ice ages, which was marked by the beginning of major land glaciation in the Northern Hemisphere 2 mya.

The PETM was accompanied by a significant number of extinctions, including loss of perhaps half of all foraminiferan species in the oceans, and the sudden appearance of several land groups, including primates. The thermal maximum may have been caused by volcanism or release of methane hydrates from the sea floor (Figure 9.5). Whatever the cause, the planet heated rapidly and dramatically. Redwood trees and subtropical algae were found at the poles. Major warming of the oceans robbed deep ocean waters of oxygen and many marine extinctions resulted. As in the Permian–Triassic mega-extinction, the loss of oxygen in deep waters was driven by changes in ocean circulation. High levels of CO_2 in the atmosphere may have also caused changes in ocean pH that affected organisms with calcium carbonate skeletons. The thermal maximum resulted in extinctions on land as well. Many lineages of mammals became extinct at this time, and several mammalian orders appeared suddenly, with no obvious precursors in the fossil record. The net result was the replacement of primitive mammals with the ancestors of modern mammals.

PROBABLE CAUSE

The PETM was caused by a combination of orbital forcings and release of methane gas from sea sediments. The orbital forcings contributed to a generally warm climate at the time (55 mya), approximately 4–6 °C warmer than today's climate. On top of that warmth, eccentricity maxima and minima initiated several spikes in warming that may have caused oceans to warm enough for methane solids trapped in sediments to gasify and enter the atmosphere (Figure 9.5). Biogenic methane is produced in huge quantities by natural metabolic processes in continental shelf marine sediments. At moderate or cool temperatures, it is stored in sediment as methane clathrates, which are solid. At high temperatures, however, methane clathrates are released as gas. Because methane is a potent greenhouse gas, once the planet warmed enough to turn methane clathrates to gas and large amounts were released into the atmosphere, the warming spike of the PETM was inevitable.

Approximately 34 mya, a cooling event associated with the onset of Antarctic glaciation resulted in some loss of species. This event at the end of the Eocene geologic period marked another extinction episode not usually included in the big five but nonetheless important and strongly linked to climate change. Global temperatures plummeted as the Earth experienced a very rapid cooling

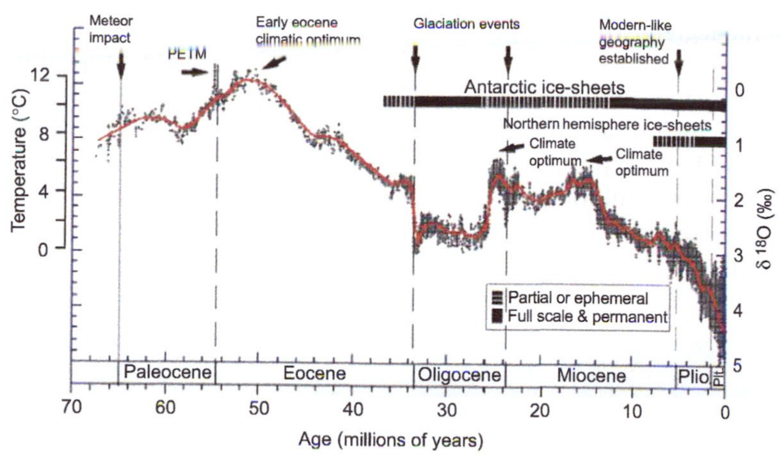

FIGURE 9.4 End-Eocene global cooling.

The initiation of the first permanent Antarctic ice sheets in the past 100 million years coincided with end-Eocene cooling, indicated by an arrow at approximately 34 mya. *Source: Zachos et al 2001. Reproduced with permission from Yale University Press.*

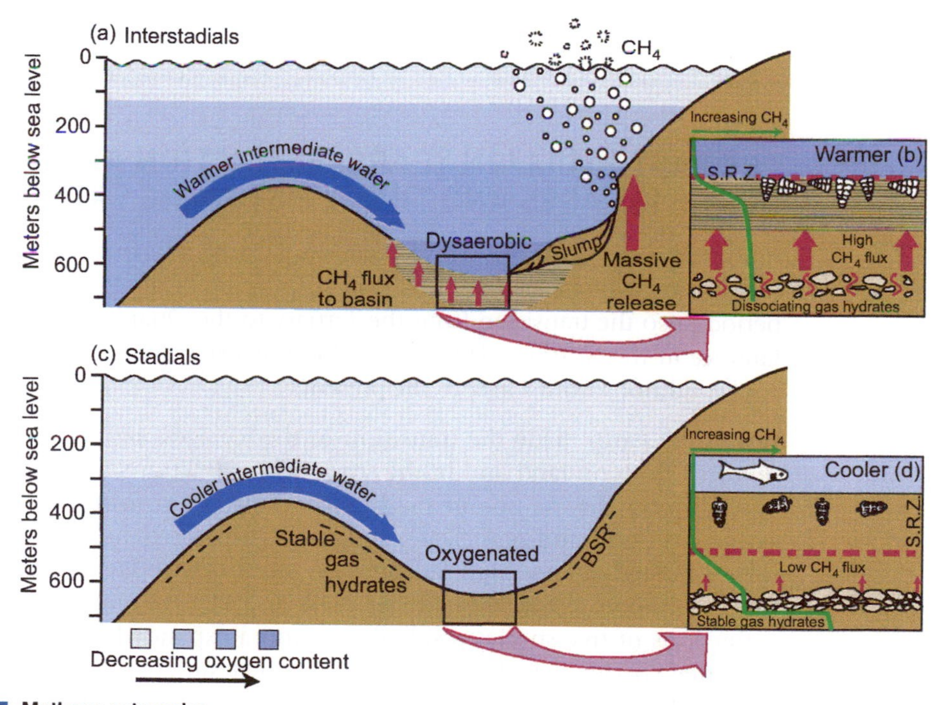

FIGURE 9.5 Methane outgassing.

Methane trapped in sediments as clathrate may be released during periods of warming - inter-glacial periods (interstadials, a), because warm methane hydrates are unstable (b). But in cool, glacial periods (stadials, c), hydrates are stable and no methane is released into the atmosphere (d). Outgassing from sediments in the Santa Barbara channel during interglacial periods is illustrated in this drawing. Release of methane from clathrates has also been implicated in the rapid warming at the Paleocene–Eocene thermal maximum. *Source: Reproduced with permission from AAAS.*

of several degrees. The biological result of this big chill was surprisingly small. Extinctions in the oceans were significant but minor relative to the big five mega-extinctions or the PETM. Terrestrial extinctions were modest as well, with species lost including some camel-like mammals and ancient rodents in North America and loss of primates and rodents in Europe. Viewed over a longer time frame of several million years, these end-Eocene extinctions were the culmination of a series of extinction events that in sum were of major significance (see box).

MID-EOCENE EXTINCTION

From approximately 40–32 mya, a major set of extinctions occurred associated with global cooling. The Earth cooled after the Paleocene, and by the mid-Eocene, ice sheets were beginning to form. Antarctic circumpolar currents and cold bottom waters from the opening of the Greenland–European ocean floor resulted in major climatic changes worldwide. Associated with these changes was a series of extinctions and biological changes. In the oceans, many planktonic foraminifera became extinct. On land, wholesale changes in plant and animal composition were recorded on all continents, with extinctions in major groups of vertebrates. These extinctions may have paved the way for cooler adapted species and species associations, accounting for the relative absence of extinctions during the full ice ages of the past 2 million years.

THE PAST 2 MILLION YEARS: EXTINCTION AT THE DAWN OF THE ICE AGES AND THE PLEISTOCENE EXTINCTIONS

At the onset of the ice ages 2 mya, significant extinctions once again took place in sync with climate change. This is the beginning of the Pleistocene geologic period (also the transition from the Tertiary to the Quaternary). At this time, land ice in the Northern Hemisphere became extensive, leading to primarily cool conditions from 2 mya to the present.

Just as the Earth made the transition to the ice ages, significant extinctions occurred both on land and in the sea. These extinctions were not abundant enough to qualify as one of the big five, but they constitute an important extinction event and one clearly linked to climate change. On land 2 mya, forests existed at very high latitudes, in areas that today are tundra. These forests disappeared as the climate cooled, and local and global multiple extinctions of tree species resulted. As tundra expanded in cooler climates, even tree species that did not become extinct suffered major range reductions. Extinctions occurred in the oceans as well, particularly in corals and mollusks of the Caribbean and the Atlantic. Extinctions in the Caribbean are linked to pronounced cooling of 5 or 6 °C. However, there is little evidence of extinction from the northern Pacific, perhaps because organisms in these oceans were more cold-adapted.

At the other end of the ice ages, there was another major extinction. Perhaps the most controversial of the lesser extinction events are the extinctions that occurred as the planet left the last ice age and the climate warmed to current interglacial temperatures. The geologic name for these extinctions is the end-Pleistocene extinction because they occurred at the end of the Pleistocene (ice ages) and beginning of the Holocene (current warm interval). The cause of these extinctions has been the subject of fierce debate among researchers.

SPOTLIGHT: A MODEST PROPOSAL

Climate change and the arrival of humans caused widespread extinctions in North America at the close of the last glacial period. Now a group of scientists wants to bring back the dearly departed. The species that went extinct included large cats, elephants, and horses. No species currently exist in North America to fill many of the niches left behind. So why not bring in the relatives of these species to restore many of the vacant ecological roles?

This is exactly what Donlan (2005) propose. Introducing lions, cheetahs, and elephants into Iowa may sound far-fetched, but the problems may be more practical than ecological. These authors argue that restoring large mammals would restore ecological functionality to grassland systems that have been without keystone species and ecosystem engineers for millennia.

This idea may not be as ecologically disastrous as it sounds. These large mammals have been extinct for only approximately 13,000 years, so in evolutionary time they have been gone only an instant. They might restore important ecological functions, and they might well draw tourists to areas that currently have limited appeal for tourism. Furthermore, several species threatened with extinction in their native ranges, such as Bactrian camels (*Camelus bactrianus*), could be given safe haven in relatively well-policed areas of the United States and Canada. Although the species are not exactly those that went extinct, they are close relatives. The net effect would be to restore the North American plains to something closer to their condition prior to human arrival, and that would make natural land use much more economically competitive through tourism. Crazy? Maybe—or maybe not.

Source: Donlan, J., 2005. Re-wilding North America. Nature 436, 913–914.

The species that were lost include giant ground sloths in South America and camels, horses, the mammoth, and saber-tooth cats in North America (Figure 9.6). These are large mammals, easily identifiable in the fossil record. There is no question they went extinct in a very short time frame.

There was a pattern to the extinctions. More occurred in North and South America than in Europe, Asia, or Africa. Large mammals were most strongly affected. Where animals existed in large and small forms, it was always the larger relative that disappeared.

These extinctions coincided with rapid warming, but they also were coeval with the arrival of humans. The first humans arrived in North and South America just as the ice age ended. A likely scenario is that climate warming facilitated human arrival, human hunting greatly reduced populations, and loss of habitat due to climate change finished off any survivors (Figure 9.7).

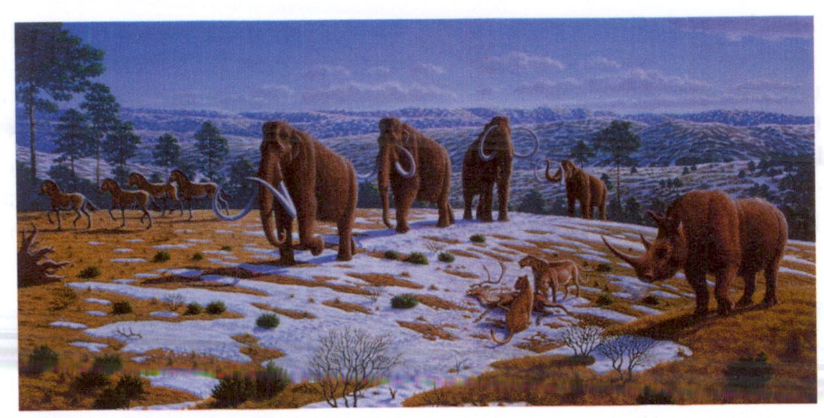

FIGURE 9.6 Species lost at end-Pleistocene.
Some of the dozens of species lost in North and South America at the end of the Pleistocene are illustrated, including saber-tooth cats (*Smilodon*) and wooly mammoth (*Mammuthus primigenius*). *Source: From Wikimedia Commons.*

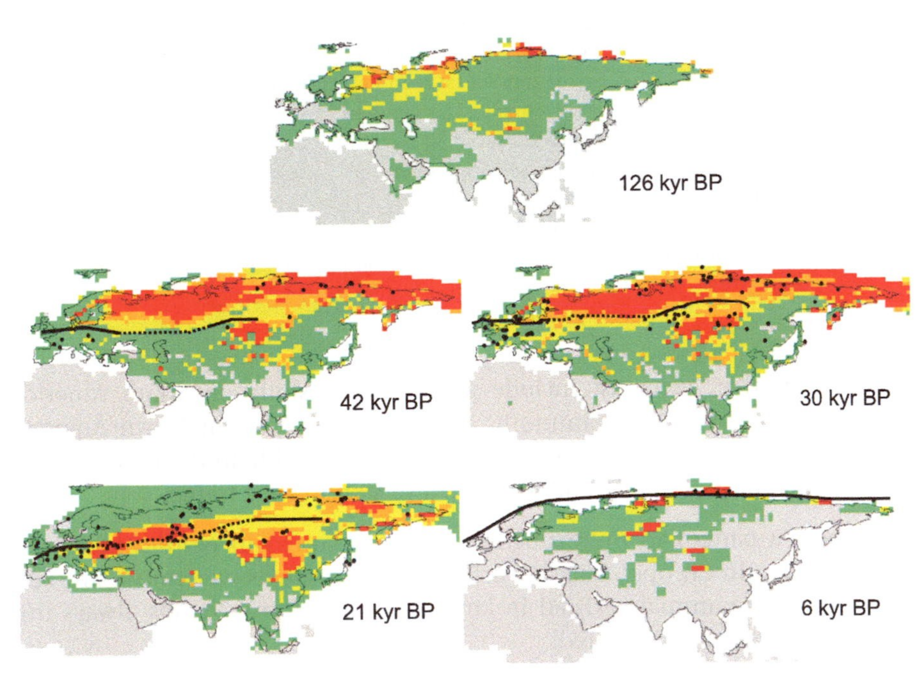

FIGURE 9.7 Declining mammoth range.
Modeled loss of mammoth range due to human expansion (dark line) and climate change (color ramp; red most suitable), from 126,000 years ago (126 kyr BP) to the end of the Pleistocene. *Source: Nogues-Bravo et al. (2008).*

The Pleistocene extinctions are therefore an example of the consequences of combined human impacts and climate change rather than the consequences of climate change alone. They may therefore be an apt analogue to the impacts of anthropogenic climate change on a heavily populated planet this century.

THE MISSING ICE AGE EXTINCTIONS

If the onset of the ice ages and retreat of the last glaciers caused extinctions, what about the other ice ages? The Earth has cycled in and out of ice ages for the past 2 million years, yet there have not been large extinction episodes with the onset and retreat of each ice age. These missing extinctions illustrate that even abrupt, major climate change is not always accompanied by massive extinction.

One answer to the riddle of the missing extinctions may be that the length of the warm periods (interglacials) between the ice ages was relatively short. An average interglacial lasted on the order of only 10,000–20,000 years. This means that for most of the past 2 million years, the Earth has been in an ice age, with only short 10,000- to 20,000-year bursts of warm climate. Species may be able to hold on through unfavorable climate, as long as favorable climate returns within a few thousand years. Once the extinctions at the onset of the ice ages removed species vulnerable to cold, the species that were left may have been relatively cold-adapted. These cold-tolerant species may then have been able to persist through the relatively brief interglacials.

PATTERNS IN THE LOSSES

Marine extinctions have tended to occur in times of cooling, especially among cold-blooded creatures in warm waters. Land extinctions in deep time (more than 100 mya) have also occurred in times of cooling or ice ages such as the snowball Earth.

Biodiversity increases over time. After each extinction event, the number of species slowly recovers, a process that takes tens of millions of years. Over time, and interrupted by minor and major extinction events, biodiversity returns to previous levels and slowly increases. The long-term trend is a marked increase in the number of species on Earth. The extinction events slow the process of species accumulation—but only temporarily.

Some species seem to be more vulnerable than others to extinction. The species in the fossil record may be divided into two types—those that are short-lived and those that persist in the record for longer periods of time. The short-lived species are seen for tens of millions of years in the fossil record, which is "short" on geologic timescales. Long-lived species are seen in the fossil record for 50 million years or more. It is the short-lived species that are affected the most by the major and minor extinction events. The more persistent species show little impact from the extinction events, which is probably why they persist longer in the fossil record. Even the long-lived species show an impact at the Permian–Triassic extinction 250 mya, showing that this event stands out above all the rest as particularly severe.

FURTHER READING

Bond, D.P., & Wignall, P.B. 2014. Large igneous provinces and mass extinctions: An update. Geological Society of America Special Papers, *505*, SPE505-02.

Gingerich, P.D., 2006. Environment and evolution through the Paleocene–Eocene thermal maximum. Trends in Ecology and Evolution 21, 246–253.

Nogués-Bravo, D., Rodríguez, J., Hortal, J., Batra, P., & Araújo, M.B. 2008. Climate change, humans, and the extinction of the woolly mammoth. PLoS biology *6* (4), e79.

Rohde, R.A., Muller, R.A., 2005. Cycles in fossil diversity. Nature 434, 208–210.

Veron, J.E.N., 2008. Mass extinctions and ocean acidification: biological constraints on geological dilemmas. Coral Reefs 27, 459–472.

Looking to the Future

CHAPTER 10

Insights from Experimentation

Theory and experimentation play important roles in estimating the biological effects of climate change, especially the direct effects of CO_2 on plants and ecosystems. Theory suggests possible direct effects of increasing atmospheric CO_2 concentrations that have been increasingly tested in laboratory and field settings. Experimentation informs understanding of complex processes during warming and of the interactions of warming, changes in precipitation, and direct influences of CO_2. This chapter briefly reviews underlying theoretical considerations and then explores both laboratory and field experiments that are shedding light on our changing world.

THEORY

Physical and physiological processes are temperature dependent, so warming is expected to have direct effects on these processes. Physical processes may be critical to plant or animal growth. Some expectations from theory would suggest that plant growth will be enhanced by warming and increased CO_2 levels, whereas other processes may in theory be expected to retard plant growth and decrease species survival.

Many of the expectations of theory based on single-factor effects have been shown to be complicated by the multifactor realities in the complex settings of nature in the wild. However, understanding underlying theory is important because many models are constructed from this theory and because multiple influences, most of them described in theory, result in the responses that have been observed in the real world.

Warming

Physiological processes limited by temperature include photosynthesis, respiration, and processes governed by protein or enzymatic action. Because they determine the quantity and quality (nutrient content) available for the entire food chain, plant physiological processes are particularly important in determining the ecosystem impacts of climate change.

Climate Change Biology. http://dx.doi.org/10.1016/B978-0-12-420218-4.00010-X

In general, plant physiological processes have a Q_{10} of approximately 2, which means that rates double with a 10 °C increase in temperature. Thus, for projected warming this century of 2–6 °C, physiological processes can be expected to increase between 20% and 60%. This may, for example, lead to increased plant growth and higher biomass production in warmer climates. However, where photosynthesis is limited by other factors, it may not be able to keep pace with respiration that is accelerating because of climate change, leading to plant stress.

Photosynthesis and respiration increase with increasing concentrations of CO_2. Rising atmospheric CO_2 levels therefore exert both a direct effect on plant physiology by increasing the concentration of a key chemical in photosynthetic and respiratory reactions and an indirect effect by raising temperatures and therefore the speed of these reactions.

Soil mineralization is perhaps the most important physical process for plant growth affected by climate change. Plant growth and biomass production are limited in many settings by nutrient availability. One key source of mineral nutrients is from weathering and decomposition of soils. As temperatures rise, the physical process of soil mineralization increases. This effect may be particularly important in the Arctic, where low temperatures limit soil mineralization.

Effects of Elevated CO_2

There are three photosynthetic pathways, each of which is affected differently by elevated CO_2. The C_3 pathway is the most common and is found in most plants. The C_4 pathway is found in many tropical grass species and is of major importance in subtropical and tropical ecosystems. Crassulacean acid metabolism (CAM) is a specialized pathway that stores photosynthetic products at night, allowing stomata to remain closed during the day. CAM is found mostly in desert succulents, in which closure of stomata during the day offers huge advantages in water conservation.

In C_3 plants, photosynthesis is catalyzed by the enzyme rubisco. Rubisco is affected directly by changes in CO_2 because it is subsaturated at current atmospheric CO_2 levels (280–380 ppm). Increases in atmospheric CO_2 concentration may therefore increase saturation of rubisco, enhancing photosynthetic rates.

At higher temperatures, the affinity of rubisco for CO_2 increases, independent of Q_{10}, further enhancing photosynthetic rate. The combined action of elevated CO_2 and temperature increase expected in a human-created greenhouse world therefore leads to a theoretical expectation of substantially elevated photosynthetic rates.

Enhanced photosynthesis helps plants conserve water. CO_2 entry into the leaf is controlled by stomata, which are pores in the leaf guarded and controlled by twin cells (Figure 10.1). Stomata open to allow CO_2 for photosynthesis into

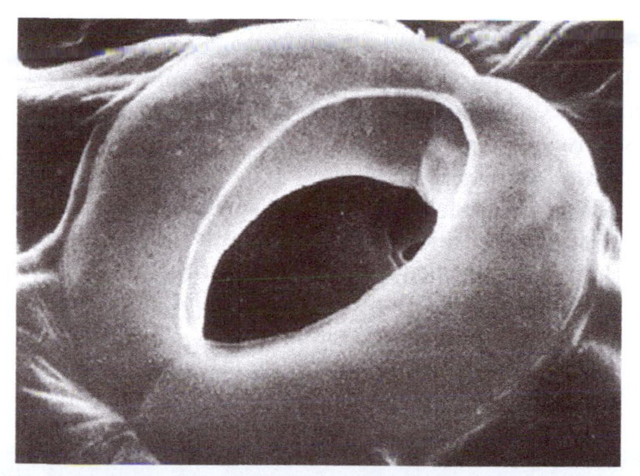

FIGURE 10.1 **Photomicrograph of a stomata on the underside of a leaf.**
Source: Dan Hungerford.

the leaf, but water vapor escapes the leaf through this same path through transpiration. If CO_2 concentrations rise, stomata can be open shorter periods of time to admit the same amount of CO_2, which reduces water loss from the leaf. However, prolonged stomatal closure can lead to overheating in the leaf, so regulation of stomatal opening and closing is a complex physiological balancing act. A second major theoretical prediction is therefore that elevated atmospheric CO_2 will reduce transpiration, help plants maintain a positive water balance, and allow them to better withstand drought or daily water stress.

C_4 plants are thought to have evolved from C_3 plants in response to improved photosynthetic efficiency and reduced water loss. The C_4 pathway uses PEP-carboxylase in primary carboxylation. This is a more efficient pathway, which means that the stomata of C_4 plants can be open less, and lose less water, than C_3 plants for a given level of photosynthesis. This gives C_4 plants an advantage in settings in which surviving water stress is important, such as tropical semiarid grasslands.

Because elevated CO_2 leads to enhanced water efficiency in C_3 plants, an important theoretical expectation is that C_4 plants will lose some of their competitive advantage under climate change. Because photosynthesis underlies production of biomass, which is in turn linked to a plant's ability to compete with other plants, changes in plant dominance might be expected owing to elevated atmospheric CO_2 in settings in which C_4 plants currently outcompete C_3 plants.

A final theoretical expectation of elevated temperature and CO_2 is that plants will do better in low-light situations. At 25 °C, a doubling of CO_2 causes the light compensation point to decrease by 40%. The light compensation point is the light intensity at which net photosynthesis becomes zero (the point at which photosynthesis exactly matches respiration).

Shading and UV Light

Shading may result from climate change as the hydrologic cycle intensifies, leading to increased cloud cover. Clouds intercept and scatter light, leading to complex responses in vegetation. Some forests photosynthesize more under light cloud cover. Cloud cover is generally parameterized at a sub-grid cell level in general circulation models (GCMs), so physical modeling can shed little light on the type of cloud cover changes that may be seen in the future. GCMs do not resolve the biologically relevant differences in cloud type. Under heavy shading, plants often respond with increased leaf internodes, which may be interpreted as an attempt to escape shading from nearby plants. It is unclear whether a similar effect might result from heavy cloud shading.

TIME AND THE BIOSPHERE

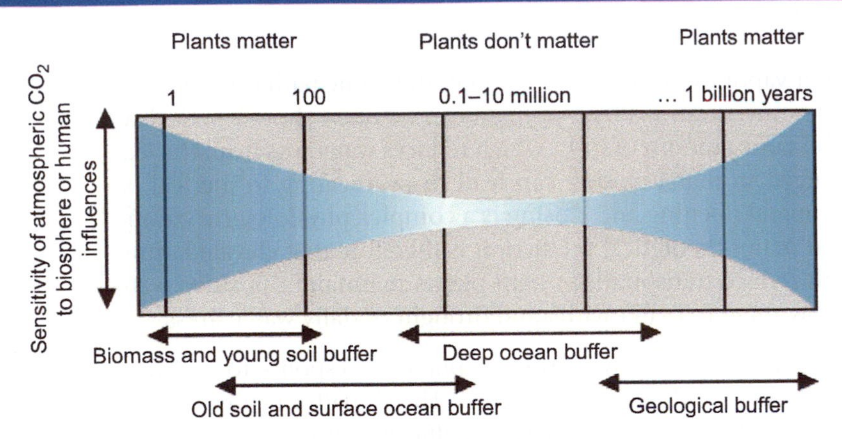

Relevance of Living Plants to Atmospheric CO_2 Concentrations.
The influence of the biosphere on the composition of the atmosphere, carbon cycles, and climate varies across timescales. On timescales of decades, human pollution and forest destruction may cause major increases in atmospheric CO_2 that will not be removed by the oceans and other natural sinks for hundreds of thousands of years. On scales of thousands to millions of years, the ocean will absorb and buffer short-term changes in CO_2. On long timescales of tens of millions of years, photosynthesis generated the oxygen content of the atmosphere. *Source: Korner (2000). Reproduced with permission from the Ecological Society of America.*

The relationship of plants to atmospheric CO_2 is most important on timescales of 1–100 years, as living matter takes up and releases CO_2 in photosynthesis and respiration (the fast carbon cycle—see Chapter 19). On longer timescales, geologic processes such as mixing of the deep oceans are more important to the distribution and disposition of CO_2. However, on very long timescales plants are once again important in determining atmospheric composition because the buildup of undecomposed plant material results in the formation of fossil fuels and a net accumulation of oxygen in the atmosphere.

Increased penetration of UV light due to damage to the ozone layer may complicate the effects of climate change. UV light can damage tissues of plants, mammals, and amphibians. Some plants make use of high-energy UV light in photosynthesis, but this is an exception. Effects of UV penetration may be particularly important at high latitudes in arctic and antarctic environments. Some plants produce compounds to protect themselves from UV radiation. The production of protective compounds or compounds that repair damage may be expected to lower plant productivity and biomass.

Artificial stimulation of plankton blooms is often mentioned as a strategy for sequestering CO_2 from the atmosphere. Adding iron to ocean surface waters can stimulate plankton blooms because iron is the limiting nutrient for many species of phytoplankton. Calculations show large productivity gains in surface waters from fertilization, whereas experimental fertilization in southern oceans has resulted in only a fraction of the calculated effect.

LABORATORY AND GREENHOUSE EXPERIMENTS

Ocean Acidification Experiments

CO_2 dissolved in seawater forms carbonate and bicarbonate, which dissociate to free H^+ ions, lowering ocean pH. This global effect of CO_2 emissions on ocean acidification is of concern because of possible effects on the physiology of marine organisms sensitive to pH changes. Experiments with corals and plankton indicate that the effects of ocean acidification may be serious for some of the most widespread and important marine organisms.

The effect of CO_2 on seawater is easily reproduced in the laboratory by infusing seawater with gaseous CO_2. As in seawater under natural conditions, less than 1% of this experimentally enhanced CO_2 remains in seawater as dissolved gas. The rest dissociates to form carbonate, an acid, which releases hydrogen ions into the seawater. This makes the seawater more acid (technically less basic, because the pH of seawater under natural conditions is around 8.2). It also changes seawater chemistry, because the hydrogen ions released combine with bicarbonate, reducing the amount of bicarbonate in the seawater.

Corals are strongly affected by these changes. The decrease in bicarbonate ion (CO_3^{2-}) reduces the calcium carbonate saturation state of seawater. Experiments in which corals are grown in tanks with elevated CO_2 show coral growth (calcification) to be directly correlated with calcium carbonate saturation state. This is expected, because calcium carbonate is the basic building block of coral skeletons. Higher calcium carbonate saturation favors faster and stronger building of the coral skeleton. Conversely, the lower calcium carbonate saturation due to addition of CO_2 to seawater is expected to decrease coral growth. Experiments confirm this effect. Experiments with

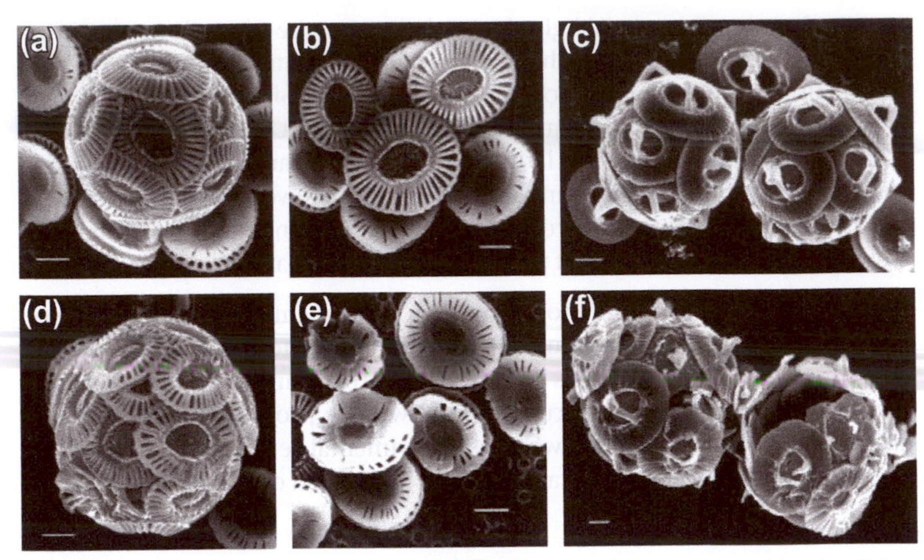

FIGURE 10.2 Plankton response to ocean acidification.
Planktonic algae have calcium carbonate skeletons (coccoliths) that are affected by the pH of seawater. In the upper panels (a-c) are plankton grown in seawater with a pH similar to current. The lower panels represent deformities of coccoliths in plankton grown in seawater with an acidity similar to that expected when atmospheric CO_2 has doubled. *Source: Riebesell et al. (2000). Reproduced with permission from Nature.*

doubled CO_2 levels show a 40% drop in coral calcification (skeleton-building). They also show reduced calcification in coralline algae, which play a major role in reef consolidation. Such reductions in growth in both corals and coralline algae would have major consequences for tropical reefs already hit by bleaching and sea level rise. Experiments suggest that one of the most important sources of ecosystem structure in the tropical oceans is at risk from ocean acidification.

Experiments on plankton demonstrate possible consequences of acidification for the base of pelagic marine food webs. Many types of plankton have calcium carbonate skeletons, which provide important structure for the organisms and are an important source of carbon sinking to the sea floor in the carbon cycle (see Chapter 19). The majority of calcification in marine organisms takes place in these plankton species. The effect of elevated CO_2 on plankton was assessed in laboratory cultures in which plankton was allowed to reproduce for several generations in beakers or tanks (Figure 10.2). These experiments showed reduced calcification in dominant phytoplankton species, for reasons similar to those responsible for reducing coral calcification (lowered carbonate saturation). This has important implications for the base of the marine food web.

FIGURE 10.3 Laboratory and greenhouse experiments.
Diffusers and enclosures may be used to maintain constant elevated CO_2 levels, whereas greenhouses or other warming devices may be used to manipulate temperature. *Source: Courtesy of Scottish Crop Research Institute.*

Experiments on Terrestrial Systems

Laboratory and greenhouse experiments artificially manipulate temperature and CO_2 to simulate elevated atmospheric CO_2 and warming, allowing us to explore many of the expectations from theory. A typical experimental apparatus may include a greenhouse for warming, various levels of irrigation for controlled water input from drier to wetter, and/or infusions of CO_2 (Figure 10.3). CO_2 infusion approaches vary from closed containers that maintain precise CO_2 levels to open-top chambers that allow entry of ambient light and rainfall while maintaining elevated CO_2 through infusers.

Most laboratory and greenhouse experiments are conducted on single species or a limited suite of species. Growth of one or more species in plots is becoming more common, whereas single-plant experiments have declined during the past several decades (Figure 10.4). Studies that capture complex interactions of multiple species are less common.

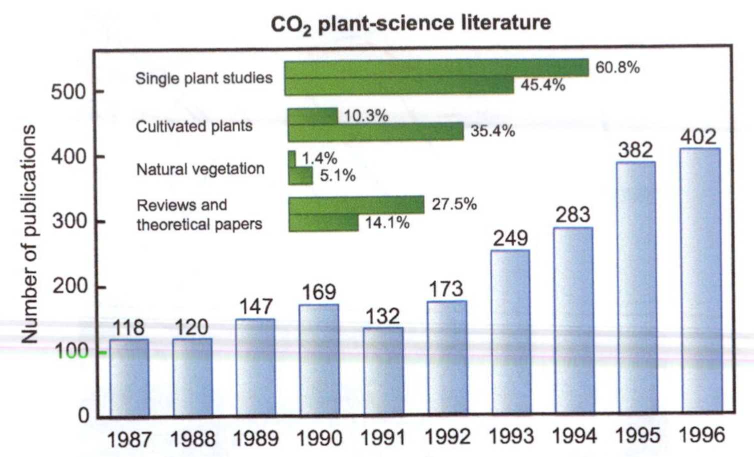

FIGURE 10.4 Decline in single-plant studies, increase in whole-vegetation experiments. The blue bars indicate total number of publications from 1987 to 1996. The green bars indicate the change in percentage publication between the first half of the time period (1987–1991; top bar) and the second half of the time period (1992–1996; bottom bar). *Source: Korner (2000). Reproduced with permission from the Ecological Society of America.*

Early results of single-species laboratory manipulations confirmed theoretical predictions of enhanced photosynthesis under elevated CO_2 and warming. Numerous studies showed increased primary productivity and biomass accumulation under experimental conditions. Other recorded responses were stimulation of growth, reduced numbers of stomata, increased performance under water stress, and changed allocation of aboveground and belowground biomass. Many of these observations were consistent with expectations from photosynthetic and physiologic theory.

Quantification of the CO_2 effect can be summed over the hundreds of published experimental results now available. Biomass enhancement is a measure that captures the net effect of stimulated photosynthesis and plant growth.

C_3 species in laboratory experiments were found to average 45% enhancement of biomass under elevated CO_2 of between 500 and 750 ppm. C_4 plants showed a lower, but still significant, biomass enhancement of 12%. CAM species were intermediate at 23% (Figure 10.5).

However, many of these early studies used simplified nutrient and competitive regimes. Many results linked to increased photosynthesis were recorded when growth conditions were very favorable, making resource constraints other than those on photosynthesis minor. In natural settings, nutrients are often limited. Under these conditions, individuals must compete with many individuals of the same and other species.

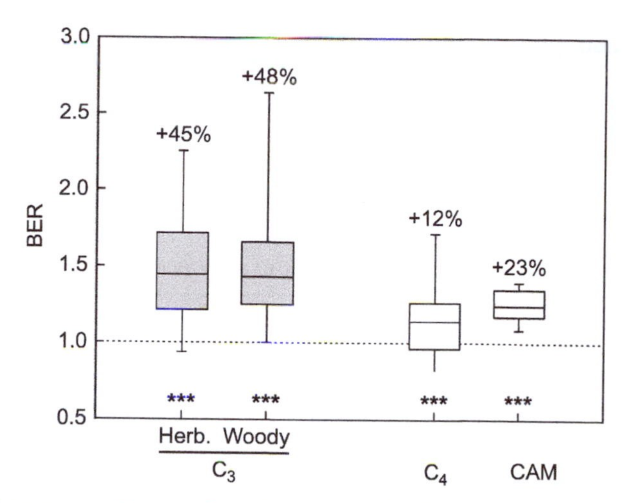

FIGURE 10.5 Increase in biomass for various categories of species (herbaceous and woody C_3 plants, C_4 species, and CAM species).
Graphs show an increase in biomass enhancement ratio, a measure of increase in biomass. Box plots such as these indicate the 5th (bottom horizontal line), 25th (bottom line of box), 50th (midline of box), 75th (top line of box), and 95th (upper horizontal line) percentiles of the distribution. *Source: Poorter and Navas (2003).*

Later studies focused on cultivating plants in multi-individual (Figure 10.6) and, increasingly, multispecies contexts with more realistic nutrient limitations. Under these conditions, much of the growth enhancement seen in single-plant experiments was reduced (Table 10.1).

These later studies led to the recognition of acclimation in plants to altered CO_2—acclimation being a reduction in response from theoretical levels under actual field conditions. A response is still noted even with acclimation, but the response is markedly less than theory or single-plant experiments might indicate.

Quantifying response in an experimental setting that includes acclimation, multiple individuals, and multiple species is more difficult. Biomass enhancement, as one index, was found to track poorly between single-plant and multiplant studies and even more poorly between monoculture and multispecies studies. These results called into question some findings in single-species trials, and they emphasized the value of field experiments (Figures 10.7 and 10.8).

Because of the significant reduction in growth and biomass effects with increasingly natural laboratory and greenhouse experiments, large-scale experiments with intact ecosystems are now a major research focus. These larger experiments are the subject of the next section of this chapter, but, their limitations notwithstanding, there are several robust results from ex situ manipulations worth emphasizing:

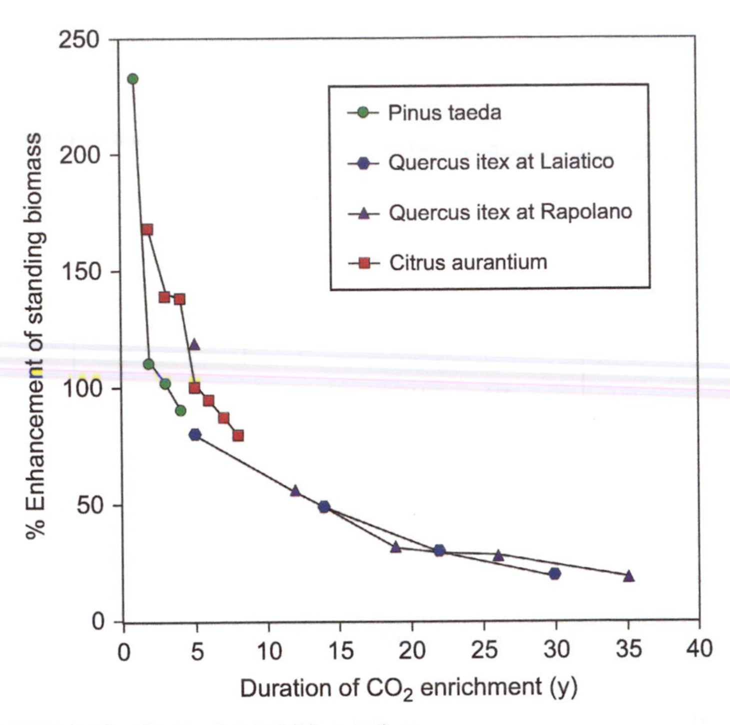

FIGURE 10.6 Acclimation to elevated CO_2 over time.
Trees grown under elevated CO_2 show a strong response initially, but the response declines strongly with time. *Source: Idso (1999).*

Table 10.1 The Effects of Environmental Conditions on Increase in Biomass under Enhanced CO_2[a]

Environmental Stress Factor	BER
None	1.47
Low nutrients	1.25
Low temperature	1.27
High UV-B	1.32
High salinity	1.47
Low water availability	1.51
Low irradiance	1.52
High ozone	2.30

[a]*Average biomass enhancement ratio (BER) of environmentally stressed C_3 plants compared to those of relatively unstressed plants. For each environmental factor, it was calculated what the BER would be if the stress factor reduced growth of the 350 plants by 50% compared with the "optimal" conditions.*
Source: Data from Poorter and Pérez-Soba (2001).

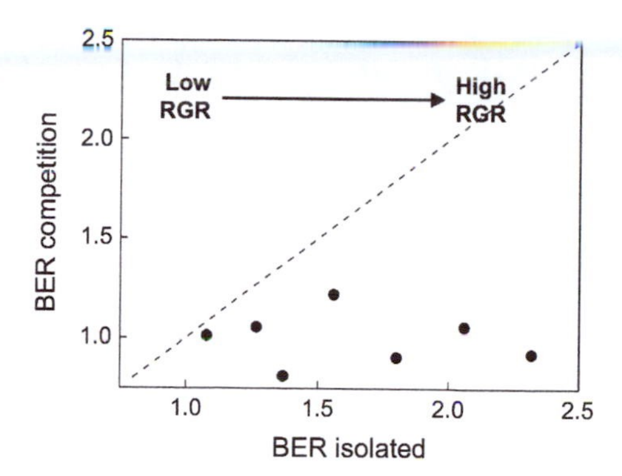

FIGURE 10.7 Biomass enhancement for seven tropical plant species grown in isolation and in a mixed community.

The CO_2 enhancement observed in the isolated trial is not evident in the mixed community. *Source: Poorter and Navas (2003).*

- Plant responses to CO_2 are mostly nonlinear. Experiments using more than two concentrations of elevated CO_2 are rare, but those that have been performed indicate that it is likely that nonlinear relationships between vegetation responses and elevated CO_2 are the rule.
- CO_2 responses acclimate in nutrient-limited and multispecies trials and with time. Initial, single-plant, and non-nutrient-limited responses are likely to be strongest. Actual responses in the field are likely to be lower, perhaps much lower, than responses seen under these idealized conditions.
- The theoretical expectation of improved resource (light and water) use and reduced vulnerability to water stress is supported by experimental evidence. In C_3 plants, and in some C_4 plants, elevated CO_2 results in increased photosynthetic efficiency, reduced gas exchange across stomata, and less water loss. This favors survival in water-limited environments or during seasonal water stress.
- Belowground responses to elevated CO_2 are important. Mycorrhizal response has been demonstrated across a wide range of plant partners and may play a critical role in plant response. Soil microbe mass and composition apparently change little with CO_2 enhancement, but soil microbial systems cycle carbon faster in high CO_2.
- Although theory suggested that C_4 plants should not be responsive to elevated CO_2, increasing evidence suggests that at least some C_4 plants benefit in elevated CO_2. Some C_4 grasses may respond favorably to elevated CO_2, which will help them maintain their competitive advantage over C_3 plants.

MYCORRHIZAE

Mycorrhizae are subterranean fungi that form mutualistic associations with many plants. Mycorrhizae associate with the roots of the plant and improve soil nutrient availability, especially phosphorus, for the plant. In turn, the fungi receive sugars and water from the plant. Mycorrhizae can help promote plant establishment and growth postclearing or in new habitats. Approximately 80% of all plant species form mycorrhizal associations and may grow more poorly when mycorrhizal symbionts are not present. This can be an important consideration during rapid range shifts in response to climate change because soil microorganisms may be less mobile than their plant hosts.

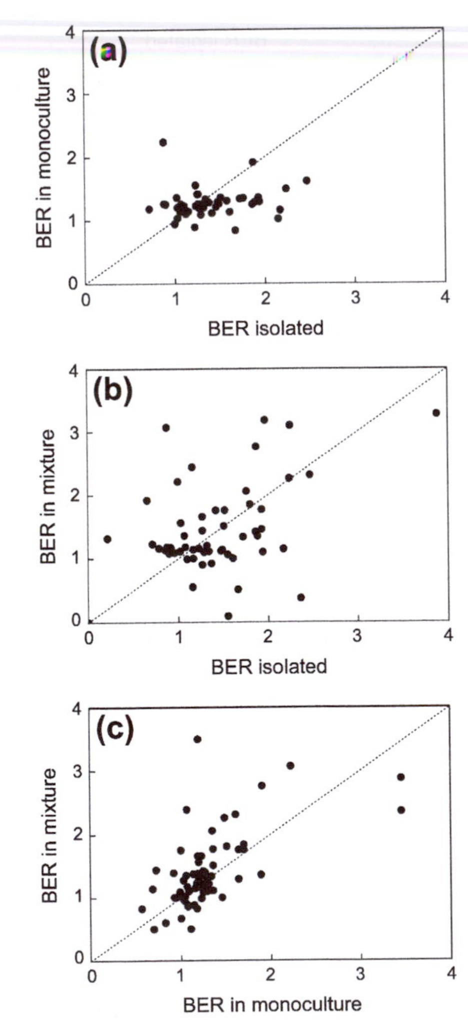

FIGURE 10.8 Effects of experimental conditions on increase in biomass under enhanced CO_2. (a) Biomass enhancement ratio of plants grown in isolation versus plants grown in monoculture ($r=-0.25$, $n=27$, $p>0.2$). (b) Isolated plants versus plants grown in a mixed culture ($r=0.04$, $n=33$, $p>0.8$). (c) Plants grown in monoculture versus plants grown in a mixture of species ($r=0.58$, $n=50$, $p<0.001$). Dotted line 5 1:1. *Source: Poorter and Navas (2003).*

FIELD EXPERIMENTS

A broad range of experiments attempt to gauge the influence of elevated CO_2 and temperature on intact vegetation. The apparatuses used in these field experiments vary from simple open-top warming chambers to vast arrays of diffusers that allow elevated CO_2 to permeate intact ecosystems.

Warming Methods

Open-top chambers, heat lamps, and buried heating coils are all used to elevate air and soil temperature in warming experiments. These and a variety of other warming constructions have been devised by imaginative investigators. The common theme of all field-warming experiments is to generate warmth to simulate anthropogenic climate change while keeping vegetation intact to allow downregulation, competition, and other processes to unfold in as natural a manner as possible. Some of the most widely used methods are now sufficiently common that fairly standard construction is used. Multistudy comparisons have found little significant difference linked to the type of warming used.

Open-top chambers are large four-walled enclosures, usually 2–4 m tall, made of Plexiglas or other plastic in metal frames (Figures 10.9 and 10.10). The clear plastic sides act like a greenhouse, warming the air in the enclosure. A thermocouple placed inside the chamber is typically used to record the elevated

FIGURE 10.9
Active (a) and passive (b) warming experiments. The active warming devices include the use of infrared warming lamps. Passive warming depends on blocking of air circulation or intensification of sunlight to create warmth. Passive warming devices are often simply circles or boxes of glass or clear plastic, which act much like miniature greenhouses but allow multispecies interactions and have minimal impact on received precipitation. *Source: (a) Courtesy of Charles Musil. (b) From the National Center for Ecological Analysis and Synthesis, University of California, Santa Barbara.*

FIGURE 10.10 Transplantation and open-top chamber experiments.
Transplantation preserves plant–plant interactions and soil properties (a). It is usually implemented with the movement of plants embedded in whole soil. Open-top chambers (b) preserve plant and soil relationships over a limited area. *Source: Finnish Forest Research Institute.*

temperature of the chamber, whereas an identical unit outside the chamber records ambient temperature. The open top allows rain to fall freely into the chamber. Closed-top chambers or chambers with partially covered tops may be used to simulate warming and drying. Each chamber encloses an area of vegetation large enough to capture an entire ecosystem, such as a grassland or forest understory. Enclosures of 1 m^2 or larger are common. An open-top chamber in combination with a CO_2 infuser may be used to study the combined effects of warming and elevated CO_2.

A popular approach in montane settings is to transplant vegetation and soil to a lower elevation. This technique is sometimes referred to as a transplantation of soils and vegetation protocol. The warmer temperatures of the lower elevation are assumed to mimic global warming. If future warming is unevenly distributed throughout the day, this assumption may be optimistic. Soil is transplanted with the vegetation to minimize edaphic effects and to lessen the trauma of transplantation. Ideally, identical plots will be dug up, potted, and left at the higher elevation as controls.

If nighttime warming dominates the warming signal, passive nighttime warmers are an appropriate research manipulation. Simply placing a sheet of plastic or metal above the vegetation will trap enough heat to raise nighttime temperatures by 1 or 2 °C or more. Theory suggests that radiative trapping by greenhouse gases will increase nighttime temperatures more than daytime temperatures. Although some early observational evidence seemed to support this theoretical prediction, recent measurements have shown mixed results, leading to questions about the role of nighttime warming.

Soil-warming experiments may be conducted in frozen or unfrozen soils. In unfrozen temperate or tropical soils, the warming affects the Q_{10} of biological processes and may affect mineralization, but it causes little change in the physical structure of soil in the short term. In cold climates, alteration of freeze–thaw cycles can result in important physical changes to soils as well.

Warming in frozen soils is often used to simulate changes in freeze–thaw cycles. Changes in freeze–thaw cycles will occur in both temperate and cold (high-latitude or high-altitude) climates. In both settings, earlier thawing lengthens the growing season.

Simulation of Drying and Increased Precipitation

Warming apparatuses are often coupled with systems for simulating changes in rainfall. Drying is easily simulated with a moving or permanent cover that excludes some or all rain. Moving covers are preferable because it is difficult to design fixed covers that do not affect temperature. With movable designs, a rainfall sensor triggers closure only during rainfall events, greatly reducing temperature artifacts at other times.

SPOTLIGHT: CHANGE—AND CHANGE AGAIN

A series of circles marks California's landscape just north of San Francisco. From the air, they look like small versions of giant circular irrigation systems used in the Midwest, and that is exactly what they are. Since 2002, Blake Suttle of the University of California has been adding artificial rain to grass plots in a natural grass–oak woodland. In some plots, rainfall is added in winter, in others in spring. The results of the spring addition experiments provide a cautionary tale for biologists trying to simulate the effects of climate change using computer models or short-term experiments (Suttle et al., 2007). Suttle's spring addition plots initially showed greater plant diversity than did controls. Forbs proliferated, and a diversity of plants thrived. However, change did not stop there. The forbs fixed nitrogen, improving the soil in ways that favored growth of grasses. Grasses outcompeted the herbaceous layer, eventually displacing most of the species that had made the plots so species rich. After 4 years, the spring addition plots were poorer in plant species than controls, and associated insect diversity was much lower as well. In the space of just a few years, competition had reversed the effects of climate change on diversity, resulting in a completely new community composition. Climate change effects may often be nonlinear and long term, making them difficult to predict in simple experiments or computer simulations.

Source: Suttle, K.B., Thomsen, M.A., Power, M.E., 2007. Species interactions reverse grassland responses to changing climate. Science 315, 640–642.

Increased rainfall is simulated using many types of irrigation apparatuses. Some experiments use water gathered during nearby precipitation events to minimize differences in water quality between treated (irrigated) and control plots. One system pairs drying (rainfall exclusion) and increased precipitation plots, capturing the rainfall from the drying plots and feeding it into an irrigation system for the increased precipitation plots, to mimic the timing and intensity of rainfall events.

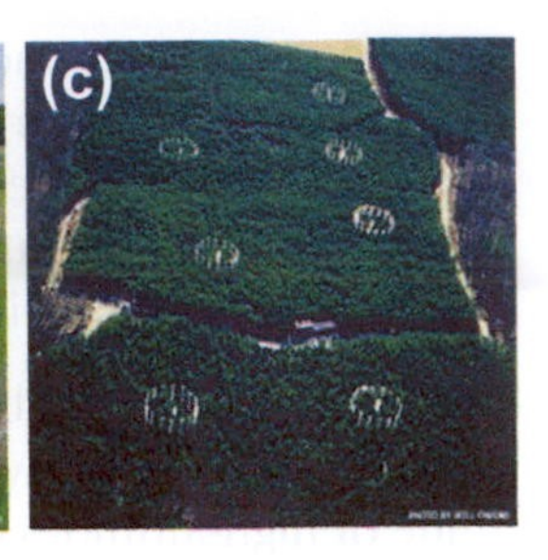

FIGURE 10.11 Free air CO_2 enrichment (FACE) experiments.
FACE experiments use massive diffusers to elevate CO_2 concentrations over a large area (a). Diffusers are often arrayed around a central measurement tower. Effects on grassland (b), forest (c), other vegetation types and crops have been measured with FACE experiments. *Source: (a) Courtesy of Jeffrey S. Pippen. (b) Courtesy of Professor Josef Nösberger, Swiss Face Experiment (ETH Zurich). (c) From Brookhaven National Laboratory.*

Free Air CO_2 Methods

The most ambitious experiments that elevate CO_2 are the free air CO_2 enrichment (FACE) experiments (Figure 10.11). These large manipulations use arrays of CO_2 diffusers to create elevated CO_2 levels in intact, large vegetation blocks. The source of the CO_2 may be natural, such as from CO_2 seeps, or artificially generated commercial CO_2 may be used. Some form of containment at the periphery of the experiment site may be needed to ensure uniform CO_2 levels.

Natural experiments occur in which CO_2 from geologic formations escapes near the surface. Plants growing in the vicinity of such seeps can be monitored for enhanced growth or photosynthesis. Experimental manipulations may be undertaken on the vegetation to assess response to elevated CO_2 under various growing conditions. For example, the vegetation may be experimentally warmed or subjected to experimental harvest.

RESULTS OF WHOLE-VEGETATION EXPERIMENTS

Among the effects of warming on whole-vegetation plots are changes in soil moisture, mineralization, plant productivity, and respiration. In general, soil moisture decreases with warming, whereas mineralization, productivity, and respiration increase. However, there is variability among sites and between vegetation types. Although almost all sites follow the general trend, some sites show little response, and a few sites even exhibit signs opposite to the norm (e.g., decreasing productivity with warming).

In a meta-analysis of multiple studies, warming increased soil respiration by a mean of 20%, organic soil horizon net nitrogen mineralization by 46%, and plant productivity (aboveground) by 19% across studies conducted at 17 sites in various areas of the world (Figure 10.12). Respiration increase was greatest in the first 3 years. This increase was less apparent in later years, but the robustness of this conclusion is limited by the small number of long-term studies.

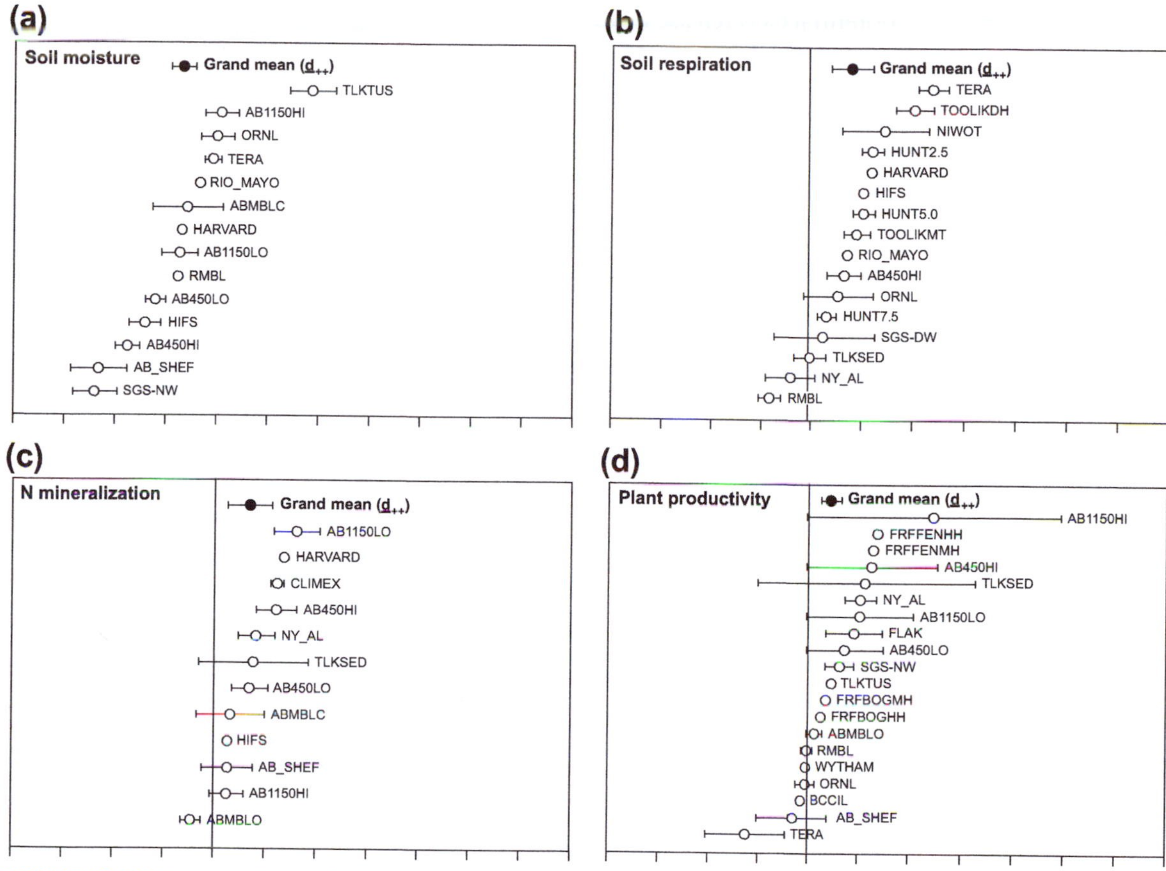

FIGURE 10.12 Response to warming.

The effects of warming on (a) soil moisture, (b) soil respiration, (c) N mineralization, and (d) plant productivity are shown for multiple studies from throughout the world. Measured mean effects at each study site are indicated by open circles; bars indicate 95% confidence intervals. The vertical line indicates no effect. *Source: Rustad (2001).*

The effects on local ecology and the global carbon cycle may be large. For instance, the soil respiration increase, if extrapolated to the world, would represent an increased release of 14–20 Pg of carbon per year, which is several times the amount released annually by human fossil fuel use and other activities. This increased release of CO_2 due to human-induced warming suggests the possibility of a runaway greenhouse effect, in which warming changes soil respiration, leading to more CO_2 release and more warming.

However, most studies measured the soil respiration effect only in the growing season, so yearlong extrapolations may be too high by a factor of 2 or 3. Furthermore, it is possible that the early high change in soil respiration may reflect the oxidation of only the most available soil carbon, in which case the soil effect may greatly attenuate with time. Even accounting for these factors, the potential CO_2 release from soils due to warming may be on a scale comparable to major sectors of human emissions, if only for a short time.

RESULTS OF FIELD CO_2 EXPERIMENTS

In one of the longest field experiments on record, two natural CO_2 seeps in Italy were used to assess the effect of elevated CO_2 on holm oak (*Quercus ilex*). In the experiment, the trunk and all aboveground biomass were removed, but the tree's root system was left intact. New recruits sprouted from the stump in a process known as coppicing. Two populations near the seeps were cut and then censused after coppicing for 30 years. The saplings benefited from the presence of a fully developed root system, demonstrating much more rapid growth than saplings established from seed.

The holm oak near the seeps grew faster than control trees for the first few years of the experiment. This early fast growth may be analogous to the growth response seen in laboratory experiments in which nutrients and water are not limited. The saplings, benefiting from an existing root system, would face few nutrient or moisture constraints in their first years.

Later, the growth response subsided. At 30 years, the CO_2-enriched trees were only 3 years ahead of the control trees growing in ambient CO_2. This corresponded roughly to the advantage in growth increment that the CO_2 trees enjoyed in the early years. Whether the CO_2 growth enhancement observed in the holm oak study would translate to normal, noncoppicing saplings remains unclear.

Experiments in Panama suggest that in the understory of tropical forests, tree recruitment is stimulated by enhanced CO_2. Because water stress may be important in limiting recruitment in tropical forests, enhanced photosynthetic efficiency under higher CO_2 may improve survival during dry periods, leading to improved recruitment. These results may correspond to findings in laboratory settings of strong promotion of growth by enhanced CO_2 in low-light settings near the photosynthetic compensation point.

SPOTLIGHT: GOING THE DISTANCE

Experimentation can shed light on processes such as dispersal and range shifts that are usually the province of paleoecology. Despite being a leading candidate for the mechanism behind extremely rapid range shifts, evidence supporting long-distance dispersal is elusive, deriving mostly from circumstantial evidence of plant colonization or recolonization of islands. A group of researchers at Duke University took a more proactive experimental approach and measured the dispersal distance of seeds released from a forest tower used to study carbon fluxes (Nathan et al., 2002). They then located the marked seeds and recorded the distance they had traveled from the tower. The measured dispersal matched model dispersal, indicating that rare, long-distance events may be one mechanism behind rapid range shifts.

Source: Nathan, R., Katul, G.G., Horn, H.S., Thomas, S.M., Oren, R., Avissar, R., et al., 2002. Mechanisms of long-distance dispersal of seeds by wind. Nature 418, 409–413.

Whole-sward CO_2 enrichment of grasslands has shown a wide range of responses in biomass, from a more than 90% increase to a small negative response. However, when fertilized grasslands are removed, the response range is greatly reduced, with an approximately 10% mean increase in biomass.

Grassland responses may be moderated by water availability. In studies in multiple regions, biomass increased with enhanced CO_2 in wet years but showed no response in dry years. This has been found in settings as different as Switzerland and California and on widely varying soil types. Whether this is a long-term or transient effect is not clear.

Grassland FACE experiments show that there are significant savings in water associated with elevated CO_2. This is to be expected because stomata need to be open less with higher ambient CO_2 concentrations. This in turn results in less time for water vapor from the interior of leaves to be lost through the stomata. The net result is less transpiration and less water requirement by vegetation. This water effect has been noted in both C_3 and C_4 grasses perfused with elevated CO_2. Evapotranspiration may be up to 20% less in elevated CO_2 plots.

Results from open-top chambers and FACE experiments suggest that soil moisture may be higher in elevated CO_2 treatments, presumably as a result of reduced transpiration. In C_4 tallgrass prairie FACE experiments, this increased soil water increased the growing season. The tallgrass prairie growing season is limited by drought in late summer, so the enhanced soil moisture delayed the onset of this limitation and prolonged the growing season.

Two FACE studies in Maryland and Florida are noteworthy because they have run for an unusually long time. Almost two decades of results from these studies indicate that elevated CO_2 may increase photosynthesis in some plants and that this increase may be sustained over years or decades, whereas in other species acclimation is strong and little growth response is observed. Acclimation was seen in these sites, but it did not totally suppress increases in photosynthesis. Thus, the increases seen were less than greenhouse experiments would suggest, but they were still substantial. This is a finding typical across a wide range of studies (Figure 10.13).

The Maryland site is dominated by wetland sedges. The increase in photosynthesis in this wetland translated primarily into increased shoot density. Increased shoot density was recorded early after initiation of the CO_2 treatment and continued throughout the study. The fate of the carbon sequestered by enhanced CO_2 was not entirely accounted for, but part of it contributed to enhanced deposition in sediments. The enhanced soil carbon in sediments increased the number of nematodes and foraminifera, altering ecosystem structure and the soil food web. Soil respiration increased in response to these changes, pointing to possible positive feedback and release of greenhouse gases under elevated CO_2 conditions.

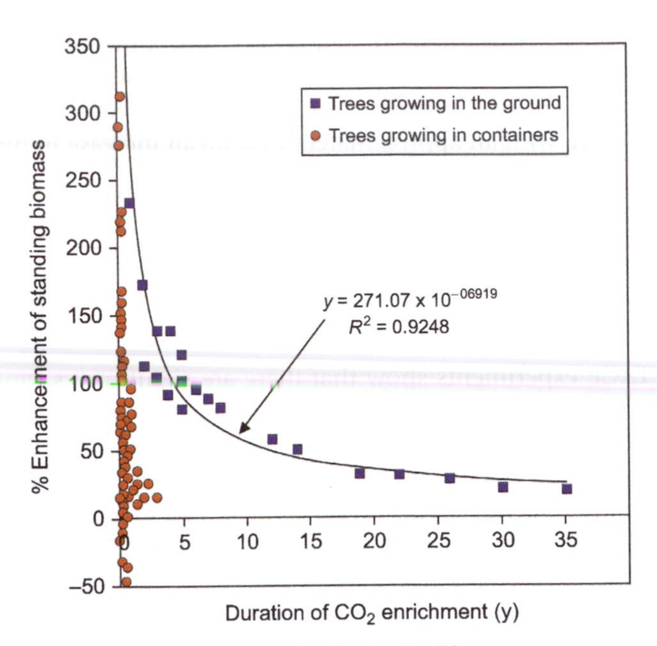

FIGURE 10.13 Acclimation in experimental and natural settings.
Single-plant experiments seldom span time frames long enough to detect acclimation. Whole-ground experiments, usually conducted over longer time frames, clearly show the effect of acclimation. *Source: Idso (1999).*

In the Florida FACE experiment, the vegetation community was scrub oak (*Quercus geminata* and *Quercus myrtifolia*), and CO_2 enrichment resulted in responses different from those in the wetland. Of the two dominant species, *Q. geminata* showed strong acclimation of photosynthesis and no growth response, whereas *Q. myrtifolia* showed no acclimation and growth was stimulated by 80% over controls. Belowground biomass increased first, whereas aboveground biomass increased later in the experiment, lagging behind the belowground effect.

In the Florida, Maryland, and other FACE experiments, a decrease in plant tissue nitrogen has been noted. The ratio of carbohydrate to protein in leaves and other structures due to elevated photosynthesis may account for some of the measured decrease. However, a more important factor is the reduced need for photosynthetic proteins under elevated CO_2. Rubisco may contain up to 25% of the nitrogen content of leaves, so reducing the need for rubisco can cut leaf nitrogen content significantly. Insect grazing decreased in both the Maryland wetland study and the Florida scrub oak study under elevated CO_2 because of reduced nitrogen content of plant tissues.

The decrease in leaf nitrogen is important because it affects the food web. Palatability to insects and decomposition by microbes are affected by tissue quality, of which nitrogen content is an important component. Thus, increased photosynthesis under elevated CO_2 may lead to less need for photosynthetic proteins in leaf tissue, which changes leaf nutrient content, affecting both insect grazers and decomposing microbes in the food web. Changes in these decomposers and primary herbivores will be passed on to other parts of the food web in ways not yet fully understood, leading to changes in the entire ecosystem.

FRESHWATER EXPERIMENTS

Experiments on freshwater fish have shown that temperature spikes lasting minutes or hours can lead to fish mortality. The mechanism of death involves decreased aerobic scope for activity. Aerobic scope is the difference between basal metabolic rate and maximum metabolic rate. When aerobic scope approaches zero, death may result. Basal metabolism increases exponentially with temperature, so it may quickly reach or exceed the maximal metabolic rate, driving aerobic scope to zero. When this happens, the ability of the circulatory system to deliver oxygen to tissues is totally occupied meeting basal (resting) metabolic needs. There is no scope for feeding or predator avoidance, and death may result from starvation or heat stress.

Experiments simulating reduced streamflow due to climate change show effects on the growth of fish. Researchers in California blocked off sections of streams to simulate 20%, 30%, and 40% reductions in flow. They found that lower flow rates meant fewer aquatic insects drifting in the water column. This reduction in food translated to lower growth rates in fish and can ultimately result in reduced fitness or death.

ARCTIC EXPERIMENTS

Temperature changes are already pronounced in high northern latitudes and can have a profound effect on soil structure and vegetation growth and composition. As a result, studies examining the effects of warming and elevated CO_2 on arctic ecosystems are worthy of special note. The results of these studies suggest that the sometimes dramatic changes already being seen in the Arctic only preface larger changes yet to come.

Arctic climate change experiments are classified into five broad categories: warming, CO_2 enhancement, fertilization, shading, and watering. Fertilization experiments are undertaken because rising temperatures are expected to increase mineralization of soils, especially as frozen soils thaw, resulting in major increases in nutrient availability. Shading studies seek to understand the possible impacts of increases in cloud cover projected for the region by many GCMs, whereas watering simulates increases in precipitation.

Increased temperature results in elevated reproduction and physiological responses in arctic plants. Biomass increased in both deciduous and evergreen shrubs, as well as in grasses. In contrast to fertilization experiments, nitrogen concentration was reduced by warming. This response probably results from increased photosynthetic efficiency. Warming increases photosynthetic efficiency even in the absence of elevated CO_2.

Relatively few CO_2 experiments in the Arctic have been performed compared to CO_2 experiments in other biomes or to other climate change experiments in the Arctic. The limited results that are available suggest a substantial effect of CO_2 on physiological process rates and some influence on biomass. However, the number of studies is so small that general conclusions across plant types and across regions are impossible. The degree of photosynthetic stimulation or acclimation in arctic plants remains a largely unanswered question.

Fertilization results in major increases in the production of seeds and bulbils. Such reproductive measures increased by nearly 300% in a number of studies in various areas of the Arctic. Phosphorus had the largest effect in most cases, indicating that it is the limiting nutrient in these settings.

Biomass is dramatically increased by fertilization of arctic grasses, but no other growth form seems to be as strongly affected. Grass biomass increases of up to 15 times have been noted, with no other plant group showing significant responses. Shrubs show increases in leaf nitrogen but not enhanced biomass production.

Little or no significant effect of watering (irrigation) or shading has been noted in arctic experiments. This is in line with the theoretical expectation that soil moisture is temperature limited rather than precipitation limited in the Arctic. This is because there is abundant precipitation as snow, which is retained into the growing season. Soils are often water saturated where they are not frozen, so water availability is not limiting in the region unless or until temperatures change dramatically. Lack of major biomass or physiological response to shading may indicate that temperature is also more important than light availability in arctic systems.

Other factors are important in arctic systems but have received little study. Most GCMs project a lengthening growing season in the Arctic, which should have a major effect on plant growth. Only a limited number of experiments have been initiated to explore this effect. Results to date indicate that advancing snowmelt may have significant effects on end of growing season cover and biomass.

Slumping as soils thaw and other physical effects of thawing may have major impacts in the Arctic because they change soil structure and free nutrients (Figure 10.14). Some of these physical changes have been investigated in freeze–thaw experiments in which time of thaw is artificially manipulated by heating elements. Results of these experiments suggest major effects on plant

FIGURE 10.14 Slumping arctic soils lead to changed vegetation composition.
Source: Kokelj et al. (2009).

growth and composition. Observations in the field bear out these expectations, with major differences in plant communities and increasing shrub biomass associated with soils that have thawed and slumped.

FURTHER READING

LaDeau, S.L., Clark, J.S., 2001. Rising CO_2 levels and the fecundity of forest trees. Science 292, 95–98.

McCarthy, H.R., Oren, R., Johnsen, K.H., Gallet-Budynek, A., Pritchard, S.G., Cook, C.W., et al., 2010. Reassessment of plant carbon dynamics at the Duke free-air CO_2 enrichment site: interactions of atmospheric [CO_2] with nitrogen and water availability over stand development. New Phytologist 185, 514–528.

Seiler, T.J., Rasse, D.P., Li, J.H., Dijkstra, P., Anderson, H.P., Johnson, D.P., et al., 2009. Disturbance, rainfall and contrasting species responses mediated aboveground biomass response to 11 years of CO_2 enrichment in a Florida scrub-oak ecosystem. Global Change Biology 15, 356–367.

Smith, S.D., Charlet, T.N., Zitzer, S.F., Abella, S.R., Vanier, C.H., & Huxman, T.E. 2014. Long–term response of a Mojave Desert winter annual plant community to a whole-ecosystem atmospheric CO_2 manipulation (FACE). Global change biology, 20(3), 879–892.

CHAPTER 11

Modeling Species and Ecosystem Response

Climate change biology has emerged as a field because of the likely impacts of human-driven changes in the atmosphere. The historical record of these changes is very short, and they can be difficult to distinguish from the effects of short-term natural variation. As a result, modeling of the future is a major tool for understanding climate effects on species and ecosystems.

Modeling is a diverse field, encompassing both conceptual models and mathematical models. When we describe how a system works, we are implicitly constructing a model of that system. If the model is rigorous, it can be tested. For instance, when we say that human CO_2 emissions are increasing CO_2 levels in the atmosphere and warming the planet, we have constructed a conceptual model that can be tested. As discussed in previous chapters, this simple verbal, conceptual model has two major testable components—the rise in atmospheric CO_2 concentrations and the resulting warming of the planet. Both of these components of the conceptual model have been tested. Atmospheric CO_2 is rising dramatically, and the mean temperature of the planet is increasing. Thus, the conceptual model has been validated by observations. Similarly, we have a conceptual model that warming will cause species ranges to shift poleward, and this conceptual model has been validated by actual observations of poleward shifts in species ranges.

As we want to understand more about the system, more detailed models may become necessary. What if we want to know how far a species range might shift, or where to place a protected area to ensure the species is

THE FIRST CLIMATE MODEL

The first climate model was a weather forecasting model. In 1922, Lewis Fry Richardson, a British mathematician, developed a system for numerical weather forecasting. Richardson's model divided the surface of the Earth into grid cells and divided the atmosphere into layers of grid cells—exactly the same format of today's weather and climate models. In each cell, equations for temperature, pressure, moisture, and other properties were executed in a series of time steps. Mass and energy were exchanged horizontally and vertically between cells. Richardson's model required hand calculation and failed badly. Decades later, the rise of computers allowed his model to be run mechanically, with increasingly useful results.

Climate Change Biology. http://dx.doi.org/10.1016/B978-0-12-420218-4.00011-1

protected under future climates? In these cases, we may want quantitative answers such as how far or to what geographic location a species' range may shift. These more quantitative answers will often require quantitative, mathematical models.

There are many different types of mathematical models. Almost all are now so complex that they are executed on computers. Early mathematical models of global processes such as planetary cooling or atmospheric chemistry were executed by hand and could take months or years of meticulous paper calculations. Since the advent of modern computers in the second half of the twentieth century, manual calculations to execute models have become obsolete.

Beginning in approximately 1950, computing power was sufficient to implement mathematical models of very complex systems, such as weather. By the 1970s and 1980s, the first reliable models of global climate were being produced. Computing power continued to increase, allowing higher resolution models of global processes and very high resolution models of local weather processes, including individual storms.

SPOTLIGHT: NOVEL CLIMATES

Climate change may result in climates that have never existed before or cause current climatic conditions to disappear. The biological relevance of these changes depends on the tolerances of individual species. A new set of climatic conditions may not affect a species if the conditions are all within its climatic tolerances. However, new conditions that exceed one or more climatic tolerances may have dramatic effects on survival and range.

Williams et al. (2001) searched for a correlation between past novel climates and novel plant associations in North America. They developed an index that showed strong correlation between past novel climates and novel species combinations. This analysis was taken a step further by Williams et al. (2007), who asked where novel and disappearing climates might be found globally owing to twenty-first century climate change. The index used was the difference from current interannual variation. This is a slightly different index than was used in the North America analysis, but it is applicable globally.

Novel climates and disappearing climates are both projected to be concentrated in the tropics. This indicates that the tropics may be vulnerable to climate change even though the magnitude of warming is considerably less in the tropics than at high latitudes. Combined with narrow niche breadth in many tropical species, the tropics may experience severe biological changes due to climate change. The impact of climate change is the product of magnitude of change and sensitivity. Sensitivity may be high in tropical organisms, whereas magnitude of change will be greatest near the poles. No region is immune to the biological effects of climate change.

Source: Williams, J.W., Jackson, S.T., Kutzbacht, J.E., 2007. Projected distributions of novel and disappearing climates by 2100 AD. Proceedings of the National Academy of Sciences 104, 5738–5742; and Williams, J.W., Shuman, B.N., Webb, T., 2001. Dissimilarity analyses of late-quaternary vegetation and climate in eastern North America. Ecology 82, 3346–3362.

Biological modeling is less complex than climate modeling, but many of the principles are the same. Biological modeling begins with conceptual models, which may be turned into quantitative mathematical models executed on computers. There are trade-offs between spatial resolution of the models and the geographic scope (domain) over which the models can be run.

TYPES OF MODELS

Three types of biological models dominate research in climate change biology: species distribution models (SDMs), dynamic global vegetation models (DGVMs), and gap models. Each of these types of models captures a different aspect of biological change along a spectrum from species to ecosystems.

SDMs, as their name implies, simulate the distribution of species—their ranges—relative to climate. SDMs create a statistical model of the relationship between current climate and known occurrences of a species (Figure 11.1). Often, these

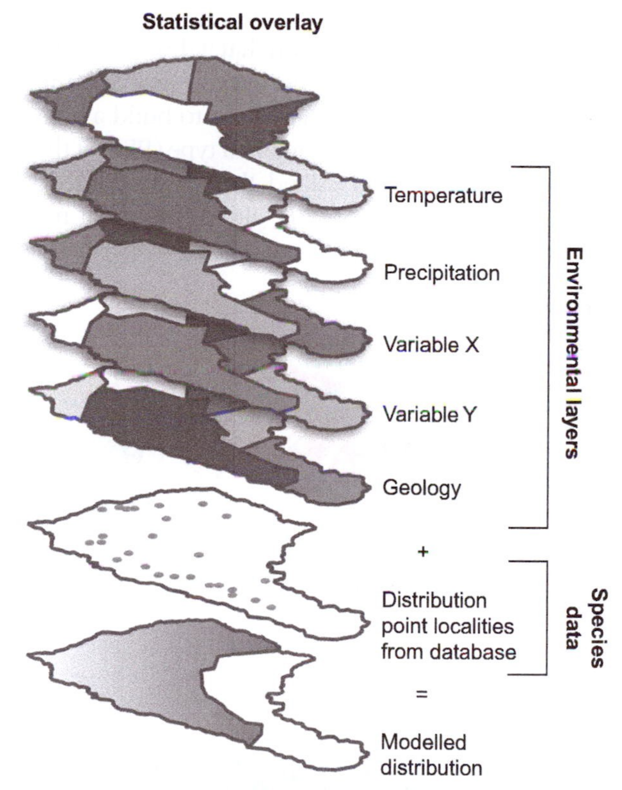

FIGURE 11.1 Schematic of an SDM.

Species distribution modeling begins with the selection of a study area (left). The study area is usually selected to be large enough to include the complete ranges of the species of interest to ensure that data sampling the entire climate space the species can tolerate are included. Climate variables and other factors constraining species distribution (shaded layers on right) are then correlated with known occurrences of the species of interest (layer with points). This statistical relationship can be projected geographically to simulate the species' range (bottom shaded area). Repeating this process using GCM-generated future climate variables allows simulation of range shifts in response to climate change. *Source: Lishke et al. (1998). Copyright Massachusetts Institute of Technology, by permission of MIT Press.*

models are used to simulate species' current ranges from a limited set of known observations. SDMs are useful in this setting when a species' distribution is not well known but needs to be included in conservation planning. The same type of model can be run using future climatic conditions, and results can be compared to current distribution to obtain an estimate of how species ranges may shift with climate change. The output of an SDM is typically a map of a species' simulated range, either in the present or in both the present and the future.

DGVMs (Figure 11.2) operate in a quite different way. DGVMs use first-principle equations describing photosynthesis, carbon cycling in soils, and plant physiology to simulate growth of and competition between vegetation types. DGVMs literally "grow" vegetation mathematically, fixing carbon from the atmosphere, distributing it to plant parts, and evolving a vegetation that it describes in terms as a number of "plant functional types" (e.g., tropical evergreen forest or temperate deciduous forest). If enough carbon is fixed and maintained to build a forest in a particular location, the DGVM registers the plant functional type (PFT) at that point as forest. If enough carbon persists for only a grassland, the PFT is recorded as grassland. The output of a DGVM is a global map of PFTs, although similar models can be run at higher spatial resolution for individual regions, such as countries or continents.

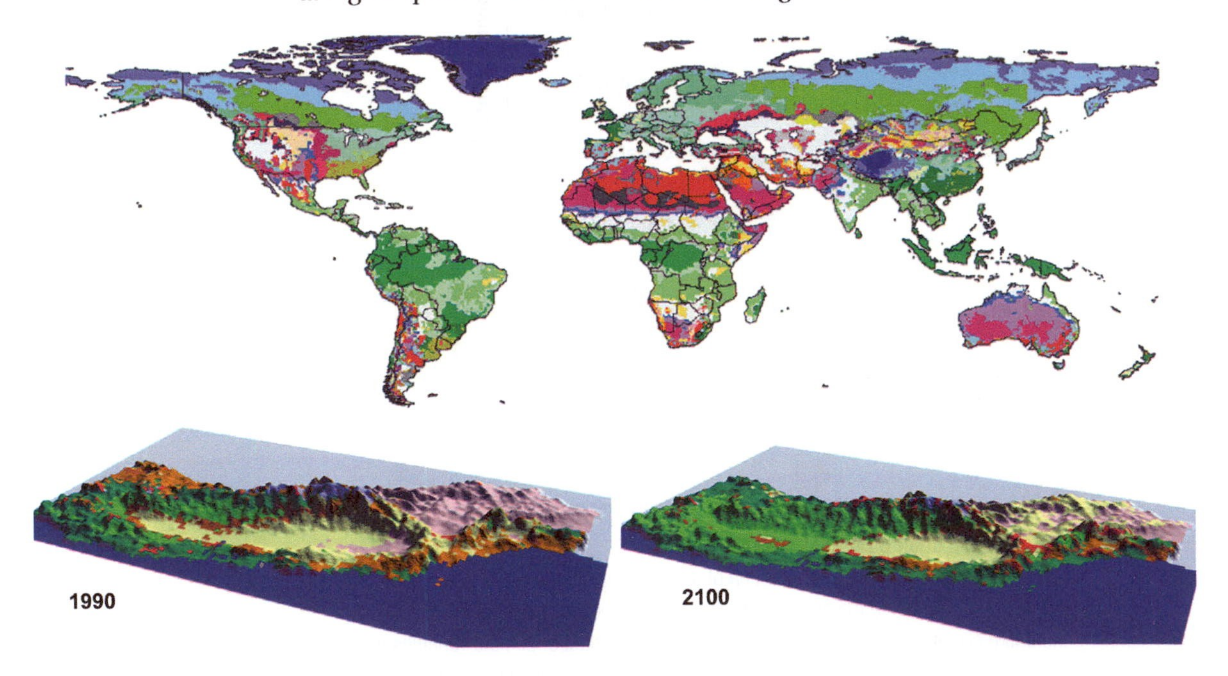

FIGURE 11.2 Global and regional vegetation simulation of a DGVM.
The global distribution of PFTs (top) can be simulated in a coarse-scale DGVM. The same DGVM run at finer resolution can simulate PFT distribution with many local features resolved (bottom left). Driving the DGVM with projected future climates from a GCM provides a simulation of the change in PFT distribution due to climate change at either global or regional (bottom right) scale. *Source: Neilson et al. (2005). Courtesy of USDA Forest Service.*

Because they run from photosynthetic equations, DGVMs are able to simulate the direct effects of elevated CO_2 on plant growth and competition, in addition to the indirect effects of the warming caused by CO_2 DGVMs are now being integrated into Earth System Models (see p256).

Gap models fall between SDMs and DGVMs on the scale from species to ecosystems. The term "gap model" derives from the attempt to simulate what happens in a forest gap after a tree falls—the growth of individual trees to fill the gap and competition between these individuals of different species. The parameters of the model are derived from known growth rates of various species of trees under different climatic conditions and spacing. This information is generally most readily available for trees of commercial importance, whose growth has been studied in various areas of their range and under various replanting spacings and combinations with other species. Because the mathematical models needed to simulate these growth and competition characteristics are fairly complex, gap models are generally run for a single forest gap. Their output is therefore a chart of species composition at that particular point rather than a map (Figure 11.3). However, as computer capabilities have advanced, it has become possible to join multiple gap models together to simulate growth and competition over small regions.

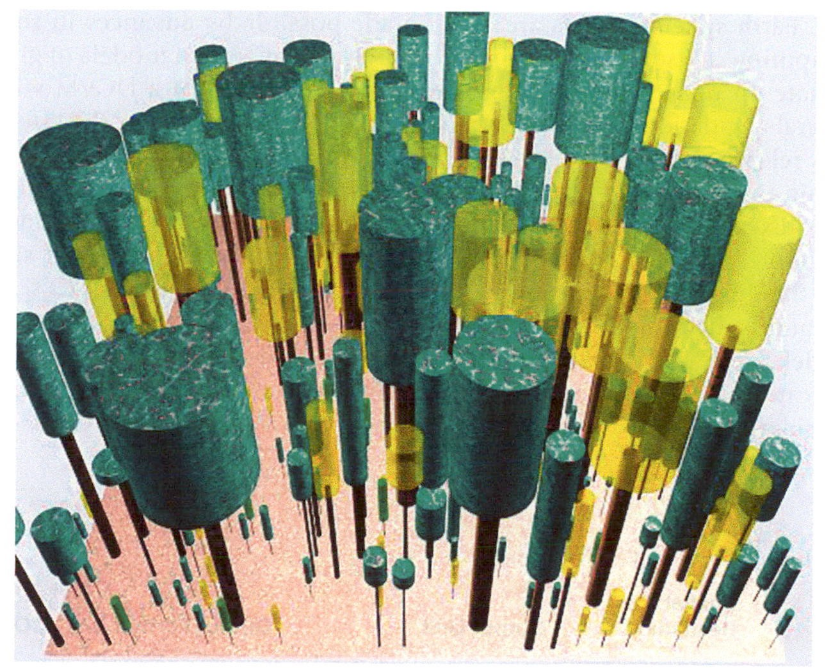

FIGURE 11.3 Gap model.
Styized simulation of trees in a 30m x 30m forest plot, similar to a gap model, is illustrated in this computer-generated schematic. *Source: Figure courtesy of H Sato, Japan Ministry of Marine-Earth Science and Technology. Reproduced with permission JAMSTEC.*

Comparison of SDM, gap, and DGVM models shows relative strengths and weaknesses of each. SDMs deliver species-denominated results appropriate for biodiversity at relatively fine scales because the statistical models involved are not mathematically complex. However, it may require modeling of hundreds or thousands of species to arrive at conclusions about changes in vegetation or ecosystems, and competition between species for novel climatic space is not explicitly addressed. Gap models simulate competition between species and provide species-denominated results. However, they address individual points rather than producing a map relevant to a study region (unless many are joined together), and the data required may be available for only a limited number of (tree) species. DGVMs address competition and produce geographically explicit results (maps) for the globe or individual regions, but they do not give information about individual species (Table 11.1).

Climate change biology studies use all three types of models. Because different modeling skills are needed for the three types of models, they are often executed by different research groups, each specializing in SDMs, gap models, or DGVMs. However, the best studies integrate findings from all three types of models to provide a more comprehensive view of possible biological outcomes.

Other types of models of biological change are being developed. In particular, Earth system models are being made possible by advances in supercomputing. These models integrate biological change into models of global climate. In their simplest form, they are the coupling of a DGVM with a general circulation model (GCM). They allow, for instance, for the effect of CO_2 release from burned forest or transitions in vegetation to be included within internal model dynamics. In GCMs, values for these biological carbon fluxes are external to the model and must be supplied by the model operator, based on simplifying assumptions. Thus, a GCM cannot simulate the dynamics of vegetation change due to climate, releasing CO_2 into the atmosphere, thus driving additional climate change. An Earth system model can simulate these dynamics. Early results from Earth system models indicate that these biological–climate interactions may be important. We consider these results in our discussion of DGVMs because DGVMs and Earth system models are related.

Table 11.1 Comparison of Gap, DGVM, and SDM Models

Model	Domain	Spatial Resolution	Output Unit	Output Format	CO_2
SDM	Species' range	0.25°21 km	Species	Map	No
Gap	Point	Point	Species	Chart (or map)	No
DGVM	Global/regional	1°210 km	Plant functional type	Map	Yes

DYNAMIC GLOBAL VEGETATION MODELS

Many distinct DGVMs now exist, developed and run by independent research groups. Several DGVM intercomparison projects have compared DGVMs run under identical conditions to compare results.

Prominent DGVMs include those developed by the University of Sheffield (SDGVM), the Hadley Centre of the UK Met Office (TRIFFID), the Leipzig group (LPJ), the University of Wisconsin (IBIS), the U.S. Forest Service (MAPSS), and others (e.g., VECODE, HYBRID, and MC1). All operate using some level of mathematical representation of plant growth, photosynthesis, and respiration, sometimes coupled with empirically driven corrections. For instance, carbon estimates from satellite images may be used to correct the model.

One strength of DGVMs is their ability to simulate the direct effects of CO_2 on plant growth, as well as the indirect effects of climate change. DGVMs may therefore help answer the question, how will the direct and indirect effects of human CO_2 pollution of the atmosphere interact? To answer this question, a DGVM is run with elevated levels of atmospheric CO_2, a simulation of climate change, or both. Each DGVM yields slightly different answers, so intercomparison projects have been formed to run several DGVMs with identical inputs (e.g., with identical emissions scenarios and GCM climate change) and compare the results. The outcome of one such intercomparison is shown in Figure 11.4.

DGVM results indicate that change in CO_2 can have a significant direct effect on vegetation distribution across the globe, but that this effect is largely neutralized by the indirect effects of climate change. DGVM runs with changed CO_2 only show changes in the major patterns of vegetation types throughout the world. DGVM runs with climate change only (the indirect effect of CO_2) show even larger changes. In some areas, the direct and indirect changes reinforce one another; in many other areas, they oppose one another. Combined CO_2 and climate change runs with DGVMs show that the climate change signal is stronger: it is difficult to distinguish between climate change only and climate change plus CO_2 runs.

SPOTLIGHT: CO_2 AND THE AMAZON

CO_2 may have played an important role in shaping speciation in the Amazon (Cowling et al., 2001). CO_2 models show that vegetation could have changed significantly at the last glacial maximum (LGM) without loss of forest cover. Forest may have been retained owing to cooling, which reduced evapotranspiration-related stress. Yet at the same time, canopy density may have changed markedly—an effect mediated by lower CO_2 levels at the LGM. Lower canopy density would result in different understory light and temperature, forcing species to adapt to new conditions and possibly driving speciation.

Source: Cowling, S.A., Maslin, M.A., Sykes, M.T., 2001. Paleo vegetation simulations of lowland Amazonia and implications for neotropical allopatry and speciation. Quaternary Science Reviews 55, 140–149.

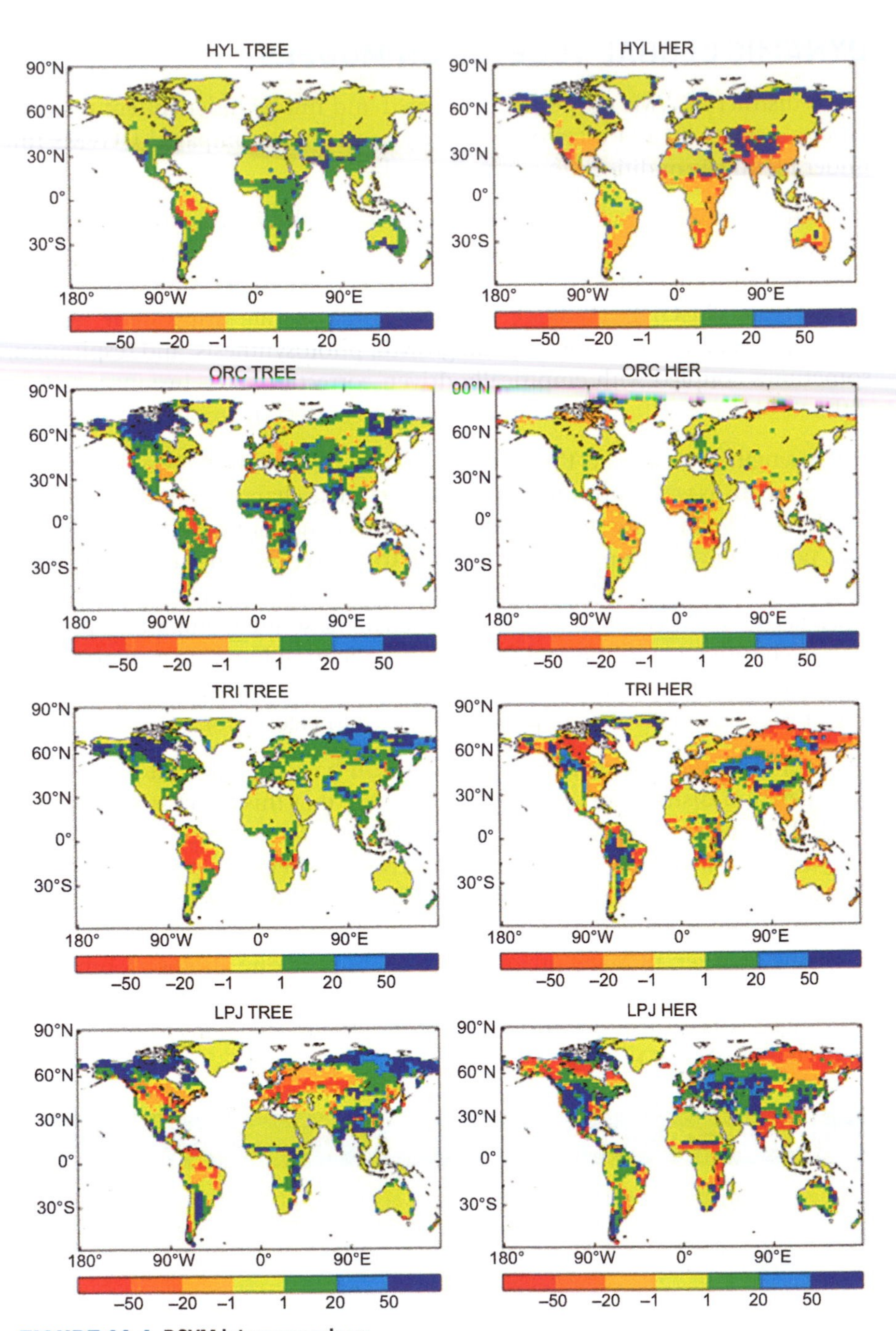

FIGURE 11.4 DGVM intercomparison.

Outputs of six different first-generation DGVMs (top six panels) compared to the composite of all six (bottom left) and the PFT distribution classified from satellite imagery (bottom right). *Source: Cramer et al. (2001).*

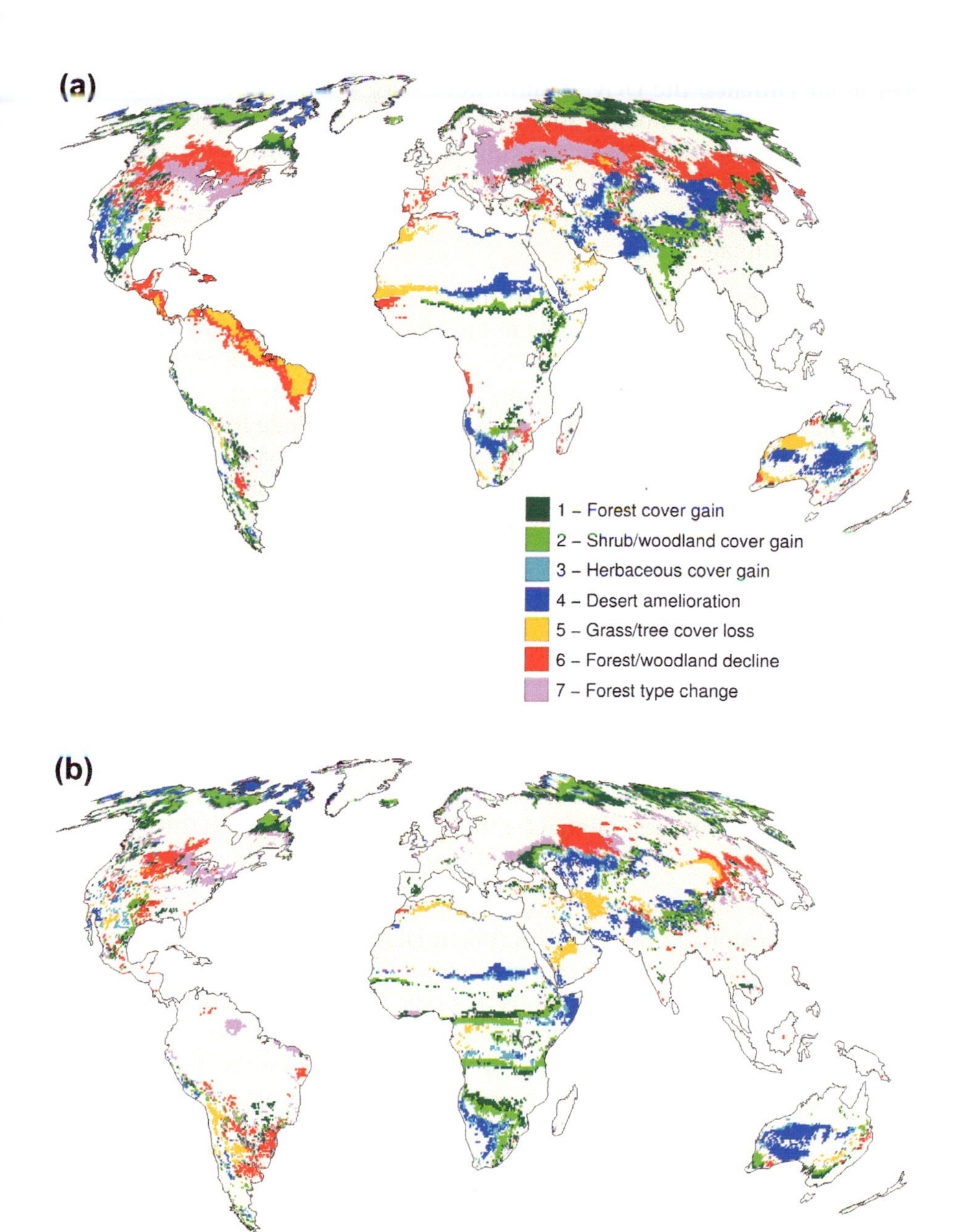

FIGURE 11.5 Climate and CO_2 DGVM simulations.
DGVM global change simulations for high emissions (A1 scenario) and high climate sensitivity GCM (Hadley) (a) and low emissions (B2 scenario) and low climate sensitivity GCM (ECHAM) (b). *Source: IPCC.*

For future climates, the DGVMs show northward expansion of boreal forest, expansion of some temperate southern forests toward the pole, expansion of several temperate forest types, and contraction of some tropical forest types, most notably in the Amazon (Figure 11.5). The direct effect of CO_2 seems especially important in damping the loss of moist tropical forests that occurs in response to climate change.

The contraction of Amazonian forest simulated by DGVMs is particularly noteworthy. The Amazon basin is home to one of the largest and most biologically rich blocks of tropical forest. DGVM results show the wet forest of the Amazon shrinking in response to climate change. DGVMs are now being combined with GCMs in Earth System Models. Earth System Models include feedbacks of vegetation on climate, such as the changing albedo of forest replacing ice. They also simulate effects of changing climate on global biome distribution and may represent the effects of moisture recycling in large tropical basins such as the Amazon.

Earth system models that incorporate feedbacks from vegetation to climate show an even stronger effect. The mechanism is moisture recycling within the basin. Moisture driven into the basin from the Atlantic Ocean is taken up by the wet tropical forest of the Amazon and then transpired. As the vegetation respires, water is lost through the stomata of leaves. This moisture enters the atmosphere, rises, condenses, and falls again as rain. Thus, forests in the eastern Amazon are the source of rainfall for forests farther west, especially in the central Amazon. If climate change results in the drying of these eastern forests, there is less water and less rainfall for the central Amazon, and it dries too—in something akin to a domino effect for forest water.

DGVMs are also useful in examining the effect of disturbance on vegetation type and the interaction of this effect with climate change. For instance, some regions have adequate photosynthesis (net primary productivity) to support forests but currently are vegetated by grasslands. Periodic burning in these systems may release carbon and maintain a savanna in an area that would revert to forest in the absence of fire. The SDGVM has been used to simulate the removal of fire from systems, and several DGVMs have been used to study what may happen to fire regimes and fire-driven vegetation types as climate changes. In many of these studies, DGVMs are implemented for individual regions rather than globally.

Although all DGVMs correctly identify the overall pattern of global vegetation, all will either overpredict or underpredict certain types of vegetation in some regions. Because there is no way to test against future, unknown vegetation types, the test of ability to reproduce current vegetation is very important. Semi-independent tests have also been developed, such as the use of evapotranspiration values from the models to predict river flows and comparing the DGVM-derived estimates to actual flow volumes. However, validation of DGVMs, and all biological models, into the future remains a challenge.

SPECIES DISTRIBUTION MODELS

SDMs are perhaps the most widely implemented of all climate change biology models. This is because their relative simplicity makes them easily implemented by individual researchers on ordinary personal computers. Several SDM software programs are available for download from the Internet. Internet sites also provide access to current climate and GCM simulation data necessary to run SDMs. SDMs are sometimes known as "niche models" because they simulate a species climatic niche in the current climate or the change in the niche as the climate changes. SDMs may be referred to in the older literature as envelope models, bioclimatic models, or range-shift models. Statistical tools that can be used to generate SDMs include GLM, generalized additive modeling, linear regression, and classification and regression trees, all of which are supported in the "R" open statistical software. Maximum entropy (Maxent) is an algorithm specifically designed for SDMs that is available for free download from the Internet. It is one of the most widely used and best performing SDM packages.

The simplest SDM uses values for climatic variables at points at which a species has been observed, compares these values to values of the same variable across a study area, and models the species as present where current climate is within the range and as absent where current climate is above or below the range—this is an "envelope" model. For instance, if a species has been observed in 50 locations, and the lowest mean annual temperature of those sites is 20 °C and the maximum mean annual temperature of those sites is 30 °C, the model will simulate species presence at all sites with mean annual temperature warmer than 20 °C but less than 30 °C. When run for only one variable, many areas will qualify as suitable, but adding additional variables (e.g., mean annual precipitation, precipitation of the wettest month, minimum temperature of the coldest month) will further constrain the simulated range of the species. By the time five or six variables have been added, the model will produce a quite reasonable estimate of the current range of a species, if that species' range is indeed limited by climate. More sophisticated SDMs use more complex statistical formulas to fit the relationship of observed occurrences to current climate, but all have conceptual similarity to this simple example.

To simulate the future range of species, an SDM substitutes climate data from a future, elevated CO_2 run from a GCM (Figure 11.6) into the statistical model developed for the current climate. The model then predicts species presence or absence based on the suitability of the future climate in a cell. In practice, climate data for both the present and the future are obtained for a series of grid cells across a study area. Each grid will have a series of climate variables associated with it, provided by interpolation of data from either weather stations (current climate) or a GCM (future climate). The model then returns a result of the probability of species presence or absence for each cell in the study area based

EXTENT OF OCCURRENCE AND AREA OF OCCUPANCY

In determining the extent of species' ranges for climate change assessment and other applications, extent of occurrence (EO) and area of occupancy (AO) are used. EO refers to the convex hull that encompasses all known occurrences of a species—in essence, a range map. AO is the area within the EO that a species actually occupies. For instance, the EO for red-winged blackbird is most of the western United States, but within that EO, red-winged blackbirds actually occur only in wetlands, which are thus the AO for the species. AO is always smaller than EO. Both AO and EO may be affected by climate change. An important case is when EO expands but AO within that range shrinks—a situation that can lead to underestimation of the threat to a species unless AO is properly recognized.

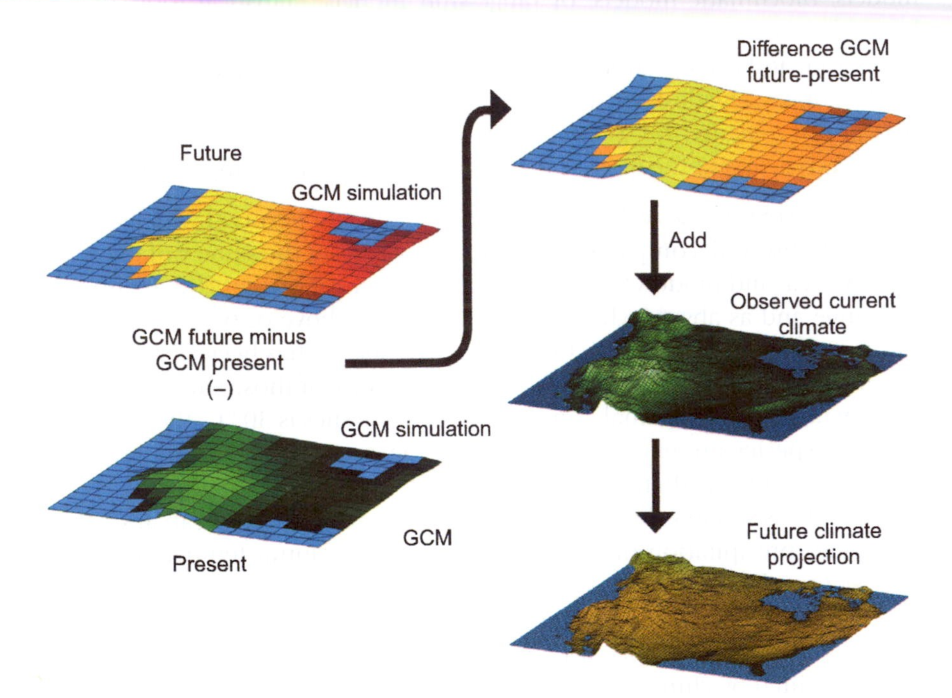

FIGURE 11.6 Generation of a future climatology for species distribution modeling. Because SDMs often require climatologies with horizontal resolutions much finer than those offered by GCMs, techniques are needed for generating downscaled future climatologies from GCM outputs. One approach commonly used applies the difference between GCM simulations of the present and the future to a current fine-scale climatology. This is done because GCM fidelity to current climate may be imperfect. The use of a historical fine-scale climatology ensures reasonable reproduction of major climatic features. The GCM difference (future–present) simulates future warming. *Courtesy of Karoleen Decatro, Ocean o'Graphics.*

on the climate variables in that cell and the statistical relationship between climate and observed occurrences of the species (Figure 11.7). The final product is a map of probabilities. The probabilities are often converted to a simple presence/absence map to simulate a traditional range map of a species (Figure 11.8).

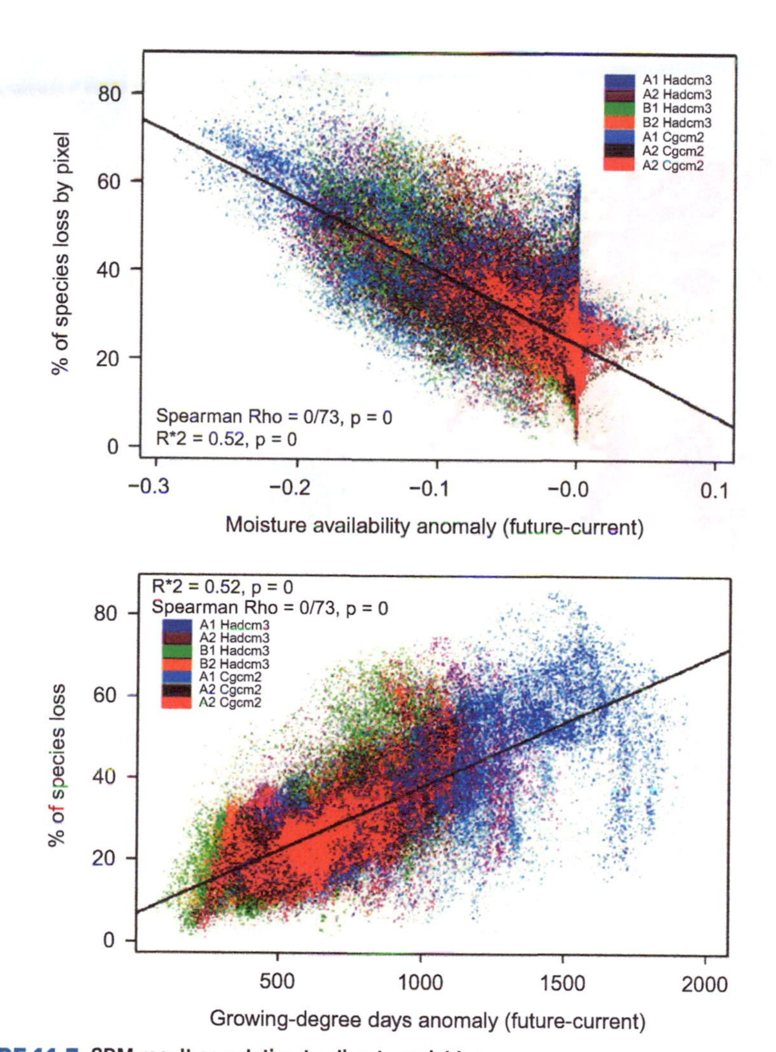

FIGURE 11.7 SDM result correlation to climate variables.

The scattergrams indicate relationships between species loss (%) and anomalies of moisture availability and growing-degree days in Europe. The colors correspond to the climate change scenarios indicated in the key. *Source: Thuiller et al. (2005). Copyright National Academy of Sciences.*

SDM applications include simulating the current range of species for protected area planning, simulating the spread of invasive species, and simulating shifts in species ranges under past and future climates. The considerable success SDMs have demonstrated in modeling current species ranges and in modeling the spread of invasives lends confidence to the application of these techniques to simulating species ranges in future climates.

For example, SDMs have been tested by trying to reconstruct the past ranges of species from their present distribution. In one such test for North America, an SDM

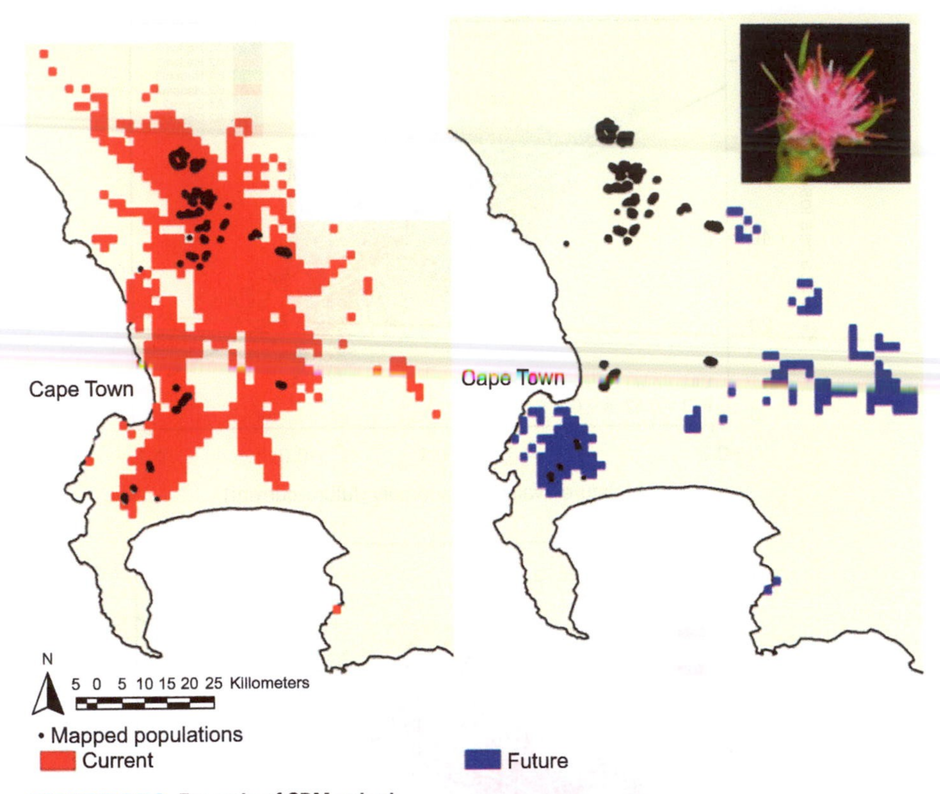

FIGURE 11.8 Example of SDM output.
SDM output for a protea (pictured) from the Cape Floristic region of South Africa. Current modeled range is shown in red, and future modeled range is shown in blue. Known occurrence points for the species are indicated by black circles. *Source: Figure courtesy Guy Midgley, Stellenbosch University.*

produced a reasonable estimate of past distribution of the eastern mole from its present distribution and vice versa (Figure 11.9). In another test, an SDM was created for the glassy-winged sharpshooter (*Homalodisca coagulata*), an insect pest that spreads bacterial disease in vineyards (Figure 11.10). The sharpshooter has invaded California from its native range in the southeastern United States, posing a serious threat to the California wine industry. The SDM correctly simulated the area of invasion based on the observed occurrence of the sharpshooter in its native range.

SDM simulations of species range shifts with climate change have been conducted for many regions and thousands of species. These studies support expectations from theory that species will move toward the poles and upslope as the planet warms. They also support the notion that species respond individualistically to climate change. This means that species move independent of one another as climate changes, resulting in the tearing apart of vegetation and animal "communities" and the reassembly of species into new assemblages.

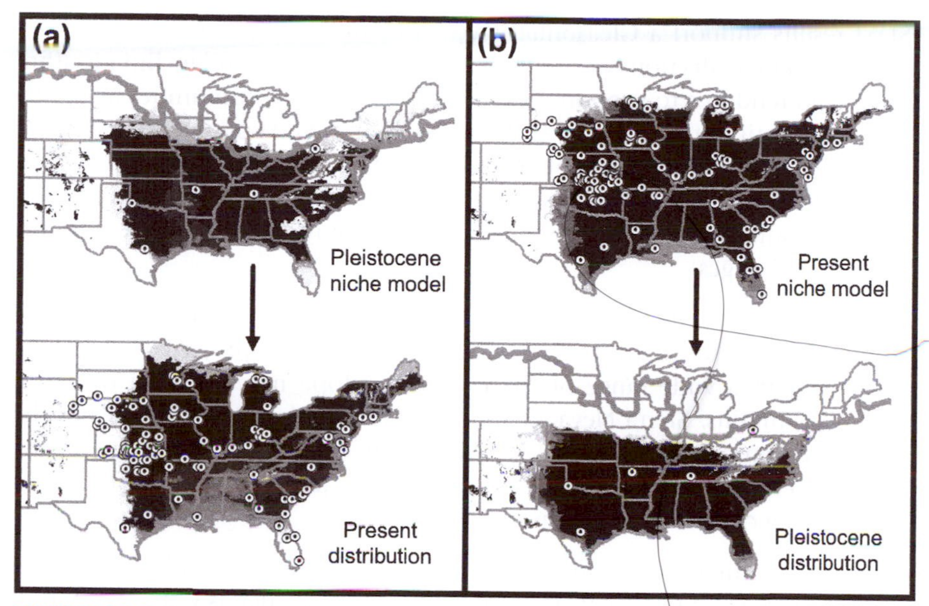

FIGURE 11.9 Backward and forward modeling of the eastern mole (*Scalopus aquaticus*). (a) SDM created from known Pleistocene occurrences predicts present distribution. (b) SDM created from known current distribution predicts known fossil occurrences. *Source: Martinez-Meyer et al. (2004).*

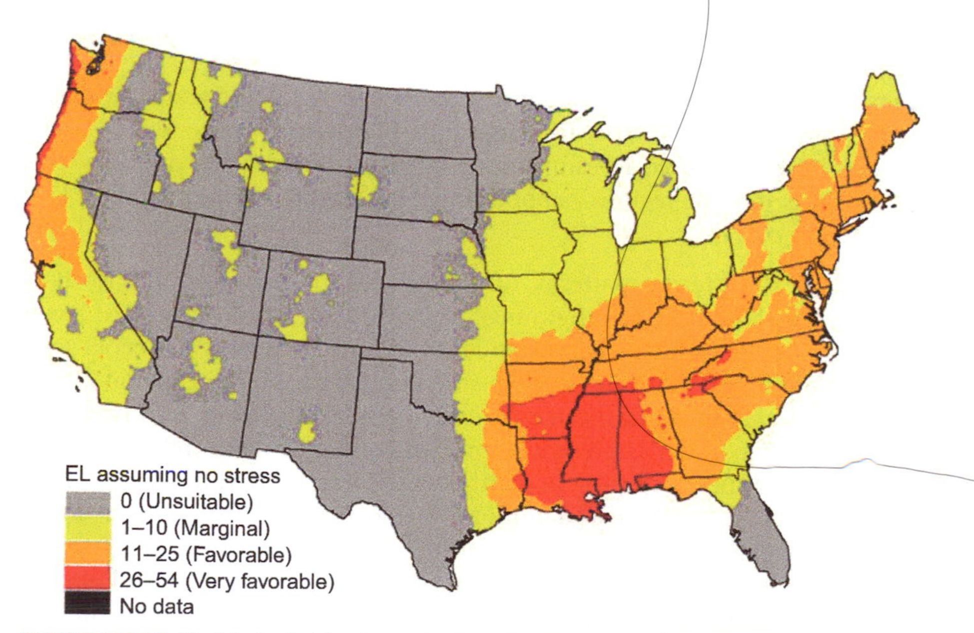

FIGURE 11.10 Model of suitable climate for glassy-winged sharpshooter in the United States. The potential spread of insect pests such as the glassy-winged sharpshooter can be predicted using SDMs. *Source: Venette and Cohen (2006).*

SDM results support a Gleasonian view of communities—that communities are ephemeral collections of species brought together by similar climatic affinities—and tend not to support a Clementsian view of communities as tightly coevolved entities that respond to climate as a unit. SDM outputs clearly show that individual species within plant and animal communities move at different paces and in different directions in response to alterations in temperature and rainfall, resulting in the disassembly and reassembly of species combinations as climate changes.

Europe and North America have been the focus of the greatest number of SDM studies. In these northern temperate settings, they have projected northward movement in a wide range of species and upslope movement in the Alps, Rocky Mountains, Sierra Nevada, and other mountain ranges. Studies in the Alps have demonstrated significant loss of range in alpine plants due to the effect of decreasing area as species move upslope, just as there is more area at the base of a cone than at the tip.

Australia was the point of origin for some of the earliest SDMs (e.g., Bioclim) and thus has had a relatively large number of SDM studies as well. These studies have shown simulated range losses in mountains and across a range of species. In a study of 92 endemic Australian plant species, most were found to lose range and 28% lost all range with only a 0.5 °C warming. Large amounts of range loss have been demonstrated in numerous montane species in Queensland (Figure 11.11).

Mexico and South Africa are other regions in which active research programs have produced SDM results for thousands of species (Figure 11.12). These studies have been used to estimate the need for connectivity and new protected areas in these regions—applications that are explored in detail in later chapters.

SDMs reveal differing patterns of range loss in montane and lowland regions. Lowland species suffer range loss over large areas because of a very shallow climate gradient in lowlands. Montane areas have a steeper temperature gradient, resulting in range shifts that are smaller in spatial extent. However, many lowland species are widespread: they begin with very large ranges and can suffer large areal range loss and still retain relatively large range area. Montane species, in contrast, are often arrayed in narrow elevational bands up the slope of a mountain. These species have small absolute range sizes so that relatively small absolute areas of range loss translate into large proportional range losses. Even where widespread lowland species suffer large proportional range losses, they may retain range sizes that are large relative to those of montane species. Thus, large absolute and relative effects in lowlands may still leave lowland species with larger ranges than those of montane species. Therefore, statistical summaries of absolute and

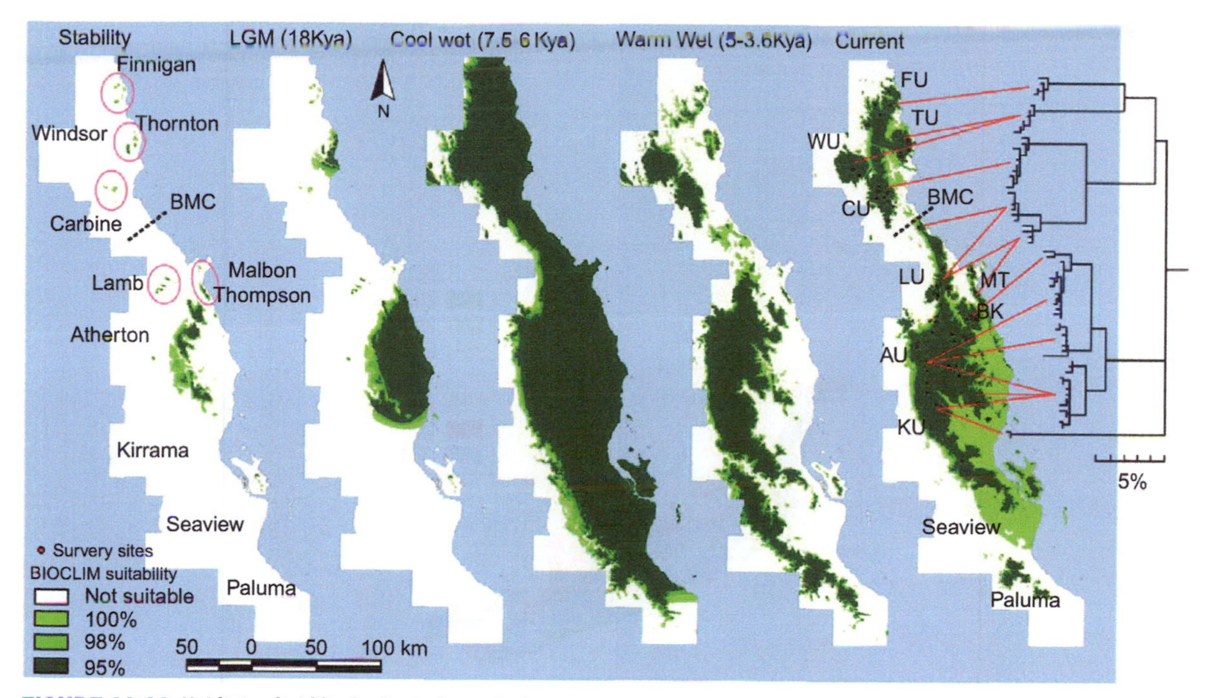

FIGURE 11.11 Habitat suitability for Australian rain forests.

Suitability for Australian rain forests is shown for three past climates and the present. The modeling tool used was Bioclim, an early SDM software. Like other SDMs, Bioclim can be used to simulate suitability for biomes or vegetation types as well as species. End panels indicate genetic similarity of forest fragments (right) and areas that are stable in all modeled time slices (left). *Source: Hugall et al. (2002).*

proportional range size must be viewed with caution: an extremely important variable is remaining absolute range size. This variable may be most important in species extinction risk, another topic that is explored in detail in later chapters.

GAP MODELS

Gap models confirm the Gleasonian view of communities that emerges from SDMs. A typical gap model shows new species appearing, existing species dropping out, and major changes in dominance among species that persist at any given location. Spliced together over an entire landscape, these changes suggest highly ephemeral communities, changing dramatically in decades or centuries during past climate changes, with more, and often more profound, changes projected for the future.

Gap modeling originated in forestry schools in the United States in the 1960s and 1970s. These models emerged from interest in optimizing individual tree growth for commercial production from plantation forests.

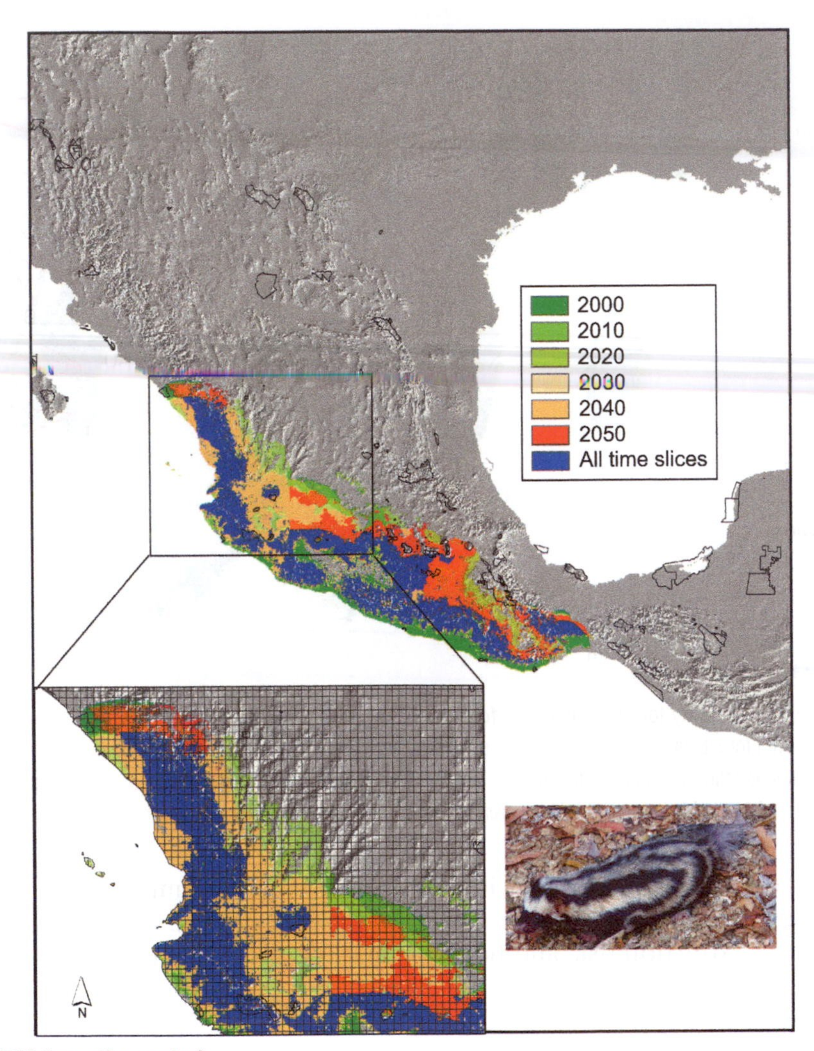

FIGURE 11.12 **SDM time-slice analysis.**
SDM simulation of present and future range for the pygmy spotted skunk of western Mexico. This model was constructed in 10-year time slices to allow pathways of contiguous habitat (present to future) to be identified. *Source: Hannah et al. (2007). Reproduced with permission from the Ecological Society of America.*

Forest ecologists realized that these models could be adapted for ecological studies. The relevant ecological unit of study was the gap left in a forest when a mature tree fell. This choice of model domain also simplified calculation of shading and nutrient competition by reducing the areas over which these processes had to be calculated. Models of gap replacement by individual trees offered insights into forest succession and the competitive ecology of forest communities.

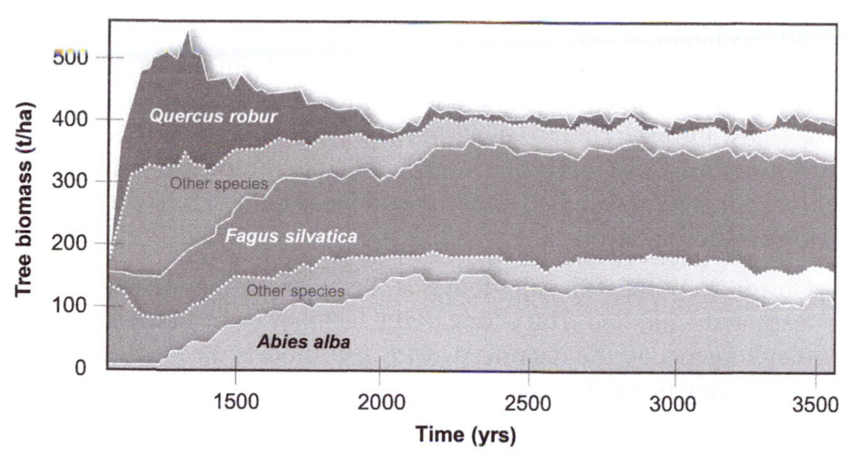

FIGURE 11.13 Gap model output.
This gap model of forest composition in Switzerland under climate change shows an early peak in oak abundance, giving way to a mixed fir–beech forest with little oak. *Source: Redrawn from Bugmann (2001).*

Many variants of gap models now exist, most of which were created to study temperate forests in Europe or North America. For example, the FORET model was developed to simulate forest in the southeastern United States, FORCLIM for European alpine forests, and FACET for the forest of the Sierra Nevada mountains of California. These models use physical equations to simulate processes such as shading, equations for biological processes, and parameters measured in forestry field trials.

Gap models have been tested for their ability to simulate known forest compositions and spatial variation of forest composition, such as altitudinal zonation of vegetation in mountains. Gap models have reliably reproduced the general altitudinal zonation features on several mountain ranges on different continents. Because altitudinal zonation is the result of a climate gradient, these tests of gap models increase confidence in their ability to simulate changes in forest composition as climate changes.

Gap models have been used extensively to model past climates and climate change (Figure 11.13). These studies have revealed state-dependent, or hysteretic, responses. These arise when a forest may have multiple stable states under a given set of climatic conditions. Which stable state actually emerges depends on the history of the forest—both the history of climate change and the biological history of the site.

The theme of state-dependent forest composition is seen in gap model studies of future climate change. The current composition of a forest may be critical in determining future trajectories of composition. In particular, forests exhibit considerable compositional inertia. Once established, mature trees may persist in climates that would be inhospitable for their establishment. Replacement of these forests happens only with disturbance such as fire or death of the trees owing to old age. These "living dead" forests may prevent the establishment of replacement vegetation and then suddenly burn or die centuries later, opening up the landscape for a completely different vegetation type. In this way, mixed conifer forest might be replaced by oak woodland as montane habitats warm, not in a gradual transition but, rather, in sudden state switches when the coniferous forest dies or burns.

A notable weak point with gap models is the relatively few studies of tropical forests using the method. This is due to the fact that many of the growth trials necessary to generate the data needed for gap models have never been conducted for many tropical trees.

MODELING AQUATIC SYSTEMS

Both biological and physical models yield important insights into the future of marine systems. Physical models of ocean chemistry and temperature suggest changes in range limits of marine organisms, most notably corals. Biological models examine ecophysiological tolerances and food web interactions, with less emphasis on geographic range shifts than in terrestrial models, perhaps because of the great existing variability in the geographic ranges of marine species. Modeling physical change with depth is particularly relevant—a challenge that has been addressed in lake models as well (Figure 11.14).

Coral reef distribution changes may be studied using models of both climate change and changes in ocean chemistry. As discussed in Chapter 3, warming sea surface temperatures are causing coral bleaching, driving tropical corals toward the poles. However, at the same time, CO_2 dissolved in seawater lowers pH and alters the saturation states of calcium carbonate, driving corals toward the tropics.

Models of the future show how these contrasting effects are likely to evolve (Figures 11.15 and 11.16). Warming waters continue to spread from the equator, making the tropics increasingly inhospitable to corals. Aragonite saturation states decline from already low values in temperate and polar waters, making these regions poorly suited to the growth of scleractinian, reef-forming corals. This "double whammy" squeezes warm-water corals between declining temperature conditions, on the one hand, and declining chemical conditions, on the other hand. Corals may literally have nowhere to go in a warmer, more acid ocean because they are unable to tolerate the bleaching in warm equatorial waters and unable to secrete their calcium carbonate skeletons in the undersaturated waters nearer the poles.

The net effect of these influences is a catastrophic reduction in the suitable range for coral reefs should human CO_2 emissions continue unchecked

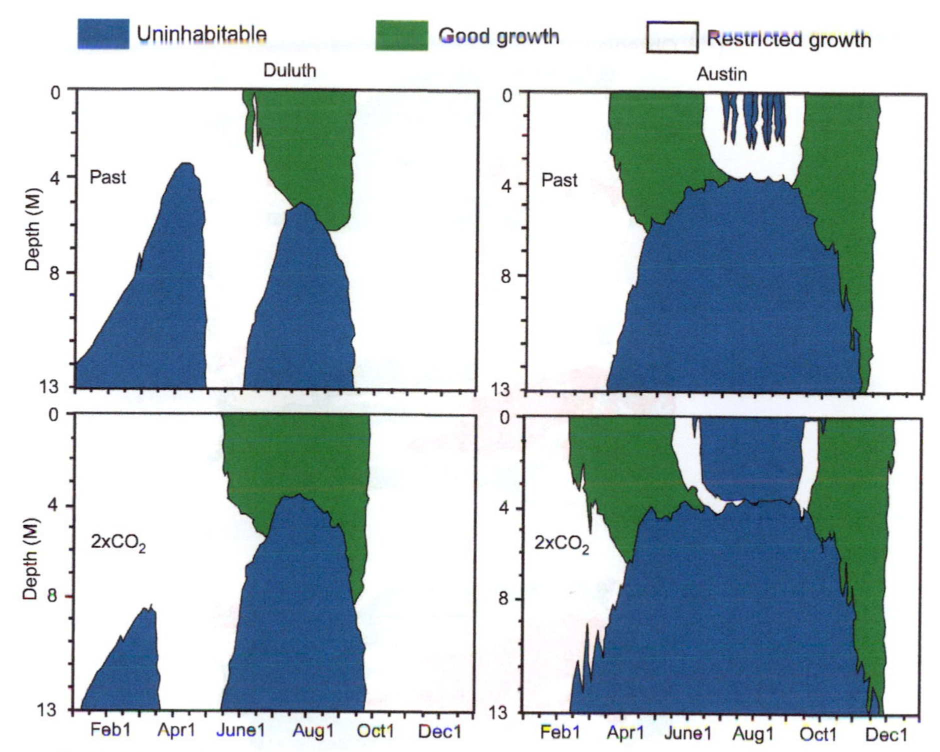

FIGURE 11.14 Lake habitat suitability profiles under double-CO_2 scenarios.
With warming, lake habitat suitability is very close with depth and over time. Simulated habitat suitability is shown for double-CO_2 scenarios for lakes in Duluth, Minnesota, and Austin, Texas. In Duluth, uninhabitable surface water extends deeper and lasts longer in the climate change scenario. In Austin, a summer window of habitability in the middepths closes, making the entire lake uninhabitable by late summer. *Source: Stefan et al. (2001).*

(Figure 11.17). Regional models are being used to refine these projections accounting for other controls on coral reef distribution, such as turbidity, and to devise conservation plans to deal with change. This is a dramatic example of possible future impact of changes already under way.

Biological models for individual marine species are less common than in the terrestrial realm. The models that do exist are generally more complex than SDMs. They typically represent food web interactions and ecophysiological changes in productivity. These marine models are the wet counterpart of DGVMs in complexity and theoretical foundation (ecophysiological rather than statistical). For example, EcoSim is a food web model that uses ecophysiological constraints to determine changes in primary productivity and has been used to assess changes in marine ecosystems surrounding Australia

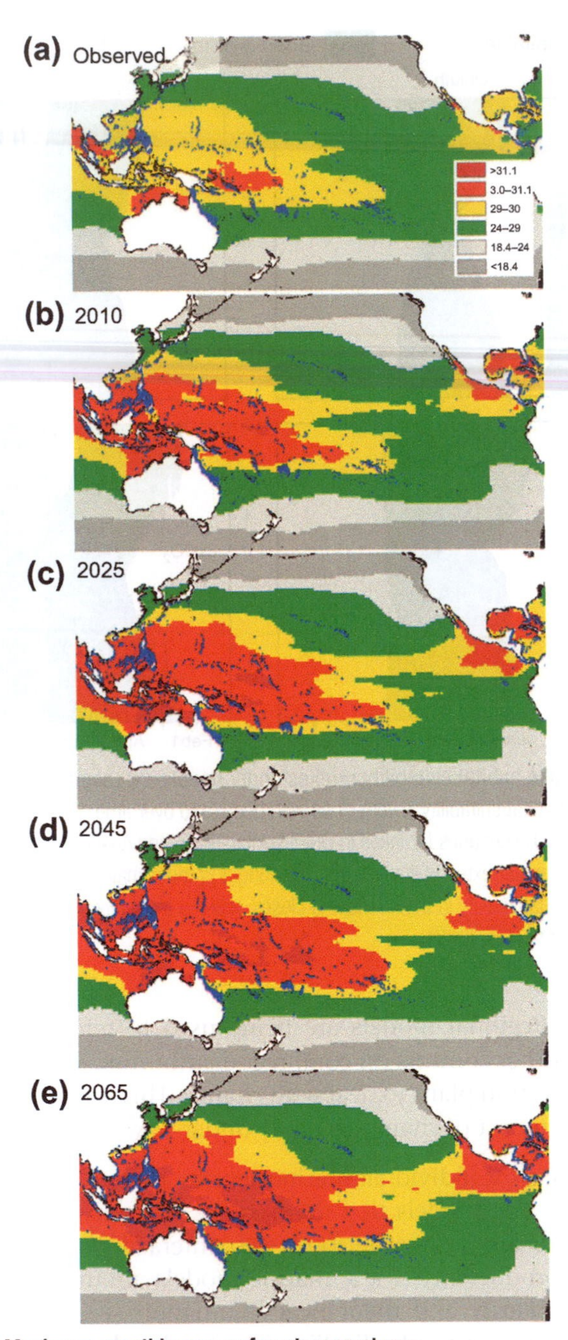

FIGURE 11.15 Maximum monthly sea surface temperatures.
(a) Observed, (b) 2000–2009 projected, (c) 2020–2029 projected, (d) 2040–2049 projected, and (e) 2060–2069 projected. Warmer temperatures cause bleaching that threatens persistence of coral reefs. *Source: Guinotte et al. (2003). With kind permission from Springer 1 Business Media.*

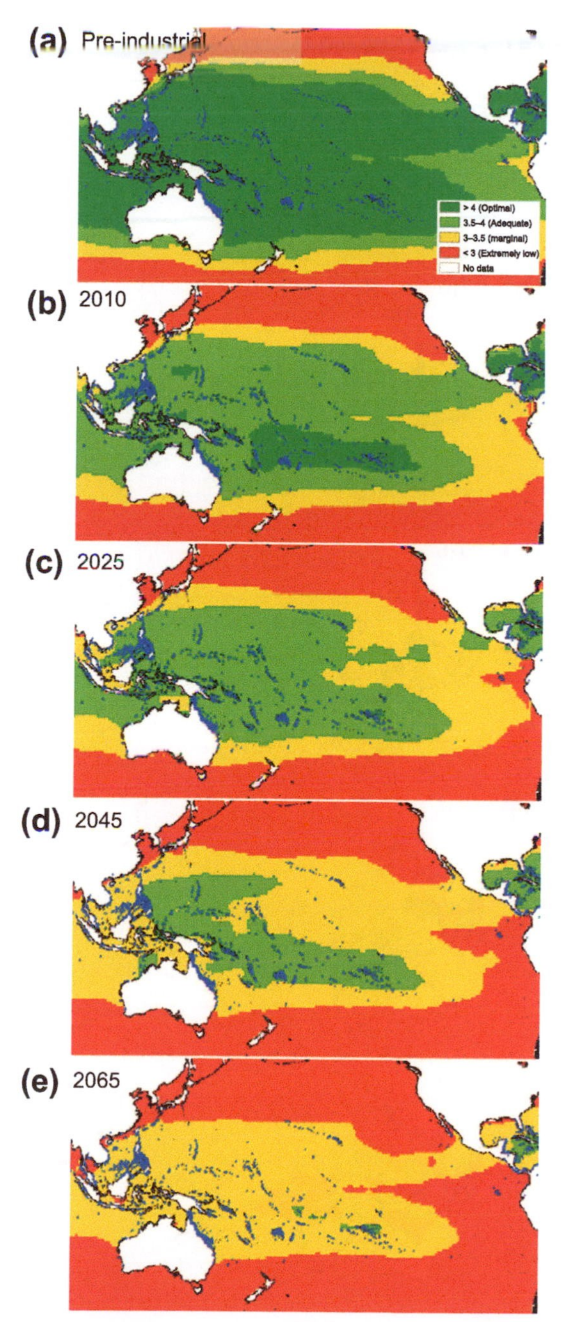

FIGURE 11.16 Aragonite saturation state of seawater.

Red/yellow, low-marginal; green, adequate/optimal. (a) Preindustrial (1870), (b) 2000–2009 projected, (c) 2020–2029 projected, (d) 2040–2049 projected, and (e) 2060–2069 projected. Low saturation states unsuitable for coral reefs collapse in toward the equator as the century progresses. *Source: Guinotte et al. (2003). With kind permission from Springer 1 Business Media.*

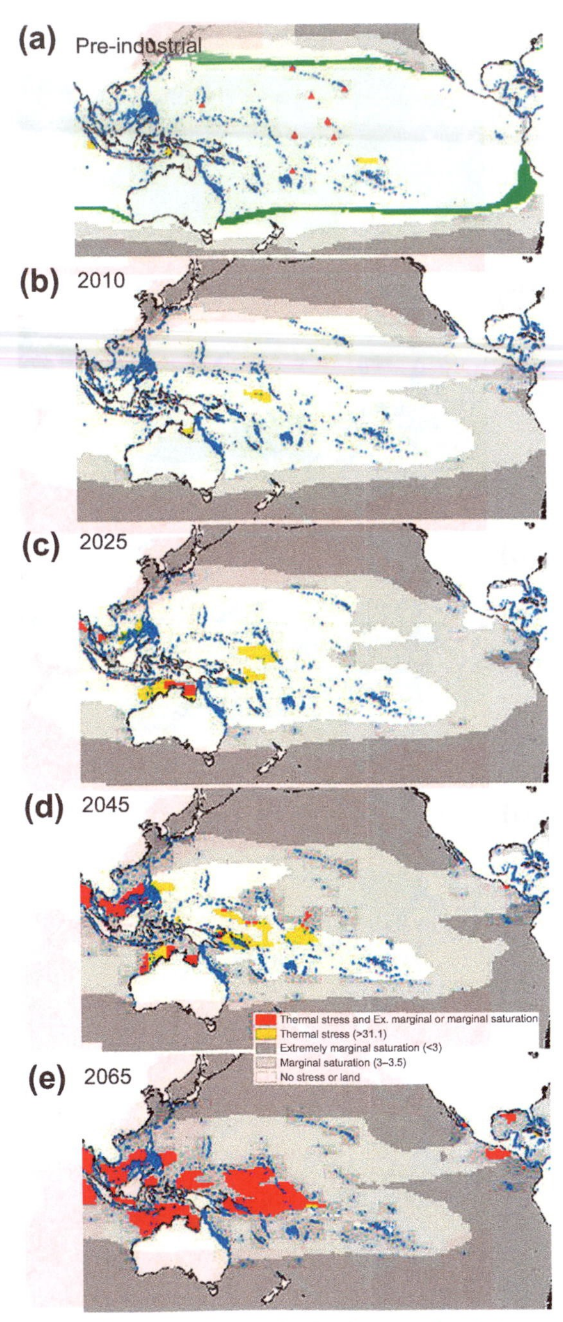

FIGURE 11.17 Areas in which temperature and aragonite saturation state combine to stress corals. (a) Observed, (b) 2000–2009 projected, (c) 2020–2029 projected, (d) 2040–2049 projected, and (e) 2060–2069 projected (see legend in panel e). *Source: Guinotte et al. (2003). With kind permission from Springer 1 Business Media.*

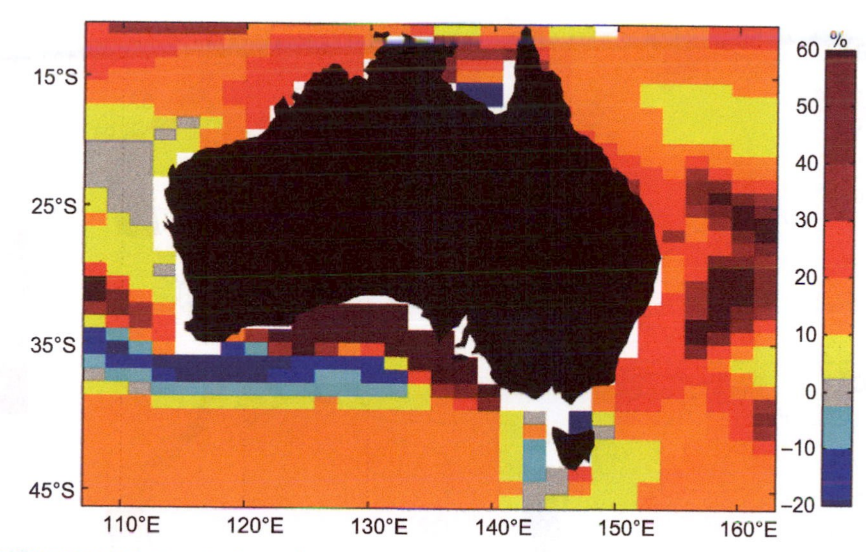

FIGURE 11.18 Output of a marine ecosystem model.
Simulated relative percentage change in phytoplankton production is shown for the region surrounding Australia, based on a nutrient–phytoplankton–zooplankton model. Production generally increases with warmer water temperatures, although note the exception in a region off southeastern Australia. The change is calculated as the percentage difference between the 2000–2004 mean and the 2050 mean. *Source: Brown et al. (2010).*

(Figures 11.18 and 11.19). Modeling of marine climate change biology is a growth field, and more tools such as EcoSim are needed, as is application of existing marine modeling tools across more geographies.

EARTH SYSTEM MODELS

Earth system models are rapidly gaining in importance as tools for integrating marine and terrestrial biology into models of atmospheric change. These models integrate vegetation and climate processes, in their simplest form, by joining a DGVM and representation of ocean primary productivity to a GCM. Changes in vegetation and marine biomass then release or sequester CO_2, which feeds back on climate. The application of these models is relatively new because the massive computing power that they require has been available only relatively recently.

Among the most important findings of early Earth system model runs has been the verification of important feedbacks between vegetation and climate, usually through CO_2 or carbon pools. These processes may have important global effects, such as CO_2 uptake by boreal forests, or effects at a regional scale, such as drying of the Amazon.

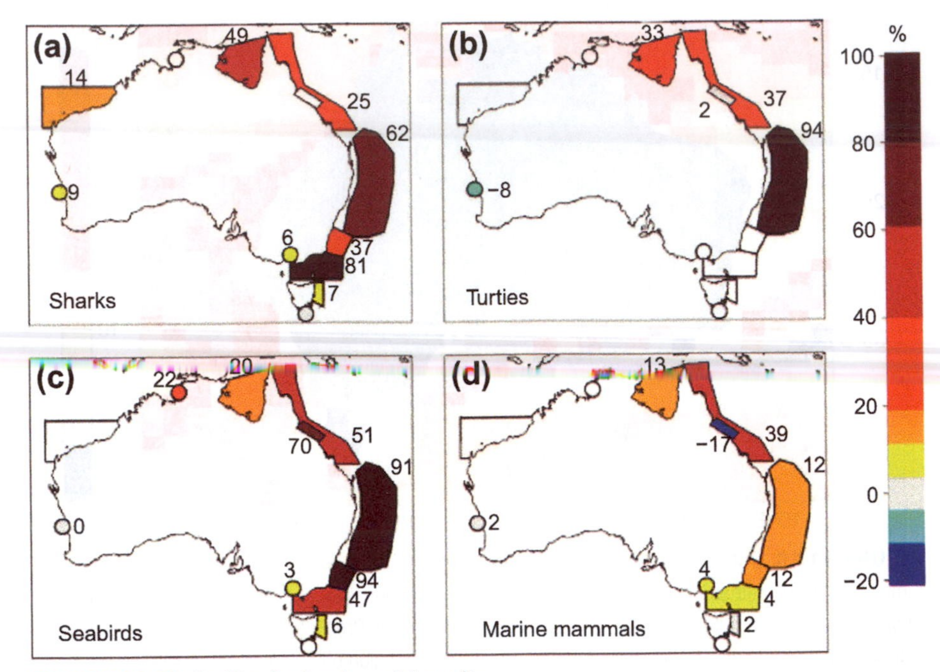

FIGURE 11.19 **EcoSim food web model results.**
Based on the production changes shown in Figure 11.18, the EcoSim food web model simulates change in abundance of species of conservation interest. Following increases in production, biomass abundance increases for most species. However, abundance of some species may decline owing to trophic effects (changed food web relationships). *Source: Brown et al. (2010).*

FURTHER READING

Bugmann, H., 2001. A review of forest gap models. Climatic Change 51, 259–305.

Midgley, G.F., Hannah, L., Millar, D., Thuiller, W., Booth, A., 2003. Developing regional and species-level assessments of climatic change impacts on biodiversity in the Cape Floristic Region. Biological Conservation 112, 87–97.

Serra-Diaz, J.M., Franklin, J., Ninyerola, M., Davis, F.W., Syphard, A.D., Regan, H.M., & Ikegami, M. (2014). Bioclimatic velocity: the pace of species exposure to climate change. *Diversity and Distributions, 20*(2), 169–180.

Sitch, S., Huntingford, C., Gedney, N., Levy, P.E., Lomas, M., Piao, S.L., et al., 2008. Evaluation of the terrestrial carbon cycle, future plant geography and climate-carbon cycle feedbacks using five dynamic global vegetation models (DGVMs). Global Change Biology 14, 2015–2039.

CHAPTER 12

Estimating Extinction Risk from Climate Change

How many species face extinction due to climate change? This is one of the most pressing questions of climate change biology. It is featured in front-page stories and on the evening news. "More than a million species wiped out by climate change" and "One third of all of the world's plants and animals gone due to climate change" are examples of some of the headlines. What is the science behind these sensational claims, and how do we know whether climate change poses a major extinction risk? This chapter provides answers to these questions.

The history of this issue began with the birth of climate change biology in the late 1980s. A series of journal articles by researchers such as Rob Peters introduced the concept to the world. A central theme of these papers was that the static nature of protected areas might not be sufficient to protect species whose ranges are becoming dynamic because of climate change. The threat of extinction if protected areas no longer functioned to protect species loomed large. However,

SPOTLIGHT: EXTINCTION RISK

The possibility of losing hundreds of thousands of species because of climate change was raised by Chris Thomas and a group of coauthors in a paper published in *Nature* in 2004. This *Nature* cover story was a synthesis of multiple modeling efforts that had been conducted on the effects of climate change on species' ranges in six regions throughout the world. The researchers leading these regional analyses were all concluding that significant species losses were likely owing to climate change. The idea for the paper emerged when the leaders of many of these studies were invited to a meeting on extinction risk from climate change organized by the International Union for Conservation of Nature. When these researchers put their results side by side, Thomas suggested applying consistent methods to all of the modeled data to derive common extinction risk estimates. The resulting assessment found 18–34% of species at risk of extinction due to climate change in midrange scenarios (Thomas et al., 2004). Applied to the 5–10 million species estimated to exist, these percentage losses represented more than 1 million species. The "million species at risk" estimate was widely reported in headlines and television coverage throughout the world. This estimate was in the press release for the paper, not in the research itself, which generated controversy. The policy and public interest generated by this paper and its media reporting ensures that extinction risk from climate change will remain a major area of research for decades to come.

Source: Thomas, C.D., 2005. Recent evolutionary effects of climate change. Chapter 6. In: Lovejoy, T.E., Hannah, L. (Eds.), Climate Change and Biodiversity. Yale University Press, New Haven, p. 398.

Climate Change Biology. http://dx.doi.org/10.1016/B978-0-12-420218-4.00012-3

interest in climate change-induced extinction did not burst into full popular attention until another paper appeared almost two decades later, in 2004.

The research paper titled "Extinction Risk from Climate Change" created front-page headlines throughout the world when it appeared as the cover story of *Nature* in January 2004 (Figure 12.1). The notion that climate change could drive more than 1 million species to extinction captured popular imagination and the attention of policymakers. The story was covered by CNN, ABC News, NBC News, NPR, and major newspapers and magazines in Europe and the United States (Figures 12.2 and 12.3). It was the subject of debate in the House of Commons and in the U.S. Senate.

An unprecedented round of scientific critique quickly followed the huge popular interest in the story. *Nature* published three articles challenging fundamental

FIGURE 12.1 Cover of the January 8, 2004 issue of *Nature*.
The Thomas et al. (2004) research (see *Spotlight*, previous page) appeared in this issue. *Source: Reprinted by permission from Macmillan Publishers Ltd.*

Last act
The inside story of a stage flop
In G2

Craig Johnston
Footballer to design genius
In G2

2004 – a hectic year in space
In Life
Plus Online & jobs

The Guardian

An unnatural disaster

- Global warming to kill off 1m species
- Scientists shocked by results of research
- 1 in 10 animals and plants extinct by 2050

Under threat

FIGURE 12.2 Front-page headlines in Europe accompanied the *Nature* paper, including this full-page color front page in *The Guardian* (United Kingdom).
Source: Copyright Guardian News & Media Ltd., 2004.

USA TODAY
Cost of global warming: 1 million species

The Washington Post
Warming May Threaten 37% of Species by 2050

San Jose Mercury News
Study: Global warming to doom many species
IT MAY BE WORSE THREAT THAN HABITAT LOSS

THE SUN
Broad study on climate envisions extinctions

San Francisco Chronicle
Dire warming warning for Earth's species

FIGURE 12.3 Headlines in the United States.
Note that some headlines (e.g., *Washington Post* in this sample) misrepresented the timeline of the estimated extinctions (the method used could not discriminate time of extinction; 2050 was the year of the emissions scenario used).

points of the paper, and publications refining or debating the underlying science continue to appear in top research journals. This welter of publications makes for a diverse literature not easily synthesized or accessed, despite the critical policy implications of the research. This paper and the resulting critiques focused on only one of several possible methods to assess risk. It is necessary to examine several lines of evidence across several disciplines to get a deeper view of extinction risk from climate change.

EVIDENCE FROM THE PAST

One way to assess extinction risk associated with climate swings is to examine biological responses in past times of rapid climate change. Association between extinction spasms and past climate change would increase the prospects for heavy extinctions associated with human-induced climate change.

It is clear that major extinctions have resulted from past climate change, as reviewed in Chapter 9. Most, if not all, of the major extinction events in the Earth's history have been directly or indirectly associated with climate change. Yet some major climate changes have not resulted in extinctions, and the conditions under which such changes took place may not be analogous to the current climate. If today's climate is much warmer or much cooler than climatic starting points for past changes, would a similar change still cause extinctions?

SPOTLIGHT: EXTINCTION AND WARMTH GO TOGETHER

Examination of long-term extinction trends indicates that extinction risk may be greater in warm climates (Mayhew et al., 2008). During the past 500 million years, high-CO_2 warm periods have had a significant correlation with higher rates of extinction. This deep-time perspective is important because the record of the more recent past provides poor analogies for future, human-driven warming. During the glacial cycles of the past 2 million years, warming events have typically occurred from cool ice age climates. No sustained warming has occurred from a warm climate, and few warm climates have existed for more than 10,000 years. However, in deeper time, warm climates have existed for millions of years, sometimes with pronounced warming during already warm spells, such as occurred in the rapid warming of the Paleocene–Eocene thermal maximum. The record of Mayhew et al. shows that in these deeper-time warm periods, extinction was elevated—a discomforting trend for human-induced warming in an already warm climate and with extensive habitat destruction standing in the way of natural adaptation.

Source: Mayhew, P.J., Jenkins, G.B., Benton, T.G., 2008. A long-term association between global temperature and biodiversity, origination and extinction in the fossil record. Proceedings of the Royal Society B: Biological Sciences 275, 47–53.

There is no good analogue for today's climate in the past because the Earth has undergone a long cooling period (tens of millions of years), culminating in the ice ages and, most recently, in a strong warming. Thus, only one or two climates in the past 50 million years have had a mean global temperature similar to that of today for an extended period of time, and none have had a similar climatic history, save for perhaps the last interglacial period.

Therefore, there are almost no relevant periods of the past with which to make a comparison about extinction and human-induced climate change. The last interglacial period seems a close approximation, but there is no record of massive extinction from that time. Another approximation would be the warming that led to the current interglacial period. As discussed in Chapter 9, there were many extinctions associated with this warming, especially in the Americas, but this was also a time when human effects were becoming strongly evident, so it is difficult to sort out the exact causes of these extinctions or explain their regional bias.

Thus, the past tells us that extinctions and climate change go together, but not always. This does not shed much light on the probability of extinctions from human-induced climate change during the next few centuries. There is some indication that rapid, large climate changes are more likely to lead to extinctions and that climate change in concert with human activity leads to extinctions—evidence that should raise alarm bells about possible future impacts of climate change. However, for quantitative estimates of how many species may be lost or which species may be most vulnerable, we have to turn elsewhere.

ESTIMATES FROM SPECIES DISTRIBUTION MODELING

The most prominent method for assessing future extinction risk from climate change is that used in the 2004 *Nature* paper that brought so much attention to the issue. The authors of that study used species distribution models (SDMs) coupled with the species–area relationship (SAR) to estimate extinction risk in six regions representative of different biomes. The resulting extinctions risks are presented in Table 12.1.

The methods behind SDMs were described in Chapter 11. SDMs allow estimates of future species' range size. SDMs were constructed for hundreds of species in each of the six regions of the study. Some showed species' ranges disappearing altogether by 2050, whereas others showed species' ranges declining by 2050 and probably headed for extinction. Most showed some decline but not enough to definitively state that a species was headed for extinction. However, some of these declining-range species might become extinct. What was needed was a method for assessing extinction risk in multiple species over the long term. The method used for this was the SAR.

Table 12.1 Extinction Risk Estimates for Multiple Regions and Taxa[a]

Taxon	Region	With Dispersal			No Dispersal		
		Minimum Expected Climate Change	Midrange Climate Change	Maximum Expected Climate Change	Minimum Expected Climate Change	Midrange Climate Change	Maximum Expected Climate Change
Mammals	Mexico *n*=96	2, 4, 5 **5**	2, 5, 7 **8**	—	9, 14, 18 **24**	10, 15, 20 **26**	—
	Queensland *n*=11	10, 13, 15 **16**	—	48, 54, 80 **77**	—	—	—
	South Africa *n*=5	—	24, 32, 46 **0**	—	—	28, 36, 59 **69**	—
Birds	Mexico *n*=186	2, 2, 3 **4**	3, 3, 4 **5**	—	5, 7, 8 **9**	5, 7, 8 **8**	—
	Europe *n*=34	—	—	4, 6, 6 **7**	—	—	13, 25,38 **48**
	Queensland *n*=13	7, 9, 10 **12**	—	49, 54, 72 **85**	—	—	—
	South Africa *n*=5	—	28, 29, 32 **0**	—	—	33, 35, 40 **51**	—
Frogs	Queensland *n*=23	8, 12, 18 **13**	—	38, 47, 67 **68**	—	—	—
Reptiles	Queensland *n*=18	7, 11, 14 **9**	—	43, 49, 64 **76**	—	—	—
	South Africa *n*=26	—	21, 22, 27 **0**	—	—	33, 36, 45 **59**	—
Butterflies	Mexico *n*=41	1, 3, 4 **7**	3, 4, 5 **7**	—	6, 9, 11 **13**	9, 12, 15 **19**	—
	South Africa *n*=4	—	13, 7, 8 **0**	—	—	35, 45, 70 **78**	—
	Australia *n*=24	5, 7, 7 **7**	13, 15, 16 **23**	21, 22, 26 **33**	9, 11, 12 **16**	18, 21, 23 **35**	29, 32, 36 **54**
Other invertebrates	South Africa *n*=10	—	18, 15, 24 **0**	—	—	28, 46, 80 **85**	—
Plants	Amazonia *n*=9	—	—	44, 36, 79 **69**	—	—	100, 100, 99 **87**
	Europe *n*=192	3, 4, 5 **6**	3, 5, 6 **7**	4, 5, 6 **8**	9, 11, 14 **18**	10, 13, 16 **22**	13, 17, 21 **29**
	Corraco *n*=163	—	—	—	38, 39, 45 **66**	48, 48, 57 **75**	—
	South Africa Proteaceae *n*=243	—	24, 21, 27 **38**	—	—	32, 30, 40 **52**	—
All species		9, 10, 13 **11** *n*=604	15, 15, 20 **19** *n*=832	21, 23, 32 **33** *n*=324	22, 25, 31 **34** *n*=702	26, 29, 37 **45** *n*=995	38, 42, 52 **58** *n*=259

[a]*Projected percentage extinction values are given based on species area (for z=0.25) and Red Data Book (bold) approaches. The three species area estimates are ordered in each cell with method 1 given first, followed by method 2 and then method 3. Values for "all species" are based on both these raw values and estimates interpolated for the empty (—) cells. In each instance,* n *is the number of species assessed directly.*
Source: Thomas et al. (2004).

SPECIES–AREA RELATIONSHIP

The SAR is derived from island biogeography theory. It is an empirically measured relationship between the size of an area and the number of species it contains. The larger the area, the greater the number of species present. This relationship has been found to hold in virtually all terrestrial systems studied, from island archipelagos to large continental areas. Furthermore, the rate at which species accumulate with area seems to follow one of two paths—one curve for islands and another, different, curve for continental areas (assuming the sampling of the continental area is done in a nested manner). The main difference between the curves, when the data are log-transformed, is the slope of the curve. The SAR exponent (z value) ranges between 0.15 and 0.39, with 0.25 being taken as typical for continental ensembles.

The idea of applying the SAR to future climate is that areas of declining climatic suitability are comparable to decreasing area of suitable habitat. As a species' suitable climatic space declines, its population should decline, rendering it more vulnerable to stochastic extinction. In the long term, this chance of extinction should be reflected in a relationship between area and numbers of species: just as larger islands hold more species than smaller islands, smaller suitable climatic space should hold fewer species than larger suitable climatic space. Because climatic space was declining for most of the species in the six regions studied, this could be translated, via SAR, into an estimate of extinction risk. This is how the extinction values in Table 12.1 were derived.

The analogy of loss of land area to loss of climate space has one important difference: land areas are shared by many species, but suitable climate space is unique to each species. Thus, to use SAR for estimates of climate change-based

STOCHASTIC EXTINCTION

Populations and species go extinct stochastically, not deterministically. Fluctuations in population size occur in all species. When mean population size is lower than the size of fluctuations, extinction is highly likely to ensue.

The magnitude of fluctuations varies, so extinction is a matter of chance. Climate change can lead to extinctions by decreasing suitable habitat and lowering overall population size, by increasing population fluctuations, or both.

extinction risk, some way has to be found to combine the suitable climate area estimates for each species, so they can be fit to the SAR relationship. Several approaches have been tried, and all yield similar results. An important criticism of the SDM/SAR approach is that these methods cannot be tested: whereas the SAR is derived from many careful measurements in the real world, reassembling SDM range losses into a SAR extinction risk estimate cannot be confirmed by real-world measurements, at least not until the range shifts and extinctions have taken place, at which point it will be too late. In addition, it is not entirely clear that SARs built up over evolutionary time by the interaction of natural forces and geography will decompose on the same line owing to climate change. As a result, some scientists accept the SDM/SAR approach as the best available, whereas others reject it as a theoretically flawed and untestable extension of SARs.

SPOTLIGHT: EXPOSURE

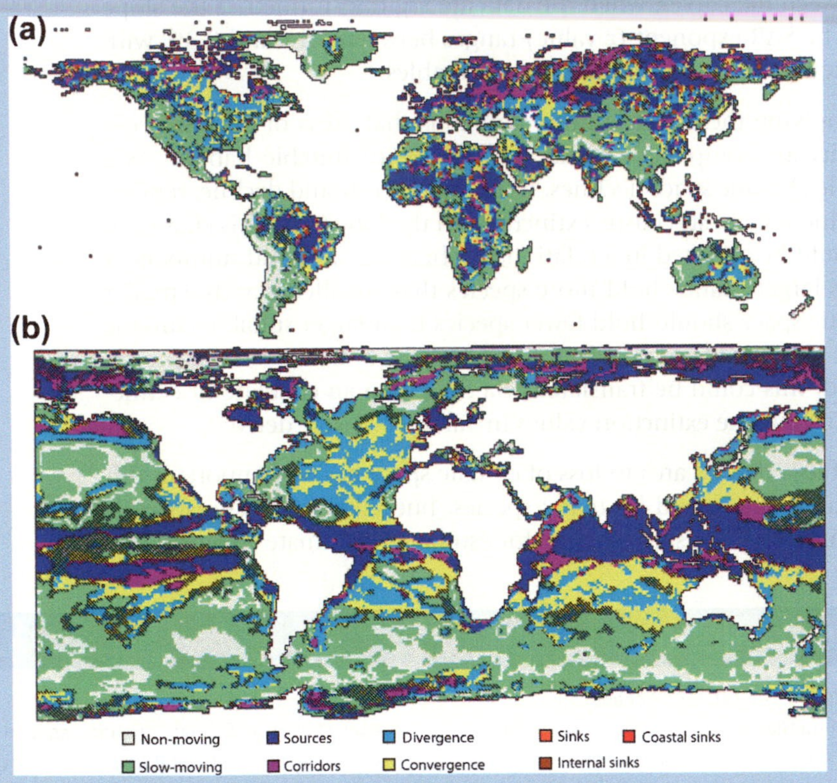

Velocity of climate change trajectory classes in both terrestrial (above) and marine (below)realms. Trajectory classes have been used to characterize the type of species range shifts dominant in an area. Sources (blue) sinks (red), areas of slow shifts (green) and convergence (yellow) are indicated.

Extinction risk, like other climate change impacts, is the product of exposure and sensitivity. Research into the exposure side of the equation has used the velocity of climate change (see Fig. 2.24) to estimate the type and magnitude of effects on species. This method allows identification of possible sources, sinks and dead-ends as species move in response to climate change.

Source: Burrows, M.T., Schoeman, D.S., Richardson, A.J., Molinos, J.G., Hoffmann, A., Buckley, L.B., Poloczanska, E.S., 2014. Geographical limits to species-range shifts are suggested by climate velocity. Nature, 507 (7493), 492–495.

A QUESTION OF DISPERSAL

What if a species cannot move? A species that cannot move, such as a plant species, will inevitably find its suitable climate shrinking: its suitable climate in the future will simply be the overlap of its suitable present climate with suitable future climate. On the other hand, a species that is well dispersed will be able to move with changing climatic conditions, fully occupying its suitable future climatic space.

The combination of SDM and SAR, clever as it is, cannot account for species dynamics. SDMs can state only what areas have suitable climate for the species in the future; they state nothing about whether the species can reach those places. Thus, the extinction risk estimates derived by combining SDM and SAR can make only two assumptions: either the species cannot disperse at all and is limited to ever-shrinking suitable climate within its present range or the species is perfectly well dispersed and can occupy all areas with suitable climate in the future. These are not very satisfactory assumptions because almost all species will fall somewhere in between these two extremes. These two assumptions do bracket the range of possibilities, however, so it is possible to state that the perfect-dispersal assumption puts a lower bound on the extinction risk of a species and the no-dispersal assumption puts an upper bound on the estimate. The results in Table 12.1 are therefore presented as two alternative values—one for no dispersal and one for perfect dispersal.

THE PROBLEM WITH ENDEMICS

One criticism of the SDM/SAR method for estimating extinction risk is that most of the species modeled were endemic to the regions being studied, so the results apply to only a subset of species of the region. Because species area curves are calculated for all species in a region, limiting the study to endemics may make the use of SAR inappropriate. There are good reasons for limiting the species modeling to endemics, however.

It is best practice in SDMs for future climates to model only species whose entire range falls within the study region. This is because a species that has range outside of the region modeled has climatic tolerances that cannot be captured in the statistics of SDMs. For instance, Anna's hummingbirds occupy both dry lowland areas and mountains in California. If you created an SDM for Anna's using data only from a lowland region, the model would "see" only relationships with relatively dry habitats. Such an SDM would be wrong for the current range: it would miss the montane distribution of the species. Worse, if the SDM were applied to future climates, it would also miss all of the future climatic space similar to the montane conditions. Thus, creating SDMs for only part of a species' range can lead to errors.

The solution to the problem of endemics lies in changing the type of SAR applied. Because it is not wise to expand the SDM to nonendemics, a different kind of SAR is needed—one that applies particularly to endemics. Fortunately, at approximately the same time that the first climate change extinction risks were being calculated, a new type of SAR was being derived—one that applied especially to endemics.

SPOTLIGHT: SENSITIVITY

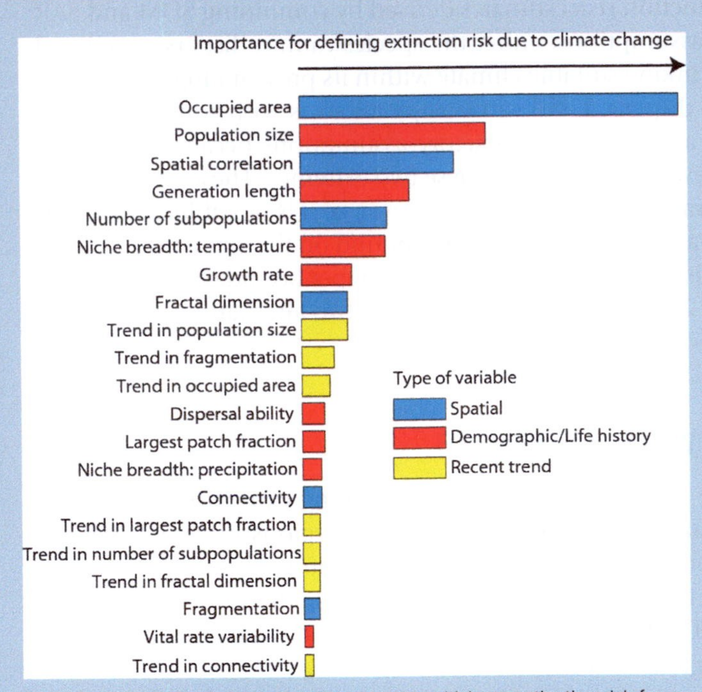

Relative importance of predictors of species' sensitivity to extinction risk from climate change. The top six predictors dominate, including niche breadth (see Fig. 12.4).

The sensitivity side of the exposure+sensitivity equation may be even more important in determining species risk of extinction. Sensitivity to extinction may be estimated from life history traits such as niche breadth and population size. Probabilities of extinction estimated from sensitivity factors have been compared to estimates from complex coupled distribution-population models. The modeled estimates could be approximated by the sensitivity factors, with a few factors dominating the results (see figure).

Source: Pearson, R.G., Stanton, J.C., Shoemaker, K.T., Aiello-Lammens, M.E., Ersts, P.J., Horning, N., Akçakaya, H.R., 2014. Life history and spatial traits predict extinction risk due to climate change. Nature Climate Change, 4 (3), 217–221.

The endemics–area relationship (EAR) is similar in principle to a SAR, but it is derived especially for endemics. Like the SAR, the EAR is also a curve with an exponent that determines the rate of species accumulation. However, species accumulate more slowly using the EAR because it counts only endemics. It is this slower accumulation curve that is best applied to extinction risk estimates using species distribution modeling. The EAR was not available for the calculations used for Table 12.1, so these values are slightly inflated.

CHECKING THE ESTIMATES

The species-based estimates of extinction risk have been confirmed by two independent methods. The second approach used dynamic global vegetation models (DGVMs) and EAR to estimate extinction risk for biodiversity hot spots. The global biodiversity hot spots have been defined based on areas of high species endemism. Because endemism rates are known for hot spots, it can be assumed that loss of habitat type for a hot-spot endemic is a global loss of habitat for that species. Based on this assumption, projection of future extent of vegetation types for a hot spot from a DGVM can be used to estimate extinction risk for hot-spot endemics. If a habitat type decreases within a hot spot, it is as if a small habitat island for the endemics occupying that habitat has just gotten smaller.

The DGVM hot-spot approach applied EAR to areas of lost habitat within hot spots to estimate extinction risk. This approach therefore had the double advantage of taking an independent approach and applying the correct SAR for endemics (the EAR). The results obtained using this approach are presented in Table 12.2. They are lower than the species-based estimates, as is expected using the slower-accumulating EAR. However, they are not negligible: they give strong support to the notion of significant extinction risk associated with climate change.

The second test came in a study of extinction risk in lizards. Lizard populations in Mexico have become extinct in areas that have warmed significantly. A model of these population extinctions then predicted population extinctions accurately on other continents. The population extinction projections were used to estimate species-level extinction risk. This ecophysiological method projected lizard extinctions of 6% in 2050 scenarios and 20% in 2080 scenarios. The hot-spots analysis and the lizard analysis suggest that the low end of the range of original estimates may be appropriate for 2050, whereas the lizard analysis suggests that the high end of the original estimates may be more similar to end-century risks.

Table 12.2 Projected Species-Area-Based Percentage of Extinctions of Endemic Species in 12 Hot Spots Judged to be Especially Vulnerable to Global Warming[a]

Hot Spot[b]	Global Vegetation Model	Perfect Migration				Zero Migration			
		Broad Biome Definition		Narrow Biome Definition		Broad Biome Definition		Narrow Biome Definition	
		Broad Specificity	Narrow Specificity	Broad Specificity	Narrow Specificity	Broad Specificity	Narrow Specificity	Broad Specificity	Narrow Specificity
California Floristic Province	BIOME3	2.4	30.9	2.5	27.8	4.5	46.4	5.2	41.9
(2125; 71)	MAPSS	3.0	4.0	7.8	40.9	5.3	6.0	14.	53.5
Cape Floristic Region	BIOME3	2.4	2.3	4.5	5.8	2.7	2.8	7.5	8.0
(5682; 53)	MAPSS	11.8	28.6	17.4	52.4	15.4	43.9	21.9	68.0
Caribbean	BIOME3	3.1	2.7	3.1	2.8	4.0	3.6	4.4	3.8
(7000; 779)	MAPSS	7.2	12.1	10.0	15.5	9.0	25.3	19.0	48.5
Indo-Burma	BIOME3	1.9	17.8	5.1	18.	2.7	27.1	7.2	31.2
(7000; 528)	MAPSS	5.5	23.8	6.7	29.6	6.2	33.6	11.9	40.
Mediterranean Basin	BIOME3	1.9	10.6	2.9	9.7	3.9	16.4	6.4	24.5
(13,000; 235)	MAPSS	3.7	4.4	5.6	26.6	8.1	9.9	12.4	44.3
New Caledonia	BIOME3	0.0	0.0	0.0	0.0	0.0	0.0	0.0	0.0
(2551; 84)	MAPSS	0.0	0.0	18.8	75.0	0.0	0.0	18.8	75.0
New Zealand	BIOME3	2.5	5.3	2.5	5.3	2.8	5.5	2.8	5.5
(1865; 136)	MAPS	4.6	24.8	4.6	29.1	6.1	40.7	10.4	38.7
Polynesia & Micronesia	BIOME3	2.2	16.6	3.0	17.7	2.2	16.6	4.1	27.8
(3334; 223)	MAPSS	3.8	42.9	8.1	55.3	5.1	43.8	14.1	58.2
Mountains of South Central China	BIOME3	4.3	3.1	4.3	12.1	8.0	8.9	8.2	28.9
(3500; 178)	MAPSS	3.5	27.3	8.8	21.6	9.5	54.6	17.3	43.5
Succulent Karoo	BIOME3	2.4	19.1	3.0	22.5	3.2	27.9	4.1	30.2
(1940; 45)	MAPSS	7.0	30.1	10.1	46.7	8.8	34.4	19.3	70.6
Southwest Australia	BIOME3	2.3	9.8	5.3	18.4	2.8	10.1	7.3	22.6
(4331; 100)	MAPSS	15.2	32.2	17.2	38.7	18.1	41.8	28.2	66.1
Tropical Andes	BIOME3	2.7	10.6	4.0	13.9	6.4	31.0	10.5	32.2
(20,000; 1567)	MAPSS	3.5	13.0	3.7	13.9	9.8	29.7	13.8	47.0

[a]*Percentages are shown for two migration scenarios, two biome breadth definitions, two levels of biome specificity, and two global vegetation models (BIOME3 and MAPSS).*
[b]*Numbers of endemic plant and vertebrate species, respectively, are shown in parentheses below hot-spot names.*

NOT JUST ABOUT POLAR BEARS ANYMORE

Many concerns about climate-related extinctions have been focused on arctic species (such as polar bears) occupying areas where the most severe changes in climate have so far occurred. The hot-spots study showed that even moderate amounts of climate change could translate into major extinctions risk and that the largest extinction risk in the long term may lie in the tropics. Other studies have confirmed this view.

A review of bird species sensitive to climate change found that most resided in hot spots. This study searched for bird species that occupy narrow elevational bands, predominantly on the slopes of mountains. Such species are likely to have small temperature tolerances because the top and bottom of their ranges are divided by relatively small temperature gradients up the mountain slope. More of these sensitive bird species occupied hot spots than the rest of the planet combined. The tropical Andes hot spot was especially rich in climate change-sensitive birds, with more than the rest of the hot spots combined (Figure 12.4).

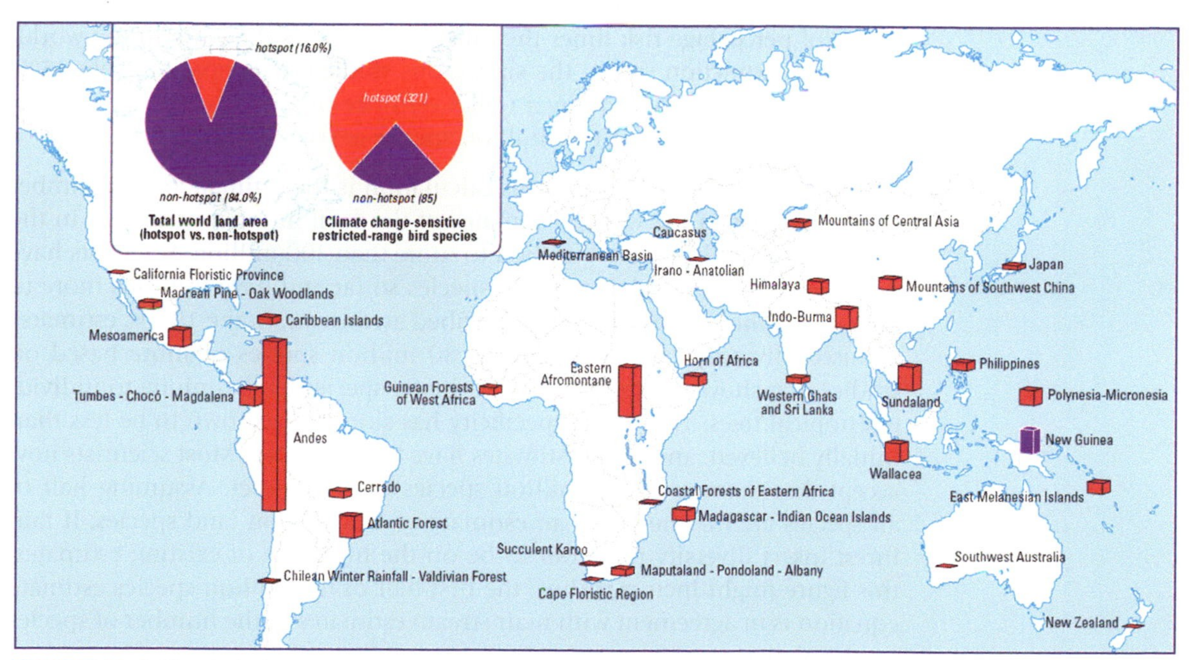

FIGURE 12.4 Global distribution of climate change-sensitive bird species.
The proportions of restricted-range endemic birds with elevational ranges of less than 1000 m are shown by region. Elevation range is indicative of temperature niche breadth, one of the most important predictors of sensitivity to extinction due to climate change. The inset shows the proportion of global land area in biodiversity hot spots (left chart, in red) and the number of climate change-sensitive bird species in global biodiversity hot spots (right chart, in red).

The first examples of extinctions linked to climate change have come from the tropics, further strengthening the view of the hot spots in particular and the tropics in general as being vulnerable to climate change. The extinctions of the golden toad in Costa Rica and dozens of other amphibians in Central and South America have been linked to mortality due to a fungal outbreak caused by climate change.

ARE A MILLION SPECIES AT RISK?

The most cited claim about extinction and climate change is that 1 million species are at risk. The million species number comes not from any of the studies of extinction risk from climate change but, rather, from the press release of the 2004 study. In that press release, the lead author of the study took the average percentage extinction risk from the six regions studied and applied it to the Earth as a whole. This extrapolation was widely picked up by the media and reported by television, magazines, and newspapers throughout the world (Figure 12.5).

The math behind the 1 million species estimate is a straightforward multiplication of percentage risk times the number of species there are in the world. The mean extinction risk in the six regions studied was approximately 25%; multiplying this proportion by a rough estimate of 5 million land species in the world yields a total of 1.25 million species at risk.

The most controversial part of this calculation is the estimate of the number of land species in the world. Estimates of the total number of species in the world vary from less than 4 million to more than 100 million. Scientists have described approximately 2 million species so far and have millions more to go. Most of the species still to be described are insects. In the 1990s, estimates of insect diversity ranged as high as 80 million species or more based on studies that showed large numbers of insect species being unique to individual tropical trees. This insect specificity has since been shown to be less than initially believed, and these estimates have been reduced. Most scientists now accept that there are 8–10 million species on the planet. Assuming half of all species are marine yields an estimate of 4–5 million land species. If rain forest insect diversity turns out to be on the high side of existing estimates, this figure might increase. Thus, the first half of the million species estimate equation is in agreement with mainstream estimates of the number of species on the planet.

The estimate of the proportion of species at risk is based on the assumption that the six regions studied in the 2004 paper are representative of the whole planet

FIGURE 12.5 **Graffiti in a Washington, DC, Metro stop, 2004.**
Release of extinction risk estimates corresponded with an increase in interest in climate change in the United States—a trend that would spike with the release of the film *An Inconvenient Truth*. *Source: Photo courtesy Steven Schneider.*

(Figure 12.6). These six regions are distributed on five continents in more than 12 biomes. They were not systematically selected to be representative, but they cover a wide enough variety of terrestrial settings to be reasonably considered representative. One bias in the regions was overrepresentation of biodiversity hot spots. However, because hot spots represent more than two-thirds of the world's species as endemics, inclusion of hot spots should increase the confidence in the result.

The number of species at risk of extinction owing to climate change is therefore almost certainly in the hundreds of thousands and probably more than 1 million. Using the most conservative estimates of numbers of species on the planet (approximately 3 million) and the lower percentage risk estimates from the DGVM/EAR study yields a lower-bound estimate of approximately 250,000 species. Using upper-bound estimates of number of species (approximately 100 million) and the SDM/SAR risk proportions gives an upper-end estimate of 20 million species or more at risk, most of them insects. Because both the SDM/SAR and the DGVM/EAR studies used 2050 climate change scenarios, the numbers of species at risk would be expected to increase significantly later in the century. Although there is much uncertainty on both sides of the equation, "a million species at risk" seems to be a reasonable first estimate.

FIGURE 12.6 Extinction risk.
The first estimates of the extinction risk from climate change and subsequent research inspired art from political cartoons to technical report covers. This 2009 image appeared on the cover of a report on Central American biodiversity and climate change prepared by The Water Center for the Humid Tropics of Central America and the Caribbean (CATHALAC). *Source: Image courtesy Luis Melillo, copyright CATHALAC.*

WHY THE FUTURE MAY NOT BE LIKE THE PAST

Unusual future conditions darken all estimates of extinction risk from climate change. Future climate and levels of habitat destruction will be very different from anything in the recent past. The present climate is warm and stable relative to the previous 2 million years, whereas levels of habitat destruction are the highest in the history of the planet. Hence, the set of conditions that species will have to navigate as climates warm will be unlike those any have faced for much of their evolutionary history.

Warming on an already warm climate will push species into uncharted climatic territory. Montane species in particular may find that they have nowhere to move. Other species will find that conditions for which they evolved no longer exist. Rare long-distance jumps to suitable climate elsewhere will be less likely because more than half the natural habitat of the planet has been replaced by human uses.

The interaction of climate change and human land use is particularly worrying. If this relationship is synergistic, actual climate-related extinctions may well outnumber the estimates, all of which fail to take into account threat synergies. Present estimates also fail to include human population dynamics, so simple expansion of human land uses and human responses to climate change may be more deleterious than current estimates reflect.

Overall, the past gives us much less insight into the consequences of future change than would be desirable for impact prediction and response. As a result, much of our understanding of extinction risk comes from modeling studies with inherent uncertainties of their own. The little independent evidence available gives some reason to believe that the modeling studies may be underestimates of the extinction risk from climate change.

FURTHER READING

Hoegh-Guldberg, O., Mumby, P.J., Hooten, A.J., Steneck, R.S., Greenfield, P., Gomez, E., et al., 2007. Coral reefs under rapid climate change and ocean acidification. Science 318, 1737–1742.

Kolbert, E., 2014. *The Sixth Extinction: An Unnatural History*. Henry Holt and Company.

Malcolm, J.R., Liu, C., Neilson, R.P., Hansen, L.A., Hannah, L., 2006. Global warming and extinctions of endemic species from biodiversity hotspots. Conservation Biology 20, 538–548.

Sinervo, B., Mendez-de-la-Cruz, F., Miles, D.B., Heulin, B., Bastiaans, E., et al., 2010. Erosion of lizard diversity by climate change and altered thermal niches. Science 328, 894–899.

Thomas, C.D., Cameron, A., Green, R.E., Bakkenes, M., Beaumont, L., et al., 2004. Extinction risk from climate change. Nature 427, 145–148.

CHAPTER 13

Ecosystem Services

Ecosystems provide goods and services that help sustain human livelihoods. Provision of ecosystem services may be altered by climate change, because all are made up of species whose growth, productivity, and location are determined by climate.

Ecosystem services can be grouped into provisioning, supporting, regulating, and cultural services. Provisioning services include water and food taken directly from ecosystems, such as drinking water and fish or wildlife captured for food. Cultural services are the values ecosystems play in human cultures, such as recreational or spiritual values. Regulating services help maintain planetary processes in ways beneficial to people, especially in regulating climate. Supporting services help maintain other ecosystem services and include nutrient cycling and primary production.

The Millennium Ecosystem Assessment (MEA) helped define ecosystem services. The MEA was a large collaborative scientific synthesis about the knowledge of the state of the world's ecosystems at the turn of the millennium. It provided the framework of the four ecosystem service categories that is widely used today. The role of ecosystems in supporting human well-being and poverty alleviation has received much more attention since publication of the MEA.

What happens to ecosystem services as climate changes is the subject of this chapter. Some ecosystem services may increase because of climate change, whereas others will decline. As maintaining ecosystem services is one goal of conservation, understanding these changes is an essential ingredient in designing conservation responses that are robust to climate change. Ecosystems may also help people adapt to climate change, enhancing ecosystem services in the face of change. Five key examples of ecosystem services changing are explored below.

FOOD PROVISION—MARINE FISHERIES

Marine fisheries provide food directly from ecosystems and are sensitive to climate change. Most human food production comes from heavily modified agricultural ecosystems. But fisheries are not generally manipulated and

Climate Change Biology. http://dx.doi.org/10.1016/B978-0-12-420218-4.00013-5

supplemented (e.g., with fertilizer) but rather produce food from ecosystems that are minimally managed. The direct link from ecosystem to human food is especially strong in marine fisheries and the impact of climate on that production is clear.

Fluctuations in marine fisheries catch is correlated with water temperature for many species (Figure 13.1). Sardine, salmon, mackerel, and other species show catch numbers that fluctuate closely with indices of atmospheric circulation and water temperature (see Chapter 5, Fig. 5.19). Species comprising more than half the catch in the Atlantic and Pacific oceans are influenced by such climate indices. Because climate change is expected to influence global atmospheric circulation patterns, these catches will respond strongly to climate change.

Water temperature at various depths influences the distribution of commercially important fish (Figure 13.2). Cod populations fluctuate in response to deep

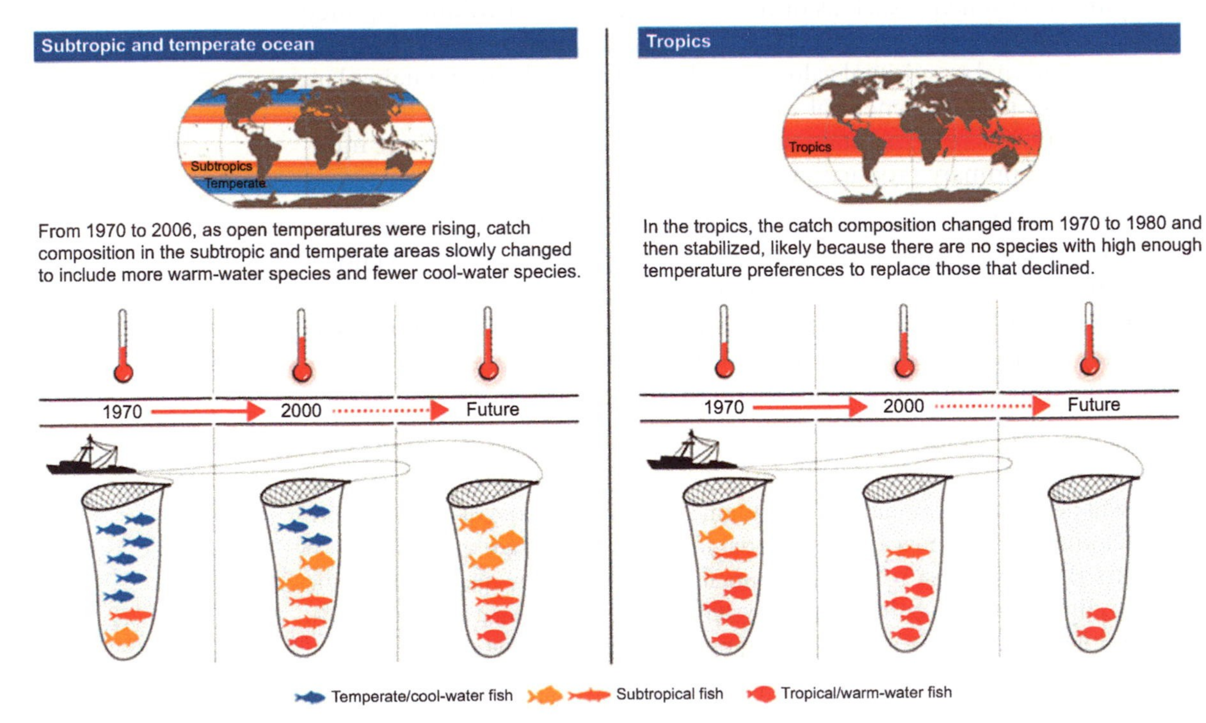
Subtropic and temperate ocean
Subtropics
Temperate
From 1970 to 2006, as open temperatures were rising, catch composition in the subtropic and temperate areas slowly changed to include more warm-water species and fewer cool-water species.
1970
2000
Future
Tropics
Tropics
In the tropics, the catch composition changed from 1970 to 1980 and then stabilized, likely because there are no species with high enough temperature preferences to replace those that declined.
1970
2000
Future
Temperate/cool-water fish
Subtropical fish
Tropical/warm-water fish
These shifts could have negative effects including loss of traditional fisheries, decreases in profits and jobs, conflicts over new fisheries that emerge because of distribution shifts, food security concerns, and a large decrease in catch in the tropics.

FIGURE 13.1 Warming oceans are reshaping fisheries.
Tropical fish species are moving toward the poles from the equator, changing catch composition in many areas and decreasing catch in some areas. Models project this trend continuing through the end of this century. *Source: Pew Oceans Center.*

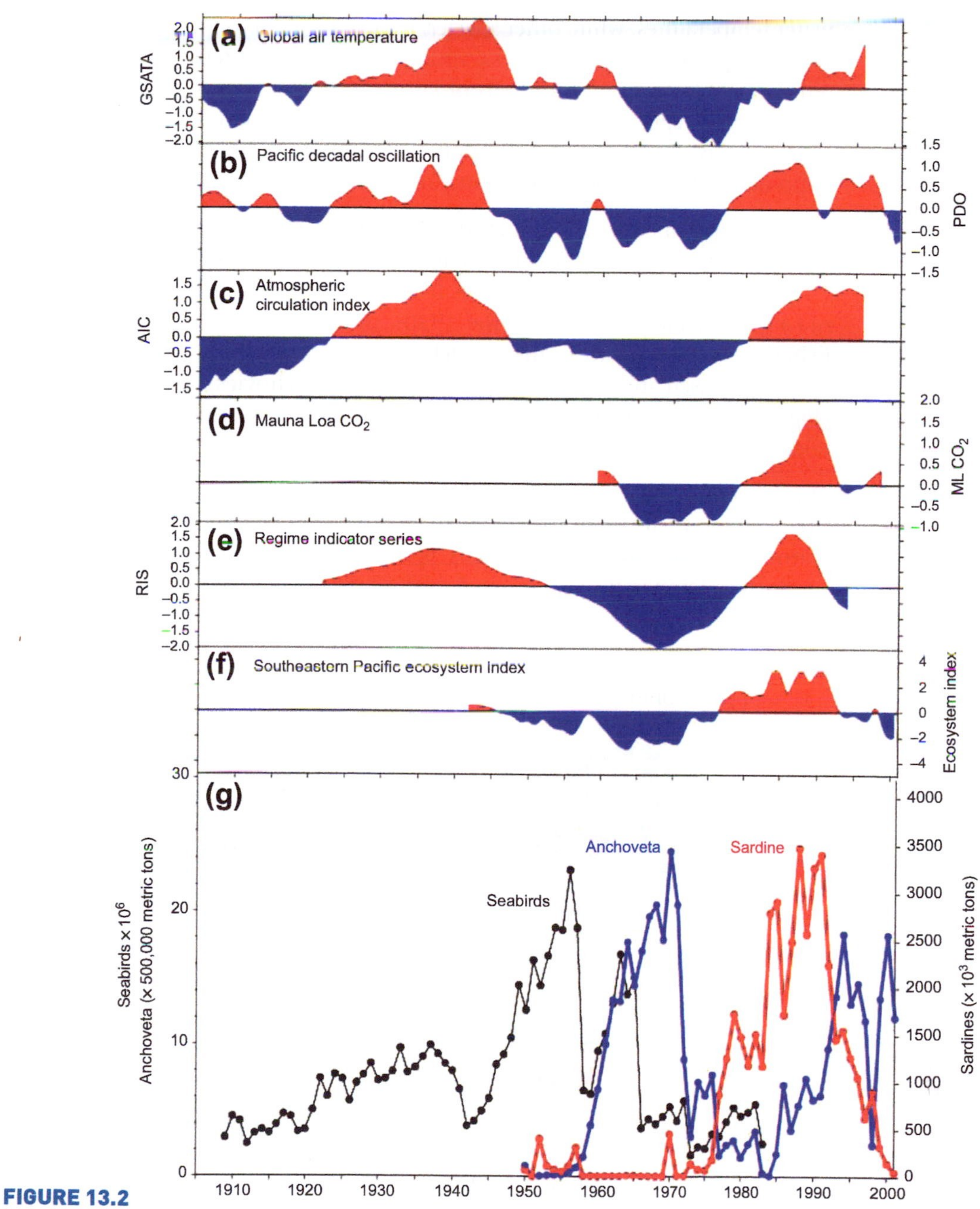

FIGURE 13.2

Sardine and anchovy catches (g) track global temperature trends (a) and local atmospheric circulation indices (b–f). Peruvian anchovy (anchoveta) catches are high in cool periods, sardine catches are high in warm periods. Cool and warm periods oscillate on an approximately 25-year cycle. In (g) seabirds are used as a proxy for anchoveta in years prior to reliable catch records being kept. *Source: Chavez et al. (2003). Reproduced with permission from AAAS.*

water temperatures, while other species, especially plankton, respond to surface water temperatures. Some species respond rapidly to short-term changes in water temperature, turning up hundreds of kilometers outside of their normal range, whereas other species respond to the evolution of water temperatures over years or decades as climate changes. Overall range shifts in marine fish are projected to approach 50 km per decade, resulting in major displacements of fisheries.

Sardines and anchovies provide an illustration of climatic controls at work on large fisheries. The Peruvian anchovy (anchoveta) fishery at its peak in the 1960's was the largest single-species fishery in the world, whereas the California sardine fishery boomed in the 1930s and 1940s. Peak years in these two fisheries alternate, with bumper catches of anchovies when the sardine fishery is depressed and large sardine catches when anchovies are relatively scarce (Figure 13.3). The alternation between "sardine regime" and "anchovy regime" takes about 25 years and corresponds to large-scale changes in climate.

The Pacific Ocean experiences a shift in water temperatures and nutrients in the "sardine cycle" about every 25 years. The anchovy fishery is high in the warm phase and the sardine fishery is high in the cool phase. Alterations in sea surface temperatures and wind strength result in water pooling and temperature gradient differences across the Pacific. The conditions that favor the

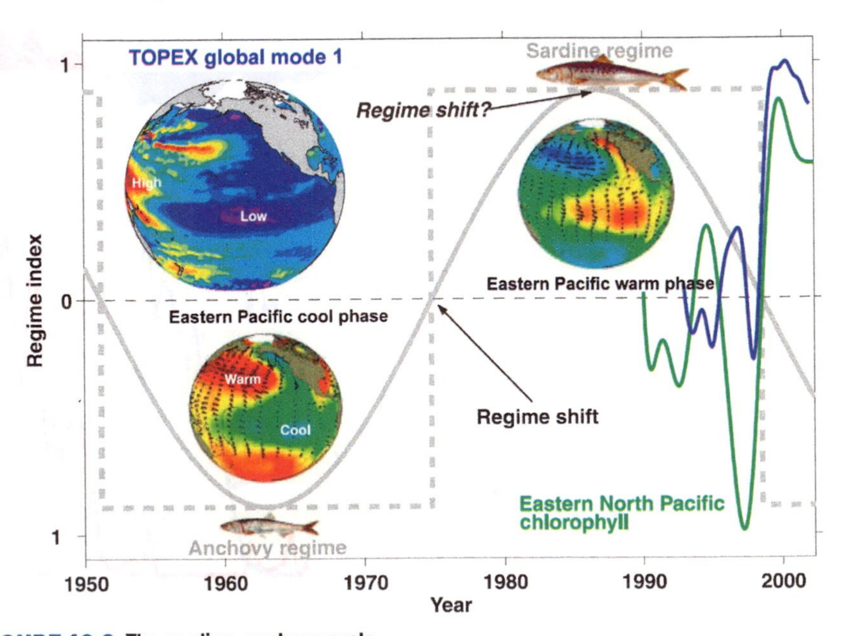

FIGURE 13.3 The sardine–anchovy cycle.
Idealized representation of the catch cycle between anchovy and sardine, showing regional water temperature patterns characteristic of each phase of the cycle. *Source: Chavez et al. (2003). Reproduced with permission from AAAS.*

anchovy fishery are associated with cooler global temperatures once the long-term trend of warming from human pollution is removed.

Many other species are affected in these Pacific regime shifts. Whole-scale ecosystem rearrangements accompany the temperature and circulation changes. In areas such as the Galapagos, these changes affect the sea life that is the basis of the tourism industry. In other regions fisheries other than sardine and anchovy are affected.

The human consequences of these ecosystem service changes are immense. In California, the entire sardine fishery collapsed when the Pacific regime switched from warm to cool (sardine to anchovy) around 1950, resulting in the loss of an entire processing industry. Changes in the Peruvian anchoveta fishery associated with these regime changes are measured in millions of tons and thousands of jobs.

Not all of the mechanisms involved in the fishery changes are understood, but it is highly likely that climate change will result in additional changes with large implications for fisheries. For instance, the sardine catch in Japan increases when local waters cool and become more productive, whereas sardine catches off California and Peru increase when local waters warm and become less productive. The changes that take place in these fisheries are clearly linked to climate change, but the exact mechanisms are still being unraveled.

WATER PROVISIONING

Water provisioning is affected by climate change when rainfall decreases, lowering streamflows, or when warmer temperatures increase the need for water without an accompanying increase in supply. This can accentuate water stress, the net balance between water availability and use. Climate change can alter the amount of water captured by cloud forests and it can cause glaciers to melt, with massive consequences for downstream water availability in both the short and the long term.

Water stress becomes severe when the proportion of surface water used for domestic, industrial, and agricultural purposes exceeds about 40%. About 2 billion people worldwide live under conditions of high water stress, and climate change is projected to increase these numbers. Population growth also influences future water stress.

Climate change is projected to accentuate water stress in eastern Africa, eastern North America, northern South America, and southeast Asia (Figure 13.4). Climate change eases water stress in some models, primarily through increase in precipitation in western North America and central Asia. Population growth will drive increasing water stress in Canada, Mexico, and Brazil and across Africa, Asia, and Europe. The combined effects of climate change and population increase mean that increasing water stress will be experienced in most parts of the planet.

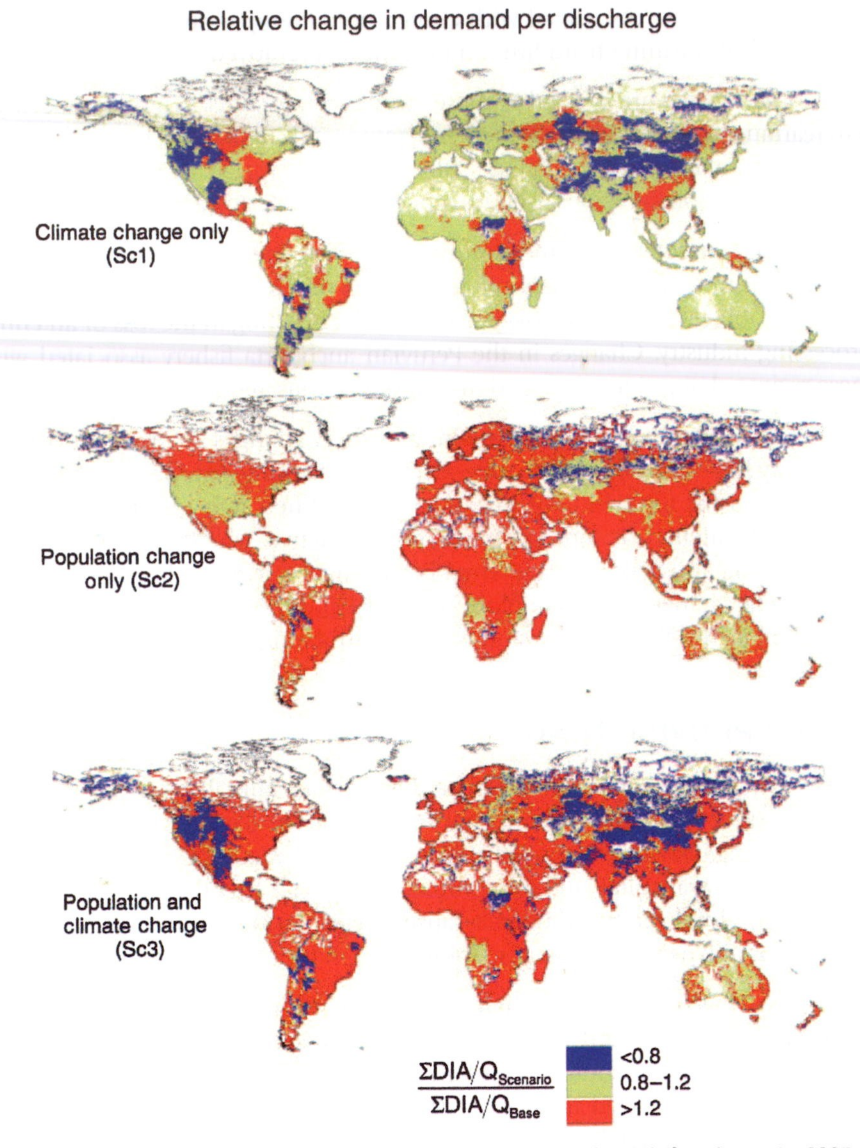

FIGURE 13.4 Change in water stress due to climate change and population change by 2025. Water stress (demand (DIA)/discharge (Q)) was calculated using a combined general circulation model-water basin model (CGCM1-WBM). Climate change effects are pronounced by 2025. Red indicates areas of increasing water stress. *Source: Vorosmarty (2000). Reproduced with permission from AAAS.*

Mediterranean climates may be particularly affected by increasing water stress due to climate change. Mediterranean climate regions include California, Chile, southwest Australia, the Cape of South Africa, and the areas bordering the Mediterranean Sea in Europe. These are naturally water-scarce areas that receive

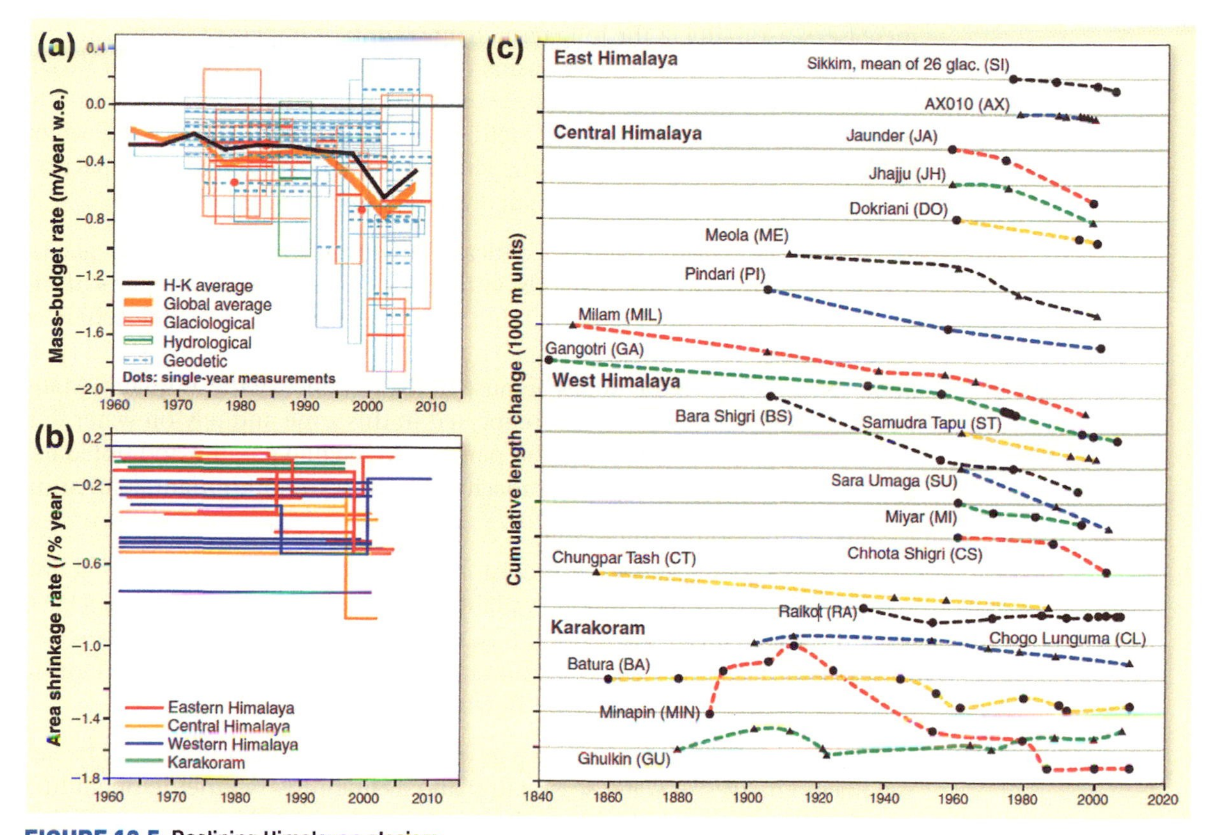

FIGURE 13.5 Declining Himalayan glaciers.
Himalayan glaciers that are the water towers for the great rivers in Asia are declining in mass (a), area (b) and length (c). Decline in length is charted in annual increments. *Source: Bolch et al. (2012). Reproduced with permission from AAAS.*

their rainfall in winter rather than summer. They are centers of wine production and have good agricultural potential, leading to high population densities and high water stress. Climate change is projected to lead to drying in most Mediterranean climate regions, with the most decrease coming in the dry summer months. The decreased summer precipitation is accompanied by pronounced warming, further compounding water stress.

In Europe, water stress increases are greatest in the south. Nearly 40% of Mediterranean populations in Europe are projected to be in watersheds with increased stress due to climate change by 2050. Water scarcity in these areas is likely to be accentuated by increased water use in agriculture and tourism sectors as temperatures warm.

Melting glaciers will affect water availability for tens of millions of people downstream, especially in the Andes and in Asia (Figure 13.5). Glacier melting increases water availability in the short term and decreases availability and increases seasonality in the long term. Large numbers of people on the Andean altiplano and in the great river basins of Asia are at risk. Loss of water provisioning due to climate change in Asia may be particularly acute in the Brahmaputra and Indus river basins. Melting of high-latitude and polar glaciers, although of great importance in the climate system, will have less impact on freshwater availability because of small population sizes in these areas.

Tropical alpine vegetation plays a critical role in moderating changes in glacial meltwaters. High-altitude wetlands, known locally as bofedales in parts of the Andes, soak up and hold water, helping to compensate for some of the changes in timing as glacial melt accelerates. These wetlands and alpine grassland (paramo) are extensive in the Andes because of the presence of the altiplano. Many of South America's major cities are located in this zone and rely on water that comes from glaciers, bofedales, and paramo. Managing paramo and bofedales to maintain their water-holding capacity is therefore an important adaptation strategy for the Andes.

Cloud forests play a key water-provisioning role by trapping water in fog and translating it into surface runoff available for human uses. Warming associated with climate change causes cloud bases to lift, decreasing the area where fog intersects with cloud forest. This ultimately decreases the extent of cloud forest, reducing the water availability for downstream populations.

Snowpack-dependent systems, such as in the western United States, will be facing major changes in water seasonality and availability. Warmer temperatures mean earlier snowmelt, higher and earlier peak flows, and less water availability late in the water year. These changes can spell trouble for agriculture, which depends on the extended water availability provided by slow snowmelt in the mountains. A 14-fold increase in counties with extreme water sustainability risk is projected for 2050 in the United States. The counties at greatest risk are in the west, southwest, and Great Plains regions. Other temperate regions dependent on runoff from snowpack, especially in arid or semiarid areas, will be at similar risk.

CARBON SEQUESTRATION

Vegetation helps regulate climate by holding carbon in plant roots, branches, and leaves. Forests are particularly important because trees hold relatively large amounts of carbon both above- and belowground. But other types of vegetation, from scrub to grasslands, all sequester carbon. Wetlands and peat soils store large amounts of carbon.

This regulating ecosystem service is itself affected by climate change. Type conversion of one biome to another may greatly change carbon storage potential per unit area. Even where biome turnover is not occurring, climate change may reduce growth rates or ultimate sizes of trees and other vegetation, reducing carbon storage.

Changes in land use can release this carbon to the atmosphere, which is why land use change is a major category of greenhouse gas emissions. Conversion of forest to annual cropland is the largest type of land use change causing increased release of carbon into the atmosphere. Slowing greenhouse gas emissions due to land use change is a major thrust in international efforts to combat climate change, but feedback in the other direction—climate change impacts on carbon storage in forests and other vegetation—is an important impact on ecosystem services.

The magnitude of climate change impact on carbon storage varies by region. Net primary productivity is increasing because of climate change in central Africa, in parts of the southern Amazon, and in high-latitude forests (see Chapter 19). But it is decreasing in the central Amazon owing to climate interactions with fire and drought, as well as across large parts of the Southern Hemisphere.

SPOTLIGHT: CARBON AND BIODIVERSITY WORK TOGETHER

Carbon storage and biodiversity conservation are globally important goals that are mutually compatible. Carbon storage reduces greenhouse gas emissions and mitigates climate change. Biodiversity conservation preserves unique species, untapped and largely unknown genetic information and ecotourism of growing economic importance. The REDD program of the climate change convention (UNFCCC) creates incentives to reduce deforestation and store carbon in natural systems. The forests conserved can be critically important reservoirs of biodiversity. Busch et al (2011) explored the factors that would make REDD implementation most effective and those that would maximize co-benefits to biodiversity. They found that the same factors needed for effective and equitable REDD - greater finance, reduced leakage and effective participation of countries with both low and high historical deforestation rates - were the same factors that mattered most for maximizing biodiversity conservation.

Source: Busch, J., Godoy, F., Turner, W.R., Harvey, C.A. 2011. Biodiversity co-benefits of reducing emissions from deforestation under alternative reference levels and levels of finance. Conservation Letters, 4(2), 101–115.

Europe and North America are likely to act as net carbon sinks, as some land cleared for agriculture in the past two centuries reverts to forest and as high-latitude forest productivity picks up owing to warming temperatures. In the western United States, however, increasing fire frequency may result in local and regional release of carbon from vegetation into the atmosphere, illustrating the regional and subregional diversity of climate effects on carbon sequestration and climate regulation.

FIRE

Wildfire degrades many of the ecosystem services provided by forest systems. Carbon storage for climate regulation is released into the atmosphere by fire. Timber is destroyed by fire. Postfire landscapes may have strongly altered water-runoff characteristics. Recreational uses are much lower in recently burned forests. As a result, tens of billions of dollars annually is spent on global efforts to fight wildfire.

Climate change will alter fire regimes in almost all parts of the globe, more strongly in some regions than in others. Fire is likely to increase in temperate, Mediterranean, and boreal systems (Figure 13.6). It is likely to decrease in tropical forests and grasslands. The Northern Hemisphere will have the greatest area of increased fire due to climate change, thanks to larger temperate areas in the north. In the tropics, most areas except the Amazon will see decreases in fire.

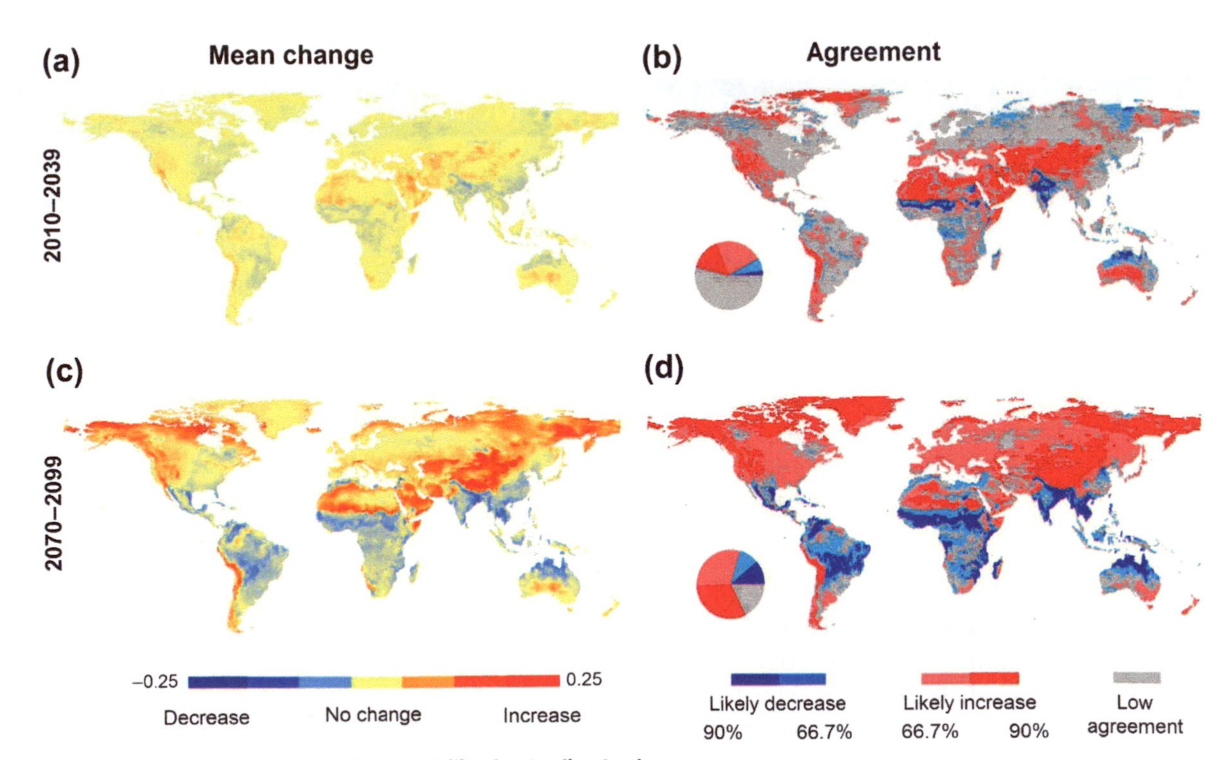

FIGURE 13.6 Global change in fire probability due to climate change.
Mean change (left panel) and multimodel agreement (right) are shown for increasing fire probability (red) or decreasing fire probability (blue). General circulation model agreement on precipitation change is low in some regions, accounting for the difference between mean and model agreement metrics. *Source: Moritz et al. (2012). Reproduced with permission from ESA.*

In the Amazon, the interplay of deforestation and climate produces a different result compared to other tropical systems. In the Amazon, drying associated with land clearing leads to drought and increased fire. See Chapters 5 and 19 for additional discussion of causes and consequences of Amazonian drought.

Many mountainous areas will see increases in fire associated with earlier snowmelt (Figure 13.7). In the western United States, for instance, increases in large wildfires from 1970 are strongly linked to climate change. Frequency of large fires increased dramatically in the 1980s, in association with warmer temperatures. The mechanism for this change is climate warming leading to earlier snowmelt, which results in drier forests in the fire season. In years when warmer spring temperatures led to early snowmelt, the frequency of large fires was more than six times higher. Forest vulnerability to this increased fire risk is concentrated in mountains, where snowpack effects are greatest.

TOURISM

Tourism is the world's largest service-sector industry and nature tourism is one of the fastest growing sectors of this industry. Changes in forests, in snowpack, in species of recreational interest (e.g., bird-watching, hunting), in streamflow, and in the oceans all affect potential tourism revenue and jobs. Major areas of concern are the ski industry and forest recreation in fire-prone areas. But perhaps the largest impact to date has been on reef tourism.

Coral bleaching from high water temperatures associated with climate change results in high coral mortality. Previously colorful reefs are reduced to white rubble and populations of fish species important in dive tourism plummet. Thailand closed 18 dive sites because of bleaching in 2011, with an estimated cost in lost tourism revenue of $2.5 million per year. Surveys of divers indicate that decline in coral quality and associated fish affect plans to visit a location. Dive sites in all oceans of the world have already been affected by coral bleaching.

Ski tourism is strongly vulnerable to climate change because of possible changes in snowpack. Earlier snowmelt results in fewer ski days and lost revenues to ski resorts and associated lodging and services. For instance, in Europe, snowlines are projected to rise 200–400 m by the end of the century owing to climate change. This movement of the snowline from its current 1300 m average to over 1700 m will reduce the number of European ski resorts with adequate snow by almost one-third. Some loss of snow can be offset with artificial snowmaking, but this added cost reduces profitability and thus ultimately forces shorter seasons as well. Some ski areas may experience higher snowfall

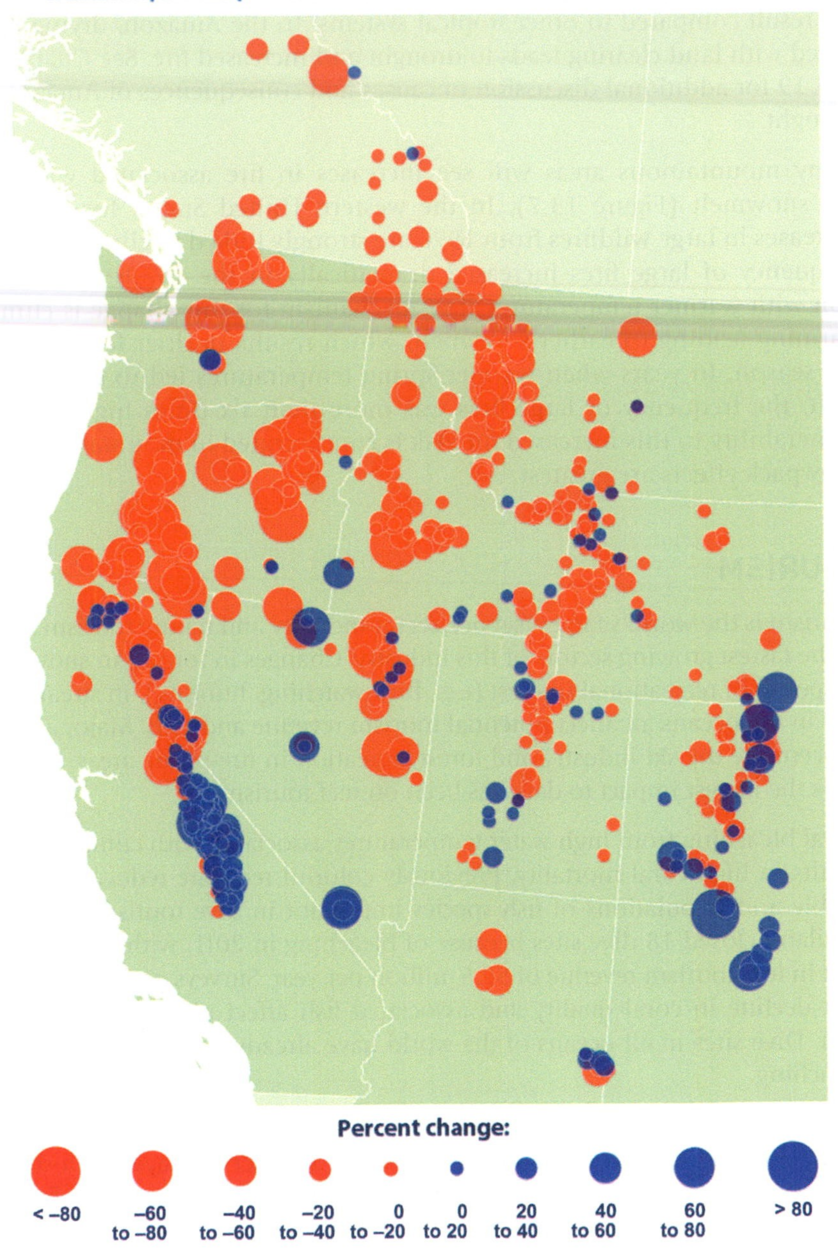

FIGURE 13.7 Snowpack fire index for the western United States.
Declines in snowpack are strongly correlated with increases in large fires. Red indicates declining snowpack, with size of circle representing strength of decline. *Source: U.S. EPA. See also Westerling et al. (2006). Data source: Mote, P.W. 2009 update to data originally published in Mote et al. (2005).*

due to climate change, so the overall impact to the industry is mixed and calls for careful local planning.

ADAPTATION AND MITIGATION

Adaptation and Mitigation are words with special meaning in climate change policy. Mitigation means reducing greenhouse gas emissions. Adaptation refers to actions to reduce the impacts of climate change. Ecosystems play a role in mitigation by storing carbon, which is why reducing deforestation is an important mitigation strategy. Ecosystems can help people adapt to climate change in many ways, especially in intentiionally designed Ecosystem-based Adaptation initiatives.

Forest recreation such as backpacking, bird-watching, and camping is affected by climate change in both temperate and tropical settings. In the tropics, two-thirds of restricted-range bird species, typically among the most sought after by bird-watchers, are projected to undergo range shifts owing to climate change. In temperate areas, especially western North America, huge forest diebacks due to drought and bark beetle kills are affecting recreation. In Rocky Mountain National Park, campsites formerly in shady glens have had to be stripped of their beetle-killed trees, leaving barren landscapes less attractive to, and less used by, visitors to the park.

Ecosystems can be managed to reduce all of these impacts of climate change on services people receive from nature. Ecosystems can also be employed to help reduce the physical impacts of climate change. These mechanisms for ecosystems to help in adaptation to climate change are known as ecosystem-based adaptation (EbA).

ECOSYSTEM-BASED ADAPTATION

Ecosystem services can help people adapt to climate change, a growing practice known as EbA. EbA includes managing ecosystems services to reduce the human impacts of altered ecosystems, such as fire or coral bleaching. But it also includes using ecosystems to help reduce other impacts of climate change, such as rising sea levels. EbA includes the adaptation of conservation strategies to help conserve biodiversity as the climate changes.

People often want to *do* something to respond to a problem. EbA often involves *stopping doing* something—namely, stopping clearing of forests and destruction of other types of vegetation. Although this approach of stopping doing something may take a little getting used to, its benefits are clear and often much cheaper than more traditional engineering solutions or technical interventions.

FIGURE 13.8 **Ecosystem-based adaptation—a man-made marsh for flood control.**
Natural vegetation can provide valuable recreational services and protect farms and dwellings in areas of increasing inundation.

There are many types of EbA activities, some of which are already being employed and others of which are being developed to help cope with future change (Figure 13.8). Because implementing EbA has costs, activities that have some immediate payoff, as well as long-term adaptation benefits, will be most likely to succeed. Examples from several sectors help illustrate the breadth and depth of EbA potential.

COASTAL PROTECTION

Healthy coral reefs and mangroves reduce wave velocity and power and so can be a valuable component of coastal protection (Figure 13.9). Where coral reefs and mangroves are degraded, coastlines are more vulnerable to erosion.

Climate change results in sea level rise and increased storm intensities, accentuating the need for coastal protection. Low-lying coastal areas may be increasingly vulnerable to inundation as climate change progresses. Maintaining or restoring natural coral and mangrove protection in these areas can reduce vulnerability to future climate change.

Mangroves play a key coastal protection role, helping to slow storm surges that may flood low-lying towns as sea level rises. Southeast Asia in particular has abundant mangroves and large coastal populations (Figure 13.10). Many mangroves in the region have been cleared to make way for aquaculture ponds. But when large storms hit, both the aquaculture ponds and the neighboring towns are flooded. Mangrove restoration is under way in countries including the Philippines and Indonesia in an attempt to reverse this trend and foster EbA. Mangroves store substantial amounts of carbon, providing an additional ecosystem service benefit to this approach. Because restoration offers both immediate and future benefits, it meets the EbA criteria of producing both short- and long-term payoff.

SPOTLIGHT: EBA ON THE ALBEMARLE SOUND

The Albemarle Sound is a large estuary on the coast of North Carolina. Soils in the estuary are derived from peat and have been drained for over a century to improve agriculture. The peat soils are cut through with drainage channels to remove seawater and make the soil useful for growing crops. But drainage is massively maladaptive in the face of climate change.

As sea levels rise, peat soils react with the salts in the water and literally dissolve. This destroys the peat soil and greatly accelerates seawater intrusion as sea level rises, threatening ecosystems, farms, and towns.

The EbA solution for Albemarle includes buying cropland ruined by rising seas and turning it back into swamp. The Nature Conservancy is buying peatland in Albemarle, filling in the drainage channels and letting the natural wetland regenerate. This maintains the peat soils, reduces huge carbon releases to the atmosphere, slows seawater intrusion from sea level rise, and provides a healthy retreating ecosystem for biodiversity and nature recreation.

Source: Pearsall III, S.H. 2005. Managing for future change on the Albemarle Sound Climate change and Biodiversity. Yale University Press, New Haven, 359–362.

EbA for coastal protection may be undertaken in conjunction with hard engineering approaches such as seawalls to provide hybrid ecosystem–mechanical solutions. Many times it can be achieved with natural regeneration that takes place when ecosystems are protected from degradation.

For example, the reefs of the Discovery Coast of Brazil help protect one of the premier beach tourism destinations in the country. The Discovery Coast is named because this is the region of Brazil in which European explorers first touched ground. It has since become the site of a multibillion dollar tourism industry. The Discovery Coast is low-lying and sandy, highly prone to erosion. But it is protected by extensive fringing reefs. Unfortunately, overfishing, sedimentation, and pollution have severely damaged these reefs. As sea level rises owing to climate change, the beaches of the Discovery Coast are at risk if these reefs remain degraded and crumbling.

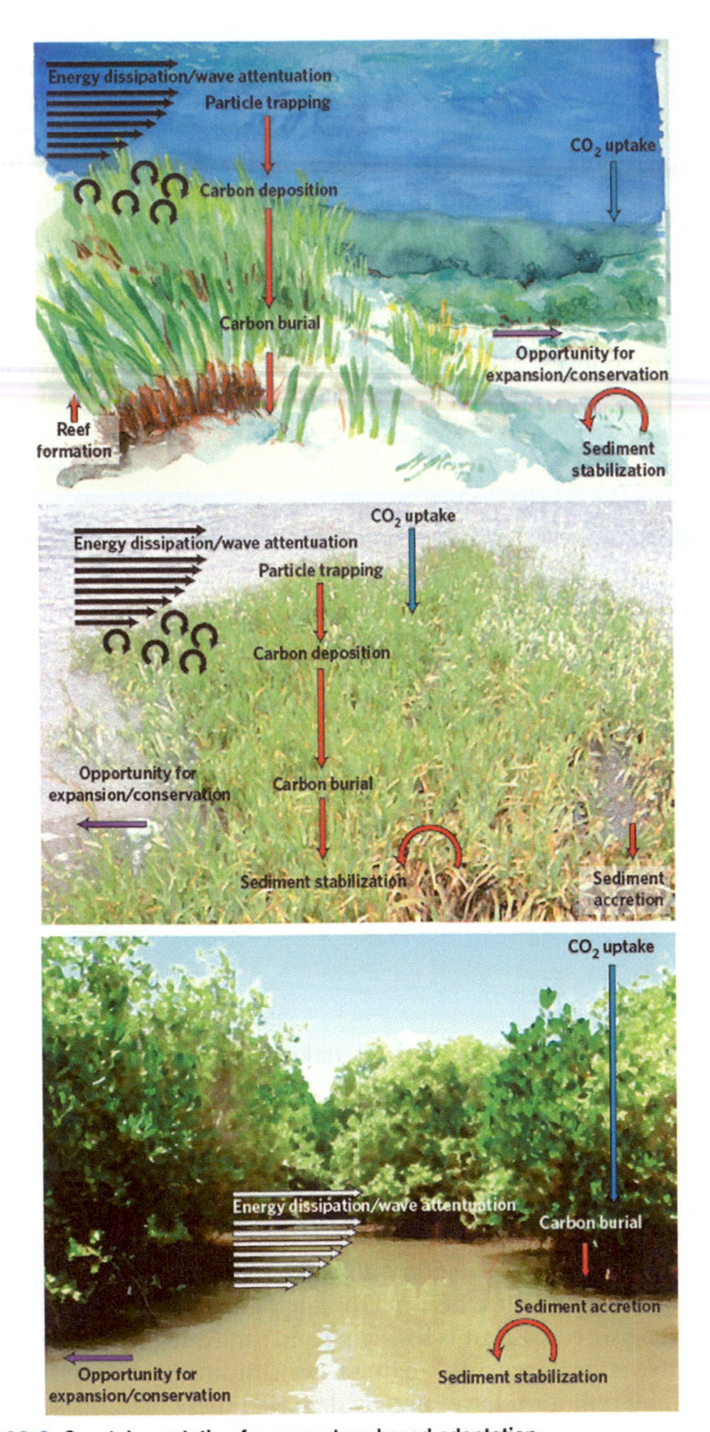

FIGURE 13.9 Coastal vegetation for ecosystem-based adaptation.

Mangroves and coastal vegetation help attenuate increased storm surges due to climate change. Restoring these vegetation types stores carbon, providing a climate regulation ecosystem service.

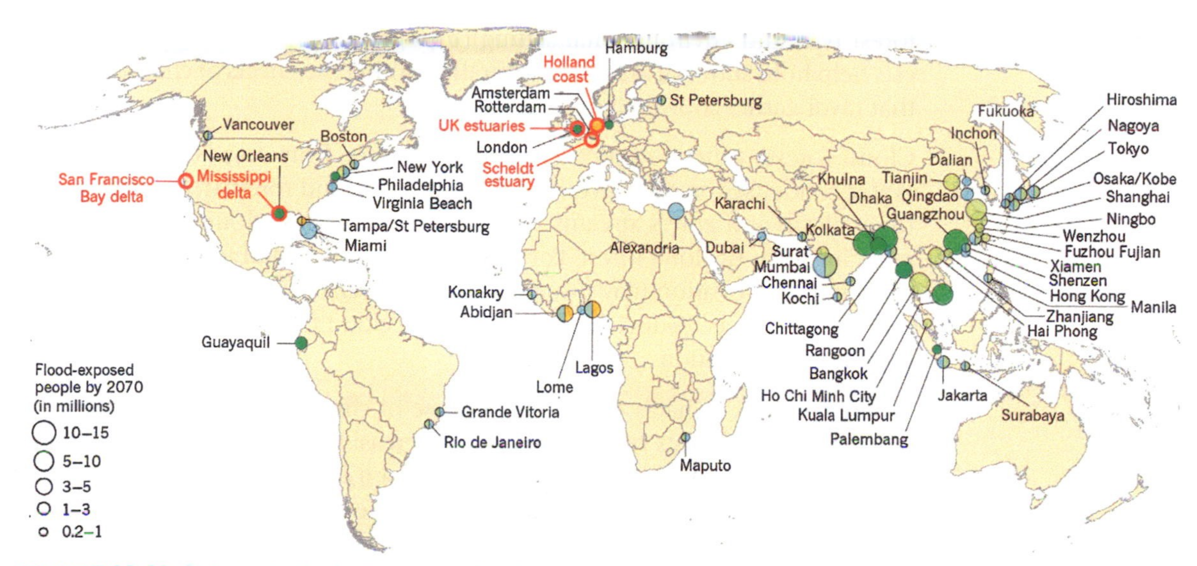

FIGURE 13.10 Global potential for flood protection from coastal vegetation.
Areas with low-lying populations are most at risk from coastal flooding, and those with extensive mangroves and other natural habitats have the most potential for ecosystem-based adaptation from coastal vegetation. *Source: Temmerman et al. (2013).*

The answer for the Discovery Coast turns out to be a fish, the blue parrotfish, *Scarus coeruleus*, known locally as "budiao." Budiao are algae-feeders and it turns out that controlling algae is the key to bringing back damaged reefs. All reefs rebuild after disturbance as corals multiply and solidify the reef surface. If algae establish on the damaged reef, corals are out-competed for space and cannot rebuild the reef. Budiao, by eating algae, let the corals win the race against algae. But budiao are delicious and they are heavily fished as a result. The EbA solution to loss of Discovery Coast beaches is therefore protection of the budiao from fishing pressure, allowing the reefs to rebuild and protect the coast from rising sea level.

Other algae-eating fish play similar key roles in other areas. Exclusion experiments demonstrate that reefs from which algae-eating fish are excluded by cages are overrun by algae and erode after disturbance, whereas areas in which algae-eating fish are plentiful see rapid reef recovery. Marine protected areas or fishing restrictions that protect algae-eating fish and increase their numbers provide an EbA solution that can be applied in many coastal areas where local protecting reefs have been damaged.

WATER SUPPLY

Forests can help regulate and increase water supply as climate changes. Cloud forests in particular generate water that can be a major component of streamflow. Other forest types may increase or decrease streamflow, depending on forest type and environmental setting (topography, riparian vegetation). Forests help hold snowpack, so they naturally combat the effects of climate change that favor earlier snowmelt.

Cloud forests generate water by capturing water from fog (surface clouds). Water condenses on the leaves and branches of cloud forest trees, drips to the forest floor, and enters streams. As the climate warms, cloud bases may lift, reducing areas of cloud forest. For this reason, where cloud forest is being lost, reducing deforestation may provide long-term benefits in water availability to downstream users. Reducing deforestation in cloud forest areas is therefore an important EbA option.

Where regional climates are drying because of climate change, the water provided by cloud forests may make up an increasing fraction of the remaining surface flow. On a regional basis, land use change can also cause lifting cloud bases. As forest is cleared for lowland agriculture, moisture entering the atmosphere from transpiration is lost and cloud formation decreases and moves to higher altitudes. This can reduce areas of cloud forest. Regional strategies that reduce lowland forest loss are another EbA option that can maintain cloud forests and the critical water they provide.

Maintaining forest cover in mountains helps combat the effects of climate change on snowmelt. As described above, warmer temperatures lead to earlier snowmelt, less water from late snowmelt, and low flows late in the water year. Forests shade snow and help retain it as conditions warm. Maintaining forests in areas of high snowpack can then help reduce the impacts of climate change on seasonality of runoff.

Where glaciers are melting, maintaining downslope tundra or paramo soils can help maintain seasonal water supplies. For instance, bofedales (see above) are compacted in many areas by grazing, which destroys their water-retention capacity. Limiting grazing allows the sponge-like properties of the bofedales to return, allowing them to hold water and release it slowly, replacing some of the water retention provided by tropical glaciers that have disappeared owing to climate change.

FOOD PRODUCTION

Crops produced on agricultural lands benefit from soil ecosystems that can maintain soil moisture and productivity. Although these systems are not natural ecosystems, maintaining more natural soil processes can help reduce the impacts of climate change.

Soils with more organic material and richer microbial ecosystems hold moisture better than heavily fertilized, industrial agricultural soils. This means that more water is available to crops as water stress increases from climate warming or decreased precipitation. Farming practices that promote agroforestry and soil organic material are good EbA options, sometimes referred to as "climate-smart agriculture."

Combining natural forest with tree crops or trees with row crops can help cool plants and soils, maintaining crop productivity as temperatures warm. For instance, shade coffee plantations are much less vulnerable to damage from high temperatures than is coffee grown in monoculture with no shade trees. Maintaining more natural options, even in human-dominated landscapes, is an EbA approach.

For instance, in Humba, Ethiopia, farmers are allowing trees to regrow in fields from stumps left after forest clearing. This low-cost method holds soil, improves soil moisture-holding capacity, and provides shade that will help keep crops cool as temperatures warm. This farmer-assisted natural regeneration lets farmers take advantage of ecosystem services provided by forest trees, without having to do anything other than let natural regrowth take its course. It provides immediate benefits and long-term adaptation to climate change.

DISASTER RISK REDUCTION

Trees can hold soil to prevent landslides or hold water to reduce flooding. For these reasons, forests on steep slopes are often protected by forestry laws. In landslide or flood-prone areas, maintaining or restoring forest is an EbA option. Laws promoting reforestation or protection have been passed after major landslide or flooding disasters in countries such as China, Brazil, and Thailand.

BOTTOM-UP CLIMATE MEETS TOP-DOWN COD

Cold water from the melting of Arctic ice is flooding into the northern oceans. Off the east coast of Canada, this results in wholesale reorganization of the food chain. At the same time, fishing pressure has wiped out cod, a dominant species in the bottom waters of these ecosystems. The influx of cold water from the north is suppressing cod production and favoring species such as snow crab and shrimp. Despite long-term reduction in fishing pressure for cod, populations are not rebounding because the top-down influence of overfishing has been combined with a bottom-up change in climate (see Chapter 5, p. 125).

As warmer temperatures intensify the hydrologic cycle, more and more intense storms are expected in many regions. Adaptation to these impacts can include restoring forests on deforested hillsides to hold landslide-prone soils. In flood-prone areas, upstream forests can hold water and reduce flood pulses, whereas lowland forest can act to slow advancing floodwaters.

Carbon stored in forests planted for disaster risk reduction contributes to carbon sequestration, offering a double benefit. Because high-intensity storms are infrequent, the income from selling the carbon stored can be an important short-term payoff that helps generate support for maintaining forests until the long-term benefits of flood or landslide reduction are realized.

FURTHER READING

Schröter, D., Cramer, W., Leemans, R., Prentice, I.C., Araújo, M.B., Arnell, N.W., Zierl, B., 2005. Ecosystem service supply and vulnerability to global change in Europe. Science 310 (5752), 1333–1337.

Buytaert, W., Cuesta-Camacho, F., Tobón, C., 2011. Potential impacts of climate change on the environmental services of humid tropical alpine regions. Global Ecology and Biogeography 20 (1), 19–33.

Nelson, E., et al., 2013. Climate change's impact on key ecosystem services and the human well-being they support in the US. Frontiers in Ecology and the Environment 11, 483–893.

Immerzeel, W.W., van Beek, L.P., Bierkens, M.F., 2010. Climate change will affect the Asian water towers. Science 328 (5984), 1382–1385.

Implications for Conservation

CHAPTER 14

Adaptation of Conservation Strategies

Adaptation is a response to climate change that reduces impacts. Intentional actions to reduce impacts will allow social development to continue in the face of climate change. For natural systems, adaptation of conservation strategies and management will play a critical role in reducing impacts. This chapter explores the adaptation actions that may help conserve biodiversity and ecosystems as climate changes.

Protected areas play a surprisingly important role in adaptation to climate change. One might think that protected areas that are fixed in place would be a poor response to a challenge that forces species' ranges to move. However, protected areas are often large enough to accommodate range shifts of kilometers or tens of kilometers and provide largely natural conditions uninterrupted

SPOTLIGHT: HOPPING HOT SPOTS

If species move, should not conservation priorities move too? Global biodiversity hot spots are among the most widely accepted conservation priorities, so their fate owing to climate change affects millions of dollars of conservation investment annually. Biodiversity hot spots are defined both by their high numbers of endemic species and by high levels of human threat. Climate change may cause shifts in either of these factors. Hot spots of diversity are less widely used in conservation priority settings but are more easily traced through time.

There is no evidence that current hot spots are shifting, but paleoecological studies show that they have shifted in the past. Renema et al. (2008) examined marine hot spots from the Miocene (42 million years ago) to the present and found that richness hot spots moved large distances. A richness hot spot that originated in the Mediterranean migrated to the Arabian Sea and then to the coral triangle area of southeast Asia. These shifts in marine richness hot spots may not equate to movements in hot spots of endemism because endemics may be the species that do not move, whereas more generally distributed species have labile margins.

Theory suggests that endemism hot spots will not move. Terrestrial hot spots of endemism are largely located in tropical mountains and islands. Species moving upslope because of climate change are likely to become more concentrated, not less, in these settings or to become extinct. Marine hot spots of endemism are less well documented but seem to also be associated with areas of high topography that may either retain species or lose them entirely. Thus, climate change may make diversity hot spots hop, whereas hot spots of endemism only get hotter.

Source: Renema, W., Bellwood, D.R., Braga, J.C., Bromfield, K., Hall, R., Johnson, K.G., et al., 2008. Hopping hotspots: global shifts in marine biodiversity. Science 321, 654–657.

Climate Change Biology. http://dx.doi.org/10.1016/B978-0-12-420218-4.00014-7

by human land uses, across which these dynamics may unfold. This can make protected areas an effective means of accommodating species range movements in response to climate change.

The role of protected areas must be viewed in the context of an overall conservation strategy that is composed of multiple elements, which may include the following:

1. Protected areas.
2. Connectivity and conservation on productive lands.
3. Management of individual species, including species rescue and translocation.

These conservation elements are the subject of the chapters in this section. As we will see, expense and management intensity of the approaches increase as we move down the list, making protected areas one of the most cost-effective and efficient approaches.

EARLY CONCEPTS OF PROTECTED AREAS AND CLIMATE CHANGE

Among the earliest and most influential works on protected areas and climate change were groundbreaking papers and book chapters by Rob Peters of the World Wildlife Fund in the 1980s. These early looks at conservation and climate change pointed out the fundamental problem of fixed protected area boundaries and dynamic species ranges. In graphic terms, this issue was illustrated as a protected area fixed in space, sometimes an island of habitat in human land uses, either overtaken or left behind by a moving species range (Figure 14.1).

SPOTLIGHT: EARLY THEORY

Conservation theory on climate change in the first decade of the discipline has been summarized by Halpin (1997). Halpin found the early conservation recommendations to be largely vague, a finding reiterated a decade later by Halpin (1997). Near the turn of the century, Halpin listed the three top recommendations as (1) multiple representations of conservation targets, in case one is compromised by climate change; (2) working outside of reserves to encompass areas in which range shifts might occur; and (3) managing landscapes for connectivity. These same themes are explored in the current conservation literature. Multiple representations and increasing the amount of protected area specifically to compensate for climate change dynamics have been found to be solid conservation planning principles. Expanding planning domains and time frames is now accepted conservation practice for climate change. Connectivity is a central focus of development for conservation planning tools.

Source: Halpin, P.N., 1997. Global climate change and natural-area protection: management responses and research directions. Ecological Applications 7, 828–843; and Heller, N.E., Zavaleta, E.S., 2009. Biodiversity management in the face of climate change: a review of 22 years of recommendations. Biological Conservation 142, 14–32.

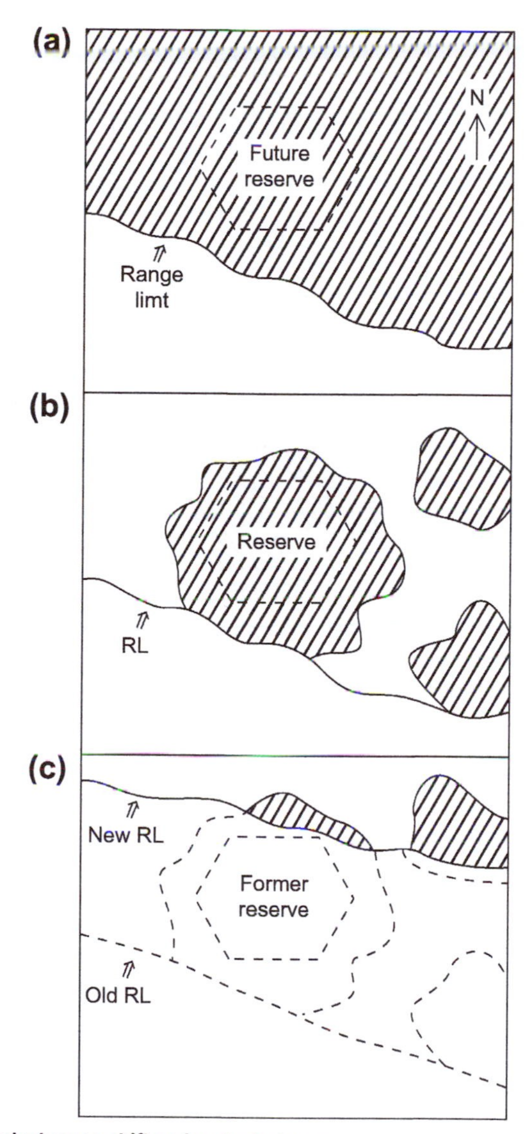

FIGURE 14.1 Species' range shift and protected area.
This figure from Peters' seminal early work on climate change and biological diversity shows the changing relationship between a species' range and a protected area. The species' range is indicated by hatching (a). In the schematic, habitat is eroded by conversion to use over time (b) and then the species range limit shifts due to climate change (c). As climate shifts, the proportion of range within the protected area changes (RL = range limit). The figure shows the reserve being lost as the species' range ceases to intersect with it, but it is unlikely that a reserve would be declassified based on the loss of only one species. The reserve would remain important for many other species, so the greater question is how to maintain protection of the species that has moved beyond the reserve. Adding a protected area within the new range of the species is one important option. *Source: Peters and Darling (1985). Reproduced with permission from Yale University Press.*

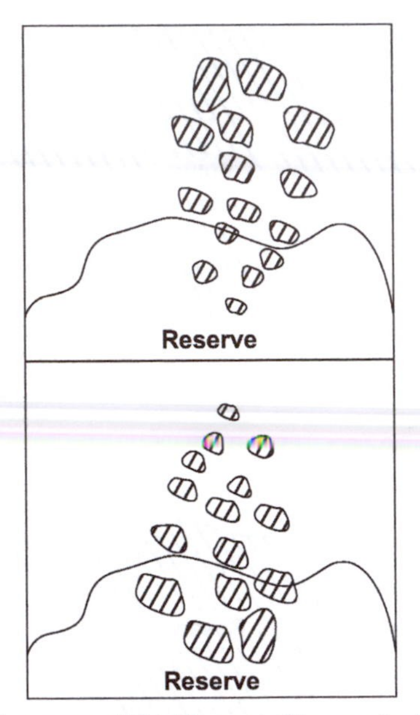

FIGURE 14.2 Metapopulation range shift with respect to a protected area.
A more sophisticated view of a species' range shift considers the area of occupancy within the species' range or the individual populations that make up the overall metapopulation of the species. Range shifts in this view involve loss or change in size of individual populations, which in turn changes the representation of the species in a protected area.

This conceptual framework remains valid for situations in which reserves are at or near the edge of a shifting species range boundary. But to capture the essence of the problem, it conceals much of the complexity of an actual range shift. In actuality, range boundaries are not monolithic entities but, rather, a collection of metapopulations that vary in time, increasing or decreasing more or less rapidly in response to climate change (Figure 14.2).

Large reserves or reserves with large elevation gradients may encompass the entire range or a subpopulation of highly restricted species. In these cases, the relationship between reserve boundary and range boundary is reversed, and the entire shift occurs within the protected area. In other cases, the challenge may be maintaining multiple healthy subpopulations across a series of reserves, even if all of the ranges are shifting or declining.

A more sophisticated view of range shifts therefore includes metapopulation dynamics, species with both large and small ranges, and a variety of range movements across the landscape. For instance, a species may have some populations moving north as a latitudinal response to warming but other populations

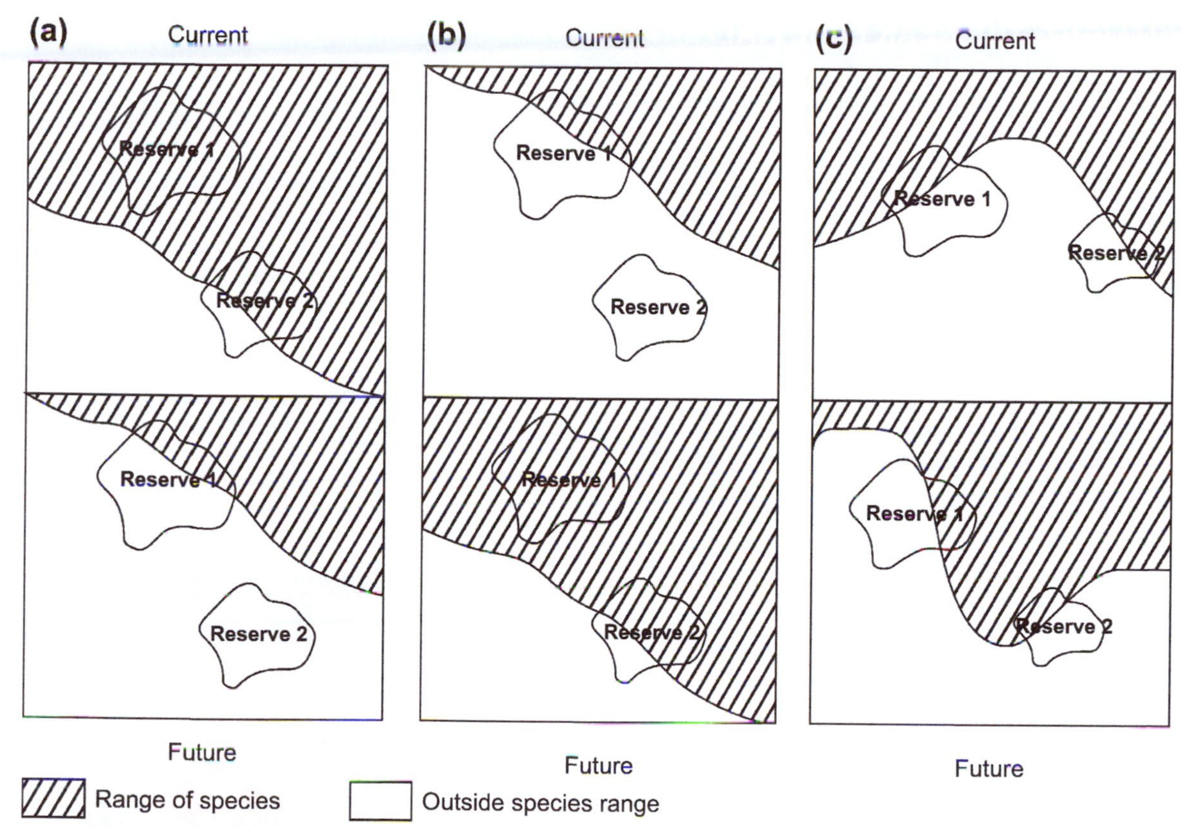

FIGURE 14.3 Range shifts relative to multiple protected areas.
Range shifts in three different species ((a), (b), (c)) are illustrated, each relative to two protected areas. This example illustrates the complexity of conserving multiple species as ranges shift. *Source: Figure courtesy of Conservation International.*

moving in the opposite direction as an elevational response if the only nearby mountains are in the south. All of these movements may take place relative to multiple protected areas: the entire suite of range responses and protected areas must be considered together to estimate the net impact on conservation of the species (Figures 14.3 and 14.4).

PROTECTED AREA PLANNING

Recent modeling research has confirmed Peters' prediction: climate change may cause species to move out of reserves. It may also cause species to move into reserves—increasing protection of some populations—if the species' dispersal abilities are sufficient.

SPOTLIGHT: THE REAR EDGE MATTERS

Much attention is given to the leading edge of range shifts, as species struggle to keep pace with climate change. However, Hampe and Pettit (2005) point out that the rear edge matters, too. The rear edge of current range shifts is the edge that has been most stable in glacial–interglacial cycles. Collapse back into glacial conditions has repeatedly wiped out leading-edge populations, whereas trailing-edge populations in warmer climes closer to the equator have endured. This means that the genetic richness of the rear edge may be greater than that of other areas of the range. It also may make rear edge populations important in resisting extinctions.

For example, cold-sensitive but drought-tolerant species have survived in Europe primarily in the south, around the Mediterranean. The most cold-sensitive species have become extinct in Europe in one glacial period or another. This means that the pattern of low global extinction in the ice ages is not replicated at regional levels. There have been many regional extinctions due to repeated glaciation, and those species that have survived have done so in trailing edges. Long-term warming may doom some rear-edge populations, but at least until now, they have persisted more than leading-edge populations.

Source: Hampe, A., & Petit, R. J. 2005. Conserving biodiversity under climate change: the rear edge matters. Ecology letters, 8 (5), 461–467.

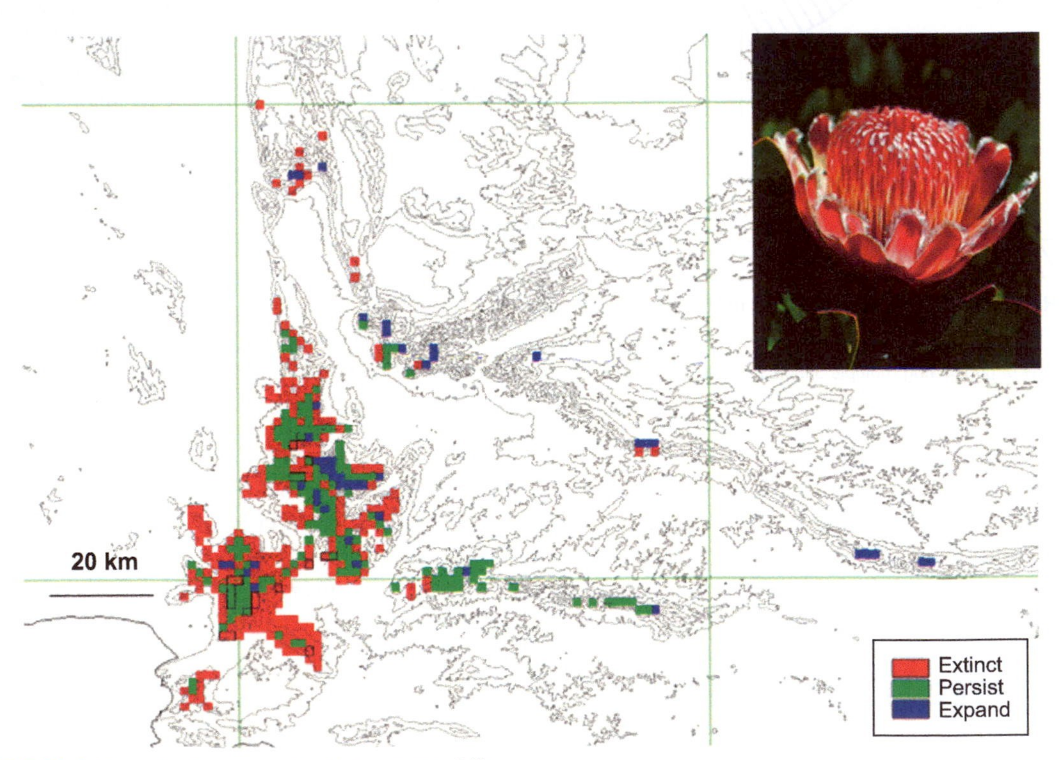

FIGURE 14.4 Diversity of movement within a range shift.
Assumptions that species will always shift poleward with warming are belied by modeling results. Here, the simulated range of a protea species in the Cape Floristic Region shifts away from the pole (blue pixels). There is no poleward landmass in this region, so this species is tracking climate upslope, moving into hills above the Cape lowlands. Blue represents newly suitable future range, red represents current climatically suitable range lost, and green represents current climatically suitable range retained.
Source: Hannah et al. (2005).

FIGURE 14.5 The king protea (*Protea cynaroides*).
The king protea is one of hundreds of protea species whose future ranges have been projected in species distribution modeling for the Cape Floristic Region. It is the national flower of South Africa. *Source: From Wikimedia Commons.*

Three of the areas in which these issues have been examined are South Africa, Europe, and Mexico. In South Africa, modeling has focused on the proteas—Mediterranean shrubs, members of the family Proteaceae, many of which have large, showy flowers (Figure 14.5). European researchers have modeled many species, including plants, birds, mammals, and amphibians. In Mexico, thousands of birds, butterflies, and mammals have been modeled.

MULTISPECIES MODELING EFFORTS

In several areas of the world, initiatives have been undertaken to model hundreds or thousands of species. In Mexico, South Africa, and Europe, these multispecies models have been used to design protected-area systems robust to climate change. By choosing areas that retain suitable climate or that allow species to migrate, new protection can help avoid extinctions due to climate change. If these new protected areas are chosen simultaneous with efforts to complete representation of species in their current ranges (many species are currently underrepresented or unrepresented in protected areas), large efficiencies result. It takes much less area to represent all species now and account for climate change than it does to complete coverage for current ranges but wait to respond to the effects of climate change.

Species move out of reserves in response to climate change. Studies of more than 1000 European species and the proteas of South Africa have shown that up to 10% of species will be lost from reserves under 2050 climate change scenarios. However, some species may increase in representation. These same studies showed some species moving into reserves, which both increased the representation of those individual species in protected areas and slowed the rate at which average representation declined owing to climate change (Table 14.1).

Additional protected area can reverse the decline in species representation. Species that lose representation because of climate change can regain some of the lost protected range if new protected areas are added in the landscape. This is accomplished by placing the new protected areas in locations where suitable climate for the species will be retained. This strategy works well in the early stages of climate change. Protection for up to 90% of species was able to be regained by adding new protected area to compensate for the effects of climate change in a study of Mexico, South Africa, and Europe. The amount of new protected area varied widely, from very small to approximately half of the existing protected-area system.

Addition of protected areas for climate change can be done at the same time that area is added to complete representation of species' current ranges. Many species are not currently represented in protected areas. When area is added to better represent these species, it can be done in places that will retain the species as the climate changes and in places that will retain other species as the climate changes. By planning additions to protected-area systems with these multiple, complementary goals in mind, the overall area requirement may be dramatically reduced (Table 14.2).

Waiting to add protected areas for climate change costs more. If protected areas were added first for current ranges and then later for climate change, the double-duty efficiency of the additions was lost, resulting in the need to add more

Table 14.1 Decrease in Species Representation in Protected Areas in the Cape Floristic Region Due to Climate Change[a]

	No. of Species (% Decline Since 2000)	
Year (Threshold)	**No-Dispersal Assumption**	**Maximum-Dispersal Assumption**
2000 (presence only)	327 (0)	–
2050 (presence only)	277 (15.3)	301 (8.0)
2050 (100-km^2 minimum threshold)	202 (38.2)	243 (25.7)

[a]*The numbers of species whose modeled ranges intersect protected areas in at least one grid cell (presence only) or at a minimum threshold of area (100 km^2) are given relative to two dispersal assumptions about species' ability to occupy newly climatically suitable areas. Source: Hannah et al. (2005).*

Table 14.2 Additional Protected Areas Needed Owing to Climate Change in Three Study Regions[a]

	A	B	C	D	E	F	G
Region	Current Protected Area (km²)	Additional Area Required to Meet Baseline Target (Current Ranges only) (km²)	Incremental Area Required to Meet Target with Climate Change, in Addition to B (Future Ranges) (km²)	Total Additional Area Required in Two Sequential Steps (B 1 C) (km²)	Total Additional Area Required in One Step (Current and Future Ranges Simultaneously) (km²)	Incremental Area Required to Meet Target Climate Change, Using One-step Approach (E 2 B) (km²)	Cost of Waiting (C 2 F) (km²)
Number of Species Modeled		Number of Species Meeting Target at Present	Number of Species Meeting Target in 2050	Number of Species Meeting Target in 2050	Number of Species Meeting Target in 2050		
Taxa							
Cape Floristic region	4681	2330	1911	4241	3487	1157	754
Plants (proteacea)		49% of A	41% of A27% of A 1 B	91% of A60% of A 1 B	75% of A50% of A 1 B	25% of A16% of A 1 B	39% of C65% of F
316		*282*	*246*	*246*	*246*		
Tropical Mexico	104,000	44,000	12,800	56,800	44,500	500	12,800
Birds and mammals		42% of A	12% of A9% of A 1 B	55% of A38% of A 1 B	43% of A30% of A 1 B	1% of A,1% of A 1 B	96% of C2460% of F
179		*178*	*160*	*160*	*160*		
Western Europe	20,850	3850	8450	12,300	7200	3350	5100
Plants (multiple families)		18% of A	41% of A34% of A 1 B	59% of A50% of A 1 B	35% of A29% of A 1 B	16% of A14% of A 1 B	60% of C152% of F
1200		*1200*	*1123*	*1123*	*1123*		

[a]*Column A indicates the area currently protected. All subsequent area amounts must be added to these existing protected areas. Column B indicates the additional area required to meet the representation target for current species ranges. Column C indicates the area increment required to meet the target for future ranges in a second, subsequent step, once the target has been met for current ranges. Column D indicates the total additional area required to meet the target for current and future ranges in two steps, or the sum of the previous two columns (B 1 C). Column E is the area required to meet the target using the alternative approach of searching for solutions for present and future ranges in a single step. Column F is the incremental area needed to address climate change when existing representation and climate change are addressed in one step, or the difference between column E and column B. Column G is the area difference between meeting the target for present and then future ranges in two separate steps (D), versus meeting it for present and future ranges at once in a single step (E). This "cost of waiting" simulates the difference between completing representation for species present and future ranges now ("early action") and completing representation for species current ranges now and then waiting until some-time in the future to complete representation for future ranges ("waiting").*
Source: Reproduced with permission from Hannah et al. (2007).

protected area. This suggests that waiting to take action (adding protected areas) until the effects of climate change on species' ranges are evident will result in significantly higher protected area addition needs.

PLANNING FOR PERSISTENCE

The design of protected-area systems for climate change is part of planning for persistence in a conservation plan. Any protected-area system plan should consider both pattern and process targets and plan for the persistence of both through time. Pattern targets are generally species or habitat types: planning for pattern means conserving a representative sample of species or habitat types, hence preserving some notion of their "pattern" in the landscape.

Planning for process means capturing temporal phenomena in the conservation plan. For instance, wildebeest migrations in the Serengeti are an important process. Any conservation plan for that region should conserve the migration, and it is clear that it is impossible to conserve the pattern of wildebeest and other wildlife on the landscape without also conserving this process.

IRREPLACEABILITY

When a species is found in only one location, that site is said to be irreplaceable. In a protected-areas network that seeks to represent all species, these sites *must* be included to meet the goal, hence they are "irreplaceable." In addition to completely irreplaceable sites, the concept of irreplaceability can be applied on a sliding scale, in which the number of rare species a site represents increases its irreplaceability score, up to a highest value that equals complete irreplaceability. Irreplaceability is a key driver for selection of cost-effective reserve systems because choosing sites with high irreplaceability minimizes the area needed in a system. Climate change can alter the irreplaceability score of sites by causing the location or the size of species' ranges to change. Reserve selection that ignores climate change will therefore be based on incomplete information about species' ranges and irreplaceability, causing it to be less effective and more expensive in the long term.

Planning for persistence refers to conservation that endures over longer time frames. Processes such as species population dynamics mean that the abundance of a species today may not be a constant in the future. For example, it may take years to establish and effectively manage new protected areas, so it is important to start the process of protection in the sites most threatened with habitat loss. This prevents the intent of preserving pattern in the landscape from being undermined by habitat destruction.

Climate change is a long-term phenomenon that affects both pattern and process. As climate changes, patterns of species distribution across the landscape are altered. Richness and endemism may shift, and, more importantly, patterns of rare sites that contribute to "irreplaceability" may be affected by climate change. Processes such as migrations may be affected by changes in phenology, as we have seen in Chapter 4.

Most area plans prioritize areas for conservation action based on both threat and vulnerability to threat (or resilience). For instance, an area far away from roads may be given a lower priority for conservation action, whereas an area laced with roads giving hunters access may need to be given priority to prevent loss of threatened populations. This is prioritizing based on threat to improve the chances of the species persisting.

Conversely, when an area is particularly resistant to a threat, it may be a lower priority for conservation. Less vulnerability or greater resilience to a threat is generally thought to lower the conservation priority of a site because it will be less affected by the threat than will other sites with similar irreplaceability.

However, this general principle is reversed when the threat is not reduced by protection. Protected areas are not effective against some threats, notably those carried in the atmosphere, such as climate change. In this case, the resilience equation is changed: it makes sense to prioritize sites with high resilience.

RESISTANCE AND RESILIENCE

Resistance or resilience to climate change may arise from many combinations of factors, and it may be a property of either landscapes or species. Resistant species or sites are less damaged by climate change; resilient species or sites are better able to recover once damaged. A species with broad physiological tolerances might be resistant, whereas a resistant site might be sheltered in a unique microclimate. A resilient site might be one in an area of high seed rain, so plants are able to reestablish easily. A resilient species might be one with high reproductive potential or good long-distance dispersal.

Conservation strategies to deal with climate change often target resilient areas, either implicitly or explicitly. Because protected areas cannot reduce the amount of climate change, resistant and resilient areas should be prioritized for adaptation, as discussed above. By prioritizing protection in areas that are resilient to climate change, conservation strategies ensure that protected area will persist. Given two areas with similar biological value (irreplaceability), protection will be more effective in the long term for the one in which target species are not wiped out by climate change. Conservation investment makes sense in that area first, although the second, higher risk, area may also warrant investment at some point.

One important exception to these general principles applies to species that are found in only one location. The sites in which these species are found have priority for protection regardless of their resistance or resilience to climate change (Figure 14.6). Because these species exist in only one location, we have no choice: we must do our best to conserve them where they are. In conservation planning terminology, these sites are completely irreplaceable and must always be protected. If they have low resilience to climate change, conservation must

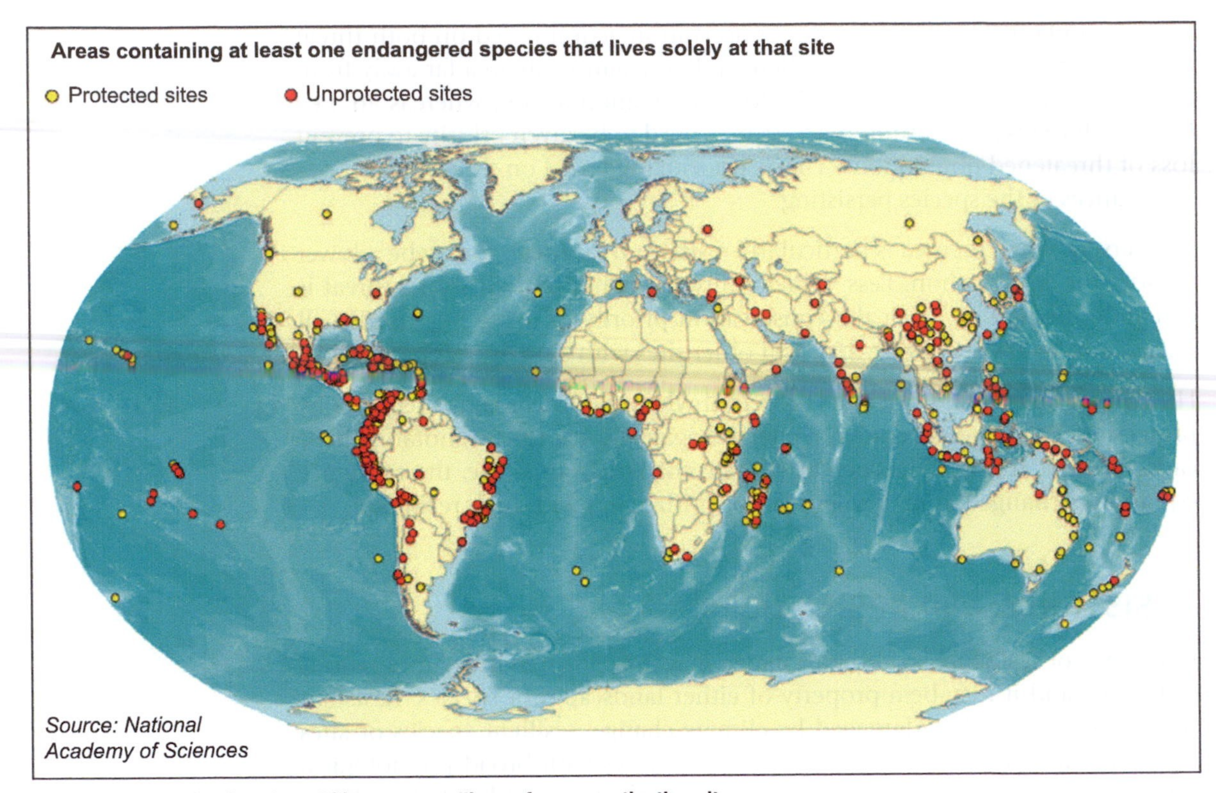

FIGURE 14.6 High-irreplaceability areas—alliance for zero extinction sites.
Alliance for zero extinction sites contain one or more species that occur only in those locations. These sites are irreplaceable: they must be conserved if species losses are to be avoided. *Source: Copyright National Academy of Sciences, USA.*

try to address that vulnerability through other actions. More than 1000 sites have been identified that harbor species found nowhere else in the world, so the number of these completely irreplaceable sites is not small.

Resilience is important in prioritizing other conservation actions as well, such as threatened species management or translocation, in ways quite different from prioritizing for protected areas. Conservation actions such as intensive management and translocation will be high priority where resilience is low. For instance, in a protected-area system in which only half the sites are believed to harbor resilient populations of a species, the other half would be prioritized for management or translocation to maintain the populations in the face of climate change.

Because our knowledge of future climate change will never be perfect, prioritization based on resilience to climate change is a risk-management exercise. We want to reduce (but not eliminate) conservation investments in sites at high risk of losing species to climate change and balance these with investment (though not exclusive) in sites with lower probabilities of losing their target species.

It is important to note that prioritization will normally need to be done for several threats at once, including both those that protected areas reduce (e.g., habitat loss) and others that they do not (e.g., climate change and acid deposition). The best conservation portfolio is one that systematically targets sites of high biological value (high irreplaceability) with appropriate emphasis on threat and resilience. For climate change, the appropriate conservation measure is to seek sites of high resistance or resilience.

"LIVING DEAD"

Suitable climate for a species may shift, leaving adult trees as "living dead." Tree mortality due to climate change is often considered to be highest in young trees. Seedlings are not deeply rooted and may be sensitive to drought or heat stress. Fully grown, deeply rooted trees may be able to tap deep groundwater and therefore be more resistant to death. Mature trees may survive under conditions that prohibit reproduction or replacement on a site. They may continue to form dominant parts of the vegetation structure until they die and are replaced by species that now find the climate at the site suitable for recruitment. Until then, the mature trees are "living dead" testaments to the former climate.

PROTECTED-AREA MANAGEMENT

Protected-area managers most often are concerned with management of threats, disturbance, species of special concern, and tourism. Climate change is an indirect threat that has the potential to alter the management of all these other factors. Reserve managers need to be aware of, and manage, both the direct and the indirect effects of climate change.

Climate change is a new type of threat that is in many ways not amenable to management through the traditional approaches used for other threats. Because it is mediated by the atmosphere and not human or biological agents, climate change penetrates all parts of a reserve and cannot be combated with staff on the ground. In this way, the direct threat of climate change is much different from other threats. Stopping climate change ultimately requires social and energy consumption changes across the planet—a change process that a reserve manager may participate in but not control. Therefore, the main management option for climate change may lie in trying to modify its effects by changing management of factors (threats, disturbance, species, and tourism) with which it interacts.

Threats that interact with climate change include encroachment of human land uses, illegal hunting, and invasive species. All of these threats may be altered by climate change. As agricultural systems are stressed by climate change, productivity may fall, resulting in demand for more agricultural land to maintain production, sometimes creating pressure for agricultural clearing inside protected areas. Invasive species may be favored by climate change, requiring more resources to combat. Anticipation of these changes may be more cost-effective than ad hoc responses. For example, it may be much cheaper to eradicate an invasive species before its population explodes. Reserves that manage threats only from their borders inward may find that climate change results in an increased number and magnitude of threats that they confront. Management that looks outward and anticipates climate change effects may greatly reduce the cost and improve the effectiveness of management responses, although in most cases requiring more management effort and budget than if climate change were not occurring.

Disturbance management includes control of fire and guiding recovery and restoration of areas affected by storms, grazing, and other disturbances. Fire frequency may be strongly affected by climate change. Fire may also "release" vegetation from existing species composition, allowing new species to colonize. For example, oak species may be favored by warmer climates but would be unable to advance into pine forests upslope—after a fire, this shade inhibition would be released and oaks could advance in, replacing the pines. Such disturbance dynamics place new demands on managers who have traditionally keyed fire control efforts on maintaining "typical" vegetation. Forest types typical of the past may no longer be typical as climate changes, and sudden turnover may occur when disturbances such as fire remove existing vegetation.

CLIMATE–HUMAN FEEDBACK

Tropical farmers often rely on forests for supplemental or emergency income. If crops fail, the sale of timber or collection of forest products can provide needed income. As climate change alters agricultural conditions, changes in crop productivity or crop failure may occur, triggering increased use of forests. Because of this human feedback on biodiversity, even where climate change is not damaging to biodiversity directly, its indirect impacts need to be anticipated to avoid threats to species.

Species of special concern include threatened species and species especially important to reserve objectives. Climate change may result in the addition of new threatened species or complicate management of existing threatened species. As with vegetation, climate change may cause shifts in species ranges or abundance, changing what constitutes a "typical" bird or mammal within a reserve—a process that may reduce the importance of some species currently targeted for management or introduce new ones. For instance, the pika is declining in high-elevation parks across western North America. This species, formerly not managed, may now need active management to avoid local extinction in many parks. In some lower elevation parks, the pika may disappear entirely, taking it in turn from a species not of special concern to a species declining and managed to prevent local extinction and, eventually, to a species that no longer exists in the park and so ceases to be a management objective entirely. Each species will be unique in its response to climate and unique in its relationship to the ecology and management objectives of a particular reserve. Managers should be aware of the impact of climate change on threatened or other managed species in their reserve and be aware that climate change may result in additional threatened species in need of management.

Tourism management includes guiding tourism to areas resilient to human use, providing facilities for enjoyment and interpretation of the natural features of a reserve, and distributing various kinds of tourism in ways compatible with multiple reserve-management goals. This can be particularly important with respect to interactions between disturbance, climate change, and invasive and threatened species. For example, areas recovering from fire may be more susceptible to introduction of invasive species carried on the boots, clothing, or vehicles of tourists. Climate change can increase the frequency of fire, altering the extent of recovering areas at the same time that it promotes establishment of invasives. Thus, reserves in which fire and invasive species are management issues may become much more sensitive to tourism impacts as climate changes. Climate change may alter the timing or extent of processes important to threatened species, changing the zoning and management needs of those species with relation to tourism. For instance, hiking in scree zones may have had negligible impact on pika populations in the past, but it might become an important threat in need of management as pika populations become stressed by climate change.

Although each threat must be assessed in the context of an individual reserve, several general principles can still be suggested for reserve managers:

- Be aware of climate change effects on current management objectives, such as threatened species or fire control.
- Be alert for nonthreatened or nonobjective species or processes whose status may change significantly because of climate change.
- Plan on longer time frames (10–20 years rather than the more typical 3–5 years).
- Consider interactions between management objectives that may be sensitive to climate change.

MARINE PROTECTED AREAS

Marine protected areas (MPAs) have been among the first to apply these principles because of the immediate nature of changes under way due to coral bleaching. MPAs serve multiple purposes and have been implemented in many areas of the planet. However, in general, MPA networks are less well developed than terrestrial protected areas, both in extent of coverage and in management

FIGURE 14.7 Healthy (top) and bleached (bottom) coral reefs. Corals that escape bleaching are resistant, those that rebound quickly are resilient.
Source: Courtesy U.S. National Oceanic and Atmospheric Administration (NOAA).

experience. There are far fewer MPAs than terrestrial protected areas, and the area of MPA coverage is a much smaller fraction of the overall ocean area of the planet than is the fraction of the land surface covered by terrestrial protected areas. However, some of the largest protected areas on the planet are MPAs, and coral reefs are especially well represented in MPAs. Here, we focus on two examples of MPA responses to climate change: MPA siting and management in response to coral bleaching and MPA needs in response to lost sea ice in the Arctic.

MPA response to coral bleaching is built around the principles of resistance and resilience (Figure 14.7). These principles are applied on different spatial scales to help guide MPA system planning and management planning of individual MPAs. In this case, resistance refers to reefs that are resistant to bleaching, and resilience refers to reefs that are more likely to recover once damaged.

At the broad-scale, system planning level, bleaching resistance and bleaching resilience are used to help prioritize sites for protection. These are not the only, nor even the main, prioritization factors, but between otherwise equal-priority sites, resistant and resilient sites are selected over low-resistance or low-resilience sites. Although biological value and irreplaceability are primary criteria, unless a site harbors completely irreplaceable attributes (e.g., species found only at that site and nowhere else in the world), there is little point in protecting a reef that will be destroyed by bleaching. Hence, resistant and resilient sites get preference.

Sites may be more or less resistant and resilient to coral bleaching based on factors such as currents, upwelling, and previous exposure of reefs to bleaching (Table 14.3). Cool currents or upwellings tend to make sites resistant to bleaching because they maintain cooler surface water temperatures. Warm currents have the opposite effect. Corals that have been bleached previously and survived (mortality in bleached corals ranges from 60% to 90% in most regions) may have been naturally selected to have more resistant zooxanthellae or other natural mechanisms of survival and recovery.

At the site-management level, resilience and resistance may be applied in zoning and management decisions (Figure 14.8). Local upwellings or currents may make some parts of an MPA more resilient than others. Physical shading is important at this scale as well. For instance, reefs in the shadow of a large mountain will be more resistant than reefs in continual exposure to the sun. Sedimentation, which negatively affects reefs in heavy doses, may actually increase resistance to bleaching by shading and cooling reefs (Figure 14.9).

Once reefs do bleach, resilience is an important factor in reserve management. Reefs that are more likely to recover become a source of recolonization of corals for areas in which mortality is high. Zooxanthellae from resilient reefs may eventually colonize damaged reefs and increase their resilience. Prioritizing these areas for protection is therefore as high a priority at the local scale as it is at the regional (system planning/site selection) scale.

Management of MPAs can reinforce resilience and recovery. When an area bleaches, temporary tourism closures can help the reef recover. Where recreational fisheries or subsistence harvest is allowed, temporary closures may also help ease pressure on coral-dependent species whose populations crash after bleaching, allowing them to recover once the corals reestablish. Lack of protective management can result in the exacerbation of the damage of bleaching, thus causing a downward spiral in reef condition that may ultimately result in the replacement of the corals in the system with algae.

Nontropical MPAs must also carefully consider climate change in management. Extensive evidence from cold-water fisheries and past climate change indicates

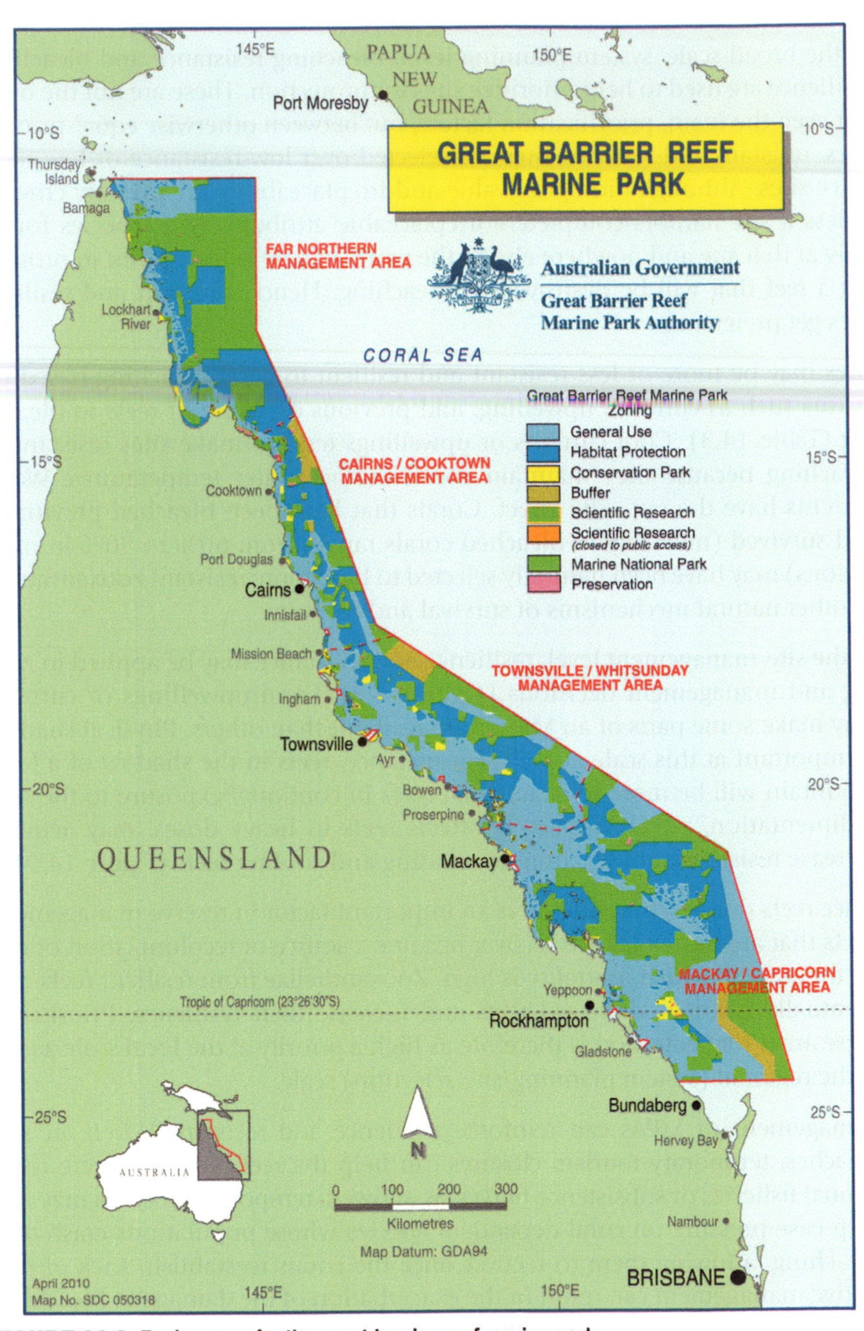

FIGURE 14.8 Zoning map for the great barrier reef marine park.

The Great Barrier Reef Marine Park is a marine protected area that has experienced extensive coral bleaching. In response to bleaching events and other management issues, a zoning plan for the park reflects permitted uses that best integrate climate change with other park management objectives. Tourism is excluded in some areas to facilitate postbleaching recovery. *Source: Map courtesy of the Spatial Data Centre, Great Barrier Reef Marine Park Authority (2010).*

FIGURE 14.9 Sedimentation and shading effects on coral bleaching.
Shading (a) and sedimentation (b) are two factors that can influence the severity of coral bleaching. Sedimentation stresses corals and may exacerbate bleaching effects, whereas shading protects corals from synergies of high temperatures and photic effects, thereby reducing the probability of bleaching. *Source: Grimsditch and Salm (2006). Reproduced with permission from IUCN.*

that pelagic fisheries and benthic communities experience major shifts linked to climate (Figure 14.10). Recent changes due to loss of sea ice in the North Bering Sea provide a compelling example of the need for additional MPAs.

Sea ice retreat in warming sea surface temperatures is driving major ecological changes in the North Bering Sea near St. Lawrence Island. The disappearance of sea ice in warmer water temperatures makes a more active ecosystem in the water column in this region. Plankton that used to fall through the water column and be deposited in the benthos now are caught up in the food web of the water column, greatly reducing the energy and nutrients reaching benthic communities. As a result, there has been a large-scale shift in the species composition of benthic communities, with mollusks decreasing in dominance and nutritional quality and being replaced by brittle stars.

FIGURE 14.10 **Marine protected areas.**
Healthy marine systems such as these can be one of the major benefits of marine protected areas (MPAs). MPAs can improve food web health and reduce chances of coral bleaching by decreasing synergistic pressures such as fishing and tourism overuse. *Source: Courtesy U.S. National Oceanic and Atmospheric Administration (NOAA).*

Walrus and spectacled eider that feed on these benthic communities are therefore faced with greatly declining nutrition. At the same time, the retreat of sea ice means that the species spend more time resting in water than on ice. Thermal losses in water are much greater than those in air, so the switch from ice resting to water resting means that eider and walrus are burning much more energy at rest at the same time that they are getting less bivalve food from the bottom. If the species try to follow sea ice to maintain their thermal balance, they find themselves in much deeper waters farther from the continental shelf, which means they must dive deeper to feed.

Walrus and eider populations have declined dramatically because of these deteriorating conditions. Spectacled eider populations have declined by up to 96% in areas of Alaska and Canada. Young walrus are being separated from their mothers as sea ice literally melts out from under them. Pups drift away while their mothers are on deep feeding dives. Lost walrus pups have been observed by research vessels in open ocean thousands of miles from normal walrus breeding grounds.

To further complicate the situation, there is a sill of cold water behind a ridge on the ocean floor off St. Lawrence Island that excludes groundfish from the North Bering Sea. This cold-water sill is breaking down as water temperatures warm. As it disappears, groundfish will enter the North Bering Sea and compete with walrus and eider for mollusks.

Bottom-trawling fisheries will probably follow groundfish when they enter the North Bering Sea, disrupting the bottom and further reducing mollusk food sources. This human disturbance may be the coup de grâce for declining walrus and eider populations if fisheries are allowed to enter the area unchecked.

An MPA in these waters is needed to protect the bottom habitat for mollusk populations to maintain food sources for walrus and eiders. This protection cannot reverse the loss of sea ice, the warming of North Bering Sea waters, or the breakdown of the cold-water sill. An MPA can, however, prevent the decline of bottom habitat due to trawling and so minimize further losses of diving species.

PROTECTED AREAS FOR CLIMATE CHANGE

Protected areas are an important component of adaptation responses to climate change. The addition of new protected areas can help compensate for protection losses caused by species' range shifts in response to changing climatic conditions. For many areas, such as the North Bering Sea and tropical reefs affected by coral bleaching, current scientific understanding is sufficient to suggest where new protected areas should be placed and how they should be managed. For other areas, awareness of risk factors and climate change planning and management principles are starting points for improved conservation. Table 14.3 summarizes important risk factors, Table 14.4 reviews general management principles, and Table 14.5 lists coral bleaching resistance factors for MPA design.

Table 14.3 Coral Bleaching Resistance Attributes and Factors

Attributes	Determining Factors
Promotion of water mixing	Proximity to deep water and regular exchange with cooler oceanic water
	Localized upwelling of cool water
	Permanent strong currents (tidal, ocean, eddies, gyres)
Screening of corals from damaging radiation	Deep shade from high land profile
	Shading of some coral assemblages by complex reef structure, multilayered coral communities, or steep slopes
	Orientation relative to the sun (north-facing slopes in Northern Hemisphere, south-facing slopes in Southern Hemisphere)
	Presence of consistently turbid water
Indication or potential preadaptation to temperature and other stresses	Frequent exposure of corals at low tides
	Highly variable seawater temperature regime (pond effect in shallow back-reef lagoons)
	History of corals surviving climate-related bleaching events
	High diversity and abundance of coral reef species
	Wide range of coral colony size and diversity in different reef zones, including centuries-old colonies
	High live coral cover
Survival of at least some coral communities	Stable salinity regime
	Large area with wide depth range and habitat variability
	Low risk of exposure to climate-related temperature stress at the location

Table 14.4 Criteria Affecting Vulnerability of Species Occurrences or Populations to Impacts of Climate Change

Factor	Increased Risk	Decreased Risk	References
Temperate range shifts	A temperate occurrence on the low latitude or lowland periphery of a species' historic range	A temperate occurrence at the poleward or upland periphery of a species' historic range and physiological limits	Peters and Darling (1985), Peters (1992)
Tropical range shifts	A tropical montane occurrence	A tropical lowland occurrence	Peters and Darling (1985), Peters (1992)
Ecosystem resilience	An exposed or management-dependent occurrence (e.g., forest edge ecosystem, small fire-dependent community, heavily exploited species)	An extensive, low-management occurrence (e.g., forest interior ecosystem, community in an intact catchment)	Forman (1997), Noss (2001)
Ecosystem connectivity	An isolated occurrence	An occurrence with functional connectivity to other occurrences	Markham and Malcolm (1996)
Genetic richness	A genetically impoverished occurrence	A genetically heterogeneous occurrence	Comes and Kadereit (1998)
Topography	A topoedaphically homogeneous occurrence	A topoedaphically heterogeneous occurrence	Peters (1992)
Extinction risk	An occurrence without restricted-range or extinction-prone species	An occurrence with restricted-range or extinction-prone species (the largest members of each feeding guild, poorly dispersing species, low reproductive rate species, and species characteristic of late-successional communities)	Diamond (1976), Terborgh (1976), Pimm (1991)
Sea level change	An occurrence on coastal wetlands unable to migrate inland	An occurrence on coastal wetlands adjacent to low-lying natural areas	Titus (1998)
Montane geography	An occurrence on steep upper mountain slopes where upward dispersal is limited	An occurrence on gentle lower mountain slopes	Halpin (1997)
Disturbance regime	An occurrence smaller than the minimum area necessary to accommodate natural disturbance cycles	An occurrence large enough to accommodate more frequent, severe, or extensive disturbances than have historically occurred	Pickett and Thompson (1978), Forman (1997)
Landscape ecology	An occurrence in a highly fragmented landscape	An occurrence in a little-fragmented landscape that promotes dispersal and reduces invasive species establishment	Hannah et al. (2002)

Table 14.5 Elements of Protected Area Management for Improved Responsiveness to Climate Change

Site planning	Climate evidence explicitly incorporated through scenario-building
	Multiple time horizons to represent uncertainty and possible long-term future conditions
	Refinement of regional scenarios
Management actions	Coordinated with other reserves in region
	Planned using scenarios of climate change and range shifts
	Based on iterative monitoring feedback
Monitoring	"At-risk" species (from climate change evidence, threatened species, management targets)
	Structured, taxon-stratified sample of all species
	Enhanced collection of climate/weather data
	Biotic survey
	Iterative feedback to management planning and action

Protected areas are only one part of the total conservation response. Adaptation to climate change will also require management of species, increases in connectivity, and landscape management. These elements of a complete program of conservation adaptation and response to climate change are outlined in the following chapters.

FURTHER READING

Hannah, L., Midgley, G., Andelman, S., Araujo, M., Hughes, G., et al., 2007. Protected area needs in a changing climate. Frontiers in Ecology and the Environment 5, 131–138.

Mawdsley, J.R., O'Malley, R., Ojima, D.S., 2009. A review of climate-change adaptation strategies for wildlife management and biodiversity conservation. Conservation Biology 23, 1080–1089.

Hannah, L., 2010. A global conservation system for climate–change adaptation. Conservation Biology 24(1), 70–77.

Green, A.L., Fernandes, L., Almany, G., Abesamis, R., McLeod, E., Aliño, P.M., Pressey, R.L., et al., 2014. Designing marine reserves for fisheries management, biodiversity conservation, and climate change adaptation. Coastal Management 42(2), 143–159.

CHAPTER 15

Connectivity and Landscape Management

Connectivity provides habitat between protected areas to further conservation objectives. Connectivity may be total, such as when a corridor of natural forest connects two parks, or partial, such as when shade coffee provides connectivity for forest birds by leaving canopy trees intact even though the understory is devoted to growing coffee. In parallel to being total or partial, connectivity may be intensive or extensive. A narrow corridor between parks is intensive, whereas a broad landscape of conservation-friendly human land use (e.g., shade coffee) connecting multiple protected areas is extensive.

There are many reasons for establishing connectivity. It may be important in providing large carnivores passage between protected areas that allows them to maintain large home ranges. Connectivity can improve gene flow between protected populations. It is important in many settings to allow species responses to climate change.

Connectivity for climate change helps accommodate species range shifts and species dispersal to track climate, promotes gene flow as species go through population bottlenecks due to climate shifts, and serves other purposes in adaptation to climate change. Perhaps the most discussed role of connectivity for climate change is that of providing avenues for species range shifts between static (fixed in space) protected areas.

The value of connectivity for promoting range shifts faces some practical limitations. Because most landscapes are dominated by or affected by human use, connections over very large distances are unlikely (Figures 15.1 and 15.2). If climate change is severe enough to require long-distance connectivity, it will also affect a large number of species. This further reduces the chances of finding available avenues through human land uses. The best way to deal with a multitude of long-distance connection needs is to avoid them, which is why stopping climate change is a top priority for the conservation of biodiversity.

Climate Change Biology. http://dx.doi.org/10.1016/B978-0-12-420218-4.00015-9

The appropriate role of connectivity is therefore in dealing with more modest connectivity needs. Even if climate change is stopped early, many species range shifts will result. Designing connectivity for climate change is about accommodating these range shifts and other effects of climate change on species and ecosystems-effects that are already underway and must be dealt with even if action is quickly taken to stop climate change.

SPOTLIGHT: FRONT MOVING IN

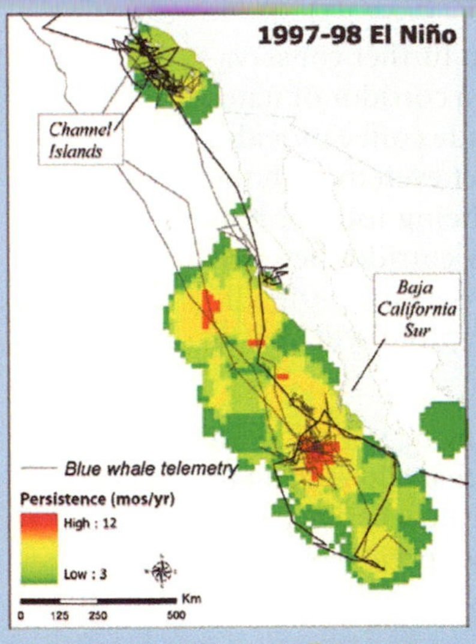

Sea surface temperature fronts (colors) off the southwest United States and northwest Mexico, shown with telemetry tracks from a blue whale (lines), illustrating the heavy use of these sea surface temperature features by large marine vertebrates (including billfish, turtles, and marine mammals). From Etnoyer et al. (2006).

In the oceans, important biological sites are not fixed in place. Large marine animals such as sea turtles and whales congregate along ocean temperature fronts. Like weather fronts, these sea surface temperature gradients are constantly shifting. Marine life follows the front, feeding on fish and other organisms that collect along the boundary between the two water masses.

Telemetry of turtles and whales clearly shows this effect. Blue whales and sea turtles follow trajectories that track sea surface temperature fronts (see figure). The radio-tagged animals prefer the interface of warm and cold water.

Conserving these rich sea surface temperature fronts requires protection that moves. Staking out a stretch of ocean above a fixed piece of sea floor cannot capture the fronts: they simply move in and out of such fixed areas. A new conservation mechanism is needed that can exclude exploitation wherever the front moves. Owing to the nearly universal use of GPS on fishing vessels, such mobile protection is now possible.

Source: Etnoyer, P., Canny, D., Mate, B., Moran, L., 2004. Persistent pelagic habitats in the Baja California to Bering Sea (B2B). Ecoregion Oceanography 17, 90–101.

Connectivity is not all positive, however. Connectivity generally involves corridors that have large amounts of edge habitat relative to core habitat or land uses that involve disturbances. These characteristics of connectivity may favor invasive species that prefer edge habitats and human systems. Connectivity may also provide avenues of dispersal for disease: Protected areas that might be spared a disease outbreak by isolation may become vulnerable to spreading diseases if connected to other parks. Similar concerns may apply to pests and weeds. Therefore, the benefits of connectivity for climate change must be weighed against possible negative effects.

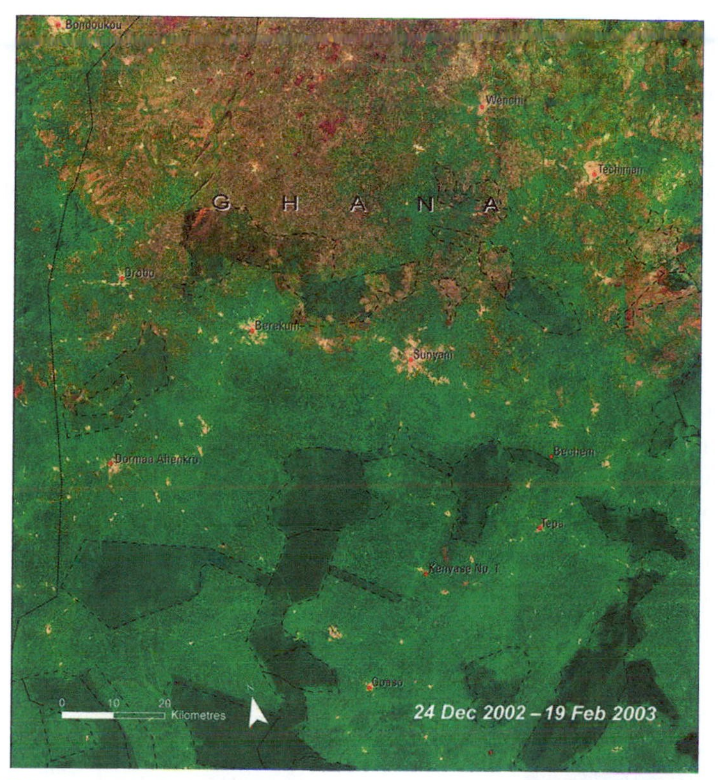

FIGURE 15.1 Low-connectivity landscape.
Satellite image of forests in Ghana. The influence of protection is visible; the irregular shapes of remaining forest fragments correspond exactly to the outlines of forest reserves (dotted lines). The large areas in which connectivity has been lost are plainly visible in almost all areas not protected. These areas are now agricultural landscapes, making reconnection difficult. *Source: From United Nations Environment Programme.*

FIGURE 15.2 A high-connectivity forest landscape.
Planning for broad-scale connectivity is still possible in this Canadian forest. *Source: Courtesy of WRI Features.*

AREA-DEMANDING SPECIES

Large species such as top carnivores often require extensive home ranges for hunting or reproductive territory (Figure 15.3). Such area-demanding species have home ranges that span multiple habitats, each of which may undergo transitions or deterioration due to climate change. Because of these climate-driven alterations in habitat, the area needs of such species may increase. Connectivity is often needed to maintain area-demanding species in the present climate, and it will be increasingly needed for area-demanding species to survive under future climates or to survive the transition between present and future conditions.

Connectivity for large species may be important in maintaining entire ecosystems during climate change. When keystone species are lost, food-chain reverberations may affect ecosystem structure and process. This general principle is particularly relevant to climate change.

Predators may influence prey population fluctuations with climate change, stabilizing the food chain. Without top predators, herbivore populations fluctuate with food availability, which is in turn dependent on climate-driven

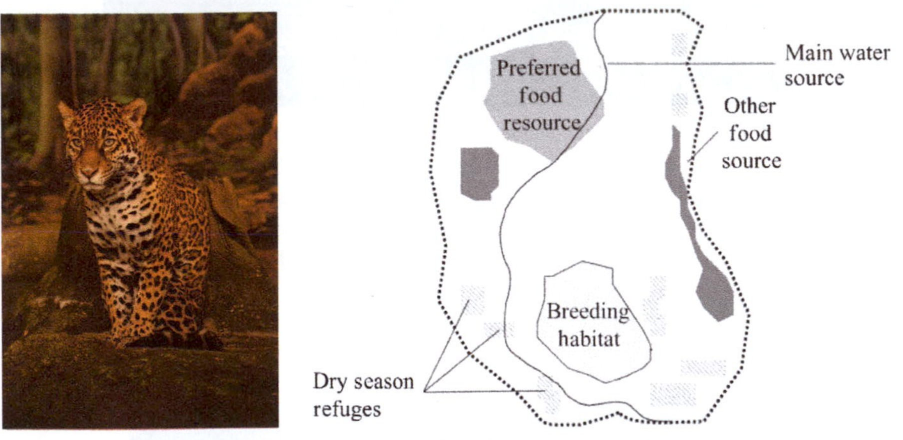

FIGURE 15.3 Jaguar and the landscape species concept.
The jaguar, *Panthera onca,* is a large carnivore that has been used as an icon for establishing landscape connectivity (left). Paseo Panthera is a network of connected protected areas intended to allow passage of jaguar across Central America. This provides important connectivity for range shifts, even though it was not specifically designed for climate change. The landscape species concept (right) can be used to design connectivity for iconic species. Species' needs at different times of the year or different stages of live history are mapped in the landscape and connectivity between these key resources can be a goal of landscape conservation. The map illustrates landscape species needs for a hypothetical large mammal. Because different elements in the landscape can be affected differently by climate change, the landscape species approach is a useful concept for climate change planning. *Source: From Wikimedia Commons and Sanderson et al. (2002).*

productivity. Prey populations in predator-free settings may go through large "boom and bust" cycles, building in times of good food availability only to crash when food is exhausted or productivity changes, reducing food availability. With top predators, herbivore populations are modulated by the presence of the carnivore rather than by food availability. They go through smaller cycles of population increase and decrease.

When climate changes, systems modulated by predators will be less prone to population crashes, decreasing the risk of local extinction of the herbivore population. Ecosystems in which large predators are missing will be more prone to crash and more vulnerable to population extinctions due to climate change.

These effects have been observed in lake systems and modeled for terrestrial systems with large predators. In a lake system experimentally manipulated to

SPOTLIGHT: PREDATOR–PREY POPULATION CYCLES

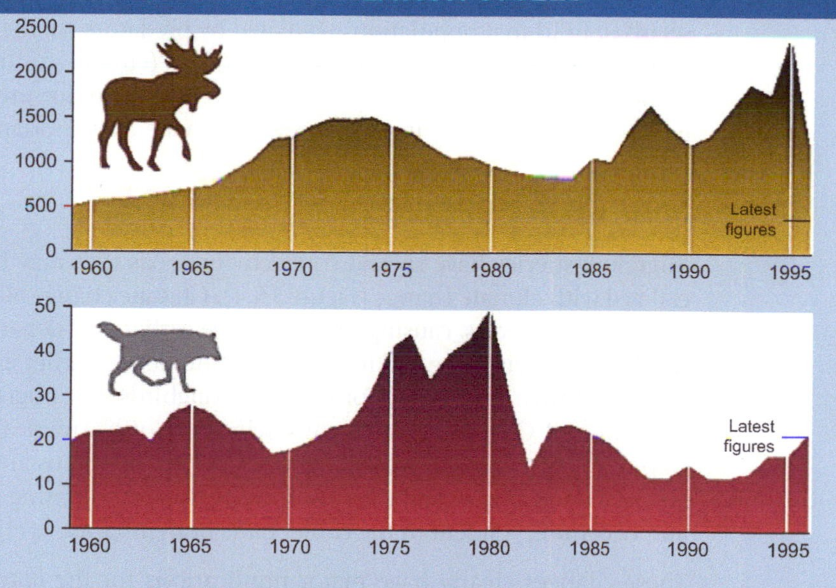

In systems in which predators and prey are not very diverse, population cycles often develop. Particularly when single species of prey and predator are involved, these cycles may be pronounced. For instance, lynx and hare populations in boreal forests may show strong cycles. When hare are abundant, lynx populations rise, and as predator density increases, hare populations cycle down. Lynx populations decrease soon after because of the lack of food, and then hare proliferate once predators are less abundant, starting the cycle anew. Similar effects are seen in wolf and moose populations on Isle Royale, Michigan (see figure). Prey populations (top panel) cycle with food availability that may be mediated by climate change. Modeling has shown that prey populations may be less vulnerable to crashes during climate change if healthy predator populations (bottom panel) exist to help keep them in check.

Source: Post, E., Peterson, R. O., Stenseth, N. C., & McLaren, B. E. 1999. Ecosystem consequences of wolf behavioural response to climate. Nature, 401(6756), 905-907.

simulate climate change, phytoplankton were more likely to crash to extinction when zooplankton predators were absent. A similar effect has been suggested for top carnivores—wolves—in terrestrial systems.

Predators such as wolves may thus be good for prey populations as climate changes. For instance, in the greater Yellowstone ecosystem, deer and elk populations fluctuate widely where wolves are absent, suggesting that additional fluctuation due to climate change could lead to "bust" parts of the cycle that would be devastating to local populations. Where wolves are present, deer and elk fluctuate in sync with wolf populations rather than food availability, and so they fluctuate much less widely. In these systems with wolves, changes in primary food availability will be less devastating because deer and elk populations are limited by wolves and not entirely by food availability.

Intact ecosystems, especially those retaining their top predators, may therefore be more robust to climate change. From microscopic lake phyto- and zooplankton systems to large terrestrial carnivores such as wolves, the general conclusion seems to hold that populations regulated by predators will be less vulnerable to climate change than populations regulated by (climate-controlled) food availability or direct climatic factors. Building connectivity for predators may therefore be a sound management strategy for climate change adaptation.

MIGRATORY SPECIES

Migratory species have special connectivity needs that may become more specialized with climate change (Figure 15.4). Climate change alters the phenology of migratory species, causing them to arrive earlier and depart later in the poleward part of their range and to arrive later and depart earlier in the more equatorial end of their migration. For instance, songbirds that migrate between North America and Central America are staying in North America longer as climate warms, meaning they leave for Central America later, thus arriving in their winter range later. Some populations or individuals are ceasing migration entirely and becoming resident in the cooler parts of their range as climate warms.

These changes clearly have major implications for the connectivity needs of migratory species. Where seasonal closures are used to protect migrants, the length and timing of these closures may have to be altered. Newly resident populations may need expanded protection in their new year-round range, whereas portions of the population that still migrate will need continuing connectivity of protection all along their migratory route.

Connectivity for migratory species is typically a series of "stepping stone" protected areas that allow feeding and rest for migrating individuals. As habitat loss to human land uses has progressed, these stepping stones have become increasingly managed to provide specific resources at specific times in the

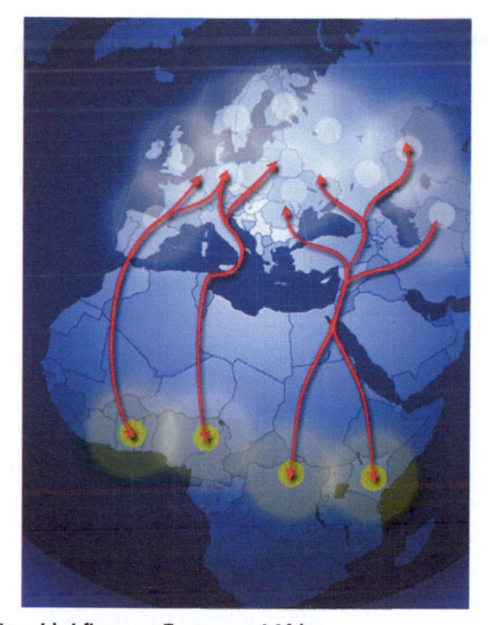

FIGURE 15.4 Migratory bird flyways, Europe and Africa.
Migratory pathways, such as these flyways, are highly likely to shift temporarily or spatially owing to climate change. Conservation responses to anticipate and track these changes are needed. *Source: Courtesy of Born to Travel Campaign, BirdLife International.*

migration. For instance, wildlife refuges in the United States now often grow grain for migratory species to feed on because food availability in surrounding landscapes is becoming more uncertain, for instance, as natural wetlands are replaced by agriculture or grain fields are replaced by commuter housing.

As climate change affects the timing of migration, it may affect the resources needed by the species at particular locations. For example, geese arriving early in their arctic summer feeding grounds are finding that the grasses they feed on are still dormant. The geese eat the rootstock of the grasses, undermining the reproduction of the very resource they need for survival. In this case, a system that required no management previous to climate change may require management to deal with a timing mismatch in the system and avoid crashes in the grass, and subsequently goose, populations. The assessment and maintenance of connectivity for migratory species will need to take changing needs such as these into consideration.

SPECIES RANGE SHIFTS

Range shifts pose a third and very large need for connectivity as climate changes. Such shifts may include area-demanding or migratory species but will include hundreds or thousands of other species as well. Many of the species whose range shifts will have to be accommodated will be plants; for these species, the appropriate connectivity will occur on much finer scales than those for area-demanding or migratory species.

For plants and other limited-dispersal species that can maintain healthy populations in relatively limited areas, connectivity requirements over broad scales would be an indication that climate change has gotten too far out of control for its biological consequences to be manageable. Instead, connectivity for these species needs to be considered on a much more local level—that compatible with existing protected areas or expansion on the periphery of individual protected areas.

These more modest connectivity needs will involve large numbers of species, so in aggregate they may represent a piece of the connectivity puzzle as significant as that for area-demanding or migratory species. Carefully placed, such connectivity can meet the needs of multiple species and provide secure futures for many species until climate change is checked.

Connectivity for these smaller scale movements needs to be added systematically to maximize the cost-effectiveness of actions. There will typically be multiple options for each species, so selecting connections that overlap with the needs of other species can greatly reduce the overall area required.

Small-scale connectivity will often occur on the periphery of existing protected areas to complement range already protected. This is in contrast to the more intuitive view of connectivity between protected areas. Whereas intersite connectivity is dramatic and easily visualized, it is connectivity within single protected areas or between the unprotected periphery and the protected area that will be most widely applicable because this can be accomplished with relatively much lower area requirements. Competition with human land use dictates that these small-scale connections will be most available and most strategic in the conservation battle to maintain species whose ranges are shifting because of climate change.

The key principle for cost-effective connectivity is to connect present and future populations and present and future attributes in a landscape that a species needs to survive. Although connectivity between protected areas is often discussed as a response to climate change, in practice such connectivity is needed only for top predators and large mammals. Many other species will need connectivity on smaller scales. Plants will most often need connectivity between areas of suitable climate in the present and future that are within dispersal distance of current populations. Small mammals and birds may require connectivity between landscape elements that are shifting as climate changes, for instance between fruiting trees affected by changes in phenology.

PLANNING FOR CONNECTIVITY

Planning fine-scale connectivity for range shifts is fundamentally different from planning connectivity for area-demanding and migratory species. Connectivity already exists for many area-demanding species and migratory species, providing a base on which planning for climate change can be built. Planning for multispecies range shifts, on the other hand, has no planning analogue and is the more difficult conservation challenge.

The fine-scale needs of individual species may be visualized as habitat "chains," linking current suitable climate to future suitable climate (Figure 15.5). Each chain of present-to-future suitable habitat is composed of multiple links, each maintaining a viable subpopulation of the species. Multiple chains may then be selected that will protect enough subpopulations to protect the entire population of the species from extinction.

Ideally, protected populations will be geographically distributed—as widely spaced as possible—to minimize the chances of extinction. By choosing widely separated populations, the odds of a chance event such as a large fire or isolated storm destroying the whole population are minimized. Distributed connectivity is also a way of dealing with uncertainty about future climates: if

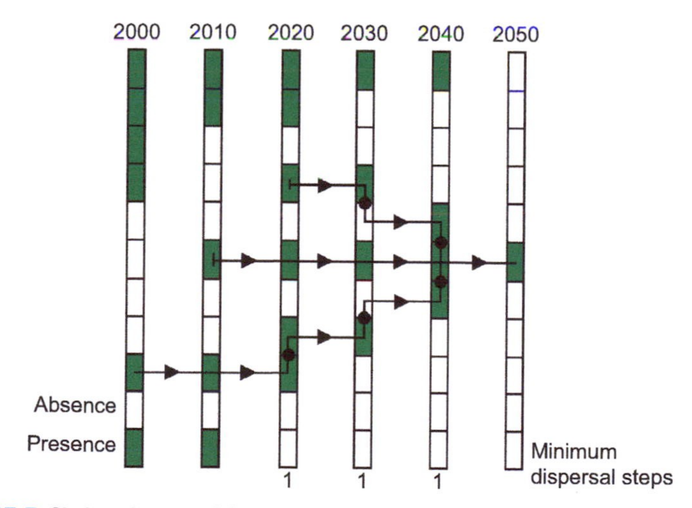

FIGURE 15.5 Chains of connectivity.
Protecting chains of connectivity is one way to ensure species representation in protected areas as climate changes. Green rectangles indicate planning units that have suitable climate for a target species at each time step. Arrows indicate "chains" of suitable habitat through time. Dispersal is limited to one 2-km cell per 10-year time step in this ant-dispersed plant species. Only one combination of cells provides a complete "chain" of habitat from 2000 to 2050. *Source: Williams et al. (2005).*

changes do not proceed as anticipated in one area, having connectivity in other areas increases the chance that multiple subpopulations will survive.

The first step in planning fine-scale multispecies connectivity for climate change is a sound protected-area planning process. Much of the fine-scale connectivity needed will be available in existing protected areas. The need for additional connectivity will be most cost-effectively achieved on the periphery of existing protected areas, adding just the increment of connectivity needed to complete species range shift requirements already largely met by existing protection.

A good protected-area planning process will provide a target population size for each species. This serves as the basis for estimating how many chains of subpopulations will be needed for a species. For instance, if the target is 100 km^2 of habitat for a species, climate change connectivity must maintain 10 10-km^2 chains from present to future or 100 1-km^2 chains. Without a clear current target, it is difficult to plan connectivity, which is why a protected-area planning process is so important: it provides the target and the framework within which planning connectivity for climate change can proceed.

Once the target has been set, expert opinion or climate modeling can be used to determine which areas will maintain suitable climate for the species, which areas will see fine-scale migration of suitable range, and which populations will undergo complete range loss or large-scale range migration. Solutions that maintain protection for populations from present to future are then selected. Areas that will retain suitable climate and are currently protected are selected first, followed by chains that occur wholly in current protected areas. Then, areas in which fine-scale connectivity can provide chains from nearby populations to future suitable habitat in protected areas are added until the target is met. As discussed in Chapter 14, such a process is most cost-effective when done for multiple species simultaneously. It is likewise most cost-effective to plan new protected areas and connectivity for climate change simultaneously (Figure 15.6).

Conservation planning software can be used to solve such problems for multiple species. With more than a few species, the trade-offs and choices to meet the target become quite complex, so it is helpful to use a computer-generated solution. Many conservation planning software programs contain a "reserve selection" algorithm that helps planners efficiently choose new protected areas to meet the needs of multiple species. Reserve selection algorithms are easily adapted to solve the multiple-chains problem of climate change connectivity. For large numbers of species, other, more sophisticated, optimization programs may return a more efficient answer. In either case, the challenge is to put enough climate change chains in place to conserve all species in the planning process, using the conservation targets set in the protected-area planning process and the most area-efficient approach possible to minimize the cost of connectivity.

CONSERVATION PLANNING SOFTWARE

Computer programs that help select sites for new protected areas are called conservation planning software. Marxan and Worldmap are examples of software packages with conservation planning applications. Most conservation planning software uses algorithms that help represent all species in an efficient (low area requirement) system. These algorithms may be modified to help plan for climate change, for instance, by requiring representation of all species in both present and future climates in chains of connectivity.

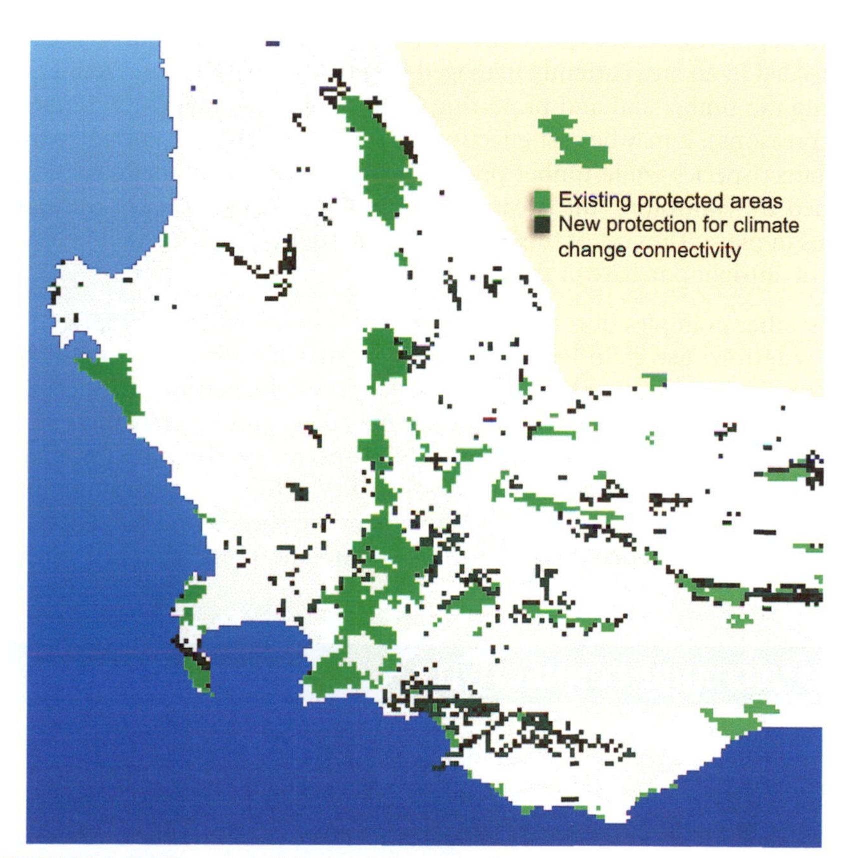

FIGURE 15.6 Building connectivity to existing protection.
Choosing chains of suitable habitat that occur partly or entirely within existing protected areas helps create cost-effective solutions by minimizing the need for new protection. This map of the Cape Floristic Region shows existing protected areas (light green) and areas needed to conserve 300 species of proteas as climate changes (dark green), selected by conservation planning software. The planning software was programmed to represent complete chains of habitat, such as that illustrated in Figure 15.5, for all species. Where chains could not be represented in existing protected areas, new areas were selected for connectivity. Note that most new protected areas connect to existing protected areas. *Source: Figure courtesy Steven Phillips.*

MANAGING CONNECTIVITY IN HUMAN-DOMINATED LANDSCAPES

Once connectivity needs are established, protection must be secured and land use managed for climate change. Where total connectivity is being established, this simply entails acquiring new protected area. In many cases, however, total connectivity will not be possible, and establishment of connectivity will involve some combination of management actions in human-dominated landscapes.

For example, climate change connectivity for an understory plant species may be needed in an area currently managed for production of timber. Rather than buying the timberland and protecting it (although this may be desirable for other reasons), it may be cost-effective to manage the land for connectivity for the target species while timber production continues. This might be accomplished by controlling the area cleared during logging to leave understory plants in place. The cost of this management activity would be well below the cost of outright purchase of the land.

Many other examples of connectivity exist in human-used landscapes. Migratory waterfowl use grain left on the ground after harvest, and tropical birds use forest canopies over shade-grown coffee. These management tools can be appropriated to meet climate change connectivity needs, even though they provide only partial connectivity. Matching species needs with cost-effective options available in target landscapes is a key activity in planning climate change connectivity, whether for area-demanding species and migratory species or for fine-scale connectivity to accommodate range shifts in multiple plant species.

MANAGING FOR CLIMATE CHANGE CONNECTIVITY

The most efficient connectivity is generally the shortest. Where managers wish to conserve species whose ranges are shifting, finding and protecting populations near an existing protected area that will become suitable habitat under future climates is more efficient than building connections between (often widely separated) protected areas.

Once such connectivity is in place, climate change poses the additional challenge of temporal variability. Unlike stable climates, in which management can be consistent year to year, management for climate change requires constant monitoring and updating of species needs and landscape uses.

Both species needs and the human use of the landscape may change with climate change, and not always in complementary ways. Management of partial connectivity in climate change therefore requires consideration of changes in land use, monitoring, and revision of management. Such management is referred to as "adaptive management" because management actions are adapted based on the results of monitoring.

PLANNING FOR CLIMATE "BLOWBACK"

Climate change will influence human land use, with potentially large consequences for conservation. In many areas, the loss of habitat inflicted by changing human land uses may be of greater conservation impact than the direct effects of climate change on species. This secondary effect, or "climate blowback" is only beginning to be explored.

Shifts in agriculture to track changing climate are the largest single concern. As agricultural conditions shift, farming may change location to track optimal conditions. Farmers adapting to climate change may put more pressure on natural systems, using more water as conditions warm or clearing forests for new fields to compensate for declining production.

Wine production provides an iconic example of possible blowback effects on conservation. Wine grape growing is famously keyed to climate. As climate changes, vineyards may expand into new areas, creating new potential for habitat loss to land clearing (Figure 15.7). Important conservation areas such as the Yellowstone to Yukon connectivity initiative will find new vineyards establishing by midcentury. Because vineyards may attract wildlife such as bears, fencing of previously open range may result, with serious consequences for connectivity. In existing areas of production, vineyards may increase water use to combat heat stress in vines (red in Fig. 15.7).

Conservationists need to prepare to work with agricultural stakeholders to avoid unintended blowback consequences of climate change. In several Mediterranean wine-producing regions, coalitions of wine producers and conservationists are planning together to allow change in vineyards while conserving high-priority areas for biodiversity and ecosystem services. Similar planning efforts can help avert conservation conflicts with other crops on the move because of climate change.

One example of anticipating change is a global effort to model crop shifts and species range shifts with climate change, looking for points of intersection (Figure 15.8). These are areas in which conservationists and agricultural interests need to work together to find mutually beneficial responses to climate change. Similar forward planning can be done at regional and local scales to anticipate and plan for changes in both food production and biodiversity. Such planning can be an important element in regional coordination of adaptation efforts.

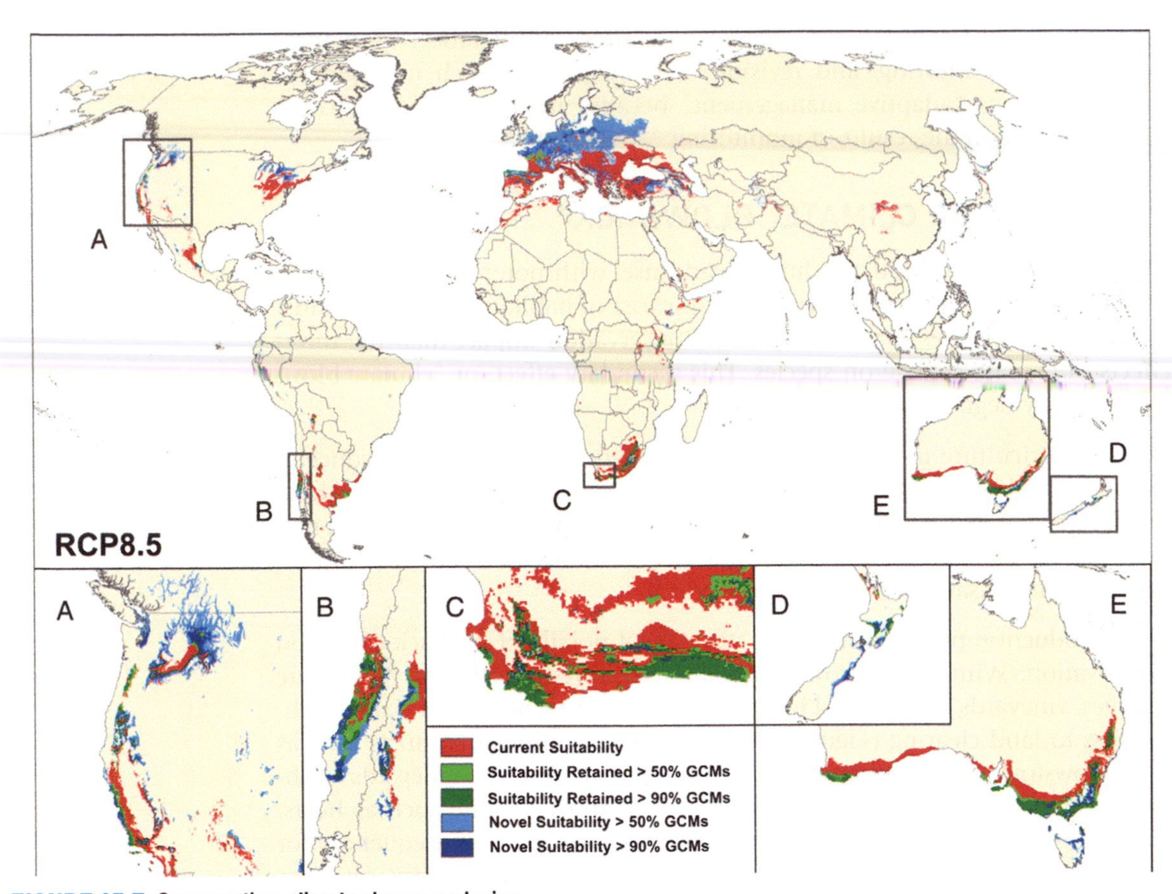

FIGURE 15.7 Conservation, climate change and wine.
Shifts in suitability for wine grape growing by 2050, shown as level of agreement in an ensemble of general circulation models. The map shows model agreement on new areas (blue) becoming suitable and on existing areas losing suitability (red). Newly suitable areas include the Yellowstone to Yukon conservation area in North America (Inset A), where open grazing range is the most common land use. Vineyards may present serious new barriers to wildlife movement in these areas. In areas of declining suitability (e.g., Insets C and E), vineyards may use water to adapt to deteriorating growing conditions, placing pressure on water resources and riverine habitats. *Source: Hannah et al. (2013).*

REGIONAL COORDINATION

All conservation action will require more regional coordination in the face of climate change, but this is especially important for connectivity. Regional coordination currently ensures harmonization of resource management objectives across multiple management units. With climate change, the scope and level of integration required will deepen. For example, if one protected area promotes a species range shift and another attempts to suppress the shift, the two management units may end up working at cross-purposes across the range of the

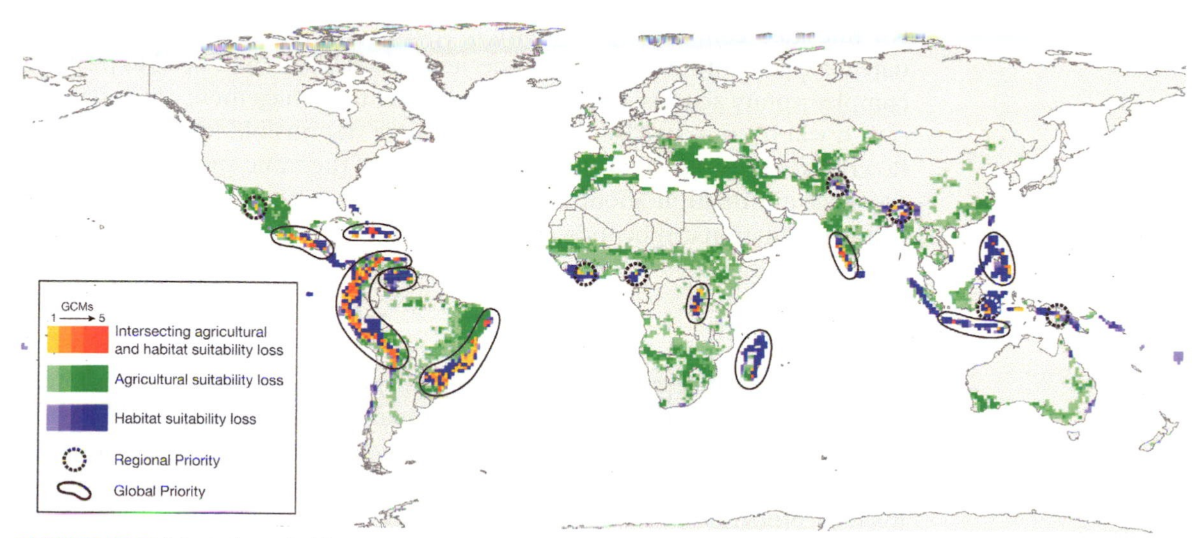

FIGURE 15.8 Adaptation priorities defined by intersecting change in agriculture and biodiversity. Climate change-driven losses in crop suitability (green) are shown with decline in climatic suitability for restricted-range birds (blue) by 2050. Intensity of color indicates level of agreement among multiple models and general circulation models. Areas of intersection (yellow–red) indicate areas in which agricultural change and decline in rare bird habitat suitability co-occur, areas in which collaborative planning between conservationists and agriculturalists may help avoid conservation blowback from shifting cropping patterns. Areas outlined in solid black show strong intersection by 2050, areas outlined with broken line show moderate intersection by 2050. *Source: Hannah et al. (2013).*

species. Because connectivity is expressly designed to promote interlinkages between protected areas, enhanced regional coordination with climate change is especially important.

Connectivity to maintain area-demanding species takes place on broad scales and inherently requires regional coordination. Thus, coordination already in place for area-demanding species can be capitalized upon for consideration of climate change. Similarly, large-scale connectivity established to deal with climate change should have regional coordination already built in.

Connectivity for migratory species already requires regional coordination, but this need will be intensified by climate change. Changing phenology of migratory species will mean that interrelationships long taken for granted will have to be reexamined. Information about such factors as arrival time, condition of individuals, and mortality in populations will have to be passed between reserves in the connected migratory network to facilitate appropriate compensatory management actions. Multiyear data will inform management actions such as planting times, whereas annual data will be important to determine when needed resources for the species should be in place and ready for changing arrival and departure times.

For fine-scale connectivity of multiple species, reserves will need to coordinate management on a species-by-species basis. Reserves will need to promote complementary actions, especially with respect to management of similar species. It will need to be clear which reserves are responsible for which species. Reserves responsible for different populations of the same species will have to exchange information about the progress of expected range shifts so that the whole population can be managed effectively.

MONITORING

Monitoring is an essential element of adaptive management for climate change. Effective monitoring requires clear management targets, readily monitored variables that are linked closely to the management targets, and a system of data collection and analysis that allows the results of monitoring to be effectively incorporated into management decisions.

For example, connectivity might be established to provide enough connected habitat for the reintroduction of wolves into two protected areas to improve ecosystem resilience to climate change. As climate change progresses, monitoring shows that wolf presence damps variability in deer populations and crashes are less frequent, as expected. However, after an El Niño event, monitoring reveals that deer and wolf populations in the two reserves are in synchrony (one of the possible negative effects of connectivity), raising the possibility of a simultaneous crash in both reserves, wiping out the wolf population. Management is adapted, temporarily breaking the connectivity between the two reserves until the populations are again asynchronous. Once monitoring shows that the prospect of a simultaneous population crash has been avoided, connectivity is restored for the long-term health of the wolf population.

As in this example, there is significant uncertainty about the magnitude of climate change and its biological effects, which can be effectively probed with monitoring. When expected changes are observed, monitoring can help ensure that the planned management responses are having their intended effect. Changes other than those expected can be detected by monitoring, providing the opportunity of early response to unforeseen impacts. A good monitoring system reduces uncertainty over time and greatly increases the cost-effectiveness of management responses in the face of uncertainty.

Monitoring systems require significant investment, which needs to be factored into the overall cost of management of connectivity. Monitoring climate variables will require temperature and rainfall recording stations and other equipment. Climate change will increase management costs in existing protected areas and connections, as well as requiring acquisitions of new protected-area connectivity. The costs of responding to climate change in biodiversity conservation are therefore substantial and need to be viewed as an additional investment that will be needed to protect conservation outcomes. The expense of monitoring should not be seen as an impediment but, rather, as an important part of the overall cost of dealing with climate change.

FURTHER READING

Hannah, L., Roehrdanz, P.R., Ikegami, M., Shepard, A.V., Shaw, M.R., Tabor, G., Hijmans, R.J., 2013. Climate change, wine, and conservation. Proceedings of the National Academy of Sciences 110(17), 6907–6912.

Hodgson, J.A., Thomas, C.D., Wintle, B.A., Moilanen, A., 2009. Climate change, connectivity and conservation decision making: back to basics. Journal of Applied Ecology 46, 964–969.

Knowlton, J.L., Graham, C.H., 2010. Using behavioral landscape ecology to predict species' responses to land-use and climate change. Biological Conservation 143, 1342–1354.

Williams, P., Hannah, L., Andelman, S., Midgley, G., Araujo, M., Hughes, G., et al., 2005. Planning for climate change: identifying minimum-dispersal corridors for the Cape Proteaceae. Conservation Biology 19, 1063–1074.

CHAPTER 16

Species Management

When protected areas and connectivity cannot adequately conserve a species, management to prevent extinction and restore populations to healthy levels is required. International, national, and state/provincial mechanisms exist to identify species at risk. Some of these mechanisms also trigger required actions to halt the decline of species at risk.

Climate change will require modification of threatened species management because it will alter ranges and phenology of threatened species. It will also move new species to threatened status. Assessing which species are seriously threatened by climate change is a major focus of climate change biology.

A general approach to conservation in the face of climate change is to manage multiple species in protected areas and managed landscapes (discussed in previous chapters) and, where protected areas alone are insufficient, identify individual species as threatened and manage them through:

- Removing other stressors
- In situ management
- Assisted migration
- Rescue (ex situ management)

This chapter discusses threatened species and then outlines conservation actions in response to climate change in each of these categories.

THREATENED SPECIES

Internationally, the Red List of Threatened Species of the International Union for Conservation of Nature (IUCN) is the accepted authority for identifying species as threatened. The Red List process involves groups of taxonomic experts from throughout the world. Each group assesses the conservation status of all of the species in a taxon (e.g., primates). For example, the primate specialist group will assess all primates based on the existing literature and knowledge of experts on the status of the species in the field. If the population

Climate Change Biology. http://dx.doi.org/10.1016/B978-0-12-420218-4.00016-0

of a species is small or rapidly declining, it may be classified into one of a series of categories indicating progressive threat to the species:

- Near threatened
- Vulnerable
- Endangered
- Critically endangered

In the United States, the national-level legislation is the Endangered Species Act (ESA). This legislation designates species by threat category, but it also limits certain federal actions once a species is listed. Threat categories under the ESA are threatened and endangered. Listing is performed by the U.S. Fish and Wildlife Service with advice from biologists rather than by biologists themselves. The process of listing is slow, and there is a large backlog of species waiting to be listed.

SPOTLIGHT: EUROPEAN UNION BIRDS DIRECTIVE

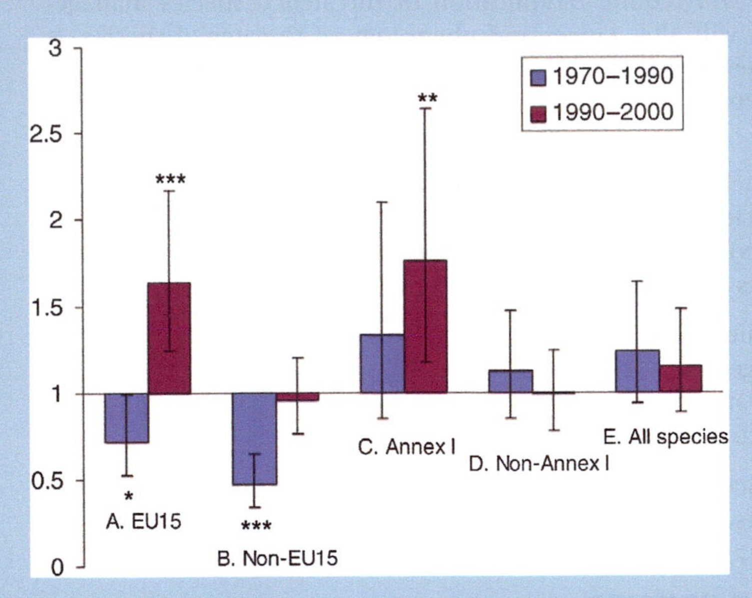

The EU Birds Directive (see text) has had measurable positive impact on species status. In the figure, the bars indicate the probability of a positive population trend for species in the 15 countries of the European Union (A), species outside the European Union (B), threatened and endangered species (C; Annex I), nonthreatened species (D; non-Annex I), and all species (E). Species within the jurisdiction of the EU Birds Directive went from a negative trend prior to the directive to a positive trend after the directive. Threatened species went from a mild positive trend to a stronger positive trend under the directive. *Figure reprinted with permission from AAAS.*

Source: Donald, P. F., Sanderson, F. J., Burfield, I. J., Bierman, S. M., Gregory, R. D., & Waliczky, Z. 2007. International conservation policy delivers benefits for birds in Europe. Science, 317(5839), 810–813.

In most other nations, there is no comprehensive endangered species legislation, but there is a European Union Birds Directive, which seeks to ensure conservation of threatened birds, and a European Union Habitats Directive that seeks to protect habitats. In addition, all countries that are signatory to the Convention on Trade in Endangered Species agree not to allow trade in wildlife or wildlife products from endangered species.

The EU Birds Directive provides evidence that protected areas and conservation measures can be successful, even in heavily populated settings. Issued in 1979, the directive provides protection and habitat preservation for birds under threat. By the turn of the twenty-first century, species that had been in decline prior to implementation of the directive had reversed their declines and showed population growth. The recoveries were due to protection of habitat in some of the world's most densely populated countries (see box p340).

Appropriate conservation action for endangered species depends on the species' biology, the nature of the threats it faces, and the extent of its distribution. A surprisingly high number of threatened species are found in only one location in the world. Others are found in multiple locations, whereas still others are area-demanding or migratory.

A review of all 4239 mammals, birds, tortoises, and amphibians considered by the IUCN to be threatened in 2006 found that their distribution relative to sites broke down as follows:

- 20% single sites
- 62% multiple sites
- 18% sites and landscape-scale conservation
- 1% landscape-scale conservation

For effective protection of these species, conservation of sites, landscapes, or seascapes is essential and may need to be supplemented by species management. Species found only in single sites often have populations so small that manipulation of habitat may be necessary to reduce the probability of extinction. Maintaining ex situ populations of these species in zoos or captive breeding centers may be desirable. Species present at multiple sites are less likely to need aggressive management, but many will still require some type of species-level management.

MANIPULATING HABITAT TO HELP A POPULATION RECOVER

Kirtland's warbler inhabits openings in jack pine forest in the central United States. Fire control programs have limited burning of the jack pine stands, reducing openings. To bolster flagging warbler populations, biologists now burn jack pine forest in prescribed areas to create openings that will become good warbler habitat. Similar manipulations may be required for some species as climate change alters fire frequencies and other disturbance regimes.

CLIMATE CHANGE IMPACTS ON THREATENED SPECIES

For species that are already threatened by other factors, climate change is an unwelcome complication that must nonetheless be addressed. Protection of key habitat combined with habitat manipulation or other measures to rebuild populations is a tool often employed in recovery efforts for threatened species. Climate change influences these efforts in ways unique to each species, including influencing sites for protection and altering habitat or population recovery efforts.

SEA TURTLES AND CLIMATE CHANGE

The world's seven species of sea turtles are all threatened, and all are threatened by climate change. Rising sea levels will inundate turtle nesting beaches. In many areas, beachfront development means that there is no inland space for beaches or turtles. Warming temperature may alter turtle sex ratios as well. The gender of sea turtles is determined by egg temperature, with warmer conditions producing more females. Consistently warming conditions may bias populations toward females, hindering population growth.

SPECIES THREATENED BY CLIMATE CHANGE

Species are threatened by climate change when the combination of physical habitat space and suitable climate is no longer sufficient to maintain a viable population. This can happen when a species already in decline owing to habitat loss suffers loss of suitable climate or when an otherwise healthy population is affected by a large loss of suitable climatic space (Figure 16.1).

Loss of suitable climate can occur in the absence of other threats because of the geography of climate. Montane species moving upslope with rising temperatures may simply run out of space to move as their preferred climate reaches the peaks of mountains (the 'escalator effect') and then literally disappears into thin air. Species at the tips of continents may similarly find nowhere to move poleward as temperatures warm. Because of local geography and climate interactions, some species may run out of climate space as they hit local dead ends.

Climate change can also interact with other stressors in ways that put species at risk. These synergistic interactions may prove to be much more sudden causes of species endangerment and extinction than either climate change or the interacting threat alone. The disappearance of toads of the genus *Atelopus* in Central and South America, for example, is due to a complex interaction of climate with chytrid fungal disease (Figure 16.2).

ASSESSING SPECIES THREATENED BY CLIMATE CHANGE

Climate change poses a special challenge for assessing the conservation status of species. The Red List is specifically designed to identify species in immediate

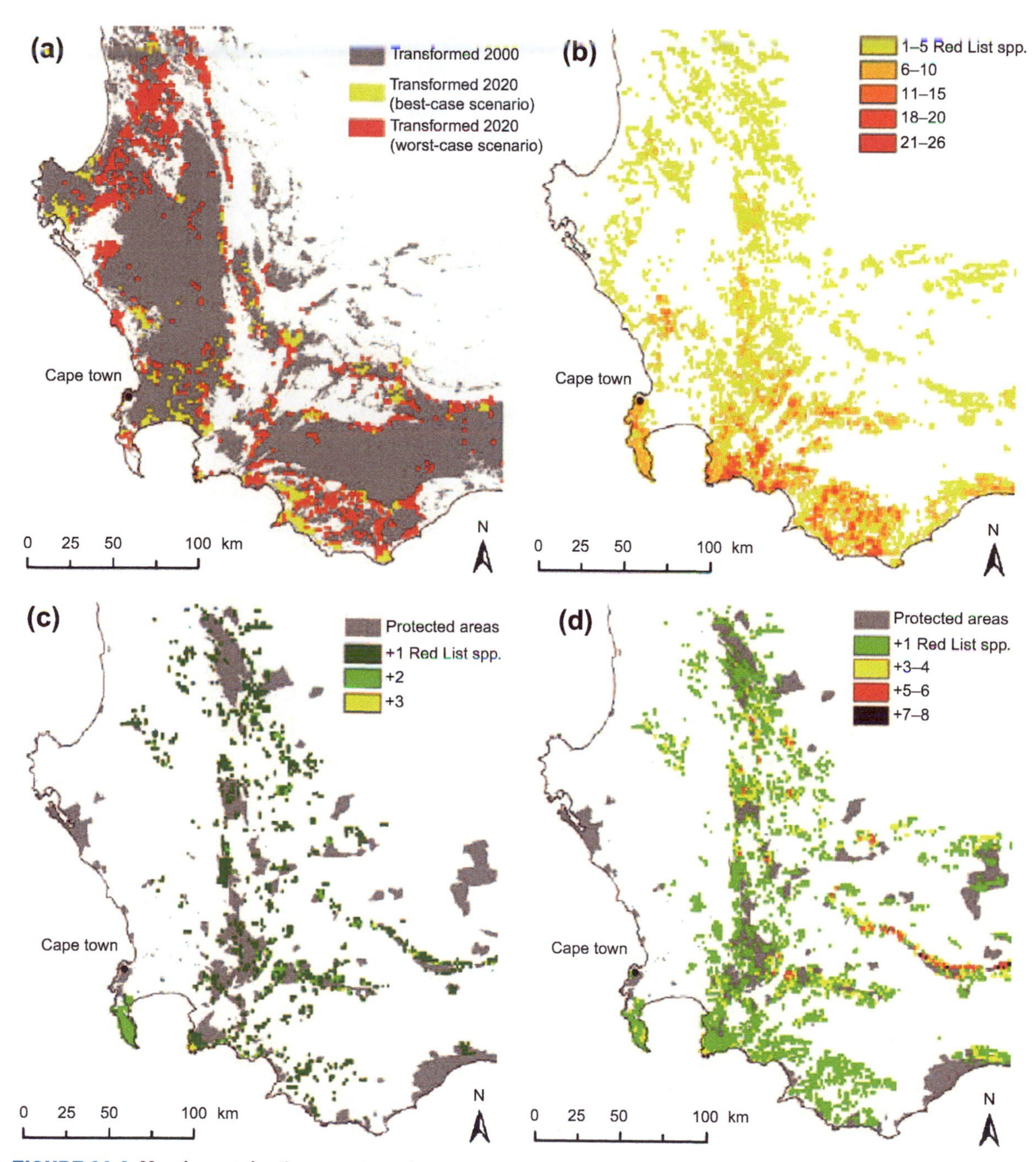

FIGURE 16.1 Mapping species threatened by climate change.
Risk of extinction due to the interaction of land use change and climate change by 2020 is illustrated in these maps of the Cape region of South Africa. Panel a illustrates land transformation trends. Panel b shows the current spatial distribution of Red List endangered and threatened species. Panel c shows the expected distribution of additional species threatened by land use change in 2020, and panel d shows the distribution of additional Red List species in 2020 when both land use and climate change are considered. The number of species threatened in 2020 by the combination of land use and climate change is more than twice the number threatened by land use change alone. *Source: Hannah et al. (2005).*

FIGURE 16.2 ***Atelopus* toads.**
Photos of a toad species, *Atelopus zeteki*, that disappeared from the forests of Central and South America before conservation measures could be put in place to protect them. *Source: (a) From Wikimedia Commons. (b) Courtesy of U. S. Fish and Wildlife Service. Photo by Tim Vickers.*

danger of extinction. Red List criteria are therefore keyed around short-term measures: has the species declined in just a few years? Is its population right now so small that extinction is imminent? Climate change, in contrast, threatens species years or decades in the future, but if conservation status and action await the effects of climate change on a species (e.g., rapid range loss), it will be far too late to do anything about the cause of the decline. How, then, can conservationists identify species at risk from climate change before it is too late?

One approach is to assess the life history characteristics that make species vulnerable to climate change. An IUCN task force took this approach, providing a detailed assessment of climate change vulnerability for 17,000 species of birds, amphibians, and corals. The study found that up to 50% of the species assessed were highly vulnerable to climate change. Less than a fifth were classified as potential persisters or adapters in each taxon. The study found that climate

IUCN THREAT CATEGORIES AND CRITERIA

The IUCN Red List is the authoritative source on endangerment of species globally. Its role in identifying species at risk from climate change has come under fire. The Red List categories are critically endangered, endangered, vulnerable, near threatened, and least concern. Species can move into one of the higher categories of threat if expert panels assess that their condition has changed based on several possible criteria. Small range size and rapidly declining range size are among the trigger criteria.

All of the Red List criteria were originally developed to respond to immediate declines due to causes such as habitat destruction or hunting. Climate change poses a different challenge because its effects may not appear for decades but are irreversible once they do. The original IUCN categories and criteria are not sensitive to climate change because immediacy of threat is central to their trigger criteria. The IUCN is now working to improve methods of red listing for climate change so that climate change-threatened species can be placed on the list before it is too late. A major advance has been the identification of species that are sensitive to climate change because of life history traits such as narrow thermal tolerances.

Source: Foden, W.B., Butchart, S.H., Stuart, S.N., Vié, J.C., Akçakaya, H.R., Angulo, A., Mace, G.M., 2013. Identifying the world's most climate change vulnerable species: a systematic trait-based assessment of all birds, amphibians and corals. PLoS One 8 (6).

change posed a significant increase in risk for species already on the IUCN Red List; 608–851 bird (6–9%), 670–933 amphibian (11–15%), and 47–73 coral species (6–9%) were found that were both highly climate change vulnerable and already threatened with extinction from other factors.

Another approach utilizes the existing Red List criterion that permits the use of population models to be considered in setting threat status; such models can be used to assess risk from climate change. Habitat suitability maps from species distribution models, based on future climate projections, can be used to define habitat patches in population models to simulate likely effects of climate change on species of concern. The modeling criterion exists because some species undergo large fluctuations in population, making measures at any individual point in time a poor index of the species' likelihood of extinction. For these species, modeling the population fluctuations can generate probabilities of extinction that in turn can be used to determine if a species is threatened by climate change.

AN ICONIC EXAMPLE

Listing of the polar bear under the ESA stands as a clear example of the challenges of assessing species endangerment due to climate change and of the complex forces of science, economics, and policy that must be resolved in real-world adaptation solutions. The ESA requires species whose populations may become endangered in the "foreseeable future" to be given protection as threatened species. The scientific basis for listing the polar bear under U.S. law was the decline in arctic sea ice, which is well under way and is projected to intensify this century, undermining the most critical of polar bear habitat needs. Economic forces at play include planned oil and gas development in polar bear habitat and also the possibility of legal challenges to force the United States to regulate greenhouse gases based on the first listing of a species due to climate change.

This case highlights the difficulties encountered in bringing species threatened by climate change under the umbrella of existing threatened and endangered species protocols. Most existing protocols, including laws and international agreements, are geared toward protecting species that will decline in the next few years, not species that are expected to decline rapidly several decades in the future. Like the international red listing process of the IUCN, the ESA has struggled with this issue.

A petition to list the polar bear as threatened under the ESA was filed by the Center for Biological Diversity and other conservation groups in April 2006. This obligated the U.S. Fish and Wildlife Service to assess the merit of the petition, which they did, finding that sufficient evidence existed to justify research into the scientific merit of the threat to polar bears. This year-long period of research and public comment attracted much attention because the U.S. administration of the time opposed international action on climate change.

The state of Alaska was concerned that listing of the polar bear would be used to prevent North Slope oil and gas development. The ESA prevents killing ("taking") of listed species even indirectly through habitat destruction. Oil and gas exploration and development activities in polar bear habitat might face legal challenges as a result. Canada opposed the listing because it would deter high-paying U.S. tourists from traveling to hunt polar bears in Canada.

Political pressure against the listing came from conservative groups opposed to action on climate change. They believed that conservation groups might file legal challenges attempting to force the United States to regulate greenhouse gases if the polar bear was listed because of threats from climate change. If the polar bear was threatened because of climate change, emissions of greenhouse gases might be interpreted as "taking" of the species under the ESA. The U.S. did subsequently regulate CO_2 as a greenhouse gas, but not because of the listing of polar bears as threatened.

The scientific basis for listing was controversial as well. Polar bear populations were at a 40-year high, so the species clearly would not qualify for listing under ordinary (non-climate change) criteria. Approximately 25,000 polar bears existed in 19 populations worldwide, only six of which were declining. Therefore, the case for listing was based entirely on climate model (GCM) projections that showed that melting of sea ice would lead to decimation of this critical polar bear habitat. Polar bears spend much of their lives on sea ice and raise their young in dens dug into the ice, and their main source of calories is seals, which they hunt on ice floes.

The ESA requires that species listings be based solely on scientific evidence. The act also specified that species could be listed as threatened only when their endangerment could be anticipated in the "foreseeable future." The GCM used for the listing assessment showed that sea ice in the Arctic would disappear this century. The modeling science was clear that major trouble awaited the polar bear in several decades, but was this the foreseeable future as intended in the act? After lengthy internal discussion and intense external pressure, the U.S. Fish and Wildlife Service decided to list the polar bear as threatened in December 2007. This triggered a 6-month public comment period.

A total of 670,000 comments were received on the proposed listing, including detailed scientific critiques and mass postcard mailings from conservation group members. Many of the science comments received were critical of the reliance on a single GCM. Others questioned the resilience of polar bears to climate change because polar bears had endured multiple climate swings during the past half-million years.

In response to these critiques, research was commissioned to examine paleoecological data and a more robust suite of GCM results. The paleoecological research showed that polar bears evolved 250,000–400,000 years ago, and during the ensuing period, sufficient sea ice existed (even in previous interglacial periods) to support substantial polar bear populations. At the same time, multiple GCMs were in agreement that sea ice would decline steadily through 2050–2060. Most models showed an ice-free Arctic by the end of the century. This degree of sea ice loss would be devastating to polar bear populations throughout the world. In January 2009, the decision to list was upheld.

The polar bear listing was a triumph of rule of law over political pressure and of scientific evidence over economic concerns. It required an interpretation of the ESA for a threat that the act was never envisioned to address. It involved modeling across a century to create a "foreseeable future." This listing is therefore a model of adaptive policymaking for those struggling with climate change in endangered species assessments in other nations and internationally.

MANAGING SPECIES THREATENED BY CLIMATE CHANGE

Once a species is identified as threatened by climate change, appropriate management actions should be implemented to reduce extinction risk. The goal of management is to stabilize the population and, where possible, recover a population to the point that its threat status can be downgraded. This sequence is identical to that undertaken for species threatened by other factors. For example, in the United States, the bald eagle was threatened by pesticides that damaged eggs and by habitat loss, leading to it being listed as endangered under the ESA when the population dropped below 900. Action was taken to ban DDT and restore habitats, and the species was removed from the endangered list in 2007 when the population had recovered to more than 20,000 individuals.

Management of species threatened by climate change is more complex. Effective management often seeks to address root causes of endangerment. However, the root causes of climate change are energy consumption and pollution that may occur far from the range of the threatened species. Conservation action for climate change therefore focuses on two fronts: adaptation to the expected impacts of climate change and reducing other threats.

Removing Other Stressors

A general management strategy for species expected to be influenced by climate change in the future is to aggressively move to reduce existing stressors (Figure 16.3). This allows the population of the species to grow, providing a more robust population when the effects of climate change begin to occur.

FIGURE 16.3 Threat synergies.
Climate change impacts may be exacerbated by other stressors, such as (a) overpopulation or (b) human-caused fire. *Source: Conservation International.*

Stressors that are particularly important to remove are those that affect areas of the species range expected to be resistant or resilient to climate change. Examples are areas that may be resistant because of a lower magnitude of change in climate variables important to the species and areas of high genetic variability that might confer resilience.

Species modeling or expert opinion may provide the knowledge necessary to prioritize. Areas in which species models show major loss of suitable climatic range may be poor choices for investment in reducing current threats because the species may not persist in these areas regardless of initial population size or removal of other stressors. Expert opinion may be particularly important in identifying synergies between climate change and other threats that may be poorly represented in models.

Because adaptation actions can only reduce, not eliminate, the impact of climate change, restoration of populations may be dependent on reversing the effects of other stressors. For example, a species might be found to be threatened if encroaching human development was expected to reduce its habitat by 30% and climate change to reduce it by another 50% during the next three decades. The climate change impact can be reduced but not eliminated, whereas direct habitat loss may be eliminated. A cost-effective management strategy would then involve a mix of reduction in both climate change stress and habitat stress, even though climate change was the major contributor to species threat.

In Situ Management

Where removing other stressors is not enough to ensure a healthy population, individual species may require special management. Examples of situations requiring in situ management include phenological mismatches and extreme range shifts. Phenological mismatches may be particularly challenging to manage because manipulation of the timing of predator, prey, food, or migrations is unlikely to be a viable long-term conservation solution. However, manipulation may provide a short-term solution or insurance policy while longer term solutions are put in place.

One biological solution for the long term is to use natural selection to improve the fitness of populations faced with phenological mismatches. Traits such as photoperiod response are critical to phenological response and are known to evolve very rapidly in response to change. Evolutionary processes may not occur quickly enough to keep pace with very rapid human-induced climate change when populations are highly fragmented by human land uses. This is where manipulation may play a short-term role. Some subpopulations may be manipulated to provide a secure core population while other subpopulations are allowed to fluctuate freely to allow natural selection to operate.

Once the evolving population is substantial, the artificial manipulation can be discontinued.

For joint artificial manipulation/natural selection to work as a strategy, the phenological transition must be limited in duration. The natural evolving population must be able to reach a self-sustaining population level at low risk of extinction, or other management options should be pursued. A subpopulation that goes through continuous or repeated bottlenecks owing to phenological mismatches is highly vulnerable to extinction and unlikely to rebound to become a self-sustaining population.

Similar intervention may be required for species projected to undergo extreme range shifts. For example, some protea species in South Africa are projected to undergo range shifts across long distances in the Cape Fold Mountains as their suitable climate jumps to progressively higher peaks along the chain (Figure 16.4). Such extreme shifts in climate space are likely to outpace species niche-tracking abilities. In this example, a species that is wind dispersed may have some probability of colonizing adjacent mountaintops, but an ant- or rodent-dispersed species would be expected to be unable to colonize distant mountains.

SPOTLIGHT: FAIRY SHRIMP AND COWS

The interaction between human land use and climate change is not always negative. Pyke and Marty (2005) examined the potential impacts of grazing and climate change on vernal pools in California. Vernal pools are ephemeral habitats that fill with rain and then slowly dissipate. They support a unique and highly endemic biota, including several species of the fairy shrimp genus, *Branchinecta*. Pyke and Marty found that cattle grazing compacted soils, which aided in water retention. This could help offset the drying effects of warmer climate. Although fairy shrimp habitat is poorly represented in protected areas and climate change would reduce the number of vernal pools (Pyke and Fischer, 2005), a combination of expanded protection and managed grazing could greatly increase the prospects for fairy shrimp survival as climate changes.

Source: Pyke, C.R., Fischer, D.T., 2005. Selection of bioclimatically representative biological reserve systems under climate change. Biological Conservation 121, 429–441; and Pyke, C.R., Marty, J., 2005. Cattle grazing mediates climate change impacts on ephemeral wetlands. Conservation Biology 19, 1619–1625.

When a species' dispersal ability is exceeded, management intervention becomes appropriate. If a species has a very low probability of tracking its suitable climatic conditions, conservation of the species may depend on artificial manipulation. One option is to maintain the species in its current range through in situ cultivation. For example, the mountaintop protea may be limited by orographic rainfall and temperature-driven water balance limitations. Simple watering might maintain a population that would otherwise perish. Seedlings may be most sensitive to climate variables such as drying, so watering may be needed only for some portion of the life cycle and only enough to supplement the population to robust levels. For species for which in situ

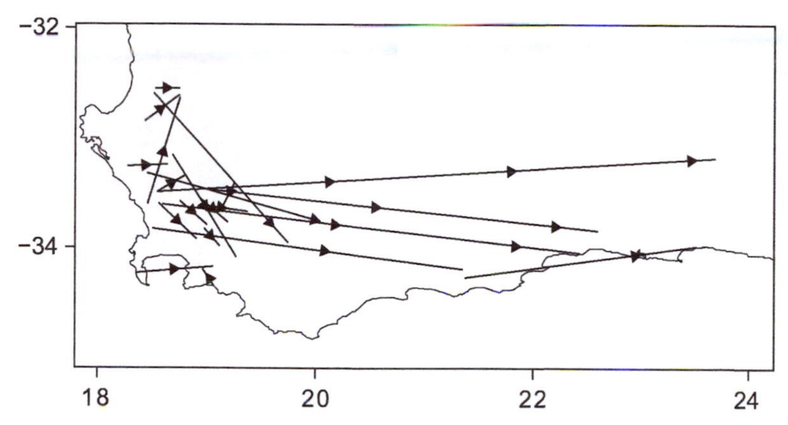

FIGURE 16.4 Mountain-hopping protea.
Arrows indicate present-to-future vectors of protea range shifts in the Cape Floristic Region. Many species move to the Cape Fold Mountains and then track suitable climate following the mountain crest.
Source: Midgley et al. (2003).

cultivation requirements are too intensive, a second management option is assisted migration.

Assisted Migration

Assisted migration is a management option for cases in which in situ manipulation would fail or become prohibitively expensive. The advantage of assisted migration is that it is a limited-term management intervention that, if successful, allows species population recovery by natural processes.

The principle of assisted migration is simply that a population unable to naturally reach suitable climatic space can be translocated artificially and then will flourish without further intervention. In practice, assisted migration may be as simple as carrying a few seeds to a suitable site, as in the Cape protea example, or as complex as moving and caring for an entire population until it is out of danger (Figure 16.5).

Although simple in practice, assisted migration may be complex to implement because it is difficult to establish the ecological framework for making decisions about when and where assisted migration is appropriate. When a species is translocated to a new location, it may out-compete species native to the location. Is this effect the desired outcome, or does it pose a new threat to yet another species? Because translocated species may wind up in association with other species for which there is no historical analogue, how do we define desirable ecological goals for the species and its new neighbors? Should only threatened species be translocated, or should large-scale ecological experiments be undertaken involving hundreds or thousands of species?

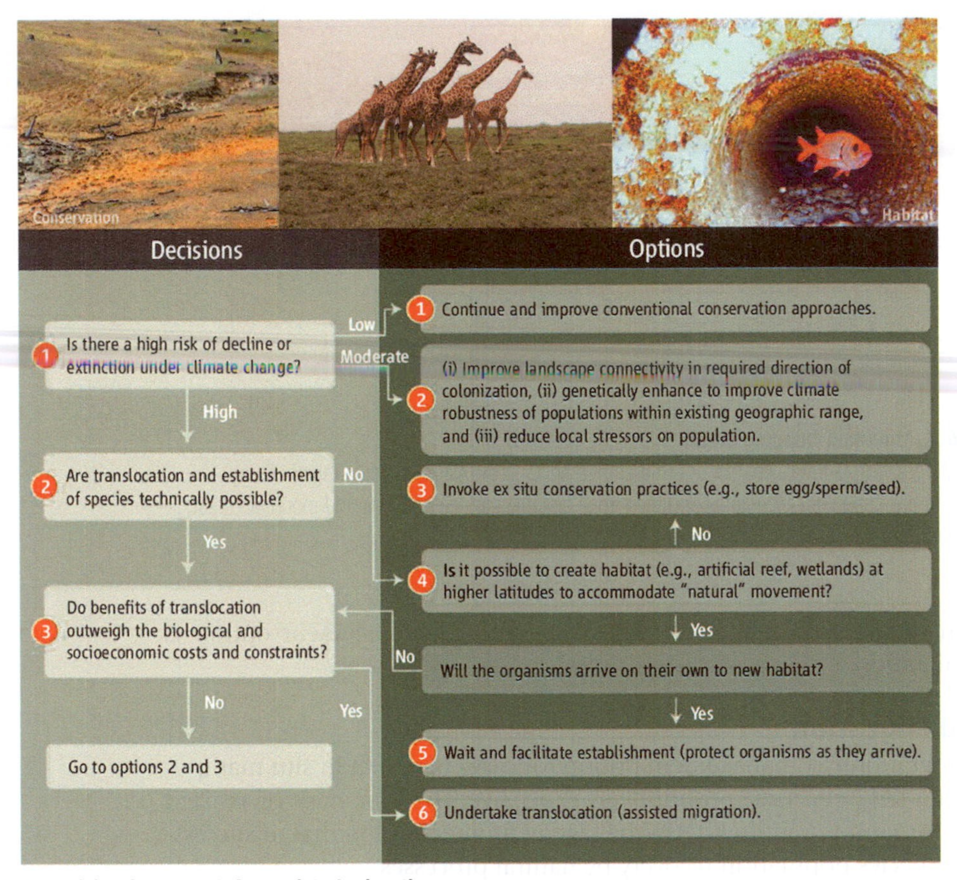

FIGURE 16.5 Decision framework for assisted migration.
A sequence of decisions can help determine when species should be considered for assisted migration and when natural processes should be allowed to prevail. Because human intervention may have unintended consequences, assisted migration should be considered a last resort option for conservation under climate change. *Source: Hoegh-Guldberg et al. (2008). Courtesy of Sue McIntyre. Copyright by CSIRO.*

Several factors must be assessed when weighing a translocation (assisted migration) effort. Accurate estimates of the species population size and distribution are required so that translocation is not undertaken based on faulty information. Estimates of suitable future climate through species distribution modeling and consideration of interactions with other species at the new location should be taken into account. The species long-distance dispersal capacity needs to be estimated so that translocation resources will not be wasted on species that could make the trip themselves. Finally, possible source populations should be studied to determine possible genetic advantages or disadvantages, such as disease transmission, associated with each.

Assisted migration is therefore desirable only after the other management strategies have been exhausted. Carefully considered translocation will require

CHOOSING A SOURCE POPULATION

When artificial assistance is needed to maintain species, source population is important. For instance, if a species is unable to keep pace with its suitable climate, in some instances human-assisted migration may be necessary. Individuals from the existing range might be transplanted ahead of the range edge into newly suitable climates. Individuals from the range margin are most likely to be adapted to climates similar to those in the area of expansion. Trailing edge populations would be expected to be most genetically adapted to the warm conditions and first to be lost as climate changes, so transplants taken from the trailing edge may facilitate range shifts and help retain trailing edge genetic adaptations.

planning, monitoring and public consultation. Poorly considered translocation may have serious deleterious ecological consequences. For these reasons, maintenance of populations that can shift range naturally is top priority, followed by actions to manage individual species in situ. Only when these options are insufficient should translocation or species rescue be considered.

Species Rescue

When all else fails, species rescue through ex situ conservation such as captive breeding should be implemented to avoid extinction. This is the last-ditch conservation response for species threatened by climate change, just as it is for species threatened by other stressors.

At best, rescue actions are temporary and allow recovery of a population that can be reintroduced in the wild. At worst, rescue prevents complete loss of a species that has become extinct in the wild. Ex situ conservation can also complement conservation in the wild, providing an "insurance" population that will survive even if unanticipated events lead to decline or loss of the wild population.

Climate change affects the rescue landscape by causing the addition of potentially large numbers of species to the pool of candidate species. Estimations of extinction risk due to climate change place the number of threatened species in the hundreds of thousands or millions. If even a fraction of these species ultimately require rescue operations and ex situ management, the cost and management burden will be large. Ex situ programs require captive-breeding facilities, full-time staff, food, and materials (Figure 16.6).

This is one reason conservation of biological diversity is best served by limiting climate change as soon as possible. Less climate change will reduce the numbers of species placed at risk, the number of species that must be rescued or maintained ex situ, and therefore the cost of conservation. The ecological cost of losing species in the wild is incalculable.

At the same time, climate change biology can inform adaptation strategies that respond effectively to the extinction risk that does arise because of climate

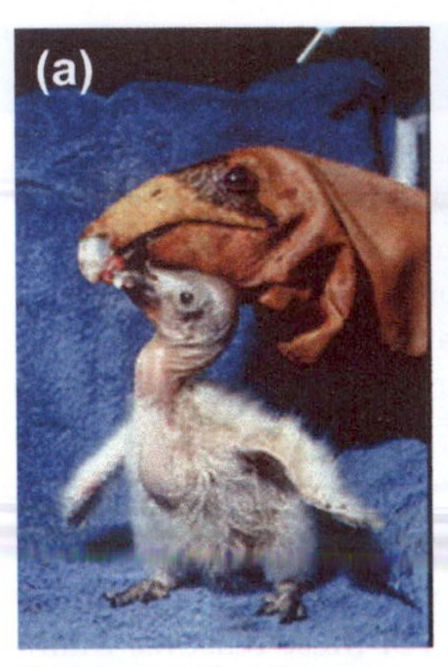

FIGURE 16.6 A Species rescued.
All 22 wild California condors were rescued from the wild in 1983. Successful captive breeding has allowed reintroductions into the wild. Captive breeding included feeding with a glove that simulated an adult condor (a). More than 200 released birds (b, c) now survive in several populations. Similar intensive management may be needed for species threatened by climate change. *Source: From Wikimedia Commons.*

change. Species at risk must be identified and conserved in the wild or through species management actions where possible, and contingency plans must be made for species for which conservation in the wild or even assisted migration may fail. It is likely that combined with the wide array of existing stressors already at play in the world, climate change will require a major new investment in conservation to maintain the planet's biodiversity.

RESOURCES FOR THE JOB

Climate change biology suggests that large additional resources for conservation will be needed to safeguard species from extinction due to climate change. It is one of the responsibilities of climate change biologists to carefully and reasonably document the risk associated with climate change and the level of resources required to respond. Fortunately, many of the tools explored in previous chapters can be brought to bear on this task.

As discussed in Chapter 14, new protected areas will be required to protect all species in their present and future ranges. Not all species have adequate protection of their current ranges, making the completion of protected-area systems an opportunity to incorporate climate change into protection planning. The incremental cost of adding protected areas for climate change will be substantial. New resources for acquisition and management of protected areas will be required because of climate change in most nations and regions.

Connectivity and management of landscapes used for human production will have to be included in strategies for adaptation to climate change. Connectivity on the periphery of existing protected areas is especially cost-effective, as discussed in Chapter 15. Species dynamics across landscapes used for agriculture and grazing will need increasing consideration, as will connectivity in landscapes used for renewable energy solutions to climate change.

Finally, the number of species requiring intensive conservation management is likely to increase dramatically because of climate change. Early identification of species at risk can help keep costs low and avoid extinctions. In some special circumstances, assisted migration may keep species from needing rescue and ex situ management. The cost of managing ex situ conservation of species will be relatively large, making in situ adaptation options such as expansion of protected areas a more cost effective option.

All of the increased costs of conservation due to climate change need to be quantified and articulated. Funds for protected-area acquisition and management will be needed first. Costs of connectivity can be incorporated simultaneously, creating protected-area networks that maximize conservation with climate change but also interface with surrounding landscapes effectively. Later, as species are placed in jeopardy as climate change advances, funds will be needed to manage individual species and, where necessary, to rescue them in ex situ conservation programs.

These estimates are best made on a regional basis. Some rough global estimates are already available, but exact needs depend on the extent of climate change and the sensitivities of individual species best assessed for individual regions. It is the job of climate change biologists to both advocate for effective measures to limit these costs by stopping climate change and to state clearly what the costs are in the face of expected climate change, so that resources can be made available and adaptation action taken to safeguard the species and ecosystems of all regions.

FURTHER READING

Foden, W.B., Butchart, S.H., Stuart, S.N., Vié, J.C., Akçakaya, H.R., Angulo, A., Mace, G.M., 2013. Identifying the world's most climate change vulnerable species: a systematic trait-based assessment of all birds, amphibians and corals. PLoS One 8(6), e65427.

Grebmeier, J.M., Overland, J.E., Moore, S.E., Farley, E.V., Carmack, E.C., Cooper, L.W., et al., 2006. A major ecosystem shift in the northern Bering Sea. Science 311, 1461–1464.

McLachlan, J.S., Hellmann, J.J., Schwartz, M.W., 2007. A framework for debate of assisted migration in an era of climate change. Conservation Biology 21, 297–302.

Finding Solutions: International Policy and Action

CHAPTER 17

International Climate Policy

UNITED NATIONS FRAMEWORK CONVENTION ON CLIMATE CHANGE

International climate policy on climate change is rooted in the United Nations Framework Convention on Climate Change (UNFCCC). The UNFCCC was agreed upon at the Earth Summit in Rio de Janeiro in 1992. The convention has now been ratified by 195 countries, including the United States.

The UNFCCC was developed in response to an evolving understanding of human-induced change in climate systems. Measurements at Mauna Loa showed rising atmospheric CO_2 concentrations, and new computer models of global climate pointed to rising levels of the burning of fossil fuels as a culprit. By the time of the 1992 Earth Summit, there was global recognition that the problems associated with burning fossil fuels required action.

The UNFCCC is a statement of the need for action, but it does not contain agreement on specific emissions reductions. The Kyoto Protocol is a supplement to the UNFCCC that sets more binding targets for greenhouse gas emissions reductions. Countries that ratified the protocol made commitments to reduce greenhouse gas emissions. The first commitment period of the protocol ran from 2008 to 2012. The second commitment period runs from 2013 to 2020. Unlike the UNFCCC, the US has never ratified the Kyoto Protocol, leaving the world's largest emitter of greenhouse gas pollution out of the bargain.

The ultimate goal of the UNFCCC is to "stabilize greenhouse gas concentrations in the atmosphere...at a level that would avoid dangerous anthropogenic interference with the climate system." Since stabilizing greenhouse gas concentrations in the atmosphere can only be achieved if there in no net release of greenhouse gases, the ultimate goal of the convention is ending greenhouse gas pollution. This is an ambitious goal that has proved hard to realize.

Nature conservation, food security and poverty reduction are included in the goals of the UNFCCC. The convention states that stabilization of greenhouse gas concentrations should be achieved "...within a timeframe sufficient to allow ecosystems to adapt naturally to climate change, to ensure

Climate Change Biology. http://dx.doi.org/10.1016/B978-0-12-420218-4.00017-2

that food production is not threatened, and to allow economic development to proceed in a sustainable manner.."

A central research question in climate change biology is what levels of change natural systems can withstand, both because it is an intrinsically interesting question in biology and because of the wording of the UNFCCC. It is difficult to define what the term "adapt naturally" might mean in the context of rates and magnitudes of change unprecedented for millions of years, but it is clear that extinctions would be a sign of natural systems breaking down (see Chapter 12).

The UNFCCC structure includes mechanisms for meetings of the ratifying countries and for gathering scientific and technical input for convention decisions. The

FIGURE 17.1

Policymakers and the public around the world recognize that the United States has been the greatest source of greenhouse gases in the atmosphere. Growth in emissions from China means that China has now passed the U.S. as the number one annual emitter of greenhouse gases, so moving forward, the U.S. and China will play central roles in determining climate change solutions. *Source: Figure courtesy Santa Barbara Independent.*

meetings of the participating countries are known as the Conference of the Parties, or COP. The advisory bodies include the Subsidiary Body on Scientific, Technical, and Technological Advice, or SBSTTA. There are many advisory subgroup meetings organized under the SBSTTA. The COP is an annual event; the first COP was held in 1995. Each COP is numbered successively, with the 1995 COP labeled as COP1. SBSTTA meets roughly once a year, but occasionally meets twice in one year.

All binding decisions under the convention are made at the COP. With 195 parties, this means that decision making can take years. Some COPs are more important than others, as major decisions evolve and a COP in which major decisions are to be made becomes a watershed event. This was the case for COP15 which took place in Copenhagen in 2009. Decisions were expected at that COP on what emissions reductions commitments would look like after the Kyoto Protocol second commitment period runs out in 2020. In the end, only tentative progress was made at that COP.

Developed countries and developing countries have different roles under the UNFCCC and Kyoto Protocol. As the Kyoto Protocol was being developed, it was recognized that most of the greenhouse gas pollution currently in the atmosphere was from developed countries, and most of that was from one source—the US (Figure 17.1). It was felt that it was not fair that developing countries, who had played little role in creating the problem, be required to reduce emissions and perhaps delay their economic growth. As a result, only developed countries were required to agree to emissions reductions. The countries that were required to reduce emissions if they agreed to be part of the Kyoto Protocol were listed in Annex 1 of the protocol, and so they are sometimes referred to as Annex 1 countries.

In the time since the Kyoto Protocol was signed developing countries' emissions have grown. Countries such as China, India, and Brazil have become major greenhouse gas emitters (Figure 17.2). China now exceeds the US in greenhouse gas pollution. So the playing field is shifting in greenhouse gas emissions with developing countries not included in Kyoto Protocol now making up a significant portion of current and future emissions. International mechanisms need to change to keep up with these trends.

RESEARCH AND INDEPENDENT NON-GOVERNMENTAL ORGANIZATIONS

Scientists and researchers participate in discussions of the United Nations Framework Convention on Climate Change (UNFCCC) too, but not directly. Only country delegates can participate in official UNFCCC deliberations. But researchers can participate in side events, observe, and discuss science developments with country delegates. A collaboration of researchers, Research and Independent Non-Governmental Organizations (RINGO) coordinates these researcher activities. RINGOs have played important roles in moving ahead the science behind UNFCCC positions and in influencing the development of programs such as REDD and the Adaptation Fund.

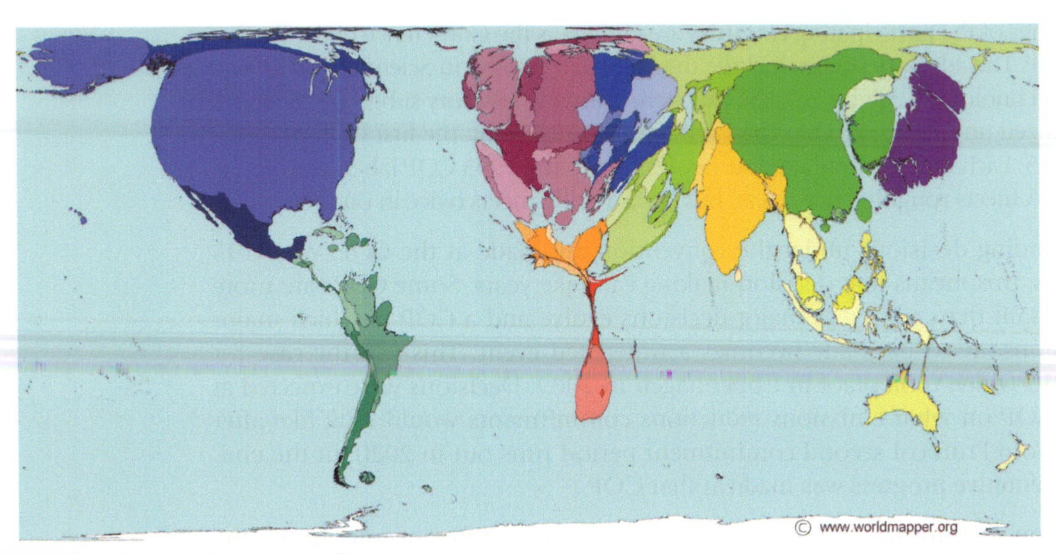

FIGURE 17.2 Global greenhouse gas emitters.
Global greenhouse gas emissions by country are shown in this proportional diagram, with the area of each country corresponding to its contribution to global greenhouse gas emissions. *Figure courtesy of Worldmapper.*

INTERGOVERNMENTAL PANEL ON CLIMATE CHANGE

The Intergovernmental Panel on Climate Change (IPCC) is independent from the climate convention, but provides strong complementary work on the science of climate change. The purpose of the IPCC is to provide state-of-the-art science summaries for the public and global policy makers. The IPCC is a body of the United Nations, housed under the United Nations Environment Program (UNEP).

The IPCC produces an assessment of climate change science on an approximately six-year cycle. Each report is numbered, starting with the first assessment produced in 1990. The IPCC reports are:

1. 1990 (First Assessment)
2. 1996 (Second Assessment Report—SAR)
3. 2001 (Third Assessment Report—TAR)
4. 2007 (Fourth Assessment Report—AR4)
5. 2013–2014 (Fifth Assessment Report—AR5)

Each assessment is carried out by three working groups, each composed of hundreds of leading scientists. IPCC reports come in four volumes, one for each working group and an executive summary. The IPCC working groups are:

Working Group 1—The Scientific Basis
Working Group 2—Impacts, Adaptation and Vulnerability
Working Group 3—Mitigation

Working Group 1 reviews climate science, such as evidence of increasing CO_2 in the atmosphere, temperature records, and changes in precipitation, both observed and projected for the future. Working Group 2 examines the impacts of those climatic changes, assessing possible adaptation options and vulnerable regions and groups. In IPCC usage, "adaptation" means what people can do to reduce climate change impacts, while "mitigation" refers to what can be done to address the root causes of climate change, namely reducing greenhouse gas pollution. Working Group 3 reviews mitigation options.

The scientific findings of every IPCC working group are reviewed by participating governments. This can result in changes to the wording proposed by the scientists. Some of the changes have been controversial, but in the end the IPCC products are scientifically sound and agreed to by all participating governments, which makes them powerful statements about the state of climate change science.

The IPCC has consistently found that human pollution is causing changes in atmospheric CO_2 and climate, with serious consequences for people and nature. The first assessment report stated that human pollution was causing rising atmospheric CO_2 levels and that this was highly likely to result in climate change. The second assessment found that there was a discernable change in global climate already underway and that its cause was human greenhouse gas emissions. The third assessment found that the climate was changing more rapidly than any time in the past 10,000 years. The Fourth Assessment Report (AR4) stated that "Warming of the climate system is unequivocal" and that climate would continue to warm for centuries, even with rapid action to reduce greenhouse gas pollution, while the AR5 stressed the urgency of action to avoid high costs in both mitigation and adaptation.

CARBON MARKETS

Developed country parties to the Kyoto Protocol agreed to undertake emissions reductions and the 1997 agreement made provisions for the trading of carbon reductions. The European Union (EU) decided to implement a carbon market to achieve these reductions in the most cost-efficient way. The European carbon market, known as the Emissions Trading System (EU ETS), covers emissions from over 31,000 power plants, factories, and airlines in Europe. It is a cap-and-trade system, with 2020 emissions targeted to be 21% lower than 2005 and 43% lower by 2030. Industries in participating countries trade emissions allowances to meet these targets in the lowest cost way. Emissions allowances are allocated in an auction system.

The EU ETS is part of the Clean Development Mechanism (CDM) and Joint Implementation (JI) of the Kyoto Protocol. Both the CDM and JI allow emissions in one country to be reduced by trading for emissions reduction in another country. In CDM, the second country is a developing country, while under JI the second country is another Annex 1 country that has agreed to emissions reductions.

Carbon markets are a type of cap-and-trade pollution control strategy. As with any cap-and-trade system, a limit is put on the amount of pollution allowed, credits are created equal to the target amount of pollution, and polluters (emitters) trade credits in a market so that the emission reduction required to meet the target is achieved in the cheapest way possible. One polluter will buy credits (the right to pollute) from another polluter when the second polluter can clean up their emissions more cheaply than the first.

Carbon markets are a way of making pollution reduction cost-efficient. They allow one source of CO_2 to reduce another, source's emissions at a cost savings. For example, a factory that is required to reduce emissions in Germany might pay a factory in Eastern Europe to reduce its emissions. This makes sense where the factory in Germany is efficient (low CO_2 emissions per unit of energy used) and the Eastern European factory has older, inefficient equipment. Buying the Eastern European factory efficient equipment is cheaper than making the German factory superefficient. Since the equipment emitting the CO_2 is usually boilers or burners of some sort, newer equipment is much more efficient than equipment that is 30 or 40 years old. Big efficiency gains can be made at a lower cost by upgrading the old equipment.

The EU carbon market is the largest in the world (Figure 17.3). China and Australia have large national carbon markets, and California has the world's largest subnational carbon market. States in the US are beginning to institute CO_2 emissions caps and carbon markets because of a lack of national-level action.

Because the US refused to participate in emissions reductions under the Kyoto Protocol, only a small fraction of global greenhouse gas emissions come under some form of carbon market. In areas not covered by governmental carbon markets, some industries still want to voluntarily engage in emissions reductions, and many of those want to achieve some or all of their emissions reductions through carbon offsets. This has led to the development of a market in carbon offsets.

The voluntary carbon market makes up only about 2% of global carbon markets (Figure 17.4). But it is significant in pioneering emissions reductions in the US, where there is no national commitment to reduction or in testing mechanisms for reducing emissions from tropical forest destruction. The Verified Carbon Standard (VCS) helps corporations ensure that their voluntary reductions meet strict standards and help conserve biodiversity. The VCS and projects that test means of reducing deforestation help set the stage for a new generation of projects under the program for Reduced Emissions from Deforestation and Degradation (REDD).

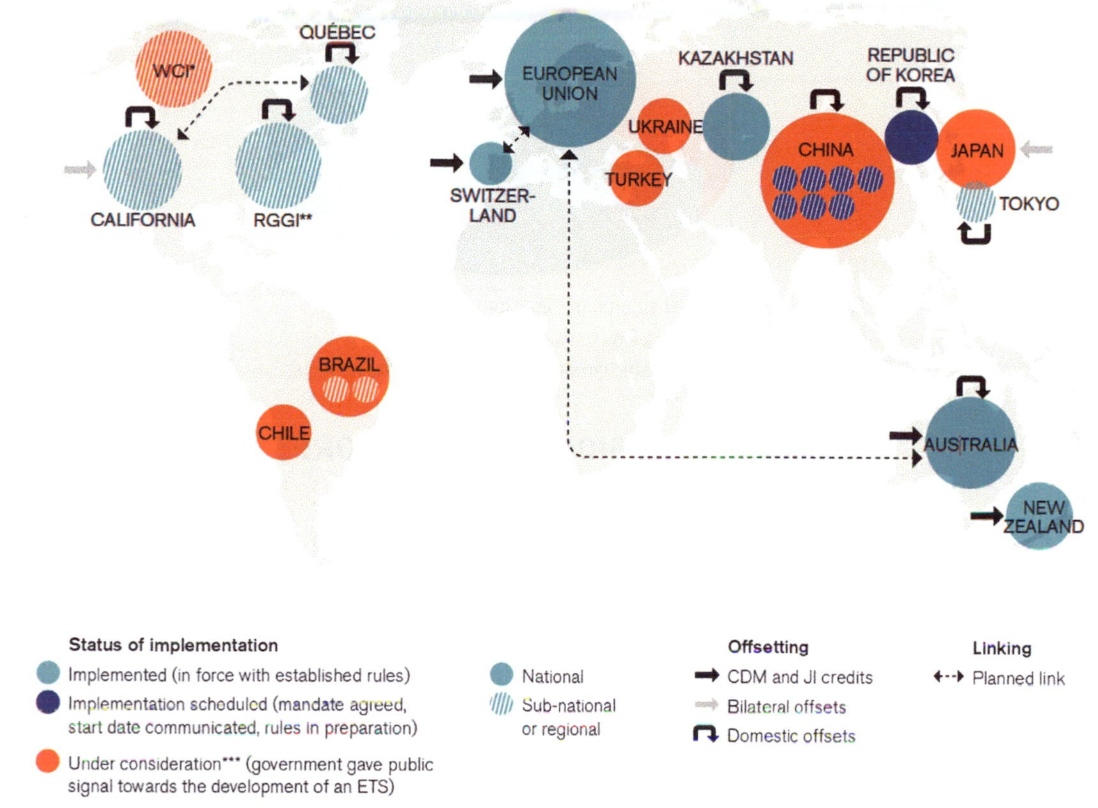

FIGURE 17.3 Map of existing, emerging, and potential emissions trading schemes.
Global emissions trading schemes are represented by their status of implementation (colored circles) and type (arrows). *Source: World Bank (2013). Figure courtesy of the World Bank Group.*

VERIFIED CARBON STANDARD

The Voluntary Carbon Standard (VCS) program was started to verify emissions reductions from voluntary offsets. It is now known as the Verified Carbon Standard. VCS provides carbon accounting methods and best practices for projects that allow individual companies to confirm that their carbon offsets are having an impact.

VCS includes standards for biodiversity protection and climate adaptation. Many corporations are concerned about overall sustainability, and would like their carbon offsets in tropical forests to also benefit biodiversity and local communities. The VCS makes this possible by providing guidelines to meet community and biodiversity goals. Some VCS projects take the adaptation of local communities into account as well. Conserving a forest which protects a watershed and harbors rare species can meet climate mitigation (carbon), biodiversity, and community adaptation needs all in the same place.

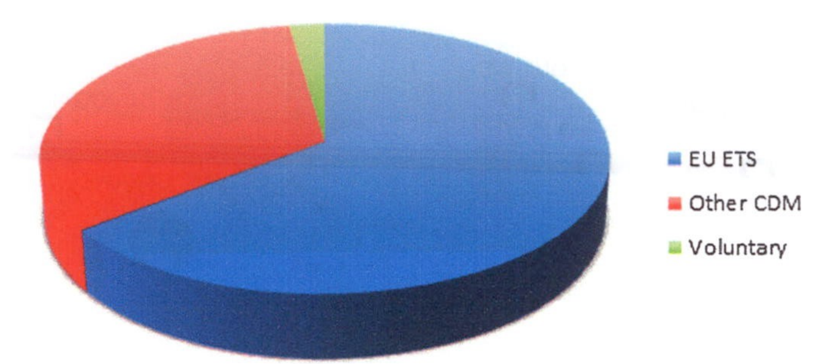

FIGURE 17.4 Global volume of emissions traded.
The global volume of emissions reductions traded is compared for the EU emissions trading scheme, other Clean Development Mechanism trading and voluntary markets.

REDUCED EMISSIONS FROM DEFORESTATION AND DEGRADATION

Deforestation makes up about one-sixth of global CO_2 emissions, which makes reducing deforestation a sensible part of an overall strategy to combat climate change. Ultimately, meeting the UNFCCC objective of stabilizing atmospheric greenhouse gas concentrations means that deforestation will have to be stopped or offset.

Most deforestation is currently in the tropics (temperate areas are increasing in forest cover), so reducing deforestation will take place mostly in developing countries. During climate negotiations in the 1990s, it was considered unfair that developing countries should be called on to play a significant role in emissions reductions when they had played a small role in creating the climate change problem. As a result, mechanisms for reducing deforestation were left out of the early emissions reductions agreements in the Kyoto Protocol.

Later, as carbon markets became established, tropical countries realized that they were left out. Billions of dollars in potential carbon market funds would never flow to reducing tropical deforestation because it was not allowed under the Kyoto Protocol. A consortium of tropical forest countries, led by Papua New Guinea, set out to correct this. In the 2005 UNFCCC discussions at COP11 in Montreal, it was agreed to begin a process for defining how reducing deforestation could be included in the international framework for emissions reduction.

Reductions in Deforestation and Degradation (REDD) are mechanisms for including both deforestation reduction and reducing forest degradation into emissions reductions systems—the result of years of discussions begun in Montreal. The REDD guidelines are intended to allow reductions in deforestation to be quantified, rewarded, and traded to help reduce global greenhouse gas emissions.

Reductions in deforestation are measured against a historical baseline, usually the historical rate of deforestation in the country (Figure 17.5). REDD

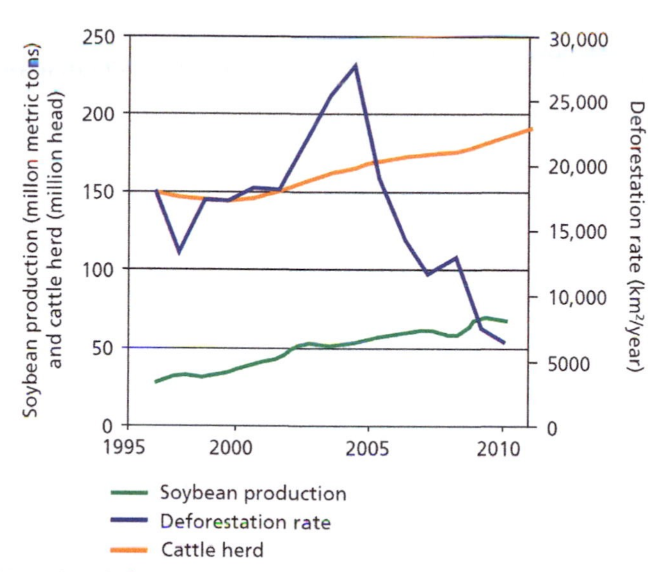

FIGURE 17.5 Deforestation and cattle and soybeam production in Brazil.
Deforestation has fallen off dramatically in Brazil since 2004, even while productive agriculture (soybeans and cattle) have increased. Much of the reduction in deforestation has been in the Amazon, while some of the increase in soy has been at the expense of dry forests. *Source: USDA.*

baselines set historical trends for deforestation in countries experiencing a net loss of forests. Deforestation dipping below these historical rates as the result of REDD programs then counts as emissions reduction.

Calculating baseline deforestation from national levels risks excluding countries that have good track records. For instance, Suriname has high levels of forest cover because of historically low deforestation. If neighboring Brazil reduces deforestation, clearing may move into Suriname if it is not part of the REDD system. A solution to this problem is to use global average deforestation rates as the baseline for countries with historically low deforestation. This acknowledges that even though deforestation rates have been low in the country, national action to prevent deforestation may be needed, and should be rewarded in carbon markets.

REDD faces two critical challenges, the first of which is called leakage. When deforestation is stopped in one area or country, it may move to a neighboring area or country, as in the Brazil-Surinam example above. If this happens, deforestation and emissions haven't been reduced, they've just been moved to another location. For instance, if a protected area protects forests from clearing, but lands outside the park are cleared instead, there has been no net reduction in deforestation and REDD fails. Or if one country cracks down on illegal loggers who move across the border to the country next door, there has been no real change in forest degradation. Both site and national deforestation reduction plans must account for leakage to prove that real reduction has taken place.

The second challenge is the equitable distribution of REDD revenues. Nations participate in the UNFCCC and REDD, but it is local people that are impacted by land-use changes that result in and reduction in deforestation. If local people do most of the work or bear most of the costs of reducing deforestation, most of the rewards from carbon payments should reach the local level. This may not be easy if carbon revenues are paid to the national government. REDD has taken so long to negotiate in part to make sure that both national level actions (such as implementing forest policy, policing, and eliminating tax incentives) and local actions (such as implementing sustainable farming practices) are rewarded.

ADAPTATION

Adaptation is what people do to deal with the impacts of climate change (Figure 17.6). For instance, when crops fail due to drought, local farmers have to switch to other less drought-vulnerable crops or rely on food aid. Or when storm surges flood a village due to a rise in sea level, the village must move, restore mangroves, or build a seawall. Adaptation and Ecosystem-based Adaptation (EbA) are key components of human response to climate change that are gaining increasing attention. EbA is described in detail in Chapter 13.

FIGURE 17.6

Adaptation innovation.

A woman tends a floating garden in Bangladesh. Floating gardens rise and fall with storm surges and rainfall, making them immune to destruction from flooding. This simple innovation protects from changes in rainfall, runoff and sea level due to climate change. Adaptation that pays immediate benefits as well as long-term climate adaptation benefits is more likely to be adopted. *Source: From https://www.flickr.com/photos/practicalaction/sets/72157624796596782/*

Like REDD, adaptation was an orphan in early climate negotiations. Many nations felt that emphasizing adaptation in UNFCCC discussions was a distraction from focusing on the root causes of the climate change problem. As a result, for over a decade, mitigation (the reduction of greenhouse gas pollution) was the major focus of international debate, with little attention paid to adaptation.

As progress on mitigation has been slow, it became increasingly apparent that significant climate change will happen before emissions are under control having major impacts on people, especially in poorer countries. International attention remained on mitigation, but adaptation was recognized as a needed counterpart to emissions reductions. At COP15 in Copenhagen in 2009, major agreements on adaptation were reached.

SPOTLIGHT: DECISION PATHWAYS

Changing decision making may be as important for climate change adaptation as are technical fixes. Decision making that anticipates and integrates change is much more likely to lead to successful adaptation outcomes. One way to improve decision-making for adaptation is to look at decision pathways - possible futures and key divergence points.

For example, in the Philippines, several coastal towns have worked together to close fisheries when fish are spawning. This replaces previous management in which closure of the fishery was fixed in time. Fisheries monitors collect fish each month and when reproducing females are detected in the monitoring system, the fishery is closed. Rangers from the nearby towns enforce the closure. This system protects the marine fishery when it is most vulnerable, and if that vulnerability shifts due to climate change, the system automatically adapts because it is built on monitoring, rather than on fixed protected areas or fixed closure times. Water temperature is one key variable in the time of reproduction for tropical reef fish, so this system of decision-making adapts as sea surface temperatures warm.

Assessing decision pathways alongside assessment of vulnerability can help produce improved adaptation outcomes. Decision-making is part of adaptive capacity and can help offset biological vulnerability (sensitivity) or physical vulnerability (exposure).

The backbone of UNFCCC adaptation agreements is a fund to compensate people impacted by climate change. This fund is supported mostly by developed countries that are the major greenhouse gas emitters, both through voluntary contributions and through a portion (2%) of CDM transactions. The Adaptation Fund has received over $300 million in contributions and has more than $100 million in projects underway. These amounts are small in relation to the total estimated costs of climate change, however.

The World Bank has estimated that climate change adaptation will cost between $70–100 billion/year in a world that is 2°warmer. Despite the large uncertainties of these estimates, they are clearly much larger than the current amounts being raised in the Adaptation Fund. The Green Climate Fund proposed in the Copenhagen COP has an overall goal of raising $100 billion/year by 2020 for both adaptation and mitigation. There is no identified source for these funds, so while this goal may be on the same order of magnitude as the need, it is unclear if funding will ever be mobilized at the scale of the goal.

WHY DOESN'T IT WORK?

From mitigation to adaptation, it is clear that international efforts to deal with climate change fall far short of the mark. Why have international climate negotiations taken so long and accomplished so little?

The climate negotiations have foundered while other international negotiations, such as the ozone layer, have succeeded. This initially seems puzzling, but the comparison with the ozone negotiations actually reveals important differences that explain much of the delay in climate progress.

Climate negotiations have struggled because there are clear winners and losers in solutions and the science is complex. The combination of these two factors have led to motivation to delay solutions, with the complex science resulting in greater uncertainty about policy responses. In contrast, the ozone layer had more straightforward science already worked out and no clear winners or losers. The causes of the ozone hole were known, and the corporations that manufactured the chemicals that created the damage were the same companies that made the alternative, nondestructive chemicals. As a result, no stakeholders had motivation to delay, and international accord on solutions were quickly reached. The ozone hole over Antarctica was discovered in the early 1980s, and the international protocol (the Montreal Protocol) restricting use of CFCs was approved in 1987. In contrast, while climate change was identified as a possible problem in the 1970s, negotiations over solutions are still continuing.

Losers exist in climate change solutions because the same corporations that produce fossil fuels are not the same corporations that produce renewable energy. End users of fossil fuels, including millions of car owners, would have to switch to different technologies to use renewable energy, at considerable costs. These corporations and individuals stand to lose billions of dollars in a transition to renewable energy, and they compose a large stakeholder group with strong motivation to delay or defeat progress in combating climate change.

The complexity of climate science provides ammunition to groups motivated to oppose international action. The basic physics of global warming are simple and have been known for over a century. But understanding the evolution of climate change into the future requires highly sophisticated models with state-of-the-art computing. This research is pushing the boundaries of modern computing power, and so is less certain than the science underlying engineering or medical technologies. Groups opposing action on climate change seize on this uncertainty to create doubt and justify inaction. The multicountry nature of the negotiations and the very different roles of developing and developed countries further slow and complicate decisions. The IPCC provides a valuable benchmark of sound science, but it is unclear when the international process will overcome the combined obstacles of solution losers and science complexity.

FURTHER READING

Busch, J., Godoy, F., Turner, W.R., Harvey, C.A., 2011. Biodiversity co-benefits of reducing emissions from deforestation under alternative reference levels and levels of finance. Conservation Letters 4 (2), 101–115.

da Fonseca, G.A., Rodriguez, C.M., Midgley, G., Busch, J., Hannah, L., Mittermeier, R.A., 2007. No forest left behind. PLoS Biology 5 (8).

Gupta, J., 2010. A history of international climate change policy. Wiley Interdisciplinary Reviews: Climate Change 1 (5), 636–653.

O'Neill, B.C., Oppenheimer, M., 2002. Climate change: dangerous climate impacts and the Kyoto Protocol. Science 296, 1971–1972.

Pielke, R., Prins, G., Rayner, S., Sarewitz, D., 2007. Climate change 2007: lifting the taboo on adaptation. Nature 445 (7128), 597–598.

CHAPTER 18

Mitigation: Reducing Greenhouse Gas Emissions, Sinks, and Solutions

STABILIZING ATMOSPHERIC GREENHOUSE GAS CONCENTRATIONS

Stopping climate change requires much more than emissions reductions: it requires stabilizing atmospheric GHG concentrations. In the absence of large new additional sinks for CO_2 and other GHGs, stabilization equates to reducing emissions nearly to zero. In turn, near-zero emissions means complete transition of the world's energy supply to renewable (non-CO_2-emitting) sources. This is a huge task—one that will need to be accomplished in stages.

Global policy discussions have focused on the stabilization of global GHG concentrations at levels that would prevent climate change from exceeding a mean global temperature increase of 2 °C. This goal corresponds to a low-end GHG stabilization target, perhaps as low as 350–400 parts per million (ppm), depending on emissions scenarios and general circulation models used to make the projection. Because global atmospheric CO_2 concentrations passed 400 ppm in 2014 and are rising by several ppm per year, this is an ambitious goal.

Some economic models suggest an optimum in trade-offs between costs of mitigation and adaptation at atmospheric GHG concentrations at or below 550 ppm. This is approximately double preindustrial CO_2 concentrations, which were near 280 ppm. Stabilization levels are expressed in CO_2 equivalents, so 550 ppm equates to the sum of CO_2 concentration at a 1:1 conversion, plus all other GHGs at their atmospheric concentration multiplied by their potency relative to CO_2. Therefore, a 550 ppm CO_2 equivalent target equates to levels of CO_2 alone of slightly below 550 ppm.

Because such low targets may be difficult to achieve, there is an increasing international discussion about allowing global GHG concentrations to rise above the target level and then bring them back down. This "overshoot" strategy is attractive where insufficient political will or technical means are available to achieve the stabilization target this century, but future technical advances will

Climate Change Biology. http://dx.doi.org/10.1016/B978-0-12-420218-4.00018-4

enable reduction to achieve the long-term target. Thus, it may be possible to have a long-term target that is more ambitious than can be achieved in the short-term.

SPOTLIGHT: REFLECTING BACK

Not all deforestation is created equal. General circulation model (GCM) simulations that couple land cover, such as forests, with atmospheric and ocean processes show that boreal and tropical deforestation have opposite effects on global mean temperature change (Bala et al., 2007). All deforestation increases atmospheric CO_2 content, as expected. However, boreal forest loss replaces dark forests with lighter grasslands or snowfields with higher albedo. This reflects more sunlight back to space, therefore cooling the planet. Despite the CO_2 release from the loss of boreal forests, GCM simulations show the world to be a cooler place when boreal forests are lost, due to the albedo effect. Tropical forest loss, on the other hand, results in replacement vegetation with similar albedo. Thus, tropical deforestation warms the planet, with the CO_2 effect dominating over the effect of reflecting sunlight. The way to a cooler planet is to emphasize decreases in tropical deforestation.

Source: Bala, G., Caldeira, K., Wickett, M., Phillips, T.J., Lobell, D.B., Delire, C., et al., 2007. Combined climate and carbon-cycle effects of large-scale deforestation. Proceedings of the National Academy of Sciences 104, 6550–6555.

The overshoot strategy has important biological implications, because it may put natural systems through two climate transitions rather than one. Is it better for a species to undergo a range shift once to a more severely-changed climate or to undergo two range shifts to a more moderate long-term change? The answer to this question is not known, and this is an important topic for research in climate change biology.

PRACTICAL STEPS FOR THE NEXT 50 YEARS

Although the stabilization of atmospheric GHG concentrations requires a formidable energy transition, there are reasons to believe that key early actions can be taken with existing technologies. Early moves to more energy-efficient and renewable technologies can provide an important head start in the needed energy transition.

One way to visualize early GHG stabilization needs is as a triangle of energy supply to keep emissions from increasing during the next 50 years (Figure 18.1). If this triangle can be met with renewable energy, efficiency, and non-CO_2 sources, the first, most important portion of a move to stabilizing GHG concentrations will have been accomplished. The total triangle corresponds to 175 Gigatons of carbon (GtC) emissions that need to be offset.

This 50-year triangle can be seen as consisting of a series of smaller wedges, each corresponding to a portion of the needed energy supply (see Figure 18.1(b)). These wedges are often referred to as "stabilization wedges." Seven such wedges, each corresponding to 25 GtC, are required to meet the entire 175-GtC triangle.

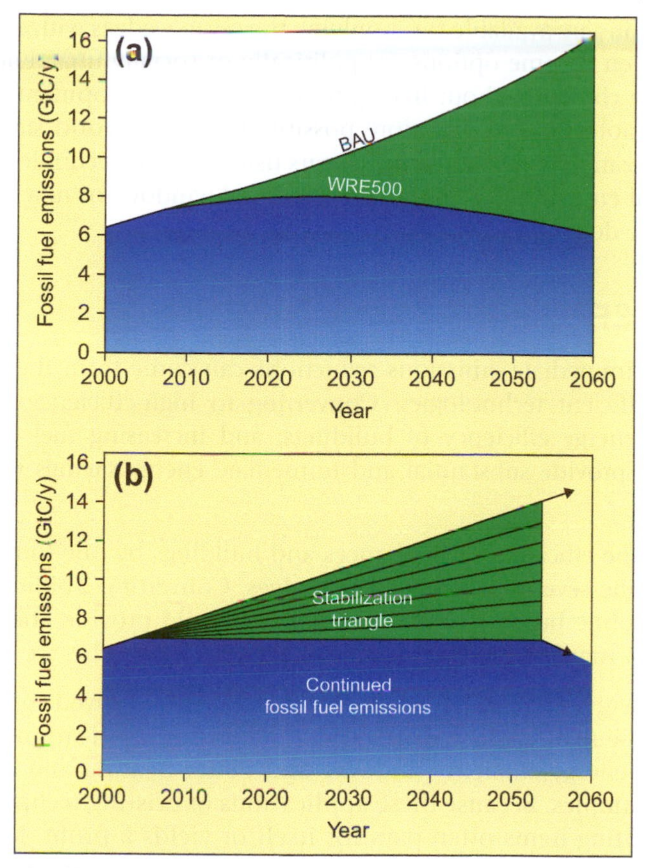

FIGURE 18.1 Stabilization triangles and wedges.
Stabilizing greenhouse gas concentrations in the atmosphere requires the leveling of emissions by 2050 and reductions thereafter. To offset new demand, a "stabilization triangle" (green) is required. One way to satisfy the energy demand in such a triangle is through existing technologies, each contributing a wedge of approximately one-seventh of the triangle (solid lines within green). More than 15 existing technologies have been identified that can supply one of 12 needed wedges, meaning that choices can still be made in energy supply (e.g., foregoing nuclear power) and still achieve the steps toward stabilization needed in the first half of the twenty-first century. *Source: Pacala and Socolow (2004). Reproduced with permission from AAAS.*

TONS OF CO_2, TONS OF CARBON

Emissions reductions are often traded in tons of carbon. Determining the amount of CO_2 equivalent to a ton of carbon requires a conversion using the molecular weights of carbon and CO_2. Carbon has a molecular weight of 12, whereas oxygen has a molecular weight of 16. CO_2 therefore has a molecular weight of $12+(2\times16)=44$. The conversion ratio of CO_2 to carbon is therefore 44/12 or approximately 3.66. Therefore, one metric ton of carbon is the equivalent of 3.66 metric tons of CO_2.

Several options are available for supplying the seven wedges with existing technologies. Even if some options are politically or socially unacceptable, there are multiple choices without having to assume the development of any new energy technologies. It is therefore possible to get the world started toward GHG stabilization during the next 50 years using existing technology. This provides critical emissions reductions and a crucial window for new energy technology to be developed to complete the job.

ENERGY EFFICIENCY

The largest immediate emissions reductions can come from the application of energy-efficient technologies. Converting to high-efficiency light bulbs, improving energy efficiency of buildings, and increasing fuel efficiency in cars can all provide substantial and immediate energy savings with existing technology.

Improving the efficiency of appliances and buildings by one-fourth can supply one of the seven needed 50-year wedges. Converting 2 billion cars from 30-mpg gasoline burners to 60-mpg hybrids would provide another wedge. Reducing car travel in nonefficient cars could provide another wedge.

Because energy efficient technologies are underemployed, gains in efficiency can be realized in a wide variety of existing applications. Improved insulation, energy-efficient windows, and better heating and cooling systems are examples. Because these applications of existing technologies save fuel, retrofitting items often pays for itself or yields a profit. Many corporations have reduced their CO_2 emissions and saved millions of dollars in the process.

RENEWABLE ENERGY SOURCES

Renewable energy sources, such as solar and wind energy, have excellent potential to provide immediate non-CO_2 emitting energy, and they are the best technologies for long-term stabilization of GHGs in the atmosphere. Renewable energy sources by definition emit no CO_2 and do not contribute to climate change. Instead, they convert energy from the sun (or, in some cases, from gravity or heat from the Earth's core) into forms that are useable for human purposes.

Existing wind energy technologies can provide at least one 50-year wedge, whereas photovoltaic solar energy can provide another. Coupled with hydrogen fuel cells or other technology, wind-electric can power two billion cars—another wedge. Ethanol from common crops, such as corn, can provide liquid fuel needs for two billion hybrid cars. Thus, the technology already exists for at least four efficiency and four renewable energy wedges—more than enough to supply the seven needed wedges. The following four sections describe the present and future potential of the major renewable energy sources in more detail.

Solar Energy

Ultimately, the sun powers all major renewable energy technologies except geothermal and tidal power. The sun drives atmospheric processes that result in wind for wind power, plant growth for biofuels, and water evaporation that makes hydropower possible. Direct energy from the sun powers solar energy systems. More solar energy reaches the Earth each minute than is consumed in fossil fuels in an entire year.

SOLAR THERMAL TECHNOLOGIES

Solar thermal energy uses heat to convert sunlight into useable energy. Where heat is the desired end energy use, direct conversion is possible. For instance, a south-facing window (in the Northern Hemisphere) may be used to heat a room in winter. Such direct uses of the sun's heat are known as passive solar systems. Active solar systems concentrate or reflect the sun's energy. Two examples of the many types of active solar systems are mirrors used to heat a central boiler for electricity generation or moving water through solar panels to heat it. Systems that use mirrors or other methods to focus the sun's rays are known as concentrating systems, and those that do not are nonconcentrating. Many passive or nonconcentrating systems are suitable for household, decentralized use. Many concentrating, active systems are used for centralized electric generation.

Solar energy may be divided into two major categories: solar thermal and solar voltaic. Solar thermal energy relies on the heating of a carrier fluid, often water. The warmed liquid may be used directly, as in hot-water heating, or used to drive another process, such as electrical generation. Solar voltaic or photovoltaic systems generate energy by capturing electrons excited by photons in sunlight. Photovoltaic systems generate electricity that finds application in a number of end uses.

PHOTOVOLTAIC CELLS

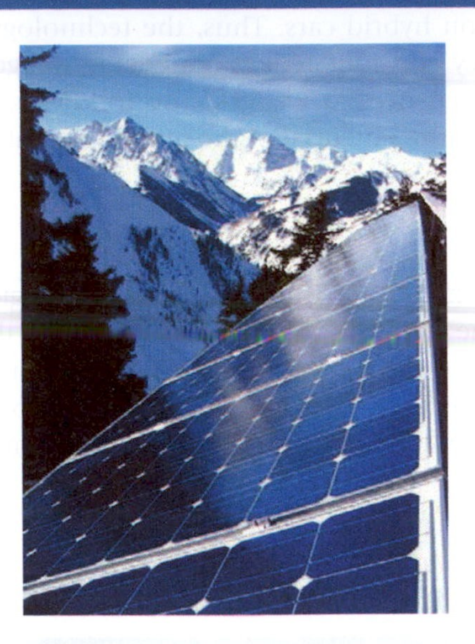

Photovoltaic (PV) cells use the electromagnetic properties of sunlight to generate electricity. Light falling on a PV array releases electrons, which are captured and channeled by the silicon structure of the PV cells. This stream of flowing electrons is electricity, which can be used to power any conventional electrical appliance or motor. Rooftop PV has the potential to provide as much as half of all of the world's energy demand, if all available rooftop space were employed. A major fraction of electricity needs could be met by rooftop PV with no land use demands that might reduce wildlife habitat.

Source: NREL.

Despite massive potential, solar power currently has limited market penetration. Even in highly suitable countries such as Australia, solar power provides less than 10% of energy demand. Nonetheless, worldwide, more than 140 million m^2 of solar thermal collectors have been installed—more than 100 GW in energy-generating potential—which is more than the total global-installed wind capacity. In areas with large amounts of available land, central-receiver solar electrical generation has major potential for expansion. Solar thermal production might supply 5–20% of all energy demand worldwide by 2050.

PV generation has strong potential for supplying a major portion of electricity demand. PV can be mounted on roofs in urban demand centers. Nanotechnology may soon provide PV paints that will allow electricity generation from any painted surface. It is possible that all residential electricity use could be provided by PV within the twenty-first century.

Wind Energy

Global-installed wind energy capacity is currently more than 60 GW. Global wind capacity increased 12-fold between 1995 and 2005. Virtually all of this wind capacity is in the form of electricity-generating turbines of different sizes.

If current growth in wind capacity continues, global-installed capacity could reach more than 1 TW by 2020, at which point global electric energy demand may be 2 or 3 TW. It is likely that mismatches between demand centers and areas with high-wind potential will limit the future expansion of wind energy. It is possible that wind energy will provide as much as 50% of electrical demand by 2050.

WIND TURBINES DIFFERENT TYPES OF WIND TURBINES

Wind turbines come in all sizes, from units the size of a bread box used to provide electricity on small boats, to giant multimegawatt turbines used in central electricity generation. Visual impacts are a major concern with large wind farms. Although the turbine towers have a small footprint, associated roads and support infrastructure may result in significant habitat loss where wind is sited in natural areas.

Sources: (a) NREL; (b) NREL; (c) Todd Spink, NREL.

Biofuels

Biofuels convert energy from the sun to plant materials, which are then processed to produce liquid fuels. Biofuels include ethanol, which may be substituted for gasoline in many recent-model automobiles. For this reason, biofuels are especially useful in providing the energy needs of the transportation sector.

Brazil has the most advanced biofuel program in the world, where approximately 40% of gasoline use has been replaced by ethanol from sugarcane and other sources. Whereas ethanol in Brazil's program can be produced from sugarcane waste, in most countries ethanol production would compete with land for food production. Hence, the 40% transportation fuel penetration of biofuels in Brazil would probably be an upper bound for most other countries by 2050.

Hydropower, Tidal Power, and Geothermal Power

Hydropower is the generation of electricity by running a turbine with water impounded behind a dam. Hydropower currently comprises slightly more than 20% of the world's electrical generation. Canada draws more than half of its electric power from hydro, whereas in the United States hydro supplies less than 10%. China has the greatest potential for growth in hydro, and is pursuing an aggressive strategy of hydro development.

Much of the potential for high dams for hydropower has already been realized worldwide. Future expansion is limited by social (displacement of communities) and environmental concerns. There is a greater potential for expansion in small- to medium-sized hydropower projects.

Geothermal energy uses heat from the Earth's core, in surface manifestations often associated with features such as geysers and hot springs, to boil water and produce electricity in a steam-driven turbine. Total world capacity is currently less than 8 GW, and even assuming a doubling by 2050, this source will remain minor in global terms. In theory, deeper earth heat can be tapped by deep drilling, which could greatly expand future geothermal potential.

A variant on geothermal production is ocean thermal production, in which the temperature gradient across the surface layers of the ocean is used to boil a highly-volatile carrier such as ammonia to drive a turbine. Ocean thermal potential is huge, but it is limited by transportation problems. Energy produced must be transported to land as electricity in huge submarine cables or converted to a liquid carrier.

Tidal energy uses the energy force of the tides to generate electricity. A dam or barrier perpendicular to the direction of tidal flow can create enough heat to generate electricity. Tidal power is largely untested but may have wider potential for expansion than ocean thermal power because it can be sited closer to end demands.

NUCLEAR POWER

Nuclear power is not renewable, but given the volume of source materials available, it is essentially inexhaustible. It produces no GHGs, making it an excellent energy source strictly from the perspective of limiting climate change. There are two types of nuclear power: nuclear fission, which is well tested as an energy source, and nuclear fusion, which is untested commercially and still in development.

NUCLEAR POWER

Nuclear power has major potential for generation without GHG emissions, but it faces major safety and environmental concerns. Among the safety concerns are potentially devastating radiation damage to humans and wildlife in the event of a major accident, and the possibility of the use of by-products to build nuclear arms. Environmental concerns include the safe disposal of very long-lived radioactive waste from nuclear reactors. Climate change provides reason to reexamine nuclear power possibilities, but social concerns seem likely to limit its overall importance as a CO_2-free energy source.

Nuclear power plant. Photo source: Marya, Wikipedia Commons.

There are approximately 450 nuclear electrical power plants in the world, approximately one-fourth of which are in the United States. Of all the countries with nuclear power plants, France receives the largest share of its electricity from nuclear power, at 78%. Current global nuclear capacity is approximately 350 GW (larger than all renewables combined), which is projected to grow little or not at all by 2020. Concern regarding the proliferation of materials for nuclear weapons and the environment will limit the growth of nuclear power in the future, unless these concerns are overcome by social momentum for action on climate change.

Nuclear fusion avoids many of the proliferation and environmental concerns of fission. Fusion is believed to produce no bomb-grade side products and far less radioactive waste than fission. However, fusion is untested technically and far from commercial viability.

THE END OF OIL

Beyond renewable and nuclear options, several modifications of CO_2-producing energy technologies have the potential to reduce global GHG emissions. CO_2 may be captured at the source or recaptured from the atmosphere and stored to reduce CO_2 emissions. The next several sections outline these technologies.

One complicating factor in the transition to low-CO_2 alternatives is the fate of the world's oil supplies. It has been predicted since the 1970s that the world is running out of oil. Many of the early predictions have been proven false by the discovery of new reservoirs and new techniques for enhancing production. However, these advances seem to be playing out, and a new wave of projections of the oil end-game are emerging.

At the core of the end-of-oil scenarios is the shape of peak oil curves. A peak oil curve depicts oil production through time. In an era of increasing use, peak production will coincide with peak consumption, meaning that peak use can only be sustained for a relatively short time. This means a precipitous fall in production once the peak in supply is passed (Figure 18.2).

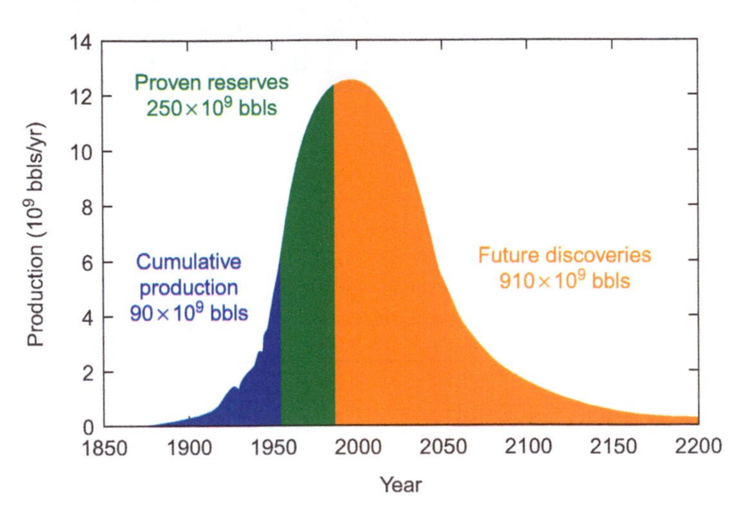

FIGURE 18.2 End of oil.
Oil production will peak and decline, with the shape of declining production roughly mirroring the upward curve of increasing production. Abundant other fossil fuels, including oil sands, oil shale, and coal, will ensure that the fossil fuel era will continue, regardless of when oil supply is exhausted (unless climate policy intervenes). *Source: From Wikimedia Commons.*

If oil peaks in the mid- to late century, renewable energy sources will be positioned to make up the lost supply, and at the same time greatly reducing GHG emissions. During the next five or six decades, renewable energy production should gain major market share and begin displacing oil, even before the peak is reached. This might help prolong oil supplies, and it would create a smooth energy transition to climate-friendly alternatives.

However, if oil peaks earlier than 2050, rapid transition to renewables may be difficult, and oil may be replaced by energy sources that do not reduce climate change emissions. Likely oil substitutes that are not renewable or climate-friendly include coal, oil shale, and tar sands, which are relatively environmentally "dirty" sources that contribute to climate change but have massive reserves.

CLEAN COAL?

World coal reserves are approximately 1000 GtC—four or five times larger than world oil reserves (Table 18.1). Coal burning can be combined with CO_2 capture at the source to make it a near zero-emission energy source. This CO_2

Table 18.1 World Coal Reserves in Gigatons[a]

	Category I: Hard Coal					Category II: Brown Coal					
	Grade A	Grade B	Grade C	Grade D	Grade E	Grade A	Grade B	Grade C	Grade D	Grade E	Total
NAM	140	0	104	97	387	12	0	3	2	11	883
									8	1	
LAM	6	1	3	7	28	0	0	0	0	1	47
WEU	18	3	14	46	185	9	1	8	1	4	289
EEU	22	22	26	9	35	8	4	2	0	2	129
FSU	88	0	22	50	202	2	0	2	4	17	288
				6	5	2			4	6	5
MEA	0	0	0	3	12	0	0	0	0	0	15
APR	37	0	37	16	64	0	0	0	0	0	153
CPA	34	40	27	16	060	1	2	2	1	47	129
			4	5		4	8	2	2		5
PAO	20	14	18	47	188	9	2	1	0	0	452
		7					3				
PAS	2	0	1	0	1	0	0	0	1	3	9
SAS	7	28	19	7	28	1	0	0	0	0	89
World[b]	372	24	51	90	361	7	5	3	8	34	624
		1	8	3	2	5	6	8	6	4	6

[a]*Expressed as oil equivalent to eliminate energy differences between different grades. Grade A, proven recoverable reserves; grade B, additional recoverable resources; grade C, additional identified reserves; grade D, additional resources (20% or remaining occurrences); and grade E, additional resources (80% or remaining occurrences).*
[b]*Totals may not add up due to rounding.*

capture technology is technologically proven but not widely dispersed. It adds considerable cost to energy produced by coal. Combined with liquefaction, this technology provides a "clean" energy source from the climate change perspective that can be used in common applications from automobiles to power plants without modification.

CO_2 capture operates on principles similar to those used to remove other pollutants from power plant streams. The high concentrations of CO_2 in the waste streams can be captured on adsorbents and disposed of. There is a capital cost and an energy input required for CO_2 capture. Capital cost varies with the process, and the energy penalty is approximately 20%.

Once captured, the CO_2 must be sequestered or stored. This is the more challenging part of the technology. CO_2 can be compressed and stored at high pressure underground in impermeable geologic formations. In practice, CO_2 injection is sometimes used to enhance oil recovery, and some oil-bearing formations will hold CO_2 without substantial leakage. Thus, although untested at large scales, the technology is available and proven. Another option includes deep ocean disposal, which must be evaluated on storage capacity and leakage (storage time).

LIFE OF A COAL-FIRED POWER PLANT

Coal-fired power plants typically have a life span of 30–35 years. This means that construction of a new coal-fired power plant will result in CO_2 emissions for the next quarter of a century or more. For this reason, climate change activists are urging the industry to discontinue plans for traditional coal-fired generation, replacing it instead with renewable energy generation or with coal-fired technology coupled with carbon capture and storage, and thus avoiding locking in long-term emissions.

TAR SANDS, OIL SHALES AND FRACKING

Far greater amounts of fossil fuel energy are found in tar sands and oil shales than in coal. Accessing these energy sources entails much higher environmental costs. Oil shales and tar sands harbor approximately twice the energy potential of coal, or 10 times the energy in remaining oil reserves. Combined, oil shales and tar sands, which are fossil fuel mixed with rock or sand, have at least 15 times the energy production potential of oil. The use of these energy sources has been constrained by the much lower recovery costs and environmental consequences of production. As oil availability declines, energy prices will rise, increasing the commercially-viable reserves of these nonoil fossil fuels.

This transition is already underway, with large areas of tar sands being developed in Canada and elsewhere. North America, a declining oil producer, may be self-sufficient in energy by 2020 due to rising production from tar sands, oil shales, and renewables. Fracking is one form of extraction of fossil fuels from sources that were traditionally too expensive to tap. It uses hydraulic fracturing to release natural gas from rocks in deep wells.

Given the great energy potential of these sources, they represent a logical energy transition as oil supplies decline. Liquefaction and gasification technologies can strip usable fossil fuels from these sources and turn them into liquid or gas fuels that can be used in most applications in which oil or gas products are currently used. Could these sources be made climate-friendly as well?

The same carbon-capture-and-storage (CCS) technologies that might be used with coal also apply to tar sands and oil shales. The equipment for CCS is costly, but components of it have been proven in other industrial applications, such as enhanced oil recovery. Chapter 19 explores natural and man-made sinks of CO_2 and explores the potentials and liabilities of CCS in more detail.

The greatest challenge in combating climate change lies in implementing a large-scale, almost total transition in energy supply away from sources that release CO_2 directly into the atmosphere. This transition can be toward renewable energy sources, CCS, or combinations of the two. Because much energy infrastructure has lifetimes measured in decades, this transition will necessarily have large built-in lead times. Early action is essential to avoid having this energy inertia turn tractable technical solutions into actions that are too little and too late. Early implementation of an organized transition can eliminate the energy inertia effect.

All the alternatives have downsides as well as climate benefits. Large land areas may be destroyed producing fuel from tar sands, even if it is coupled with CCS. Fracking has major implications for groundwater quality and public health. Most renewable energy sources require more land area than oil production, possibly impacting important habitats. The careful balancing of land use, climate change, and biodiversity impact concerns is required.

A key element in a planned-energy transition is choosing long-term energy pathways that minimize damage to the environment. Some energy pathways, benign on small scales in most settings, can be very damaging to biodiversity when implemented on large scales or in sensitive areas. Identifying these biological pitfalls by defining biodiversity-friendly and climate-friendly energy paths is a final challenge for climate change biology.

GEOENGINEERING

An alternative to reducing greenhouse gas emissions is to try to counterbalance the effects of greenhouse gases by geoengineering the global environment.

Geoengineering is also known as climate engineering. There are two broad categories of geoengineering schemes: CO_2 removal and solar radiation management.

CO_2 removal involves altering the global carbon cycle or sequestration of CO_2 and is discussed in Chapter 19. CO_2 removal schemes have been proposed that include the fertilization of the southern oceans with iron to increase phytoplankton growth and sinking of undecomposed plankton. Sequestration proposals include injecting CO_2 at a high pressure into the deep oceans.

Solar radiation management forms of geoengineering attempt to counterbalance warming by physical means. Deploying giant mirrors to space has been proposed to deflect a portion of incoming solar radiation in order to reduce atmospheric warming. Injecting aerosols to block sunlight in the upper atmosphere has been proposed with similar intended effects.

Another approach to reducing warming is to alter the albedo, or reflectance of the Earth. Replacing dark rooftops in urban environments with white rooftops is one approach to altering albedo that has multiple benefits. The reduced surface heat absorption of the white roofing results in lower urban temperatures and offsets a fraction of warming caused by greenhouse gases. Unfortunately, the area of urban rooftops in the world is tiny in relation to the amount of albedo change needed to counteract the effect of greenhouse gases.

Deploying mirrors to space is prohibitively expensive, while albedo change on scales sufficient to offset significant warming would require manipulating large areas of natural vegetation with unknown environmental consequences. Release of aerosols into the stratosphere is relatively inexpensive, but its possible effects on climate and the atmosphere is unknown. Trading one grand experiment in climate engineering (greenhouse gas pollution) for another seems a poor solution.

None of the solar management approaches address ocean acidification. Space mirrors, changes in albedo, and aerosols leave CO_2 in the atmosphere and do nothing to reduce the impacts on marine organisms caused by acidification. All solar management approaches may become unpredictable, unstable, or ineffective at large scales, because it is impossible to test them at the scales at which they would need to be deployed. Unintended side effects might cause more damage than they avoid.

While all geoengineering approaches face problems of control and predictability, as well as possible unintended side effects, CO_2 removal approaches at least address ocean acidification as well as warming, and more directly address the root causes of climate change. All of these approaches involve alterations to the carbon cycle, and are considered in Chapter 19.

EXTINCTION RISK FROM CLIMATE CHANGE SOLUTIONS

Energy technologies all carry risk to the environment. This risk varies with the type of technology, location, and environmental impact of concern. Any energy technology sited in a highly-sensitive area may have serious environmental impacts. As any technology is taken to scale, the chances of individual facilities intersecting environmentally-sensitive areas increase.

This section focuses on the biodiversity impacts of energy sources that may play important roles in combating climate change. Other important environmental impacts of energy sources include air and water pollution. Biodiversity impacts are particularly relevant to climate change biology because extinction risk from various solutions to climate change can be compared to the extinction risk from climate change.

Spatial configurations of energy sources are critical to determining extinction risk and biodiversity impacts. Extinction risk from climate change is a global phenomenon, so extinction risk from climate change solutions must also be calculated on a global scale. At this scale, many spatial variants are possible. This chapter explores the general magnitude of the land use requirements of each technology and highlights some spatial sensitivities.

Wedges beyond 50 years

The geometry of greenhouse gas (GHG) stabilization continues past the 50-year mark. The 50-year stabilization triangle stabilizes emissions at approximately present levels. Stabilizing global GHG concentrations requires an additional, similarly-sized triangle to take emissions to near zero. Because the oceans are absorbing CO_2, stabilization may be achieved at slightly more than zero emissions, but for purposes of estimation, it can be assumed that stabilizing GHG in the atmosphere requires that we stop putting GHG into the atmosphere.

Because demand will continue to grow in the second half of the century, a third stabilization triangle is needed, similar in size to the first two. This triangle supplies demand occurring in the second half of the century, primarily from developing countries.

The equivalent of three emissions stabilization triangles are therefore needed to stabilize GHG concentrations in the atmosphere by the end of the twenty-first century, assuming demand continues to rise. The first of these is the 50-year triangle, followed by a triangle to reduce existing (primarily developed country) emissions to zero, and a third to provide for growth in demand during the second half of the century (primarily in developing countries).

The second and third triangles cannot be fulfilled with efficiency wedges because they will have already been used from 2000 to 2050. This leaves more than twice as much energy to be supplied by carbon-neutral energy technologies from 2050 to 2100 as in the first half-century. Collectively, the three triangles can be referred to as "100-year" triangles.

The energy pathways developed before 2050 will have much to do with the land use requirements and biodiversity impacts of the second two 100-year triangles. If renewables are the main means of meeting these energy needs, land use requirements will be substantial. Only nuclear and fossil fuel carbon capture and storage (CCS) can meet these needs with substantially reduced land use requirements. Nuclear and CCS have lower land use needs relative to solar and wind energy, but still higher than the existing footprints of oil or natural gas.

Past Experience

Experiences in Hawaii and Brazil indicate that rapid alternative energy development may result in unexpected impacts on biodiversity. In both of these examples, progressive energy policy resulted in desired increases in alternative energy supply, but market forces caused unintended damage to natural systems. Both examples involve biofuels, but other renewable energy sources will cause heightened biodiversity impacts when taken to scale.

In Hawaii, impacts on rare native forests resulted from the development of biofuels based on sugarcane waste. In the 1980s, Hawaii developed an aggressive alternate energy program to counter the state's total dependence on imported oil. Sugarcane, a major crop in the state, generated large volumes of waste biomass after the juice had been pressed from the cane during sugar production. Bagasse, as the waste cane is known, could be used to fire steam turbines for electric generation. With state subsidies, a set of bagasse-burning power plants were built, with contracts to supply a significant amount of the state's electrical power.

However, when sugar prices dropped in the mid-1980s, cane production was curtailed, leaving less feedstock for electricity generation. The sugar mill contracts for energy production were still in place, so there was a need for an alternate biomass feedstock. Several producers bought large commercial tree chipping machines and began chipping the native forest to use in place of bagasse. An environmental outcry followed and the chipping was discontinued, but the threat of negative biodiversity consequences from rational energy policy and response to energy markets was clear.

Brazil has experienced larger negative impacts on biodiversity in response to its biofuel development program. Brazil initiated a sugarcane for ethanol program, also in the 1980s, in response to high oil prices in the late 1970s. At that time, government subsidized production, which became too expensive when sugar prices increased in the late 1980s. The program resumed in the late 1990s, however, and rapidly gained market penetration. Large areas of sugarcane were planted in response to rapidly increasing demand, often resulting in the clearing of tropical forests in the Amazon (wet forest) or Cerrado (dry forest).

The clearing of forests for biofuels, which characterized much of the Brazilian biofuel expansion, is also being observed in other regions. Clearing of tropical forests for oil palm for biofuels is a major source of deforestation in Indonesia and Malaysia. A large-scale transition to biofuels could accelerate these negative impacts, which may spread to other regions, such as Central Africa.

The balance of alternative energy sources and their spatial distribution are critical determinants of biodiversity impacts, so it is important to be aware of the impacts and spatial suitability for each source. The following sections provide overviews of some possible impacts, land use intensity, and spatial needs for major alternative energy sources.

LAND USE REQUIREMENTS OF ALTERNATE ENERGY

Renewable energy sources require land surface for the production of energy, in contrast to oil and gas, which are conveniently stored underground by natural processes. In essence, oil and gas are fossil biofuels; the land production required already took place in the past. What we generally consider the environmental and land use impacts of oil and gas production are actually only the impacts of extraction from storage and transportation. Thus, alternative energy sources have much greater potential impacts on land use and, therefore, on biodiversity.

Land use requirements of several alternate energy sources and climate change solutions can be approximated based on the 50-year and 100-year supply triangle concept. For each source, an illustrative land requirement is given for a single 50-year wedge contributing to the triangle needed to stabilize emissions and for three 100-year wedges, one for each of the triangles needed to stabilize GHG concentrations in the atmosphere.

Each wedge corresponds to 25 GtC of supply, so the two illustrative scenarios equate to 25 and 75 GtC of energy production, respectively. Because some of the technologies have the potential to contribute more than 75 GtC in the twenty-first century, and some may actually contribute much less, these benchmarks are used to facilitate comparison between the sources, not for projection of actual land use impacts. To aid visualization of the areas, comparisons are made for each source to the total amount of land currently under cultivation for food production worldwide (1.5 billion ha).

Solar and Wind

One 25-GtC solar photovoltaic wedge would require approximately 2 million ha of surface area. The equivalent in central tower solar thermal electric generation would require slightly more land area. Thus, three 100-year wedges would require between six and 10 million ha. This is a large area, but it equates to only approximately 1% of the land currently under agricultural cultivation worldwide, so the area requirements of solar energy are small in relation to those of other human land uses.

GLOBAL BIODIVERSITY HOTSPOTS

Biodiversity hotspots are areas of high endemism and high threat that are recognized as global conservation priorities. They are foremost among the areas in which siting of land-intensive renewable energy must be sensitively planned. The hotspot concept was pioneered by the ecologist Norman Myers and made popular by Conservation International. A total of 34 hotspots have been identified in areas as different as the tropical forests of the Andes and the desert vegetation of the Succulent Karoo in Southern Africa. All have in common at least 1.5% of all plants on Earth as endemics and more than 70% loss of original pristine habitat. Planning renewable energy in ways that minimizes loss of intact forest and other natural habitats can help reduce the loss of unique biodiversity.

Wind energy production requires approximately 10–20 times more area than solar energy production. One 50-year wedge of wind energy has been estimated to occupy 30 million ha. Three 100-year wedges would therefore occupy nearly 100 million ha. Nonetheless, this large area is less than 10% of the area currently in crop production. In addition to area requirements, wind energy farms are a significant source of mortality for bats and birds. Bats can be killed by the pressure drop when they fly into the low-pressure wake of a large wind turbine blade. Birds die in accidental collisions with blades, a major concern along migratory pathways. Raptor deaths can be reduced by locating attractive perching hardware, such as ladders, inside the turbine tower rather than outside near the blades.

Both solar and wind energy land use impacts can be reduced by siting on degraded land or disturbed land of little biological value. Co-location with housing, such as rooftop collectors or with agriculture help, reduce the need to clear natural habitat to generate solar or wind power (Figure 18.3). Wind conditions in cities are not especially favorable for wind energy, and visual impacts of windmills can be substantial, so siting near demand is less feasible for wind than for solar. However, whereas solar cannot easily co-locate with agriculture because it competes with crops for light, wind is often located on grazing land and can be co-located with agriculture.

Biofuels

Biofuels are the most land-hungry of the alternative energy options (Figure 18.4), and the land requirements can compete directly for land needed for food

FIGURE 18.3 **Solar and wind co-location (Top).**
Photovoltaic panels can be co-located with other land uses, such as residential roofs or open space, reducing the impact on natural habitats. *Source: From Wikimedia Commons. (Bottom) Co-location of solar and wind with agriculture. From Juwi.*

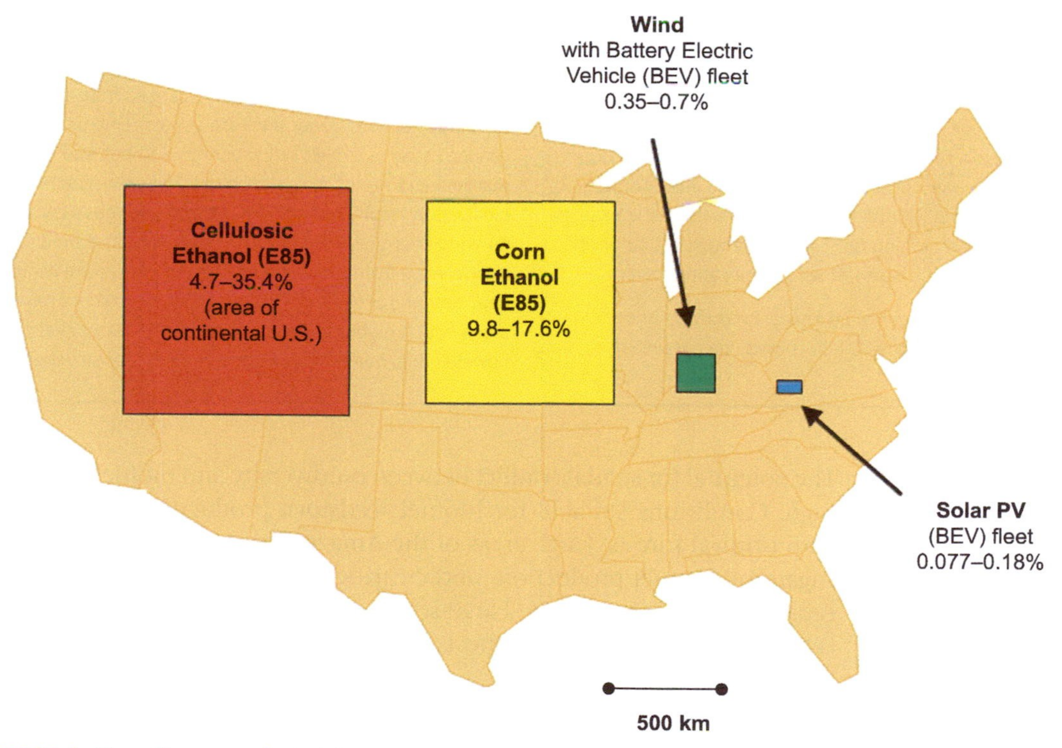

FIGURE 18.4 **Alternative energy footprint.**
The land area required to supply all transportation needs in the United States from different renewable energy sources is shown relative to the overall land area of the United States. Battery Electric Vehicles (BEV) assumed for wind and solar options. Percentages refer to the proportion of total US land area. *Source: Figure based on data from Mark Jacobson.*

production. Biofuel production has already resulted in higher corn prices in Mexico, leading to social protest. Impacts on biodiversity may be extremely high from biofuels as well because, due to social concerns, production may be pushed off of prime agricultural lands and into undisturbed habitats, where production is low and even larger areas are required per unit of production.

One 50-year wedge provided by biofuels would require 250 million ha, one-sixth of the world's croplands. Three 100-year wedges would occupy more than half as much land as is currently used for food production. Because this level of competition with food production would be unacceptable in a world with a growing population, large-scale biofuel production would have to occur on marginal lands, incurring even higher-area requirements. It is likely that biofuel production for three 100-year wedges would require more than 1 billion acres, much of it valuable wildlife habitat.

NANTUCKET OPPOSITION TO WIND

On the island of Nantucket, near Cape Cod in Massachusetts, there is strong local opposition to plans for wind energy development. Massachusetts is a traditionally liberal state with a strong history of backing action on climate change and development of alternative energy. Nantucket is a residence for well-educated, high-income families. However, local values have come into conflict with green ideals. Nantucket is known for its rustic seaside atmosphere, something not compatible with 130 futuristic, 300-foot tall, multimegawatt wind-generating towers.

Spatial conflict between wind energy production and biodiversity is moderate. The best wind sites are open, high-wind velocity areas where vegetation tends to be low. Like solar sites, however, some regions and some sites may have high-wind potential and high probability of negative impacts on biodiversity. Bird deaths caused from striking the blades of wind generators are a major environmental concern associated with wind energy. Settings such as forest margins and raptor migratory routes, where impacts on biodiversity might be high, are generally not good candidates for wind energy due to the costs of clearing and wind interference from trees. Visual impacts may make wind siting near demand problematic, pushing it toward more remote areas with a higher biodiversity value. The Nantucket project, however, has been approved and is under construction.

The potential for spatial conflict between biodiversity and biofuels is extremely high. Conditions suitable for biofuel feedstock production overlap strongly with tropical forests. Large areas of the Amazon have already been cleared for sugarcane biofuel production, and clearing for oil palm biofuels is a leading cause of forest loss in Asia. Large-scale production of biofuels would almost certainly have huge negative effects on tropical biodiversity.

The negative effect of biofuels on biodiversity may extend beyond the terrestrial realm. Increases in the anoxic dead zone in the Gulf of Mexico have already been recorded as a result of increasing fertilizer runoff from the Mississippi River due to biofuel production in the Central United States. Growth of biofuel feedstock in the oceans or harvesting of seaweed for biofuels could also have serious negative impacts on biodiversity. Clearly, despite their seeming simplicity and compatibility with current technologies, biofuels have by far the largest negative impacts on biodiversity, resulting from the combination of huge area requirements and a major spatial overlap with high biodiversity areas.

Hydropower, Tidal Power, and Geothermal Power

Hydropower provides by far the largest proportion of current energy supply of all renewable sources, and as a result, its land area requirements and biodiversity impacts are well-known. Approximately 20% of global electric demand is supplied by hydropower, and it accounts for more than half of all renewable energy production. The energy output of hydroelectric dams varies greatly with topography. Countries such as Switzerland and New Zealand get several times more energy per area than tropical countries. Using midrange yield values, an additional 25-GtC wedge from hydroelectric would require two to three million ha. This makes the land intensity for hydro dams similar to that for solar. Although only 10% of potential sites have been developed in tropical countries, these sites are lower yielding and much less climate-friendly. Decaying vegetation in tropical dams releases methane and CO_2, making them net emitters of GHG.

Spatial conflicts with biodiversity are high in tropical sites. Many tropical lowland sites are rain forest areas that are low yielding and high GHG emitting and have high biodiversity impact. Development of these sites will be destructive to terrestrial and freshwater biodiversity.

Land use impacts of geothermal energy are lower than for any other renewable technology for one 50-year wedge. Total global geothermal resources probably do not support more than one geothermal wedge. Tidal power is largely untested, and its impacts would be marine rather than terrestrial. This makes calculation of land requirements and possible biodiversity impacts difficult.

Nuclear Power

Nuclear power has a very modest footprint, much less than any renewable technology with the possible exception of geothermal. Land use is for the reactor/generation complex and for access roads. A 50-year nuclear power wedge would require about 700,000 ha, while three 100-year wedges would require about 2.1 million ha. Spatial conflict with biodiversity may exist on a site-by-site basis, but is not an overall constraint to development.

The issues that constrain nuclear energy development are fears of weapons-grade nuclear materials proliferation, consequences of accidents, and problems associated with the long-term storage of radioactive waste. These concerns make future large developments of nuclear power questionable. However, should it be developed, it could have a major beneficial climate change effect with relatively low land use and biodiversity impacts. These attributes make nuclear power one of the most polarizing and paradoxical energy sources from an environmental standpoint.

Estimating Extinction Risk

Extinction risk from climate change solutions may be calculated using the same species–area relationships used to calculate extinction risk from climate change. Smaller areas hold fewer species, in a nonlinear relationship. Because the shape of the curve is known (see Chapter 12), declining area may be equated to losses of species. The difference between the current number of species and the future number of species equates to species' extinctions.

The area needed by different energy mixes to meet global demand is a first step towards calculating extinction risk (Figure 18.5). Generally, higher land use

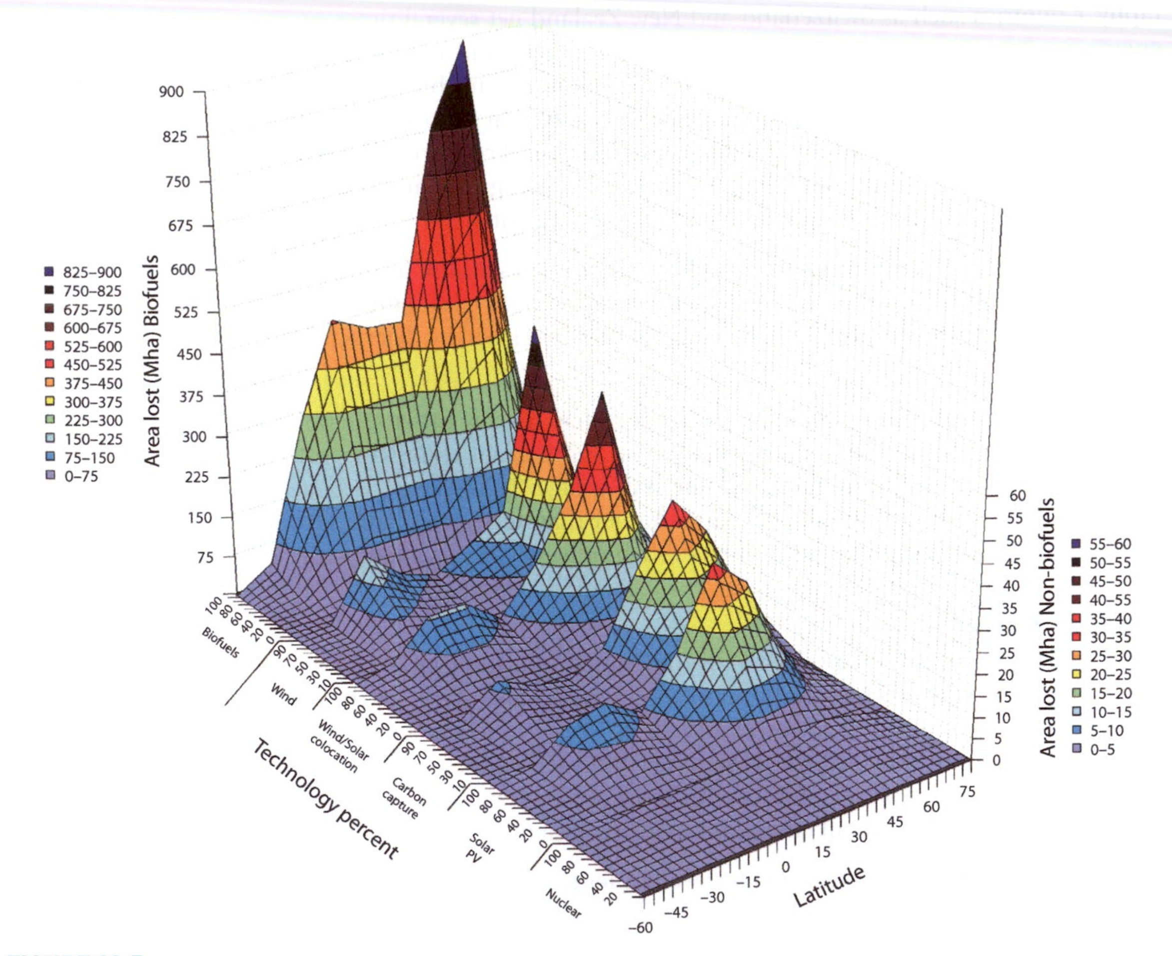

FIGURE 18.5

Energy pathways and extinction risk.Global area use of energy technologies by latitude. Average land use by latitude associated with global energy development to 2100. Land use is shown rather than biodiversity loss, because at 200 times greater impact on biodiversity for biofuels, no other technology shows on a graph of the same scale. The high-area losses for biofuels shown here combine with a high concentration of biofuel potential in the species-rich tropics to make species extinction risk from biofuels the highest of all sources by orders of magnitude. *Source: N. Snider and L. Hannah.*

demand and more tropical land use will equate to higher extinction risk. The area demand of 1000 possible alternative energy mixes to meet global demand, each in 1000 different spatial variants and assigned at random according to energy potential, has been calculated. This assessment produces a landscape of extinction risks associated with different energy pathways and spatial configurations.

Pathways with large proportions of biofuels in the mix use far more area than other energy mixes and have greater impacts in the tropics, where biodiversity is high. On average, the global area used for biofuels is 200 times greater than any other energy option. Options with high proportions of solar, nuclear, and carbon sequestration have generally low-associated extinction risks. Some high-extinction risk variants are found with most pathways, indicating that only a limited number of climate-friendly, biodiversity-friendly energy pathways exist. These must be pursued selectively and early to avoid major impacts on biodiversity.

SHORT-TERM WEDGES AND LONG-TERM PATHWAYS

The short-term potential for supply from renewable, climate-neutral sources is promising. A combination of several wedges drawn from any of a dozen or more existing technologies can provide this needed early progress in combating climate change. With careful siting and promotion of the least-damaging alternatives, impacts on biodiversity can be minimized.

Finding long-term energy pathways to stop climate change is more challenging. Up to three times more wedges are needed to meet this long-term challenge. Many low-impact sites will already have been used by the time these options are needed in the latter half of the twenty-first century, forcing new capacity toward medium- and high-sensitivity areas.

Low-impact, long-term pathways can be found, but they are not always the most readily available options. Steering economies and markets toward low-impact options requires early action: pathways not chosen now may be difficult to reenter later. Early planning and technology development can ensure that appropriate siting and technologies are available in the critical post-2050 period.

Climate change biologists concerned about the impact of both climate change and its solutions on biodiversity need to identify low-impact pathways and actions to facilitate their development. Among the technology choices that are important in these low-impact pathways are the avoidance of tropical biofuels, promotion of sequestration, co-location of solar power on existing structures, co-location of wind on existing agricultural lands, general minimization of siting in biodiversity hotspots and in high-biodiversity sites, and, perhaps, expanded deployment of nuclear energy.

FURTHER READING

Carrete, M., Sanchez-Zapata, J.A., Benitez, J.R., Lobon, M., Donazar, J.A., 2009. Large-scale risk assessment of wind-farms on population viability of a globally endangered long-lived raptor. Biological Conservation 142, 2954–2961.

Danielsen, F., Beukema, H., Burgess, N.D., Parish, F., Bruhl, C.A., Donald, P.F., et al., 2009. Biofuel plantations on forested lands: double jeopardy for biodiversity and climate. Conservation Biology 23, 348–358.

Jacobson, M.Z., Delucchi, M.A., 2009. A path to sustainable energy by 2030. Scientific American 301, 58–65.

Pacala, S., Socolow, R., 2004. Stabilization wedges: solving the climate problem for the next 50 years with current technologies. Science 305, 968–972.

Sanford, T., Frumhoff, P.C., Luers, A., & Gulledge, J., 2014. The climate policy narrative for a dangerously warming world. Nature Climate Change, 4(3), 164–166.

de Vries, B.J.M., van Vuuren, D.P., Hoogwijk, M.M., 2007. Renewable energy sources: their global potential for the first-half of the 21st century at a global level: an integrated approach. Energy Policy 35, 2590–2610.

CHAPTER 19

Carbon Sinks and Sources

Developing realistic options for reducing CO_2 in the atmosphere requires an understanding of the Earth's carbon cycle (Figure 19.1). The carbon cycle is composed of natural sinks and sources of carbon, which can be manipulated to favor removal of CO_2 from the atmosphere, or may be perturbed by human greenhouse gas emissions or attempts to sequester carbon to reduce atmospheric CO_2.

The carbon cycle plays a major role in climate and climate change. Carbon dioxide and methane are both carbon-containing compounds and major greenhouse gases. CO_2 is a major player in the global carbon cycle, as well as the largest component of human greenhouse gas emissions. The burning of fossil fuels releases carbon that has been stored for over millions of years, moving carbon from geologic stores to more rapidly-moving pools in the carbon cycle. Knowledge of these pools and movements can help in the evaluation of ways to safely remove greenhouse gases from the atmosphere and meet international mitigation policy goals.

THE CARBON CYCLE

The carbon cycle has two components: the fast carbon cycle and the slow carbon cycle. The fast cycle involves biological processes, such as photosynthesis and decomposition, while the slow cycle involves transitions of inorganic carbon, such as the weathering of rocks and soils. The slow cycle is implicated in governing climate change on a timescale of millions of years, while the fast cycle participates in decadal to millennial climate changes.

The main natural source of CO_2 in the atmosphere is from releases of volcanic gases from mid-ocean ridges and volcanoes. CO_2 is released from subduction zones at the margins of tectonic plates when carbonate rocks are subjected to intense pressure and temperature. CO_2 released into the atmosphere dissolves in the oceans, and is taken up by living organisms. Some of the major global carbon pools are thus atmospheric CO_2, dissolved CO_2 and its ions in the oceans, and carbon in living matter.

Climate Change Biology. http://dx.doi.org/10.1016/B978-0-12-420218-4.00019-6

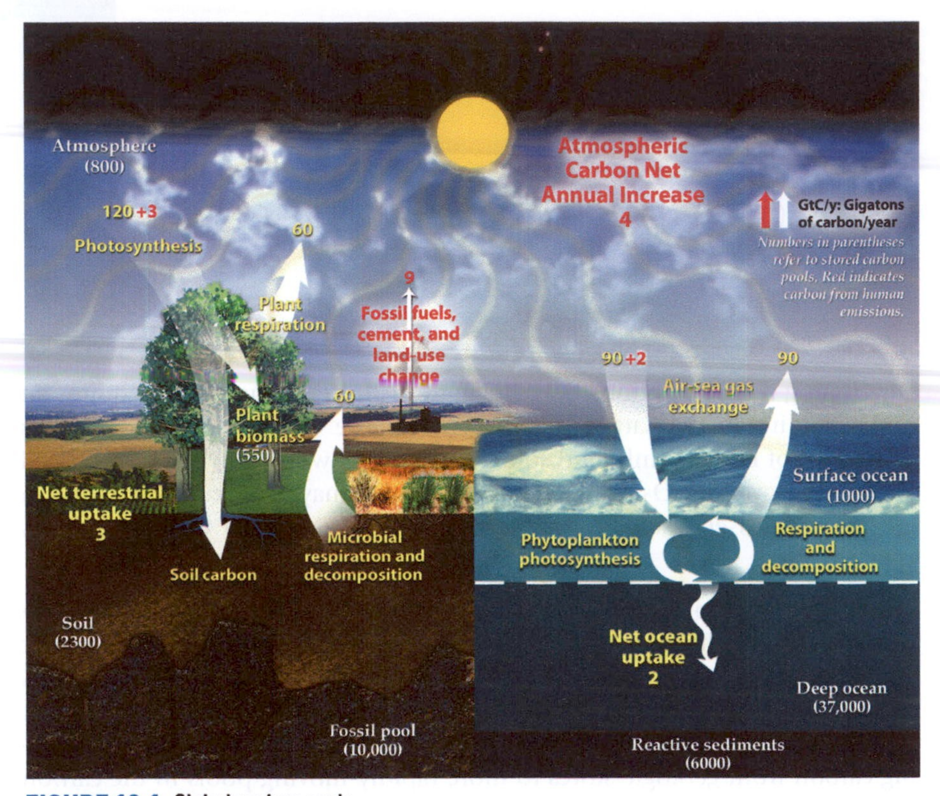

FIGURE 19.1 Global carbon cycle.
Terrestrial and marine pools and fluxes (arrows) are shown, indicating both natural sources and sinks (white) and the fate of CO_2 from the burning of fossil fuels (red). *Source: DOE.*

SLOW CARBON

Carbon is moved from fast pools into long-term storage, when carbonate rocks are formed from the skeletons of corals or the shells of marine organisms, or when sediments that contain organic material are converted into rock. Coals and black shales are especially important in storing organic carbon of terrestrial origin, while carbonate rocks, such as limestone, are particularly important in storing carbon of marine origin (from corals or shells).

These long-term processes of release and deposition unfold over millions of years, and are therefore known as the slow carbon cycle to contrast them with the higher turnover processes, such as biological uptake and decay in the fast carbon cycle, which can occur in years or seasons.

The slow carbon cycle impacts climate in important ways. As rocks weather, silicate minerals ($CaSiO_3$) combine with CO_2 to create calcium carbonate ($CaCO_3$) and silica (silicon dioxide–SiO_2). This removes CO_2 from the atmosphere, which results in a cooling effect on climate because CO_2 is a greenhouse gas. This cooling is generally offset by warming caused by increases in CO_2 from volcanic outgassing and other sources, but geologic events can be in imbalance, resulting in a net warming or cooling in climate.

For instance, the uplift of the Himalayas about 50 million years ago began a slow shift in global climate. As the Indian subcontinent collided with Asia, billions of tons of previously buried rock were exposed in developing mountain slopes and ridges. These rocks, most of them silicates, began to weather, producing silica and removing CO_2 from the atmosphere and the oceans.

The drawdown of atmospheric CO_2 caused a slow cooling of the planet until the modern northern and southern ice caps were formed. The circum-Antarctic ocean current was established about 34 million years ago, which isolated Antarctica from the global climate and accelerated ice cap formation. The cooling from the Himalayan uplift continued, and the associated cooling eventually led to the initiation of the Pleistocene ice ages, with their glacial and interglacial periods. We are in the latest and possibly the last of the interglacial periods, so the Himalayan uplift played a very direct role in the evolution of modern climate.

Other periods of past mountain building and volcanism have resulted in global cooling or warming on geologic timescales. The details of many of these events are mysteries of deep time. But the lesson that CO_2 can drive global climate change is very relevant today, as human greenhouse gas pollution is rapidly impacting the fast carbon cycle.

FAST CARBON

The fast carbon cycle is the cycle of living things. Plants use carbon to make sugar for energy and carbohydrates for structure and energy storage. Animals eat plants for energy and nutrition. Both plants and animals break up sugars and carbohydrates for energy, releasing CO_2 back into the atmosphere in the process.

Most of these actions take place in terms of minutes to decades, a blink of an eye in the geologic context of the slow carbon cycle. Some biological pools, such as large trees, may last centuries. Biological fluxes may move carbon in space, as well as time, or from the fast cycle to the slow cycle, as when plant material is washed to the sea, deposited in ocean sediments, and incorporated into sedimentary rock.

Respiration, decay, and fire are important parts of the fast cycle. Plants and animals respire organic carbon, breaking carbon bonds to get energy. Respiration takes organic carbon that plants have fixed in photosynthesis and combines it with oxygen to produce energy, CO_2 and water. Plant and animal material that is not respired decays when the organism dies, returning carbon (most often as CO_2 or

CH_4) to the atmosphere as it is decomposed by microorganisms. Fire can convert organic material directly to CO_2 without the need for any living intermediaries.

The fast carbon cycle has both terrestrial and ocean components. The terrestrial component involves grasslands, forests, and other vegetation types. The ocean component involves phytoplankton, zooplankton, and marine organisms. The greatest amount of carbon in the terrestrial part of the cycle is in plants, while the greatest amount of carbon in the marine part of the cycle is in phytoplankton and zooplankton.

The fast cycle affects climate change and climate affects the fast cycle. Enhanced rates of decay or fire may increase atmospheric CO_2, while increased growth of plants can reduce atmospheric CO_2. There are inherent feedbacks in the system, both positive and negative, because warmer climates and enhanced atmospheric CO_2 can stimulate photosynthesis, while climate can influence the role and frequency of fire.

Human actions are altering the fast carbon cycle, with potentially profound implications. Greenhouse gas pollution is altering climate and atmospheric CO_2 levels. Land use change, the clearing and burning of forests to make way for agriculture, is releasing carbon into the atmosphere as CO_2. Livestock raising and agriculture may result in carbon cycle dynamics that are very different from those of natural systems, sometimes increasing and sometimes decreasing net releases of CO_2 and methane.

For these reasons, it is important to understand the fast carbon cycle in more detail. Two major components of the fast carbon cycle—ocean and terrestrial—have very different pools and fluxes. Understanding these pools and fluxes allows us to add up the net carbon balance of the planet and understand the impacts of human greenhouse gas pollution.

OCEAN CARBON CYCLE

The oceans hold about 50 times more carbon than the atmosphere, and have taken up about 30% of the greenhouse gases emitted by human sources so far. This makes the oceans immensely important in the global carbon cycle, and in the balance of CO_2 in the atmosphere. Oceans play a key role in determining how much warming results from human pollution.

Biological processes central to the ocean carbon cycle are ecosystem photosynthesis, respiration, and the biological pump. Phytoplankton in the oceans take up CO_2 in photosynthesis and release oxygen. Respiration, both in phytoplankton and zooplankton, uses oxygen and releases CO_2. The CO_2 uptake in photosynthesis and release in respiration is balanced, except for carbon incorporated into organisms' bodies or shells. On land, the organic matter of dead organisms

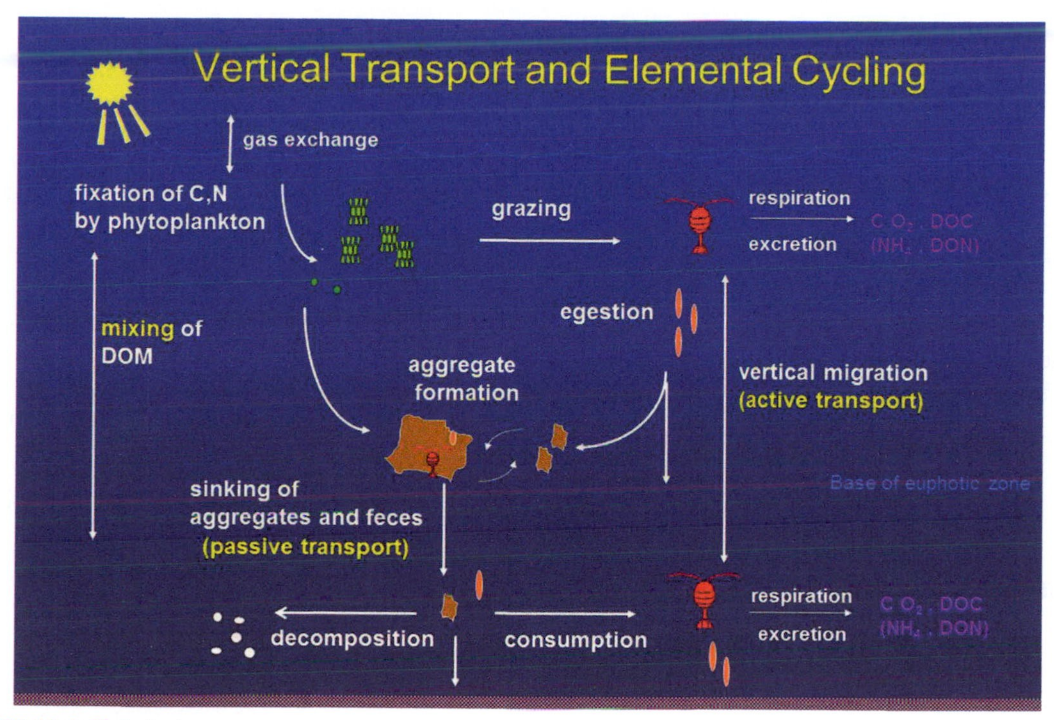

FIGURE 19.2 **Biological pump.**
The biological pump moves CO_2 from near-surface waters to depth. Phytoplankton fixes CO_2. Phytoplankton and zooplankton that are not consumed or decomposed fall through the water column to the ocean floor, where they may become incorporated into sediments and enter the slow (geologic) carbon cycle. *Source: Steinberg et al. (2012).*

is decayed, releasing carbon back to the atmosphere. In the oceans, there is less decay at the surface, and organic matter and shells sink to the depths.

The biological pump is driven by these sinking remnants of organisms (Figure 19.2). The carbon in this "rain of detritus" joins CO_2 in deep ocean waters, or enters the long carbon cycle as it is incorporated into carbonate or sedimentary rocks. This biological pump transfers carbon from the ocean surface to depth.

Important physical processes in the ocean carbon cycle are mixing and the solubility pump. Once CO_2 from the atmosphere is dissolved in surface water, its movement away from the surface plays a key role in limiting further ocean uptake from the atmosphere. If saturated waters remain near the surface, little further uptake is possible. If saturated water moves away from the surface to deeper waters, more uptake is possible. Near the surface, wind is the main force that moves surface water to depth, and this process is referred to as mixing. In zones of strong mixing, CO_2-saturated surface water is continually moved away from the surface, being replaced by undersaturated water in which more CO_2 from the atmosphere can dissolve.

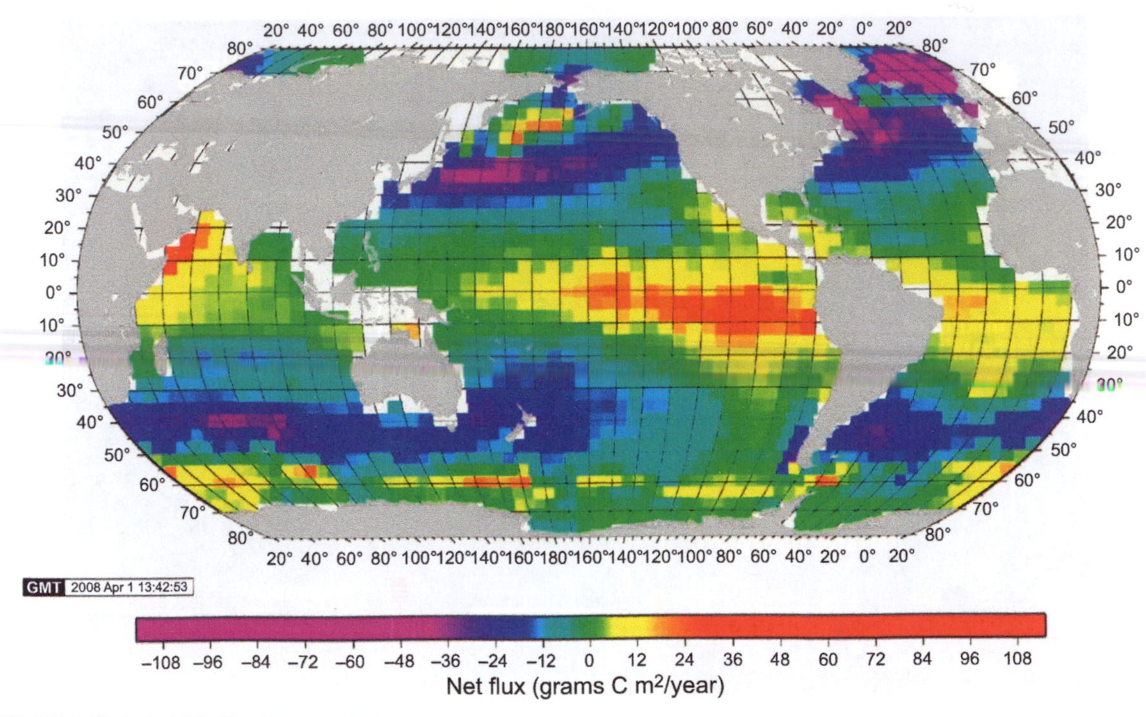

FIGURE 19.3 **Net CO_2 flux from oceans.**
The solubility pump results in CO_2 moving from cold waters near the poles where CO_2 solubility is high (blue-purple), through the thermohaline circulation to upwelling regions in the tropics. The lower solubility of CO_2 in warm water results in a CO_2 release in the tropics (yellow-red). *Source: NASA.*

The solubility pump is created by deep water mixing and involves the effect of water temperature on CO_2 solubility. Warm water holds less dissolved gas than cold water, so cold waters near the poles can hold greater amounts of CO_2 (mostly as carbonate and bicarbonate ions) than warm waters in the tropics. Thermohaline circulation (Chapter 2) then moves these waters, creating a pump. The dense, salty, and cold water sinks near the poles, circulates in the deep oceans, and rises again near the equator in a process that can take hundreds or thousands of years (Figure 19.3). Deep water rises in the tropics, where it warms and releases CO_2 because of the lower CO_2 solubility in warm water. The net effect is pumping CO_2 from the surface to deep ocean layers and from high latitudes to the tropics. Both the biological pump and solubility pump move carbon from surface waters to deep waters, so deep ocean water contains much more carbon than surface waters.

Carbon movement in the ocean is slow, even in the fast cycle, and the quantities involved are immense. The oceans contain about 36,000 gigatons of carbon; more than 90% in the form of bicarbonate. In any one year, the ocean and atmosphere exchange about 90 gigatons of carbon, equivalent to about a quarter of one percent of the oceans' carbon. The flux between the atmosphere and ocean is small compared to ocean pools because of the long residence time of carbon in the oceans. The residence time of dissolved inorganic carbon (including CO_2) is on the order of 80,000 years. The residence time of deep ocean water is between 200 and 2000 years. In contrast, the transit time of carbon in the bodies and shells of plankton is relatively rapid, taking only a few days or weeks.

TERRESTRIAL CARBON CYCLE

The terrestrial carbon cycle takes place almost entirely within the layer of living plants and animals at the surface of the Earth. This layer is very thin in relation to the depth of the oceans. Within this zone, carbon in CO_2 is taken up into living things, combined with oxygen to produce energy (releasing CO_2), and/or decomposed. Most of these processes take place in terms of minutes to decades, relatively rapidly in comparison to the long residence times in the oceans. However, some terrestrial organic matter, such as lignin, decomposes very slowly and has long residence times in the soil.

The two major carbon pools of the terrestrial carbon cycle are in biomass and soil. Biomass is further subdivided into aboveground (or standing) biomass and belowground biomass. Aboveground biomass is comprised of trunks, branches, and leaves, while belowground biomass is composed of roots (Figure 19.4). Carbon in biomass has residence times of hundreds of years in large trees and much shorter times in most other types of organisms. Soil carbon pools result primarily from the accumulation of plant matter, and generally have longer residence times than in the biomass pool.

Photosynthesis can be viewed as the beginning of the terrestrial carbon cycle. CO_2 is taken up by plants through stomata in leaves, and combined with water in one of several photosynthetic pathways, producing carbohydrates and O_2. The photosynthetic process builds up organic matter that is then acted on by the processes of respiration, decay, and fire.

Organic matter that is not respired by plants or animals decays when organisms die. Decomposition by bacteria, fungi, and insects releases carbon from dead plant and animal material. The decomposition products are mostly CO_2 or methane (CH_4), thus returning carbon to the atmosphere. Plant and animal material that is not decomposed stays in the soil pool (Figure 19.5) until it is decomposed or enters the long carbon cycle by being incorporated into rock.

Fire releases CO_2 directly back into the atmosphere from plant material. In most natural systems, fire follows a cycle in which CO_2 released by fire is taken back up by regrowing vegetation. For instance, in temperate grassland systems, summer fire releases CO_2 that is removed from the atmosphere when the grass regrows in spring, so there is no net change in atmospheric CO_2 in the long

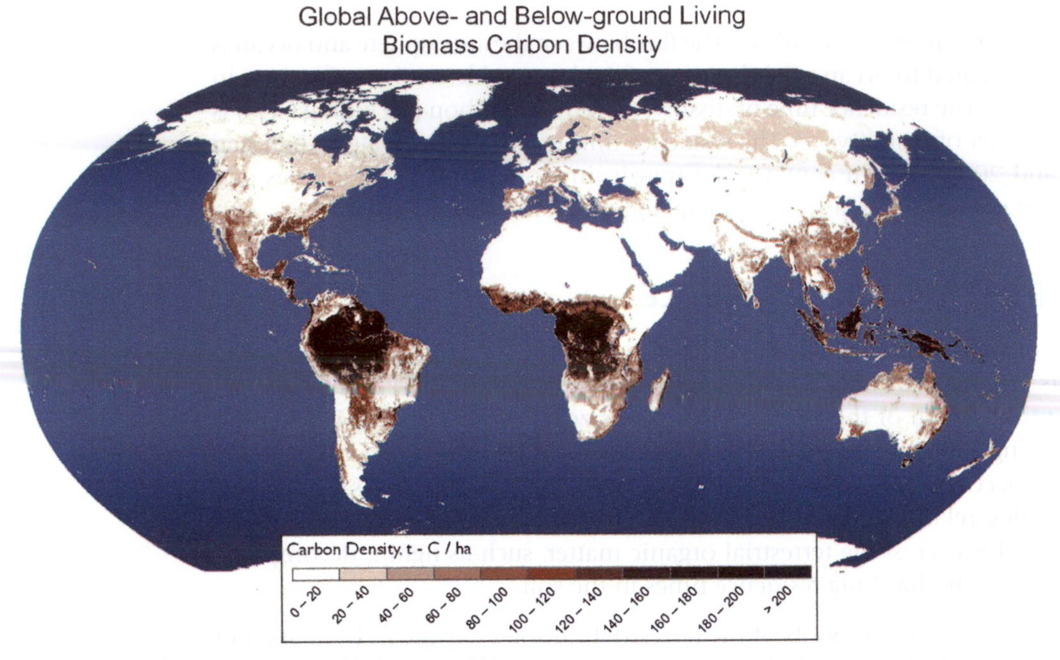

FIGURE 19.4 **Carbon in aboveground and belowground biomass.**
Carbon is fixed from CO_2 and stored in vegetation aboveground (woody material and leaves) and belowground (roots). The aboveground store is quickly released if vegetation is burned. The belowground store can be a large fraction of aboveground carbon and is released much more slowly after a fire or disturbance. *Source: IPCC.*

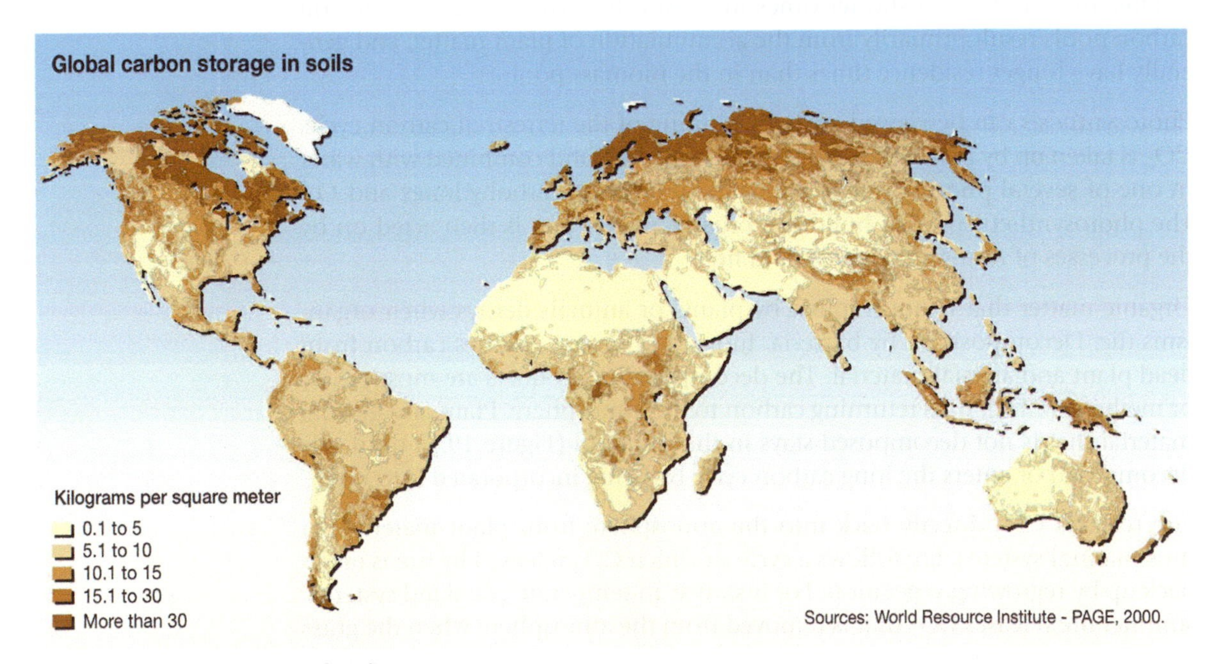

FIGURE 19.5 **Carbon storage in soils.**
Carbon stored in soils can have long residence times. Carbon density is particularly high in peat soils at high latitudes and in tropical forests. *Source: World Resources Institute.*

term. Other fire systems have longer return intervals, but still result in no net increase in atmospheric CO_2 when measured over decades.

Fire associated with human activities may result in net change in the amount of carbon in biomass and the atmosphere, however. When a forest is burned and converted to agricultural fields, for instance, the release of CO_2 into the atmosphere is no longer balanced by a CO_2 uptake in regrowing natural vegetation, so there is a net increase in atmospheric CO_2. Crops growing in place of the forest may take up some carbon, but crop plants generally store much less carbon than mature trees. This conversion of natural vegetation to human uses (including cities, agriculture, and dams) is known as land use change, and it is a major contributor of CO_2 and methane to the atmosphere.

HUMAN INFLUENCE ON THE CARBON CYCLE

Humans have released huge amounts of CO_2 into the atmospheric carbon pool by burning fossil fuels that previously sequestered carbon. Chapter 2 outlined the impact of human CO_2 emissions on the climate system. Because CO_2 contains carbon, its release into the atmosphere has implications for the global carbon cycle as well.

From the Industrial Revolution to the present, the burning of fossil fuels has added about 330 billion metric tons (330 Pg) of carbon to the Earth's atmosphere. Emissions from land use change, mostly forest clearing and burning, added about 158 million metric tons (158 Pg) over the same period. CO_2 from land use change outside the tropics was significant up until about 2000, but since that time almost all land use change emissions have come from the tropics.

Each year, current fossil fuel emissions add another 8.4 billion metric tons (8.4 Pg) of carbon to the atmosphere. Land use adds another 1.5 billion metric tons (1.5 Pg), for total annual CO_2 emissions of about 10 billion metric tons (9.9 Pg). Total emissions are increasing by about half a billion metric tons per year and have increased by 35% since 1990.

Where does all that CO_2 go? The atmosphere is the greatest sink for the added human CO_2, retaining about 45% of emissions (Figure 19.6). The ocean and terrestrial carbon pools take up about a quarter each (30% to land sinks and 24% to ocean sinks in 2006).

While these carbon additions have huge implications for climate change, their magnitude in relation to natural carbon pools and fluxes is not large. The annual human emissions of about 10 Pg of carbon are less than 10% of the annual terrestrial or ocean carbon exchange with the atmosphere, and less than 5% of terrestrial and ocean fluxes combined. The 330 Pg of total cumulative human emissions since the Industrial Revolution is small in comparison to the Earth's large carbon pools, particularly that of the

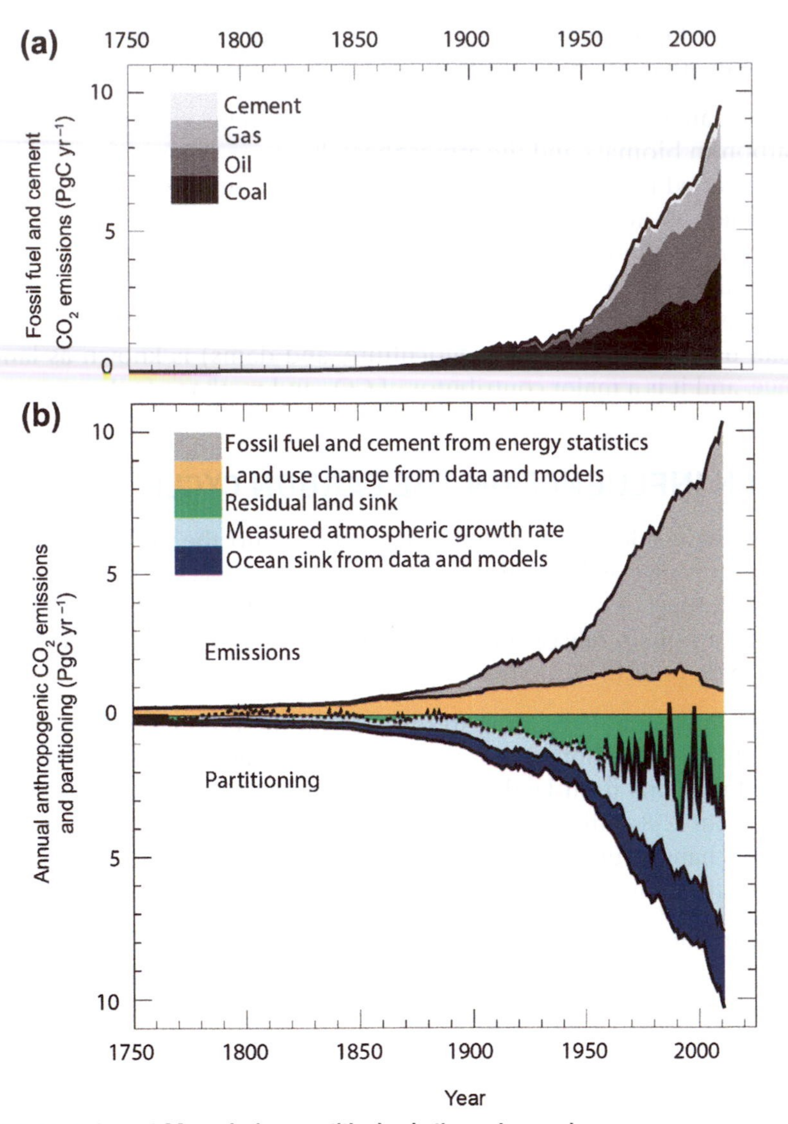

FIGURE 19.6 **Annual CO_2 emissions partitioning in the carbon cycle.**
Annual CO_2 emissions from fossil fuel use and land use change (upper half of panel b) are balanced by partitioning into the atmosphere, ocean, and land sinks (bottom half of panel b). Panel a shows breakdown of fossil fuel emissions by source type. *Source: IPCC. See also Canadell et al. (2007).*

deep ocean, which alone contains 3700 Pg of carbon. Soils house 2300 Pg of carbon. The cumulative human emissions are only about 1% of the total global combined atmospheric, ocean, and terrestrial carbon pools.

However, human emissions are large in relation to atmospheric pools, and this is where the emitted carbon is important in influencing climate. The 330 Pg of total cumulative human emissions is almost half the size of the total atmospheric carbon pool (over 800 Pg carbon). Since about 45% of human emissions remain in the atmosphere, the cumulative amount of carbon that remains in the atmosphere is about 150 Pg carbon, or about 20% of the atmospheric carbon pool.

When CO_2 from human pollution is added to the atmosphere, the fluxes, pools, and pumps of the cycle come into play. Increased CO_2 in the atmosphere is dissolved into ocean waters, where it is moved away from the surface by surface mixing and moved to the depths in thermohaline circulation in the solubility pump.

The solubility pump will take CO_2 and move it to the deep ocean water. As atmospheric CO_2 levels rise due to human causes, more CO_2 dissolves in seawater and more is moved to depth in the solubility pump. While that deep ocean water will eventually resurface and release CO_2 in the tropics, that takes hundreds or thousands of years, so the solubility pump is an important near-term sink for CO_2 added to the atmosphere by human pollution.

On the other hand, the biological pump does not play a strong role in modulating atmospheric CO_2 from human sources. This is because CO_2 is not the limiting factor for photosynthesis in the oceans. As a result, higher CO_2 due to human pollution does not change ocean photosynthesis rates, leaving the rate of the biological pump unaffected.

CO_2 that remains in the atmosphere stimulates photosynthesis in plants, enhancing the terrestrial uptake of carbon. Since CO_2 is a fundamental input for photosynthesis, increased concentrations of CO_2 in the atmosphere speed up the photosynthetic reactions. As cool, high- latitude regions warm, photosynthesis increases due to the accelerating effect of temperature on chemical reactions. Growth of trees and other vegetation is stimulated by higher CO_2 and temperature, resulting in greater carbon storage in plants. Globally, terrestrial systems have been a major carbon sink, soaking up about a quarter of the carbon emitted into the atmosphere as CO_2 from the burning of fossil fuels.

RECENT TRENDS IN TERRESTRIAL SOURCES AND SINKS

Some recent trends in carbon sources and sinks have important implications for how the carbon cycle responds to the input of CO_2 from human sources. Some regions that have been carbon sources over the past two centuries are transitioning to carbon sinks, while other areas have recently switched from sink to source.

Figure 19.7 illustrates the change in Net Primary Production (NPP) in different parts of the globe from 2000 to 2009. NPP is the amount of carbon accumulated in plant biomass through photosynthesis. Areas of increasing NPP are storing carbon and are carbon sinks. Areas of decreasing NPP are releasing carbon and are carbon sources.

Drought over large areas is responsible for turning some regions from sinks to sources. As drought continues, plants have less water to use in photosynthesis and NPP declines. This is happening across large areas in South America, Australia, Southern Africa, and Southeast Asia. The Amazon, in particular, is turning from a carbon sink to a carbon source. Overall, in the Southern Hemisphere NPP fell over 70% of the land area from 2000 to 2009.

But in the Northern Hemisphere, forests are regrowing in areas formerly cleared for agriculture, resulting in increased NPP. In the Northern Hemisphere, NPP increased on 65% of vegetated land. High-latitude forests, whose growth was formerly constrained by cold, are increasing in NPP as climates warm. These northern carbon sinks have been overwhelmed by Southern Hemisphere

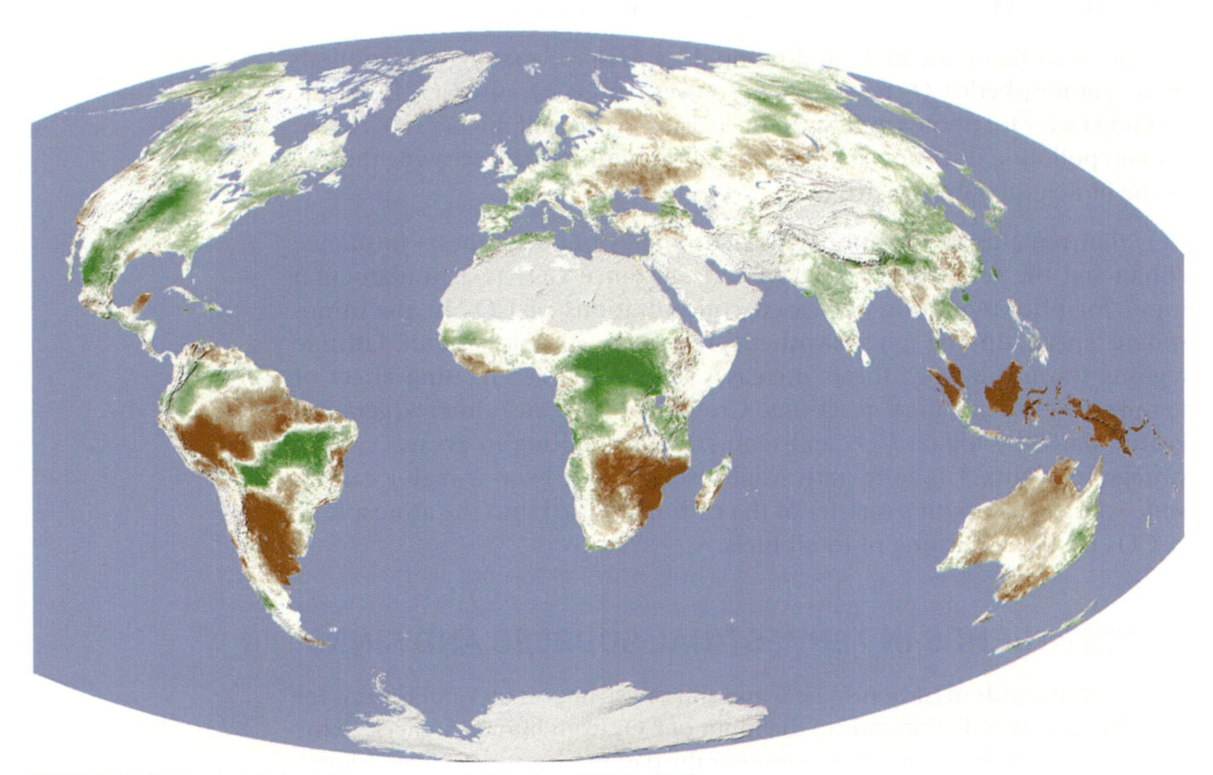

FIGURE 19.7 Trends in terrestrial sources and sinks from 2000 to 2009.
Recent trends in NPP show that the Amazon has turned from a carbon sink to a carbon source due to drought and fire. Sources are shown in brown, sinks in green. High-latitude forests and Central Africa remain sinks, while subtropical vegetation and forests in Southeast Asia are now sources. So while warming can promote greater plant growth leading to sinks, it can also promote drought and fires that convert sinks to sources. *Source: Zhao and Running (2010).*

systems, such as the Amazon, turning from sink to source so that terrestrial systems are a net source of CO_2.

CARBON CYCLE AND CARBON SEQUESTRATION

Moving CO_2 from the atmosphere to terrestrial or ocean carbon pools would eliminate its effect on global climate. Since ocean and terrestrial systems already absorb more than half of human CO_2 emissions, and ocean and terrestrial carbon pools are orders of magnitude greater than annual human emissions, moving carbon from the atmosphere does not seem too far-fetched. This movement from atmospheric carbon pools to pools that do not influence climate is known as carbon sequestration.

AMAZONIAN FOREST CHANGE

Large amounts of carbon are being lost from tropical forests around the world due to land use change. Among the leading causes of tropical forest loss are clearing for small-scale agriculture and clearing for large plantation crops, such as oil palm and soybeans. In some cases, this clearing can lead to feedback loops that result in further burning and the release of CO_2 to the atmosphere (for example, in peat soils, see below).

Perhaps the largest of these feedback loops is that affecting the forests of the Amazon basin. Amazonian forests naturally transpire large amounts of water vapor, which helps sustain rainfall deeper into the basin (see Chapter 5), a positive feedback that is disrupted when the forest is cleared.

Clearing of the Amazon has been taking place for decades, and has been recently decreasing due to new policies and the enforcement of clearing restrictions. However, the cumulative damage of years of forest loss and drying may have already triggered negative feedback involving fire, drought, and degradation.

As the Amazonian forest has been cleared, it has been converted from a nearly continuous cover of high canopy trees to large and small fragments, as well as large areas untouched by clearing. Smaller forest fragments are subject to drying from the edges, slowly deteriorating the wet forest conditions in the fragment interior.

Fire has entered the Amazonian system along with human land uses. Even low-intensity ground fires can kill wet tropical forest trees because they have thin bark not naturally adapted to survive fire. After several burning cycles, trees are girdled by fire and die.

The combination of fragmentation, fire, and climate change is converting large areas of the Amazon from continuous wet forest into remnants of degraded forest. As trees are lost to land clearing, logging, fire, and drought, their carbon is released into the atmosphere, accelerating climate change and exacerbating droughts. Whether recent advances in reducing deforestation in the Amazon will be in time to prevent this spiral of degradation remains to be seen.

To place an upper bound on the problem, we might first ask if people can keep burning fossil fuels without using up all the oxygen in the atmosphere. Atmospheric O_2 exists because organic matter produced by photosynthesis over hundreds of millions of years has not been completely recombined with oxygen in decomposition, instead remaining as hydrocarbons in rocks (trees are the lungs of the planet only on geologic timescales). If all of those hydrocarbons could be

TROPICAL PEAT

Some soils, particularly peat soils, decompose very slowly. As a result, peat accumulates dead plant matter and stores carbon. While peatlands cover only about 3% of the surface of the Earth, they store more carbon than all of the world's forests. Peatlands store as much as 500 billion metric tons of carbon, and are one of the world's richest carbon stores on an area basis, storing an average of about 1500 metric tons of carbon per hectare. Tropical peat may account for 10–20% of the total carbon stored in the world's peatlands.

This carbon remains locked in the peat until land use of climate change (or both) results in drying of the peat. When the peat dries, it becomes highly flammable. Fires in deep peat can burn underground for weeks or months, releasing the carbon stores from the soil. Intact peat swamps emit methane, a greenhouse gas, but this emission is small in relation to the huge amounts of CO_2 that are released when peat is cleared, dried, and burned.

Southeast Asia harbors large areas of peat soils that have developed in flooded or periodically inundated tropical forests. The cycle of destruction in Southeast Asia is similar to that in the Amazon, with large-scale forest clearing, drying, and burning. What is different in Southeast Asia is that the forest often sits on peat. When these forests over peatlands are cleared and burned, the peat can dry and catch fire.

Peat swamp forests cover 26 million hectares in Southeast Asia, an area the size of South Africa. Many of these forests have now been cleared and drained. When fire is used to clear adjacent forests, it may ignite the drying peat, starting long-burning, difficult-to-control fires. Peat fires can burn for months, as in Sumatra in 2014. 25 million hectares of peat forest burned in Indonesia in a single year in the 1997–1998 El Niño event. The potential annual greenhouse gas emissions from peat forests in Southeast Asia are more than the emissions from 89 million cars.

burned, the residual built-up oxygen we rely on to breathe would be consumed. As it turns out, only a small percentage of geologic hydrocarbons are in forms such as coal, oil, or tar sands that can be accessed and used for fuel. So even using generous estimates of global fossil fuel reserves, it is clear that burning fossil fuels will not make the planet uninhabitable in terms of breathable O_2.

Once it is clear that burning fossil fuels will not use all the oxygen needed for human survival, the main concern from the use of fossil fuel becomes the effect of CO_2 emissions on global climate. One way to reduce this concern is to sequester carbon released by fossil fuel use in nonatmospheric carbon pools. A big-picture consideration for sequestration is the residence time of carbon in different pools. A sink that returns carbon to the atmosphere in years or decades does little to reduce exposure to climate change. But a sink that holds carbon for centuries buys time across multiple generations. Even though the carbon may reemerge into the atmosphere, causing climate change for future generations, perhaps by that time more sophisticated sequestration methods may have been devised. A centuries-long sink is worth considering, but one must bear in mind that it has consequences for future generations. A sink that holds carbon for thousands or tens of thousands of years is roughly equivalent to a permanent sink on human timescales and is a sound long-term sequestration option.

Unfortunately, most carbon pools that can be readily manipulated are small or have short residence times (Figure 19.8). Forest carbon pools can be increased by restoring forests. But global carbon in forests is only about 300 Pg, or 30 years of annual CO_2 emissions. Since around 30% of global forest cover has been lost to clearing, even the most ambitious forest sequestration program would only store a few years of global CO_2 emissions. The potential for sequestration in restored grasslands is even smaller and the potential in soils restoration, while much larger, would take decades to accumulate. While solutions involving these terrestrial pools may provide valuable short-term sequestration, longer-term options are needed to mitigate climate change.

One proposal for longer-term storage is to prime the biological pump in the oceans. Photosynthesis is limited in the southern oceans by iron. Fertilizing

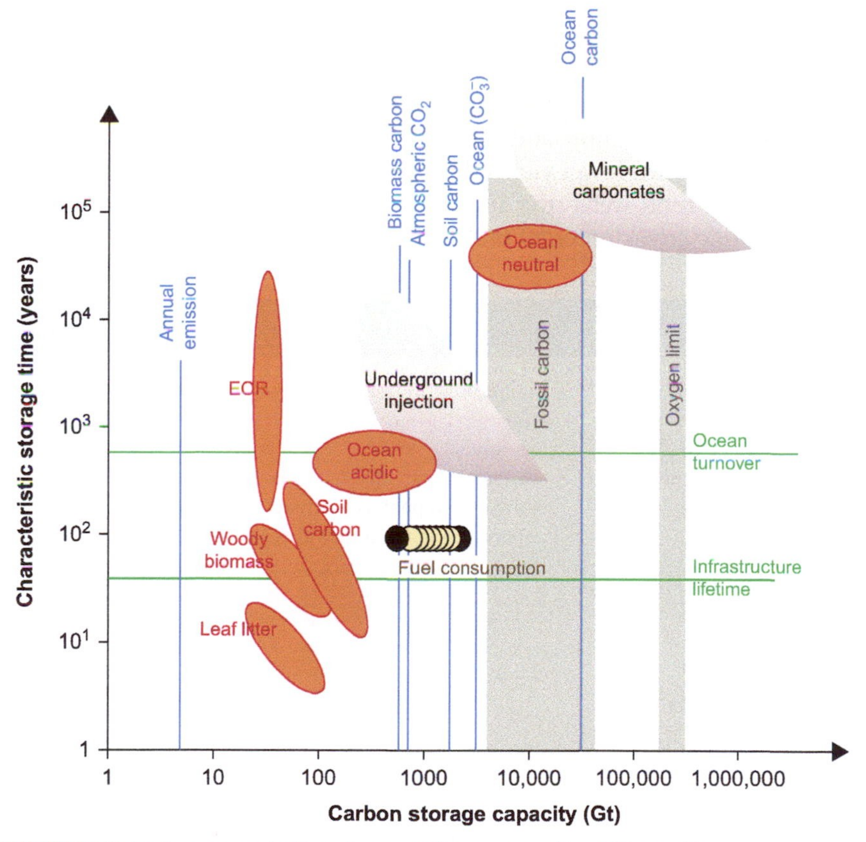

FIGURE 19.8 Carbon sequestration storage options—capacity and longevity.
On log scales, capacity and storage time are charted for each option. For instance, Enhanced Oil Recovery (EOR) injections of CO_2 into spent oil wells has a low capacity but moderate and variable storage time. Biomass and soil carbon have a low capacity and short residence times (fast carbon cycle), so are part of short-term solutions, but need to be coupled with long-term solutions with residence times typical of the geologic carbon cycle. Ocean acidic is the deep injection of liquid CO_2 into ocean bottom waters. Evolving fossil fuel consumption, total fossil fuel reservoirs (including oil sands and shales), and the total oxygen content of the atmosphere (oxygen limit on fossil fuel burning) are shown for comparison. *Source: Lackner (2003). Reproduced with permission from AAAS.*

OCEAN FERTILIZATION

One natural approach to CO_2 sequestration is not as green as it appears. CO_2 is taken up by billions of tons of algae in the Earth's oceans. Especially in southern oceans, this algal activity is limited by a lack of iron. Initial calculations indicated that adding iron to surface waters of the southern oceans could stimulate algal blooms and sequester large amounts of CO_2. However, even small-scale tests initiated major changes in ocean food webs, the full consequences of which are not completely understood. Large-scale iron fertilization would impact the base of the marine food web in one of the world's most ecologically fragile marine environments, and would cost more per unit of carbon removed than many alternative technologies.

surface waters with iron, therefore, produces algal blooms. As the algae die, their bodies would join the detritus rain of the biological pump, moving CO_2 from surface seawater into the deep ocean slow carbon cycle, where it might reside for hundreds or thousands of years. However, experiments with iron fertilization have shown lower yields than predicted by theory and interfere with natural marine food webs over thousands of square kilometers, thus risking the trade of one environmental problem for another.

SPOTLIGHT: FROZEN CARBON

Very large amounts of carbon are frozen in peat soils at high latitudes. Frozen arctic soils harbor about twice as much carbon as the atmospheric carbon pool. The thawing of frozen peat could put more CO_2 into the atmosphere than all human fossil fuel emissions.

However, as tundra thaws, photosynthesis and respiration increase in vegetation. This results in the uptake of CO_2 from the atmosphere. Early indications from studies of whole landscapes show that the CO_2 uptake of tundra plants more than balances the carbon released from thawing peat, leaving tundra as a net carbon sink.

But because the volumes of carbon involved are so large, tundra carbon is critical to future atmospheric CO_2 levels. The fate of frozen peat and the balance between carbon uptake by plants and carbon release from peat as tundra thaws is of immense importance to the future of climate change.

Source: Schädel, C., Schuur, E.A., Bracho, R., Elberling, B., Knoblauch, C., Lee, H., et al. 2014. Circumpolar assessment of permafrost C quality and its vulnerability over time using long-term incubation data. Global Change Biology 20 (2), 641–652.

If biological solutions are too risky or short-lived, are there physical means of moving carbon into long-term pools? Here again sequestration involving deep ocean pools has been suggested. CO_2 becomes liquid at the high pressures prevailing in the deep oceans, so one suggestion has been to capture CO_2 emissions and inject them into the deep oceans at high pressure. Again, impacts on ocean ecology are a concern. CO_2 from the deep ocean liquid pools would enter the thermohaline circulation, eventually bringing it back to the surface, but more importantly, altering the ocean pH and chemistry for deep ocean-dwelling organisms in unknown ways.

SPOTLIGHT: DEEP-SEA BLUES

Injection of CO_2 into the deep ocean has been suggested as a means of sequestering CO_2 from the atmosphere—a solution that has profound potential environmental consequences. CO_2 is a liquid at high pressures, and it would remain a liquid under the high pressures of the deep ocean. This prevents rapid dissolution back into the water column, which would defeat the purpose of deep ocean sequestration. CO_2 can be captured at power plants or other sources to reduce the buildup of GHG, or future technologies may allow scrubbing from the atmosphere to reduce atmospheric concentrations. CO_2 from either source might be liquefied and injected into the deep ocean, where it would be relatively stable. Seibel and Walsh (2001) studied the potential impact of this technology on life of the deep oceans. Apart from the organisms that would be killed directly in the zone occupied by the liquefied CO_2, slow leakage will result in changes in the acidity of deep sea waters because CO_2 dissociates into carbonic acid and hydrogen ions within the cells of deep-sea organisms. Deep-sea organisms may be very sensitive to changes in pH. They have metabolic rates as much as three orders of magnitude lower than those of shallow-dwelling organisms, which reduces their tolerance to acid–base changes. For example, deep-sea fishes have a low capacity for active ion regulation. Metabolic shutdown and protein synthesis inhibition are further possible complications. Models of CO_2 ocean disposal indicate that the pH of the entire ocean could be changed by 30% (0.1 pH unit) by disposal sufficient to stabilize atmospheric CO_2 concentrations at 550 ppm.

Source: Seibel, B.A., Walsh, P. J. 2003. Biological impacts of deep-sea carbon dioxide injection inferred from indices of physiological performance. Journal of Experimental Biology 206 (4), 641–650.

Geologic sequestration on land may be more promising (Figure 19.9). CO_2 is already injected into underground formations in enhanced oil recovery operations, proving that underground injection and containment is possible. Large underground geologic formations exist that can either retain CO_2 (such as salt domes) or neutralize CO_2 (as carbonates or bicarbonates). But salt domes and saline aquifers have a limited capacity, and neutralization entails additional costs.

A problem faced by all geologic sequestration options is recapturing CO_2 that has been released to the atmosphere. CO_2 can be captured at the point of emissions for large plants and point sources, the first step in CO_2 capture and storage (CCS) initiatives (see Chapter 18). But for small or mobile sources,

SPOTLIGHT: OCEAN FERTILIZATION EXPERIMENT

For ocean fertilization to work, a relatively small amount of iron must cause large changes in algal productivity, and the algal bloom generated by iron fertilization needs to sink to the seafloor and be entombed in sediment, taking the carbon from the CO_2 out of the global carbon cycle. Theoretical calculations show that the magnitude of such sequestration could be substantial.

Field tests show otherwise. Buesseler and Boyd (2003) reviewed iron-enrichment experiments in the southern oceans, showing that area requirements are much higher than shown in theory, and much less carbon reaches the seafloor than expected, despite creating blooms that covered thousands of square kilometers of ocean surface. Apart from potentially massive ecological effects on the ocean, ocean fertilization seems unlikely to be a practical or cost-effective means of sequestering human GHG pollution.

Source: Buesseler, K.O., Boyd, P. W. 2003. Will ocean fertilization work? Science 300 (5616), 67--68.

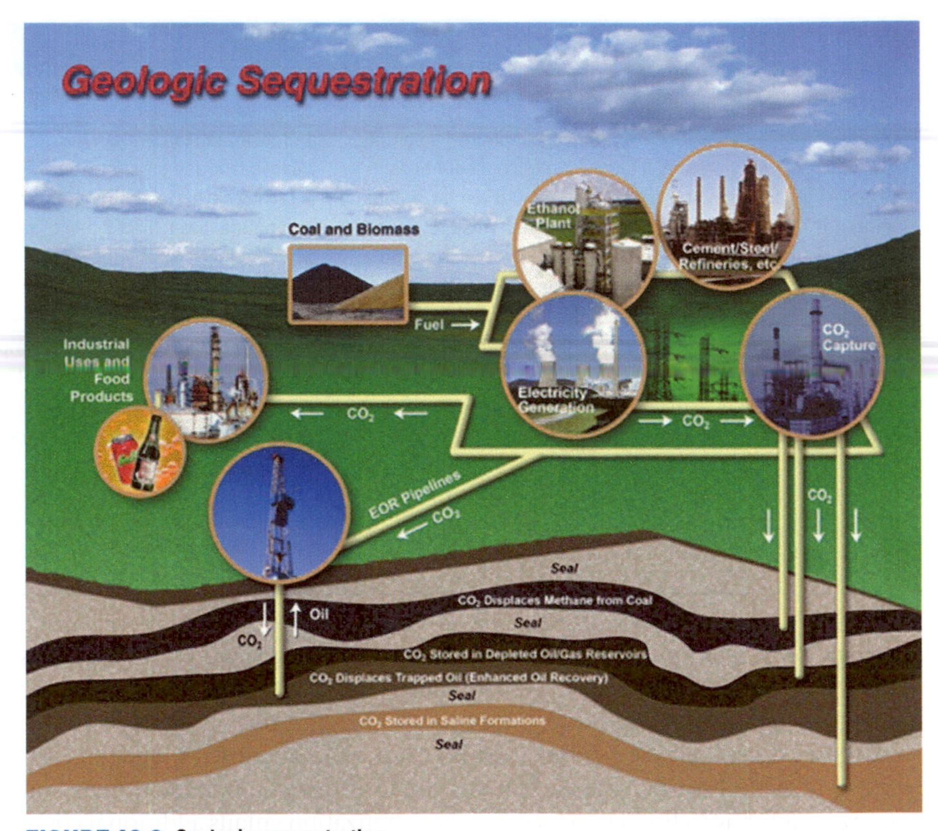

FIGURE 19.9 Geologic sequestration.
CO_2 captured at the source can be injected into abandoned oil wells or geologic formations. A small amount can be used in industrial processes, offsetting CO_2 production for those purposes. *Source: DOE.*

such as automobiles, or for land use change, capture is not feasible. This leaves large amounts of emissions that need to be recaptured from the atmosphere.

GETTING CO_2 BACK

Once CO_2 is released to the atmosphere, it is much more difficult to capture and sequester, but it is possible using adsorption techniques similar to those used to remove CO_2 at the source. Technologies that recover CO_2 already in the atmosphere are known as free air sequestration technologies (FAST).

The major technical obstacle in FAST systems is moving large volumes of air over the adsorption surface. Because CO_2 concentrations are below 400 ppm, more than 2500 l of air must be moved to remove one liter of CO_2. Removing a ton of CO_2 from the atmosphere would require passing more than one million liters of air over an adsorbent. It is simply too costly to move this volume of air mechanically for the process to be cost-effective.

FREE AIR SEQUESTRATION TECHNOLOGIES

Free air sequestration technologies (FAST) have been proposed that remove CO_2 directly from the atmosphere. FAST systems are not operational yet, but those in development use an adsorptive surface to sequester CO_2 from the air. Because CO_2 in free air is diluted, a major constraint for these systems is moving large quantities of air without using more energy than the equivalent of the CO_2 captured.

FIGURE 19.10 Artist's Conception of a FAST array.
FAST removes CO_2 from the atmosphere. This allows sequestration of past or diffuse emissions. Technologies to remove and sequester free atmospheric CO_2 are still in development. *Source: Figure courtesy Columbia University.*

As a result, FAST depends on passive air movement and require large adsorbent surfaces. This dictates that they are best placed in areas with reliable winds that are not good sites for wind energy production. Because CO_2 removal can be done on site, distance to markets is not a consideration. Therefore, sites that are poor candidates for wind energy because they are far from energy users make good FAST sites. For example, FAST arrays in the Sahara have been proposed (Figure 19.10).

FAST is in the developmental stage. Some prototypes would produce a mineral product that could be stored anywhere and would be relatively inert. Other processes recycle adsorbent to render pure CO_2, which then must be sequestered in geologic formations. No technology has moved beyond the prototype stage. The commercial viability of the devices is debatable and highly dependent on the price of carbon in world markets.

FAST provides one exciting possible future option for stabilizing atmospheric GHG concentrations without massive energy technology transitions. This option is controversial because it may seem to foster dependence on fossil fuels and undermine attempts to make deeper energy transitions. However, given the environmental costs of some renewable energy options, FAST may indeed be a rapid and relatively environmentally-favorable technological fix for climate change.

FURTHER READING

Canadell, J.G., Le Quéré, C., Raupach, M.R., Field, C.B., Buitenhuis, E.T., Ciais, P., Marland, G., 2007. Contributions to accelerating atmospheric CO_2 growth from economic activity, carbon intensity, and efficiency of natural sinks. Proceedings of the National Academy of Sciences 104 (47), 18866–18870.

Lackner, K.S., 2003. A guide to CO_2 sequestration. Science 300, 1677–1678.

Lenton, T.M., Vaughan, N.E., 2009. The radiative forcing potential of different climate geoengineering options. Atmospheric Chemistry and Physics 9, 5539–5561.

Matthews, H.D., Turner, S.E., 2009. Of mongooses and mitigation: ecological analogues to geoengineering. Environmental Research Letters 4, 9.

Schmitz, O.J., Raymond, P.A., Estes, J.A., Kurz, W.A., Holtgrieve, G.W., Ritchie, M.E., Wilmers, C.C., 2013. Animating the carbon cycle. Ecosystems, 1–16.

Steinberg, D.K., Lomas, M.W., Cope, J.S., 2012. Long-term increase in mesozooplankton biomass in the Sargasso Sea: Linkage to climate and implications for food web dynamics and biogeochemical cycling. Global Biogeochemical Cycles 26 (1).

CHAPTER 20

Assessing Risks, Designing Solutions

Like the global mitigation policy debate, the design of adaptation solutions is a social process. Biologists play an important role in this process, but they are only one part of a complete team. Climate change biologists are central in providing input to the design of adaptation for conservation strategies. They are also important in delivering input that protects ecosystem services and avoids damage to nature in the design of human adaptation. This chapter explores the process of adaptation design and the role of biologists within it.

IMPACTS, RISKS, AND ADAPTATION

Impacts are the effects that climate change has on biological and other systems. Impact is the product of the degree of change in the climate system and the sensitivity of the species or ecosystems being assessed. Vulnerability is roughly synonymous with impact, but it may also be used to refer to that portion of impact that cannot be ameliorated by adaptation. Vulnerability is especially used in social literature to refer to populations at risk that have a limited ability to adapt.

Risk is the magnitude of an impact multiplied by the probability of its occurrence. Risk assessments routinely balance low-probability, severe-impact outcomes with high-probability, low- or moderate-impact outcomes.

Adaptation is the ability of a system to modify or change its characteristics to compensate for the effects of climate change. Adaptation is an inherent property of a system that may be enhanced, sometimes greatly, by intentional human action. Biological systems have limited adaptation capacity. Thus, the adaptation of conservation systems is necessary to ensure the survival of species and ecosystems in the face of climate change.

THE ASSESSMENT PROCESS

The design of solutions begins with an assessment of impacts and risks to the system. Solutions require social responses, such as the redesign of protected

Climate Change Biology. http://dx.doi.org/10.1016/B978-0-12-420218-4.00020-2

areas, so the assessment process is inherently a social process. Climate change biologists can conduct impact or risk assessments in isolation, but unless these influence social processes, such as through the media, they will never become part of social adaptation solutions. In the remainder of this chapter, the technical aspects of biological impact assessments are discussed alongside the social process in which these assessments are usually embedded.

As participants in a broader assessment process, climate change biologists may find that their first task is to ensure that ecosystems and biodiversity are included in the assessment plan. Biodiversity and ecosystems are often left out of planning processes that focus on severe human impacts, such as loss of food or mass displacement of families. It is the role of biologists to remind planners and policy makers that some of the best adaptation solutions often involve ecosystems and their services, and that if nature is ignored in an assessment, adaptation solutions may be designed that would result in the loss of many important recreational and social values associated with nature.

A section or subgroup devoted to biodiversity and ecosystems is an important part of any impact or adaptation assessment, but the consideration of nature should not stop there. The best assessments will integrate the consideration of species and ecosystems into all facets of the assessment. Biologists can advocate for such integration and, more importantly, make it happen by devoting their time to assessment topics that do not focus directly on nature or conservation.

DOMAIN AND GRAIN

Domain and grain refer to the extent of a study area and the resolution of analysis within that area. For instance, the domain of general circulation models (GCMs) is global, and GCM grain is typically a horizontal resolution of 100–300 km. A good risk or impact assessment will define a domain and grain explicitly and early in the process. Decisions about domain and grain may be generically referred to as determining the "scale" of the assessment. Whereas not all assessment components will use spatially explicit analysis (i.e., maps), the climate scenarios used in the assessment will be spatially explicit. Therefore, domain and grain must be defined for the climate change analysis, and this is the basis for all other assessment elements.

Typically, a social or political process will begin that launches a national, regional, state, or local assessment. Biologists will be asked to participate in the assessment, or they will learn of it and ask to take part. The domain of the assessment will already generally be defined by the type of assessment (regional or national). The exact boundaries of the domain then need to be defined by participants in the assessment. This can be quite important because edge effects in climate models may degrade confidence in projections near the domain border.

The grain of the assessment must then be determined. This is an important step because finer grain assessments may require statistical downscaling of climatologies or regional climate models that require considerable lead time to prepare. It is important to allow for this preparation time when the assessment timeline is planned. Biological processes often unfold at fine scales, so biologists may have to advocate for a grain size that is appropriate to their needs and for the allocation of time and resources necessary for the preparation of fine-grain ("fine-scale") climatologies.

Once the assessment is under way, climate change biologists participating in the assessment will find that they are but one group of scientists and stakeholders in a much broader process. Climatologists and social scientists will certainly be involved; stakeholders, engineers, and policy makers will frequently be involved as well. Providing biological information to these diverse participants in accessible formats is a major challenge, particularly when a biological portion of the assessment is unfolding simultaneously.

BIOLOGICAL ASSESSMENT

The biological component of an assessment should include paleoecologists, climatologists, taxonomic specialists, and ecologists. Conservationists will also frequently be involved because the ultimate goal of most biological impact or adaptation assessments is to influence conservation policy. Depending on the nature of the assessment, stakeholders, policy makers, and social scientists may also be involved.

Where the participation of nonbiologists is left to the discretion of the biological team, the inclusion of social scientists and policy makers should be carefully considered. The natural tendency to talk biology with biologists may not produce the most useful assessment. The inclusion of social scientists in the biological analyses may provide important social context that will help products of the assessment be more readily accepted. Exposing policy makers to the biological discussions may help them appreciate the reasoning behind conservation recommendations and help them be more effective advocates for the strategies proposed.

The degree of participation of stakeholders other than scientists and policy makers should be gauged against the domain of the assessment. In national assessments, it may be difficult to have broad stakeholder participation beyond the public vetting of findings. In local assessments, however, stakeholder participation is essential. Fewer stakeholders are implicated in assessments with relatively small, local domains (e.g., a park or city), making the logistics of accommodating stakeholders less daunting. Because stakeholders will play a major role in the implementation and success of local adaptation solutions, having them involved in the assessment and design is good policy.

The biological analyses of the assessment can benefit from close involvement of climatologists. Often, biologists download future climate projections and use them to drive their models without consultation with climatologists. This is simple and straightforward, but it bypasses the wealth of knowledge about climate and future change that can be provided by climatologists. For example, the climatologists preparing future projections or historical climatologies for use in an assessment will often be aware of biases in the data or modeling that can help inform the interpretation of biological models driven by the climatologies. Any good assessment will have a team of climatologists preparing base climatologies and projections for use by other subgroups in the team, so it is often simply a question of reaching out to the climatology team to secure its participation in the biological analysis.

Modeling is an important component of every biological impact assessment. The assessment will certainly use climate models, and it should use as many biological models as is appropriate and feasible within the resources of the assessment. Because species distribution models are inexpensive and quick to implement, they are part of most assessments. Gap models may be especially important to include in assessments that involve forested lands. Dynamic global vegetation models provide a range of useful information on vegetation types and disturbance, but they may be time-consuming and expensive to implement. Where researchers are already implementing one or more of these types of models, they should be included in the assessment team. Often, modest additional resources, such as the provision of a graduate student to rerun models specifically for the domain of the assessment, can make existing modeling very relevant to the purposes of an assessment. Finally, where resources permit and there is strong need, an assessment may undertake new modeling efforts to provide critical biological data. For instance, it is difficult to envision anything more than a cursory assessment of protected area needs without species distribution modeling.

However, modeling is only one part of a complete assessment. Paleoecology can provide important insights into how biological systems have responded to past change, which is often a more reliable predictor of future response than is provided by modeling, even if precise past analogs to the future are rare. Good taxonomists and ecologists almost always think about climate change. Including their insight is vital to a discerning assessment. Therefore, although modelers often lead biological assessments, it is important to look well beyond modeling—to climatologists, paleoecologists, taxonomists, ecologists, social scientists, policy makers, and stakeholders—to produce a robust assessment result.

Integration of these diverse multidisciplinary views is a major challenge of a good assessment. Work in disciplinary subteams can be important to generate consensus among specialists—for example, paleoecologists agreeing on the interpretation of a fossil record relevant to the area. Work within a discipline can often be accomplished by e-mail, telephone, or in meetings. Integration across disciplines, in contrast, is a more lengthy process that almost always requires personal interaction and takes more than a single meeting to accomplish. Workshops of one to several days are often a valuable format to reap the full rewards of interdisciplinary analysis.

Finally, the results of an assessment must be communicated to target audiences and the public. A communication plan should be part of the assessment process. This certainly includes a report and summary for policy makers, but it should also include a communications strategy (e.g., press releases and public meetings) and discussions with key decision-makers and stakeholders. The time and resources required for these activities need to be carefully considered in planning for the assessment.

STAND-ALONE BIOLOGICAL ASSESSMENT

Although most biological assessments will be integrated into larger efforts such as national climate change impact assessments, there are situations in which stand-alone ecosystem, biodiversity, or conservation impact assessments are appropriate. Many impact and adaptation assessments have been conducted without full consideration of ecosystems and biodiversity. In these situations, it may be appropriate to conduct a stand-alone biological assessment to augment the existing work. In other settings, a specialized conservation assessment may be required—for instance, for a protected areas agency revising planning for the lands it manages or for a park needing to update its management plan.

In these settings, many of the principles of biological analyses as part of broader assessments apply. For instance, stand-alone assessments should engage climatologists and a diverse array of biological expertise. This may be challenging because a stand-alone assessment will typically have less financial resources than a broad social assessment.

A useful format for a stand-alone assessment is a series of workshops during an 18-month to 2-year time frame. A first workshop involving the key technical contributors can identify analyses that can be conducted in 6–8 months in preparation for a public workshop. These analyses are then conducted on a voluntary basis or with whatever limited funds are available. Examples include the preparation of statistically downscaled future climate projections for the region, the assessment of historical temperature trends from weather station data, the review of literature, the syntheses of paleoecological data, and the completion of other analyses within a limited time frame that will produce a useful information base for a broader workshop.

Once the preliminary analyses are complete, the main workshop of the assessment is convened. The technical team comprises the core of the workshop participants, with additional biologists, social scientists, policy makers, and stakeholders. The workshop is divided into two parts—the technical sessions and the public sessions. The technical sessions are used to generate impact scenarios integrating biological, social, and climatological insights. The public sessions are used to open and close the workshop, introduce a broader audience of stakeholders and policy makers to the issues identified at the opening, and present conclusions of the analyses of impacts at the close of the workshop.

DESIGN OF ADAPTATION SOLUTIONS

Effective impact assessment leads directly to the design of adaptation solutions. Some impact assessment processes will intentionally stop at the identification of impacts, leaving the design of solutions to a public policy process. Other impact assessments will integrate policy makers from the beginning and move seamlessly to adaptation design. In either case, the findings of the impact assessment are the building blocks for the design of adaptation activities. A clear plan for this transition to action is critical.

In most cases, participants in impact assessments will begin to envision adaptation responses. There are often preliminary recommendations in impact assessments that describe at least the broad types of measures that might be effective in reducing the impacts identified. Typically, the costs of these recommendations are not estimated, and they may not incorporate social or political considerations. The next step is to take these broad adaptation ideas and translate them into actual responses that are affordable and feasible.

Pilot activities and feasibility studies are an important part of this process. For instance, in an adaptation assessment in Madagascar, participants recommended the reconnection of the country's heavily-fragmented forests to facilitate range shifts. However, this recommendation did not specify where or how much reconnection was required, nor its cost or source of financing. Feasibility studies are therefore necessary to identify areas in which range shifts are likely, forests are heavily fragmented, and the technical knowledge of how to restore forests exists. Costing studies are needed to determine the cost of potential reforestation and, in an iterative process, how much area can be restored. Pilot studies are needed to test reforestation strategies, to gauge community participation, and to resolve any policy barriers identified. In Madagascar, funding from carbon offset programs defrayed some of the cost of reconnecting forests, benefits to communities from firewood and tourism were quantified, and the cost of resolving poorly docu mented land tenure was factored into the cost of planned reforestation.

The results of pilot activities and feasibility studies then permit the scaling and costing of full implementation of adaptation response. Major political commitment is required to mobilize the funding and create the implementation capacity needed. Where a biological assessment is part of a broad adaptation assessment, funding and implementation momentum may come from the political commitment to the overall process. For stand-alone biological assessments, a major political effort may be needed to mobilize the necessary resources to take adaptation activities from pilot to full scale.

TWO EXAMPLES OF ADAPTATION SOLUTIONS

To illustrate adaptation solutions, we use Yellowstone and Madagascar as examples. The domain of these two exercises is very different. For Yellowstone, the exercise was to design adaptation for the national park and surrounding public lands—a site-focused exercise. The Madagascar assessment examined impacts and responses across the entire island nation, with the intent of identifying adaptation actions that could benefit national-scale biodiversity conservation strategies. Both are somewhat atypical in that they are stand-alone biological assessments, but this facilitates focus on adaptation responses relevant to climate change biology.

The conclusions of the Madagascar workshop were developed in an open plenary session that included biologists, social scientists, policy makers, and stakeholders. The recommendations were endorsed and presented to a larger body of policy makers and the public. The participants recommended the protection and restoration of forests along rivers that paleoecological evidence indicated had been important range-shift corridors during past climate change. They advocated restoring connectivity to heavily-fragmented forests and protecting key remaining forests still outside the protected area estate.

Feasibility studies were initiated immediately following the workshop. The first step was to assess the cost of forest restoration. Restoration was implicated in both the riverine corridor restoration and reconnecting forest fragments. However, reforestation on large scales had not been attempted since colonial times in Madagascar. A costing feasibility study surveyed existing small-scale reforestation programs in Madagascar and estimated costs of large-scale programs. Forest restoration was found to be several times more expensive than conserving standing forest. This indicated that modeling or other evidence would be needed to focus this relatively expensive solution on the areas most critically in need. Meanwhile, a species distribution modeling feasibility study was under way to identify areas likely to be highly important for reconnecting forest fragments. A third feasibility study was implemented using satellite imagery to assess the condition of riverine corridor forests and determine areas in need of restoration. As the feasibility studies concluded, pilot studies were launched at multiple field sites to confirm the cost estimates for forest restoration, test species and replanting techniques in different soils and climates, and test different formats for community employment and management. The results of these feasibility and pilot studies will be a sound body of experience, confidence in methods, and reliable costing on which to base a nationwide program of adaptation activities.

In the Yellowstone site-level assessment, park managers, stakeholders, and conservation groups participated in a series of meetings that identified climate change under way in the region, as well as vulnerable aspects of park operations and biology. They then targeted several species and ecosystems for adaptation action. The grizzly bear is an iconic Yellowstone species targeted in the assessment. Grizzlies are a major tourist attraction and an ecosystem architect within the park, but they are also the cause of substantial conflict with humans, much of it associated with bear mortality. Grizzlies are relatively secure within the national park, but they come into conflict with humans on surrounding lands when they kill livestock or threaten humans. The killing of problem animals and animals killed in human conflicts are a major source of bear mortality. As climate has warmed in the Yellowstone region, winter has shortened and spring has become warmer. This results in a shorter hibernation period for grizzlies, greater foraging, and more conflict with humans. At the same time, these climatic conditions have led to an outbreak of bark beetles in high-elevation whitebark pines, which has killed hundreds of thousands of trees and greatly reduced the availability of whitebark pine cones and seed (which is a major food source for grizzlies, especially for young bears). This forces grizzlies from the high country earlier and reinforces the climate-induced pattern of wider foraging. Combined, these factors are bringing bears into greater contact with humans, resulting in increased mortality for bears. The pattern is expected to intensify with future warming.

Recommendations of the assessment were therefore to strengthen existing programs for avoiding human–grizzly conflict (e.g., bear-proof trash removal), to restore high-country food sources (particularly whitebark pine), and to foster the development of alternate food sources to replace those lost. Whitebark pine restoration is under way, both through natural regeneration and through management programs. The replenishment of cones will take years as trees grow and mature. In the meantime, increased bear conflict-avoidance programs and monitoring, including the addition of radio collars for potential problem animals, is being implemented. Feasibility studies are under way for the restoration of native trout populations as a food source. Native trout inhabit shallower habitats and make better prey for bears, whereas introduced trout that are outcompeting the native trout occupy deeper waters and are not important bear food sources. Restoration of the native trout has major benefits for both freshwater systems and bear populations.

AND DO IT AGAIN

Assessment of impacts and design of adaptation measures is an iterative process. Like language, it is a journey, not a destination. Once impacts have been identified and responses designed and implemented, it is time to begin the process anew. Have the expected impacts materialized? Are the responses designed appropriate? What new challenges have emerged in the 5–10 years it took to design and conduct an assessment and implement adaptation strategies?

Communicating and implementing insights from climate change biology is not an easy task. Park managers have a full plate of management problems they are already addressing with limited budgets; a new major problem is not welcome news for them. Politicians tuned in to election cycles or solving immediate crises have little aptitude for dealing with long-term problems whose worst effects may be decades away. Convincing these audiences that major effort and new expenditures are required to address the biological impacts of climate change takes time, sound evidence, and persistence.

However, the world is awake to the global threat posed by greenhouse gas pollution, and is mounting unprecedented efforts to combat the problem. Somewhat more slowly, world leaders are realizing that even the most rapid possible greenhouse gas stabilization will leave major changes to be dealt with, and that adaptation must be a part of the solution. The global framework for acting on information from climate change biology research is therefore positive and getting stronger.

This global awareness is now penetrating to regional and local levels. The world's first national park has developed a climate change impact and adaptation plan. The world's highest- priority global biodiversity hotspot is engaged in a nationwide conservation adaptation effort. If it happens in these remote and special places, it will happen increasingly in nations, ecosystems, and sites throughout the world. It is a mounting challenge, one that demands strong research and reliable data in response. A growing chorus of voices are demanding answers to guide policy and management in the face of change. This is a challenge that requires a new discipline and a new generation of scientists.

This new discipline is climate change biology.

References

Allan, J.D., Palmer, M., Poff, N.L., 2005. Climate change and freshwater ecosystems. Chapter 17. In: Lovejoy, T.E., Hannah, L. (Eds.), Climate Change and Biodiversity. Yale University Press, New Haven, p. 398.

Alsos, I.G., Eidesen, P.B., Ehrich, D., Skrede, I., Westergaard, K., Jacobsen, G.H., et al., 2007. Frequent long-distance plant colonization in the changing Arctic. Science 316, 1606–1609.

Atkinson, A., Siegel, V., Pakhomov, E., Rothery, P., 2004. Long-term decline in krill stock and increase in salps within the Southern Ocean. Nature 432, 100–103.

Axford, Y., Briner, J.P., Cooke, C.A., Francis, D.R., Michelutti, N., Miller, G.H., et al., 2009. Recent changes in a remote Arctic lake are unique within the past 200,000 years. Proceedings of the National Academy of Sciences 106, 18443–18446.

Bala, G., Caldeira, K., Wickett, M., Phillips, T.J., Lobell, D.B., Delire, C., et al., 2007. Combined climate and carbon-cycle effects of large-scale deforestation. Proceedings of the National Academy of Sciences 104, 6550–6555.

Benning, T.L., Lapointe, D., Atkinson, C.T., Vitousek, P., 2002. Interactions of climate change and biological invasions and land use in the Hawaiian Islands: modeling the fate of endemic birds using geographic information system. Proceedings of the National Academy of Sciences 99, 14246–14249.

Benton, M.J., Twitchett, R.J., 2003. How to kill (almost) all life: the end-Permian extinction event. Trends in Ecology & Evolution 18, 358–365.

Betts, R.A., Shugart, H.H., 2005. Dynamic ecosystem and earth system models. Chapter 15. In: Lovejoy, T.E., Hannah, L. (Eds.), Climate Change and Biodiversity. Yale University Press, New Haven, p. 398.

Bolch, T., Kulkarni, A., Kääb, A., Huggel, C., Paul, F., Cogley, J.G., Stoffel, M., 2012. The state and fate of Himalayan glaciers. Science 336 (6079), 310–314.

Both, C., Bouwhuis, S., Lessells, C.M., Visser, M.E., 2006. Climate change and population declines in a long-distance migratory bird. Nature 441, 81–83.

Breshears, D.D., Cobb, N.S., Rich, P.M., Price, K.P., Allen, C.D., Balice, R.G., et al., 2005. Regional vegetation die-off in response to global-change-type drought. Proceedings of the National Academy of Sciences 102, 15144–15148.

Brown, C.J., Fulton, E.A., Hobday, A.J., Matear, R.J., Possingham, H.P., Bulman, C., et al., 2010. Effects of climate-driven primary production change on marine food webs: implications for fisheries and conservation. Global Change Biology 16, 1194–1212.

Buckup, P.A., Melo, M.R., 2005. Case study: phylogeny and distribution of fishes of the characidium lauroi group as indicators of climate change in Southeastern Brazil. In: Lovejoy, T.E., Hannah, L. (Eds.), Climate Change and Biodiversity. Yale University Press, New Haven, p. 398.

Buesseler, K.O., Boyd, P.W., 2003. Climate change: will ocean fertilization work? Science 300, 67–68.

Bugmann, H., 2001. A review of forest gap models. Climatic Change 51, 259–305.

Busch, J., Godoy, F., Turner, W.R., Harvey, C.A., 2011. Biodiversity co-benefits of reducing emissions from deforestation under alternative reference levels and levels of finance. Conservation Letters 4 (2), 101–115.

Bush, M., 2002. Distributional change and conservation on the Andean flank: a palaeoecological perspective. Global Ecology and Biogeography 11, 475–484.

Bush, M., 2003. Ecology of a Changing Planet, third ed. Prentice Hall, Upper Saddle River, NJ. 477.

Bush, M.B., Silman, M.R., Urrego, D.H., 2004. 48,000 years of climate and forest change in a biodiversity hot spot. Science 303 (5659), 827–829.

Bush, M.B., Flenley, J., Gosling, W.D., 2007. Tropical Rainforest Responses to Climatic Change, Vol. 396, Springer.

Buytaert, W., Cuesta-Camacho, F., Tobón, C., 2011. Potential impacts of climate change on the environmental services of humid tropical alpine regions. Global Ecology and Biogeography 20 (1), 19–33.

Canadell, J.G., Le Quéré, C., Raupach, M.R., Field, C.B., Buitenhuis, E.T., Ciais, P., Marland, G., 2007. Contributions to accelerating atmospheric CO_2 growth from economic activity, carbon intensity, and efficiency of natural sinks. Proceedings of the National Academy of Sciences 104 (47), 18866–18870.

Chavez, F.P., Ryan, J., Lluch-Cota, S.E., Ñiquen, M., 2003. From anchovies to sardines and back: multidecadal change in the Pacific Ocean. Science 299 (5604), 217–221.

Clark, J.S., Fastie, C., Hurtt, G., Jackson, S.T., Johnson, C., King, G.A., et al., 1998. Reid's paradox of rapid plant migration—dispersal theory and interpretation of paleoecological records. BioScience 48, 13–24.

Cohen, A.S., 2003. Paleolimnology: The History and Evolution of Lake Systems. Oxford University Press, USA.

Coope, G.R., 2004. Among insect species because of, or in spite of, ice age climatic instability? Philosophical Transactions of the Royal Society of London 359, 209–214.

Cowling, S.A., Maslin, M.A., Sykes, M.T., 2001. Paleo vegetation simulations of lowland Amazonia and implications for neotropical allopatry and speciation. Quaternary Science Reviews 55, 140–149.

Cramer, W., Bondeau, A., Woodward, F.I., Prentice, I.C., Betts, R.A., Brovkin, V., Young-Molling, C., 2001. Global response of terrestrial ecosystem structure and function to CO_2 and climate change: results from six dynamic global vegetation models. Global Change Biology 7 (4), 357–373.

Davis, M.B., Shaw, R.G., Etterson, J.R., 2005. Evolutionary responses to changing climate. Ecology 86, 1704–1714.

Dansgaard, W., Johnsen, S.J., Clausen, H.B., Dahl-Jensen, D., Gundestrup, N.S., Hammer, C.U., et al., 1993. Evidence for general instability of past climate from a 250-kyr ice-core record. Nature 364, 218–220.

La Deau, S.L., Clark, J.S., 2001. Rising CO_2 levels and the fecundity of forest trees. Science 292, 95–98.

Delcourt, H.R., Delcourt, P.A., 1991. Quaternary Ecology: A Paleoecological Perspective. Chapman & Hall, New York. 242.

Deutsch, C.A., Tewksbury, J.J., Huey, R.B., Sheldon, K.S., Ghalambor, C.K., Haak, D.C., et al., 2008. Impacts of climatewarming on terrestrial ectotherms across latitude. Proceedings of the National Academy of Sciences 105, 6668–6672.

Donald, P.F., Sanderson, F.J., Burfield, I.J., Bierman, S.M., Gregory, R.D., Waliczky, Z., 2007. International conservation policy delivers benefits for birds in Europe. Science 317, 810–813.

Donlan, J., 2005. Re-wilding North America. Nature 436, 913–914.

Drake, B.G., Hughes, L., Johnson, E.A., Seibel, B.A., Cochrane, M.A., Fabry, V.J., et al., 2005. Synergistic effects. Chapter 18. In: Lovejoy, T.E., Hannah, L. (Eds.), Climate Change and Biodiversity. Yale University Press, New Haven, p. 398.

Etnoyer, P., Canny, D., Mate, B., Moran, L., 2004. Persistent pelagic habitats in the Baja California to Bering Sea (B2B). Ecoregion Oceanography 17, 90–101.

Etnoyer, P., Canny, D., Mate, B., Moran, L., Ortega-Ortiz, J., Nichols, W., 2006. Sea-surface temperature gradients across blue whale and sea turtle foraging trajectories off the Baja California Peninsula, Mexico. Deep Sea Research Part II – Topical Studies in Oceanography 53, 340–358.

Fargione, J., Hill, J., Tilman, D., Polasky, S., Hawthorne, P., 2008. Land clearing and the biofuel carbon debt. Science 319 (5867), 1235–1238.

Flenley, J.R., 1998. Tropical forests under the climates of the last 30,000 years. Climatic Change 39, 177–197.

Foden, W., Midgley, G.F., Hughes, G., Bond, W.J., Thuiller, W., Hoffman, M.T., Hannah, L., 2007. A changing climate is eroding the geographical range of the Namib Desert tree Aloe through population declines and dispersal lags. Diversity and Distributions 13 (5), 645–653.

Foden, W.B., Butchart, S.H., Stuart, S.N., Vié, J.C., Akçakaya, H.R., Angulo, A., Mace, G.M., 2013. Identifying the world's most climate change vulnerable species: a systematic trait-based assessment of all birds, amphibians and corals. PLoS One 8 (6).

da Fonseca, G.A., Rodriguez, C.M., Midgley, G., Busch, J., Hannah, L., Mittermeier, R.A., 2007. No forest left behind. PLoS Biology 5 (8).

Gingerich, P.D., 2006. Environment and evolution through the Paleocene – Eocene thermal maximum. Trends in Ecology and Evolution 21, 246–253.

Graham, R.W., Grimm, E.C., 1990. Effects of global climate change on the patterns of terrestrial biological communities. Trends in Ecology and Evolution 5, 289–292.

Grebmeier, J.M., Overland, J.E., Moore, S.E., Farley, E.V., Carmack, E.C., Cooper, L.W., et al., 2006. A major ecosystem shift in the northern Bering Sea. Science 311, 1461–1464.

Greene, C.H., Pershing, A.J., 2007. Climate drives sea change. Science 315, 1084–1085.

Grimsditch, G., Salm, R.V., 2006. Coral reef resilience and resistance to bleaching. IUCN, Gland, Switzerland.

Guinotte, J.M., Buddemeier, R.W., Kleypas, J.A., 2003. Future Coral Reef Habitat Marginality: Temporal and Spatial Effects of Climate Change in the Pacific Basin. Springer-Verlag. 551–558.

Gupta, J., 2010. A history of international climate change policy. Wiley Interdisciplinary Reviews: Climate Change 1 (5), 636–653.

Guzmán, H., Edgar, G., 2008. Millepora boschmai. In: IUCN 2009. IUCN Red List of Threatened Species, p. 2. Version 2009.

Halpin, P.N., 1997. Global climate change and natural-area protection: management responses and research directions. Ecological Applications 7, 828–843.

Hampe, A., Petit, R.J., 2005. Conserving biodiversity under climate change: the rear edge matters. Ecological Letters 8, 461–467.

Hannah, L., 2010. A global conservation system for climate-change adaptation. Conservation Biology 24, 70–77.

Hannah, L., Hansen, L., 2005. Designing landscapes and seascapes for change. Chapter 20. In: Lovejoy, T.E., Hannah, L. (Eds.), Climate Change and Biodiversity. Yale University Press, New Haven, p. 398.

Hannah, L., Salm, R., 2005. Protected areas management in a changing climate. Chapter 22. In: Lovejoy, T.E., Hannah, L. (Eds.), Climate Change and Biodiversity. Yale University Press, New Haven, p. 398.

Hannah, L., Lovejoy, T.E., Schneider, S.H., 2005a. Biodiversity and climate change in context. Chapter 1. In: Lovejoy, T.E., Hannah, L. (Eds.), Climate Change and Biodiversity. Yale University Press, New Haven, p. 398.

Hannah, L., Midgley, G., Hughes, G., Bomhard, B., 2005b. The view from the cape. Extinction risk, protected areas, and climate change. Bioscience 55, 231–242.

Hannah, L., Midgley, G., Andelman, S., Araujo, M., Hughes, G., Martinez-Meyer, E., et al., 2007. Protected area needs in a changing climate. Frontiers in Ecology and the Environment 5, 131–138.

Hannah, L., Roehrdanz, P.R., Ikegami, M., Shepard, A.V., Shaw, M.R., Tabor, G., Hijmans, R.J., 2013a. Climate change, wine, and conservation. Proceedings of the National Academy of Sciences 110 (17), 6907–6912.

Hannah, L., Ikegami, M., Hole, D.G., Seo, C., Butchart, S.H., Peterson, A.T., Roehrdanz, P.R., 2013b. Global climate change adaptation priorities for biodiversity and food security. PloS One 8 (8).

Hays, G.C., Richardson, A.J., Robinson, C., 2005. Climate change and marine plankton. Trends in Ecology & Evolution 20, 337–344.

Heller, N.E., Zavaleta, E.S., 2009. Biodiversity management in the face of climate change: a review of 22 years of recommendations. Biological Conservation 142, 14–32.

Hewitt, G.M., Nichols, R.A., 2005. Genetic and evolutionar impacts of climate change. Chapter 12. In: Lovejoy, T.E., Hannah, L. (Eds.), Climate Change and Biodiversity. Yale University Press, New Haven, p. 398.

Hodgson, J.A., Thomas, C.D., Wintle, B.A., Moilanen, A., 2009. Climate change, connectivity and conservation decision making: back to basics. Journal of Applied Ecology 46, 964–969.

Hoegh-Guldberg, O., 1999. Coral bleaching, climate change and the future of the world's coral reefs. Review Marine and Freshwater Research 50, 839–866.

Hoegh-Guldberg, O., 2005. Climate change and marine ecosystems. Chapter 16. In: Lovejoy, T.E., Hannah, L. (Eds.), Climate Change and Biodiversity. Yale University Press, New Haven, p. 398.

Hoegh-Guldberg, O., Bruno, J.F., 2010. The impact of climate change on the world's marine ecosystems. Science 328, 1523–1528.

Hoegh-Guldberg, O., Mumby, P.J., Hooten, A.J., Steneck, R.S., Greenfield, P., Gomez, E., et al., 2007. Coral reefs under rapid climate change and ocean acidification. Science 318, 1737–1742.

Hoegh-Guldberg, O., Hughes, L., McIntyre, S., Lindenmayer, D.B., Parmesan, C., Possingham, H.P., et al., 2008. Assisted colonization and rapid climate change. Science 321, 345–346.

Hoorn, C., Wesselingh, F.P., Ter Steege, H., Bermudez, M.A., Mora, A., Sevink, J., Antonelli, A., 2010. Amazonia through time: andean uplift, climate change, landscape evolution, and biodiversity. Science 330 (6006), 927–931.

Hugall, A., Moritz, C., Moussalli, A., Stanisic, J., 2002. Reconciling paleodistribution models and comparative phylogeography in the Wet Tropics rainforest land snail gnarosophia bellendenkerensis (Brazier 1875). Proceedings of the National Academy of Sciences 99, 6112–6117.

Hughes, L., 2000. Biological consequences of global warming: is the signal already apparent? Trends in Ecology and Evolution 15, 56–61.

Hulme, M., 2005. Recent climate trends. Chapter 3. In: Lovejoy, T.E., Hannah, L. (Eds.), Climate Change and Biodiversity. Yale University Press, New Haven, p. 398.

Huntley, B., 1991. How plants respond to climate change: migration rates, individualism, and the consequences for plant communities. Annals of Botany 67, 15–22.

Huntley, B., 2005. North temperate responses. Chapter 8. In: Lovejoy, T.E., Hannah, L. (Eds.), Climate Change and Biodiversity. Yale University Press, New Haven, p. 398.

Huntley, B., Webb, T., 1989. Migration – species response to climatic variations caused by changes in the earth's orbit. Journal of Biogeography 16, 5–19.

Idso, S.B., 1999. The long-term response of trees to atmospheric CO_2 enrichment. Global Change Biology 5, 493–495.

Immerzeel, W.W., Van Beek, L.P., Bierkens, M.F., 2010. Climate change will affect the Asian water towers. Science 328 (5984), 1382–1385.

Intergovernmental Panel on Climate Change, 2007. Climate Change 2007: Impacts, Adaptation and Vulnerability; Contribution of Working Group II to the Fourth Assessment Report of the Intergovernmental Panel on Climate Change. Cambridge University Press, Cambridge, UK.

Intergovernmental Panel on Climate Change, 2007. Climate Change 2007: Mitigation of Climate Change; Contribution of Working Group III to the Fourth Assessment Report of the Intergovernmental Panel on Climate Change. Cambridge University Press, Cambridge, UK.

Intergovernmental Panel on Climate Change, 2007. Climate Change 2007: The Scientific Basis;- Contribution of Working Group I to the Fourth Assessment Report of the Intergovernmental Panel on Climate Change. Cambridge University Press, Cambridge, UK.

Intergovernmental Panel on Climate Change, 2013. Climate Change 2013: The Scientific Basis;- Contribution of Working Group I to the Fifth Assessment Report of the Intergovernmental Panel on Climate Change. Cambridge University Press, Cambridge, UK.

Intergovernmental Panel on Climate Change, 2014. Climate Change 2014: Impacts, Adaptation and Vulnerability; Contribution of Working Group II to the Fifth Assessment Report of the Intergovernmental Panel on Climate Change. Cambridge University Press, Cambridge, UK.

Intergovernmental Panel on Climate Change, 2014. Climate Change 2014: Mitigation of Climate Change; Contribution of Working Group III to the Fifth Assessment Report of the Intergovernmental Panel on Climate Change. Cambridge University Press, Cambridge, UK.

Jackson, S.T., Overpeck, J.T., Webb, T., Keattch, S.E., Anderson, K.H., 1997. Mapped plant-macrofossil and pollen records of late quaternary vegetation change in eastern North America. Quaternary Science Reviews 16, 1–70.

Jackson, S.T., Webb, R.S., Anderson, K.H., Overpeck, J.T., Webb, T., Williams, J.W., et al., 2000. Vegetation and environment in eastern North America during the last glacial maximum. Quaternary Science Reviews 19, 489–508.

Jacobson, M.Z., Delucchi, M.A., 2009. A path to sustainable energy by 2030. Scientific American 301, 58–65.

Karl, T.R., Trenberth, K.E., 2005. What is climate change? Chapter 2. In: Lovejoy, T.E., Hannah, L. (Eds.), Climate Change and Biodiversity. Yale University Press, New Haven, p. 398.

Kennett, J.P., Cannariato, K.G., Hendy, I.L., Behl, R.J., 2000. Carbon isotopic evidence for methane hydrate instability during quaternary interstadials. Science 288, 128–133.

Kerr, R.A., 2000. A North Atlantic climate pacemaker for the centuries. Science 288, 1984–1986.

Kerr, R.A., 2001. Rising global temperature, rising uncertainty. Science 292, 192–194.

Klyashtorin, L.B., 1998. Long-term climate change and main commercial fish production in the Atlantic and Pacific. Fisheries Research 37 (1), 115–125.

Knowlton, J.L., Graham, C.H., 2010. Using behavioral landscape ecology to predict species' responses to land-use and climate change. Biological Conservation 143, 1342–1354.

Knutti, R., Sedláček, J., 2013. Robustness and uncertainties in the new CMIP5 climate model projections. Nature Climate Change 3 (4), 369–373.

Koh, L.P., Levang, P., Ghazoul, J., 2009. Designer landscapes for sustainable biofuels. Trends in Ecology and Evolution 24, 431–438.

Kokelj, S.V., Lantz, T.C., Kanigan, J., Smith, S.L., Coutts, R., 2009. Origin and polycyclic behaviour of tundra thaw slumps, Mackenzie Delta Region, Northwest Territories, Canada. Permafrost and Periglacial Processes 20, 173–184.

Korner, C., 2000. Biosphere responses to CO_2 enrichment. Ecological Applications 10, 1590–1619.

Korner, C., Basler, D., 2010. Phenology under global warming. Science 327, 1461–1462.

Lackner, K.S., 2003. A guide to CO_2 sequestration. Science 300, 1677–1678.

Landmann, G., Reimer, A., Lemcke, G., Kempe, S., 1996. Dating Late Glacial abrupt climate changes in the 14,570 yr long continuous varve record of Lake Van, Turkey. Palaeogeography Palaeoclimatology Palaeoecology 122, 107–118.

Lantz, T.C., Kokelj, S.V., Gergel, S.E., Henryz, G.H.R., 2009. Relative impacts of disturbance and temperature: persistent changes in microenvironment and vegetation in retrogressive thaw slumps. Global Change Biology 15, 1664–1675.

Lapointe, D., Benning, T.L., Atkinson, C., 2005. Case study: avian malaria, climate change and native birds of Hawaii. In: Lovejoy, T.E., Hannah, L. (Eds.), Climate Change and Biodiversity. Yale University Press, New Haven, p. 398.

Lau, L., Young, R.A., McKeon, G., Syktus, J., Duncalfe, F., Graham, N., et al., 1999. Downscaling global information for regional benefit: coupling spatial models at varying space and time scales. Environmental Modelling and Software 14, 519–529.

Lawton, R.O., Nair, U.S., Pielke, R.A., Welch, R.M., 2001. Climatic impact of tropical lowland deforestation on nearby montane cloud forests. Science 294, 584–587.

Lenton, T.M., Vaughan, N.E., 2009. The radiative forcing potential of different climate geoengineering options. Atmospheric Chemistry and Physics 9, 5539–5561.

Lischke, H., Guisan, A., Fischlin, A., Williams, J., Bugmann, H., 1998. Vegetation responses to climate change in the Alps: modeling studies. Chapter 8. In: Cebon, P., Dahinden, U., Davies, H., Imboden, D., Jaeger, C. (Eds.), Views from the Alps: Regional Perspectives on Climate Change. MIT Press, Cambridge, MA, p. 515.

Loarie, S.R., Duffy, P.B., Hamilton, H., Asner, G.P., Field, C.B., Ackerly, D.D., 2009. The velocity of climate change. Nature 462, 1052–U1111.

Logan, J.A., Powell, J., 2001. Ghost forests, global warming, and the mountain pine beetle (Coleoptera: Scotytidea). American Entomologist 47, 160–173.

Logan, J.A., Regniere, J., Powell, J.A., 2003. Assessing the impacts of global warming on forest pest dynamics. Frontiers in Ecology and the Environment 1, 130–137.

Lovejoy, T.E., 2005. Conservation with a changing climate. Chapter 19. In: Lovejoy, T.E., Hannah, L. (Eds.), Climate Change and Biodiversity. Yale University Press, New Haven, p. 398.

Lovejoy, T.E., Hannah, L., 2005. Global greenhouse gas levels and the future of biodiversity. Chapter 24. In: Lovejoy, T.E., Hannah, L. (Eds.), Climate Change and Biodiversity. Yale University Press, New Haven, p. 398.

Lovejoy, T.E., Hannah, L. (Eds.), 2005. Climate Change and Biodiversity. Yale University Press, New Haven.

Lowry, W.P., Lowry II, P.P., 1989. Fundamentals of Biometeorology: Interactions of Organisms and the Atmosphere. The Physical Environment, vol. I, Peavine Publications, McMinnville, OR. p. 310.

Lutgens, F., Tarbuck, E.J., 2001. The Atmosphere: An Introduction to Meteorology, eighth ed. Prentice Hall, Upper Saddle River, NJ.

Magnuson, J.J., Robertson, D.M., Benson, B.J., Wynne, R.H., Livingstone, D.M., Arai, T., et al., 2000. Historical trends in lake and river ice cover in the Northern Hemisphere. Science 289, 1743–1746.

Malcolm, J.R., Markham, A., Neilson, R.P., Garaci, M., 2005. Case study: migration of vegetation types in a greenhouse world. In: Lovejoy, T.E., Hannah, L. (Eds.), Climate Change and Biodiversity. Yale University Press, New Haven, p. 398.

Malcolm, J.R., Liu, C., Neilson, R.P., Hansen, L.A., Hnnah, L., 2006. Global warming and extinctions of endemic species from biodiversity hotspots. Conservation Biology 20, 538–548.

Malhi, Y., Roberts, J.T., Betts, R.A., Killeen, T.J., Li, W.H., Nobre, C.A., 2008. Climate change, deforestation, and the fate of the Amazon. Science 319, 169–172.

Markgraf, V., Kenny, R., 1995. Character of rapid vegetation and climate change during the late-glacial in southernmost South America. In: Huntley, B., Cramer, W., Morgan, A.V., Prentice, H.C., Allen, J.R.M. (Eds.), Past and Future Rapid Environmental Changes: The Spatial and Evolutionary Responses of Terrestrial Biota. Springer-Verlag, Berlin, pp. 81–102.

Markgraf, V., McGlone, M., 2005. Southern temperate ecosystem responses. Chapter 10. In: Lovejoy, T.E., Hannah, L. (Eds.), Climate Change and Biodiversity. Yale University Press, New Haven, p. 398.

Marshall, P.A., Schuttenberg, H., 2006. A reef manager's guide to coral bleaching. Great Barrier Reef Marine Park Authority.

Martinez-Meyer, E., Townsend Peterson, A., Hargrove, W.W., 2004. Ecological niches as stable distributional constraints on mammal species, with implications for Pleistocene extinctions and climate change projections for biodiversity. Global Ecology and Biogeography 13, 305–314.

Matthews, H.D., Turner, S.E., 2009. Of mongooses and mitigation: ecological analogues to geoengineering. Environmental Research Letters 4, 9.

Mawdsley, J.R., O'Malley, R., Ojima, D.S., 2009. A review of climate-change adaptation strategies for wildlife management and biodiversity conservation. Conservation Biology 23, 1080–1089.

Mayhew, P.J., Jenkins, G.B., Benton, T.G., 2008. A long-term association between global temperature and biodiversity, origination and extinction in the fossil record. Proceedings of the Royal Society B: Biological Sciences 275, 47–53.

McCarthy, H.R., Oren, R., Johnsen, K.H., Gallet-Budynek, A., Pritchard, S.G., Cook, C.W., et al., 2010. Reassessment of plant carbon dynamics at the Duke free-air CO_2 enrichment site: interactions of atmospheric $[CO_2]$ with nitrogen and water availability over stand development. New Phytologist 185, 514–528.

McGlone, M.S., 1995. The responses of New Zealand forest diversity to quaternary climates. In: Huntley, B., Cramer, W., Morgan, A.V., Prentice, H.C., Allen, J.R.M. (Eds.), Past and Future Rapid Environmental Changes: The Spatial and Evolutionary Responses of Terrestrial Biota. Springer-Verlag, Berlin, pp. 73–80.

McLachlan, J.S., Hellmann, J.J., Schwartz, M.W., 2007. A framework for debate of assisted migration in an era of climate change. Conservation Biology 21, 297–302.

Midgley, G., Millar, D., 2005. Case study: modeling species range shifts in two biodiversity hotspots. In: Lovejoy, T.E., Hannah, L. (Eds.), Climate Change and Biodiversity. Yale University Press, New Haven, p. 398.

Midgley, G.F., Hannah, L., Millar, D., Thuiller, W., Booth, A., 2003a. Developing regional and species-level assessments of climate change impacts on biodiversity in the Cape Floristic Region. Biological Conservation 112 (1), 87–97.

Midgley, G.F., Hannah, L., Millar, D., Thuiller, W., Booth, A., 2003b. Developing regional and species-level assessments of climate change impacts on biodiversity in the Cape Floristic Region. Biological Conservation 112, 87–97.

Milly, P.C.D., Wetherald, R.T., Dunne, K.A., Delworth, T.L., 2002. Increasing risk of great floods in a changing climate. Nature 415, 514–517.

Moline, M.A., Claustre, H., Frazer, T.K., Schofield, O., Vernet, M., 2004. Alteration of the food web along the Antarctic Peninsula in response to a regional warming trend. Global Change Biology 10, 1973–1980.

Monmier, M., 1999. Air Apparent: How Meteorologists Learned to Map, Predict and Dramatize Weather. University of Chicago Press, Chicago. 309.

Moritz, M.A., Parisien, M.A., Batllori, E., Krawchuk, M.A., Van Dorn, J., Ganz, D.J., Hayhoe, K., 2012. Climate change and disruptions to global fire activity. Ecosphere 3 (6). art49.

Morley, R.J., 2007. Cretaceous and tertiary climate change and the past distribution of megathermal rainforests. Chapter 1. In: Bush, M.B., Flenley, J.R. (Eds.), Tropical Rainforest Response to Climate Change. Springer-Praxis, Chichester, UK, p. 396.

Mote, P.W., Hamlet, A.F., Clark, M.P., Lettenmaier, D.P., 2005. Declining mountain snowpack in Western North America. Bulletin of the American Meteorological Society 86 (1), 39–49.

Nathan, R., Katul, G.G., Horn, H.S., Thomas, S.M., Oren, R., Avissar, R., et al., 2002. Mechanisms of long-distance dispersal of seeds by wind. Nature 418, 409–413.

Neilson, R.P., Pitelka, L.F., Solomon, A.M., Nathan, R., Midgley, G.F., Fragoso, J.M.V., et al., 2005. Forecasting regional to global plant migration in response to climate change. Bioscience 55, 749–759.

Nelson, E., et al., 2013. Climate change's impact on key ecosystem services and the human well-being they support in the US. Frontiers in Ecology and the Environment 11, 483–893.

Nogues-Bravo, D., Rodiguez, J., Hortal, J., Batra, P., Araujo, M.B., 2008. Climate change, humans, and the extinction of the woolly mammoth. PloS Biology 6, 685–692.

Orr, J.C., Fabry, V.J., Aumont, O., Bopp, L., Doney, S.C., Feely, R.A., et al., 2005. Anthropogenic ocean acidification over the twenty-first century and its impact on calcifying organisms. Nature 437, 681–686.

Overpeck, J., Cole, J., Bartlein, P., 2005. A "Paleoperspective" on climate variability and change. Chapter 7. In: Lovejoy, T.E., Hannah, L. (Eds.), Climate Change and Biodiversity. Yale University Press, New Haven, p. 398.

O'Neill, B.C., Oppenheimer, M., 2002. Climate change: dangerous climate impacts and the Kyoto Protocol. Science 296, 1971–1972.

Pacala, S., Socolow, R., 2004. Stabilization wedges: solving the climate problem for the next 50 years with current technologies. Science 305, 968–972.

Palliard, D., 1998. The timing of Pleistocene glaciations from a simple multiple-state climate model. Nature 391, 378–381.

Pandolfi, J.M., 1999. Response of pleistocene coral reefs to environmental change over long temporal scales. American Zoologist 39, 113–130.

Parmesan, C., 1996. Climate and species range. Nature 382, 765–766.

Parmesan, C., 2005a. Biotic response: range and abundance changes. Chapter 4. In: Lovejoy, T.E., Hannah, L. (Eds.), Climate Change and Biodiversity. Yale University Press, New Haven, p. 398.

Parmesan, C., 2005b. Case study: detection at multiple levels – euphydryas editha and climate change. In: Lovejoy, T.E., Hannah, L. (Eds.), Climate Change and Biodiversity. Yale University Press, New Haven, p. 398.

Parmesan, C., Yohe, G., 2003. A globally coherent fingerprint of climate change impacts across natural systems. Nature 421, 37–42.

Paters, R.L., Lovejoy, T.E. (Eds.), 1992. Global Warming and Biological Diversity. Yale University Press, New Haven, CT.

Patz, J.A., Olson, S.H., 2006. Malaria risk and temperature: influences from global climate change and local land use practices. Proceedings of the National Academy of Sciences 103, 5635–5636.

Pearsall III, S.H., 2005. Case study: managing for future change on the albermarle sound. In: Lovejoy, T.E., Hannah, L. (Eds.), Climate Change and Biodiversity. Yale University Press, New Haven, p. 398.

Pearson, P.N., Ditchfield, P.W., Singano, J., Harcourt-Brown, K.G., Nicholas, C.J., Olsson, R.K., et al., 2001. Warm tropical sea surface temperatures in the Late Cretaceous and Eocene epochs. Nature 414, 470.

Pennisi, E., 2001. Climate change – early birds may miss the worms. Science 291, 2532.

Perry, D.A., Borchers, J.G., Borchers, S.L., Amaranthus, M.P., 1990. Species migrations and ecosystem stability during climate change: the belowground connection. Conservation Biology 4, 266–274.

Peters, R.L., Darling, J.D.S., 1985. The greenhouse effect and nature reserves. BioScience 35, 707–717.

Peterson, R., 1995. The Wolves of Isle Royal: A Broken Balance. Willow Creek Press, Minocqua, WI.

Peterson, A.T., 2003. Projected climate change effects on rocky mountain and great plains birds: generalities of biodiversity consequences. Global Change Biology 9, 647–655.

Peterson, A.T., Tian, H., Martinez-Meyer, E., Soberon, J., Sanchez-Cordero, V., Huntley, B., 2005. Modeling distributional shifts of individual species and biomes. Chapter 14. In: Lovejoy, T.E., Hannah, L. (Eds.), Climate Change and Biodiversity. Yale University Press, New Haven, p. 398.

Pielke, R., Prins, G., Rayner, S., Sarewitz, D., 2007. Climate change 2007: lifting the taboo on adaptation. Nature 445 (7128), 597–598.

Pimm, S.L., 2001. Entrepreneurial insects. Nature 411, 531–532.

Poorter, H., Navas, M.L., 2003. Plant growth and competition at elevated CO_2: on winners, losers and functional groups. New Phytologist 157, 175–198.

Post, E., Pedersen, C., Wilmers, C.C., Forchhammer, M.C., 2008. Warming, plant phenology and the spatial dimension of trophic mismatch for large herbivores. Proceedings of the Royal Society B: Biological Sciences 275, 2005–2013.

Pounds, J.A., Fogden, M.P.L., Campbell, J.H., 1999. Biological response to climate change on a tropical mountain. Nature 398, 611–615.

Pounds, J.A., Fogden, M.P.L., Masters, K.L., 2005. Case study: responses of natural communities to climate change in a highland tropical forest. In: Lovejoy, T.E., Hannah, L. (Eds.), Climate Change and Biodiversity. Yale University Press, New Haven, p. 398.

Pounds, J.A., Bustamante, M.R., Coloma, L.A., Consuegra, J.A., Fogden, M.P.L., Foster, P.N., et al., 2006. Widespread amphibian extinctions from epidemic disease driven by global warming. Nature 439, 161.

Pyke, C.R., Fischer, D.T., 2005. Selection of bioclimatically representative biological reserve systems under climate change. Biological Conservation 121, 429–441.

Pyke, C.R., Marty, J., 2005. Cattle grazing mediates climate change impacts on ephemeral wetlands. Conservation Biology 19, 1619–1625.

Raper, S.C.B., Giorgi, F., 2005. Climate change projections and models. Chapter 13. In: Lovejoy, T.E., Hannah, L. (Eds.), Climate Change and Biodiversity. Yale University Press, New Haven, p. 398.

Raynaud, D., Barnola, J.M., Souchez, R., Lorrain, R., Petit, J.R., Duval, P., et al., 2005. Palaeoclimatology: the record for marine isotopic stage 11. Nature 436, 39–40.

Renema, W., Bellwood, D.R., Braga, J.C., Bromfield, K., Hall, R., Johnson, K.G., et al., 2008. Hopping hotspots: global shifts in marine biodiversity. Science 321, 654–657.

Retallack, G.J., 2001. A 300-million-year record of atmospheric carbon dioxide from fossil plant cuticles. Nature 411, 287–290.

Ricketts, T.H., et al., 2005. Pinpointing and preventing imminent extinctions. Proceedings of the National Academy of Sciences 102, 18497–18501.

Roessig, J.M., Woodley, C.M., Cech, J.J., Hansen, L.J., 2004. Effects of global climate change on marine and estuarine fishes and fisheries. Reviews in Fish Biology and Fisheries 14, 251–275.

Rogner, H.H., 1997. An assessment of world hydrocarbon resources. Annual Review in Energy and the Environment 22, 217–262.

Rohde, R.A., Muller, R.A., 2005. Cycles in fossil diversity. Nature 434, 208–210.

Root, T.L., Hughes, L., 2005. Present and future phenological changes in wild plants and animals. Chapter 5. In: Lovejoy, T.E., Hannah, L. (Eds.), Climate Change and Biodiversity. Yale University Press, New Haven, p. 398.

Root, T., Price, J.T., Hall, K.R., Schneider, S.H., Rosenzweig, C., Pounds, J.A., 2003. Fingerprints of global warming on wild animals and plants. Nature 421, 57–60.

Roy, K., Pandolfi, J.M., 2005. Responses of marine species and ecosystems to past climate change. Chapter 11. In: Lovejoy, T.E., Hannah, L. (Eds.), Climate Change and Biodiversity. Yale University Press, New Haven, p. 398.

Roy, K., Jablonski, D., Valentine, J.W., 2001. Climate change, species range limits and body size in marine bivalves. Ecology Letters 4, 366–370.

Rustad, L.E., Campbell, J.L., Marion, G.M., Norby, R.J., Mitchell, M.J., Hartley, A.E., et al., 2001. A meta-analysis of the response of soil respiration, net nitrogen mineralization, and aboveground plant growth to experimental ecosystem warming. Oecologia 126, 543–562.

Sabine, C.L., Feely, Richard A., Gruber, Nicolas, Key, Robert M., Lee, Kitack, Bullister, John L., et al., 2004. The oceanic sink for Anthropogenic CO_2. Science 305, 67–371.

Sanderson, E.W., Redford, K.H., Vedder, A., Coppolillo, P.B., Ward, S.E., 2002. A conceptual model for conservation planning based on landscape species requirements. Landscape and Urban Planning 58 (1), 41–56.

Schindler, D.W., 2009. Lakes as sentinels and integrators for the effects of climate change on watersheds, airsheds, and landscapes. Limnology and Oceanography 54, 2349–2358.

Schmitz, O.J., Raymond, P.A., Estes, J.A., Kurz, W.A., Holtgrieve, G.W., Ritchie, M.E., Wilmers, C.C., 2013. Animating the carbon cycle. Ecosystems, 1–16.

Schröter, D., Cramer, W., Leemans, R., Prentice, I.C., Araújo, M.B., Arnell, N.W., Zierl, B., 2005. Ecosystem service supply and vulnerability to global change in Europe. Science 310 (5752), 1333–1337.

Schuur, E.A., Bockheim, J., Canadell, J.G., Euskirchen, E., Field, C.B., Goryachkin, S.V., Zimov, S.A., 2008. Vulnerability of permafrost carbon to climate change: Implications for the global carbon cycle. BioScience 58 (8), 701–714.

Scott, D., 2005. Case study: integrating climate change into Canada's National Park System. In: Lovejoy, T.E., Hannah, L. (Eds.), Climate Change and Biodiversity. Yale University Press, New Haven, p. 398.

Sedwick, C., 2008. What killed the wooly mammoth? PLoS Biology 6 (4). e99.

Seibel, B.A., Walsh, P.J., 2001. Potential impacts of CO_2 injection on deep-sea biota. Science 294, 319–320.

Seiler, T.J., Rasse, D.P., Li, J.H., Dijkstra, P., Anderson, H.P., Johnson, D.P., et al., 2009. Disturbance, rainfall and contrasting species responses mediated aboveground biomass response to 11 years of CO_2 enrichment in a Florida scrub-oak ecosystem. Global Change Biology 15, 356–367.

Seimon, T.A., Seimon, A., Daszak, P., Halloy, S.R.P., Schloegel, L.M., Aguilar, C.A., et al., 2007. Upward range extension of Andean anurans and chytridiomycosis to extreme elevations in response to tropical deglaciation. Global Change Biology 13, 288–299.

Shackleton, N., 2001. Paleoclimate. Climate change across the hemispheres. Science 291, 58–59.

Sinervo, B., Mendez-de-la-Cruz, F., Miles, D. B, Heulin, B., Bastiaans, E., Villagran-Santa Cruz, M., et al., 2010. Erosion of lizard diversity by climate change and altered thermal niches. Science 328, 894–899.

Sitch, S., Huntingford, C., Gedney, N., Levy, P.E., Lomas, M., Piao, S.L., et al., 2008. Evaluation of the terrestrial carbon cycle, future plant geography and climate-carbon cycle feedbacks using five dynamic global vegetation models (DGVMs). Global Change Biology 14, 2015–2039.

Smol, J.P., Wolfe, A.P., Birks, H.J.B., Douglas, M.S.V., Jones, V.J., Korhola, A., et al., 2005. Climate-driven regime shifts in the biological communities of arctic lakes. Proceedings of the National Academy of Sciences 102, 4397–4402.

Soto, C.G., 2001. The potential impacts of global climate change on marine protected areas. Reviews in Fish Biology and Fisheries 11, 181–195.

Stanley, G.D., Fautin, D.G., 2001. The origins of modern corals. Science 291 (5510), 1913–1914.

Stefan, H.G., Fang, X., Eaton, J.G., 2001. Simulated fish habitat changes in North American lakes in response to projected climate warming. Transactions of the American Fisheries Society 130, 459–477.

Still, C.J., Foster, P.N., Schneider, S.H., 1999. Simulating the effects of climate change on tropical montane cloud forests. Nature 398, 608–610.

Stuiver, M., Grootes, P.M., Braziunas, T.F., 1995. The GISP2 delta O-18 climate record of the past 16,500 years and the role of the sun, ocean and volcanoes. Quaternary Research 44, 341–354.

Suttle, K.B., Thomsen, M.A., Power, M.E., 2007. Species interactions reverse grassland responses to changing climate. Science 315, 640–642.

Taylor, S.W., Carroll, A.L., October 30–31, 2003. Disturbance, forest age, and mountain pine beetle outbreak dynamics in BC: a historical perspective. In: Shore, T.L., Brooks, J.E., Stone, J.E. (Eds.), Mountain Pine Beetle Symposium: Challenges and Solutions. Natural Resources Canada, Canadian Forest Service, Pacific Forestry Centre, Kelowna, British Columbia, p. 298. Information Report BC-X-399, Victoria, BC.

Temmerman, S., Meire, P., Bouma, T.J., Herman, P.M., Ysebaert, T., De Vriend, H.J., 2013. Ecosystem-based coastal defence in the face of global change. Nature 504 (7478), 79–83.

Thomas, C.D., 2005. Recent evolutionary effects of climate change. Chapter 6. In: Lovejoy, T.E., Hannah, L. (Eds.), Climate Change and Biodiversity. Yale University Press, New Haven, p. 398.

Thomas, C.D., Bodsworth, E.J., Wilson, R.J., Simmons, A.D., Davies, Z.G., Musche, M., et al., 2001a. Ecological and evolutionary processes at expanding range margins. Nature 411, 577–581.

Thomas, D.W., Blondel, J., Perret, P., Lambrechts, M.M., Speakman, J.R., 2001b. Energetic and fitness costs of mismatching resource supply and demand in seasonally breeding birds. Science 291, 2598–2600.

Thomas, C.D., Cameron, A., Green, R.E., Bakkenes, M., Beaumont, L., Grainger, A., et al., 2004. Extinction risk from climate change. Nature 427, 145–148.

Thorne, J.H., Seo, C., Basabose, A., Gray, M., Belfiore, N.M., Hijmans, R.J., 2013. Alternative biological assumptions strongly influence models of climate change effects on mountain gorillas. Ecosphere 4 (9).

Thuiller, W., 2004. Patterns and uncertainties of species' range shifts under climate change. Global Change Biology 10, 2020–2027.

Thuiller, W., Lavorel, S., Araujo, M.B., Sykes, M.T., Prentice, I.C., 2005. Climate change threats to plant diversity in Europe. Proceedings of the National Academy of Sciences 102, 8245–8250.

Venette, R.C., Cohen, S.D., 2006. Potential climatic suitability for establishment of Phytophthora ramorum within the contiguous United States. Forest Ecology and Management 231, 18–26.

Veron, J.E.N., 2008. Mass extinctions and ocean acidification: biological constraints on geological dilemmas. Coral Reefs 27, 459–472.

Visconti, G., Beniston, M., Iannorelli, E.D., Barba, D. (Eds.), 2001. Advances in Global Change Research: Global Change and Protected Areas. Kulwer Publishers, Dordrecht, Netherlands.

Visser, M.E., Van, N.A.J., Tinbergen, J.M., Lessells, C.M., 1998. Warmer springs lead to mistimed reproduction in great tits (parus major). Proceedings of the Royal Society B: Biological Sciences 265, 1867–1870.

de Vries, B.J.M., van Vuuren, D.P., Hoogwijk, M.M., 2007. Renewable energy sources: their global potential for the first-half of the 21st century at a global level: an integrated approach. Energy Policy 35, 2590–2610.

Wall, D.H., 2005. Case study: climate change impacts on soil biodiversity in a grassland ecosystem. In: Lovejoy, T.E., Hannah, L. (Eds.), Climate Change and Biodiversity. Yale University Press, New Haven, p. 398.

Walther, G.R., Post, E., Convey, P., Menzel, A., Parmesan, C., Beebee, T.J.C., et al., 2002. Ecological responses to recent climate change. Nature 416, 389–395.

Watson, R.T., 2005. Emissions reductions and alternative futures. Chapter 23. In: Lovejoy, T.E., Hannah, L. (Eds.), Climate Change and Biodiversity. Yale University Press, New Haven, p. 398.

Weiss, H., Bradley, R.S., 2001. What drives societal collapse? Science 291, 988.

Westerling, A.L., Hidalgo, H.G., Cayan, D.R., Swetnam, T.W., 2006. Warming and earlier spring increase western U.S. forest wildfire activity. Science 313, 940–943.

Williams, J.W., Shuman, B.N., Webb, T., 2001. Dissimilarity analyses of late-quaternary vegetation and climate in eastern North America. Ecology 82, 3346–3362.

Williams, P., Hannah, L., Andelman, S., Midgley, G., Araujo, M., Hughes, G., et al., 2005. Planning for climate change: identifying minimum-dispersal corridors for the cape proteaceae. Conservation Biology 19, 1063–1074.

Williams, J.W., Jackson, S.T., Kutzbacht, J.E., 2007. Projected distributions of novel and disappearing climates by 2100 AD. Proceedings of the National Academy of Sciences 104, 5738–5742.

Willis, K.J., van Andel, T.H., 2004. Trees or no trees? The environments of central and eastern Europe during the Last Glaciation. Quaternary Science Reviews 23, 2369–2387.

Wilson, R.C.L., Drurl, S.A., Champan, S.A., 2000. The Great Ice Age—Climate Change and Life. Routledge, London. 267.

Zachos, J., Pagani, M., Sloan, L., Thomas, E., Billups, K., 2001. Trends, rhythms, and aberrations in global climate 65 Ma to present. Science 292, 686–693.

Zhao, M., Running, S.W., 2010. Drought-induced reduction in global terrestrial net primary production from 2000 through 2009. Science 329 (5994), 940–943.

Index

Note: Page numbers followed by "f" indicate figures; "t" tables; "b" boxes.

D

F

G

L

M

Q

R

S

T

Edwards Brothers Malloy
Ann Arbor MI. USA
March 11, 2015